Phaeocystis, major link in the biogeochemical cycling of climate-relevant elements

Phaeocystis, major link in the biogeochemical cycling of climate-relevant elements

Edited by

M.A. van LEEUWE
University of Groningen, The Netherlands

J. STEFELS
University of Groningen, The Netherlands

S. BELVISO
Laboratoire des Sciences du Climat et de l'Environnment, Gif sur Yvette, France

C. LANCELOT
Université Libre de Bruxelles, Bruxelles, Belgium

P.G. VERITY
Skidaway Institute of Oceanography, Savannah, GA, USA

and

W.W.C. GIESKES
University of Groningen, The Netherlands

Reprinted from *Biogeochemistry*, Volume 83 (2007)

Springer

A C.I.P. Catalogue record for this book is available from the Library of Congress.

ISBN 978-1-4020-6213-1 (HB)
ISBN 978-1-4020-6214-8 (e-book)

Published by Springer,
P.O. Box 17, 3300 AA Dordrecht, The Netherlands.

www.springer.com

Cover illustration: The cover illustration is based on Verity et al. (this volume)

Printed on acid-free paper

TABLE OF CONTENTS

Special Issue:
***Phaeocystis*, major link in the biogeochemical cycling of climate-relevant elements**

Guest Editors:
M.A. Van Leeuwe, J. Stefels, S. Belviso, C. Lancelot, P.G. Verity, W.W.C. Gieskes

Biogeochemistry (2007) 83:1-2
DOI 10.1007/s10533-007-9088-0

International Council for Science
Scientific Committee on Oceanic Research

Introduction

Maria van Leeuwe

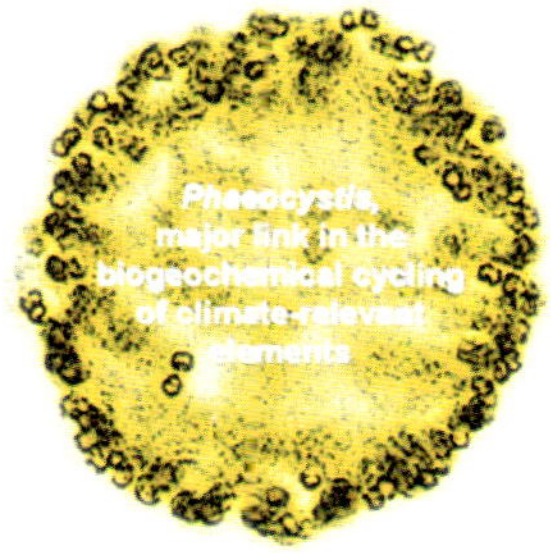

Published online: 28 April 2007

This special issue of *Biogeochemistry* comprises a selection of papers that describe the role of the ubiquitous microalgal species *Phaeocystis* sp. The papers were presented at the final meeting of Working Group no. 120, ''Phaeocystis, *major link in the biogeochemical cycling of climate-relevant elements*'', of the Scientific Committee on Oceanic Research (SCOR), held at the University of Groningen in Haren, the Netherlands, 30 August–3 September 2005. The SCOR working group embodied the desire expressed by the scientific community to give momentum to research dedicated to this microalgal species. The final meeting illustrated the success of the working group. Despite the awareness that there are still many questions left, much has also been learned over the past decade. Most importantly, we were able to define the most prominent gaps in our knowledge over the course of the meeting.

The combination of a diverse yet well-selected spectrum of scientific disciplines—varying from phylogeny to carbon degradation—with contributions by 46 registered scientists from 11 countries, made this meeting a successful and stimulating event. The opportunity to submit the contributions together in this one special volume of *Biogeochemistry* fully meets our desire to arrive at a joint publication as the outcome of that last meeting. All the presentations *taken together* tell the tale of an algal species that plays a prominent role in biogeochemical cycles but that is still surrounded in mystery.

A section on the taxonomy of *Phaeocystis* sets the stage for this issue. The review by Medlin addresses one of the most important uncertainties that trouble many attempts to improve our understanding of this organism, i.e. the crucial, yet very complex, issue of strain diversification. The intricate life cycle of *Phaeocystis* makes it even harder for one to decipher the role of this organism in biogeochemical cycles. The various stages that make up the life cycle, a phenomenon that is typical for this organism, concur with striking modifications in size and shape. These modifications may well determine the success of *Phaeocystis* as a cosmopolitan species. Moreover, they are crucial for the fate of the organic carbon represented in this species. Therefore, an ecophysiological perspective provides essential insights into a full understanding of the biogeochemical function of *Phaeocystis* in the marine ecosystem.

The biogeochemical role of *Phaeocystis* is tightly linked to ecosystem dynamics. On a small scale, viruses play an important, sometimes decisive, role in the functioning and fate of *Phaeocystis* blooms. The community of grazers, furthermore, determines the fate of *Phaeocystis*. An extensive review by Nejstgaard takes our understanding in this area a serious step forward, by clearly defining the gaps in current knowledge. In this section it is also emphasised that grazing activity and cell lysis do not necessarily

M. van Leeuwe (✉)
University of Groningen, Groningen, The Netherlands
e-mail: m.a.vanleeuwe@rug.nl

create conditions for significant carbon export into the ocean interior. Other factors that distinguish a retention system from an export system and, therewith, control carbon fluxes through the ecosystem, are discussed in various papers.

A substantive effort in studies on *Phaeocystis* is focused on its role in sulphur cycles by the production of dimethyl-sulphoniopropionate (DMSP). To put the role of *Phaeocystis* in a global perspective, Stefels has put in an effort to enumerate its dimethylsulphide (DMS) contribution. This review bridges the gap between field scientists and modellers. Models exist in a wide variety, covering a vast area, from the cellular level to global processes. The paper that describes models, by Whipple and colleagues, makes a good example of how models can expose gaps in our knowledge and how a combination with empirical studies can bring new insights.

The last treatise in this issue comprises a synthesis of all the presentations at the meeting. The aim of this chapter is not simply to present state-of-the-art knowledge, but to highlight the most relevant insights that we have gained so far and, at the same time, define urgent questions that need to be addressed. One of the stimulating outcomes of the meeting was that all the participants expressed their interest in continuation of a "*Phaeocystis* community", with potential collaboration on the level of data exchange or even joint field work. The synthesis chapter may focus future research on *Phaeocystis*.

This symposium also marked the end of the long and distinguished career of Prof. Winfried Gieskes, who chaired this SCOR Working Group jointly with Dr. Sauveur Belviso. Winfried Gieskes has devoted his career to the role of microalgae in the biogeochemical cycling of climate-relevant elements, and he has published a number of benchmark papers. His early work on pigment analysis and his contributions to the development of a method using high performance liquid chromatography have greatly influenced the use of algal pigments as biomarkers for algal distribution. In the 1970s he was one of the first to recognise the role of coastal eutrophication in *Phaeocystis* dynamics, and, since then, *Phaeocystis* has always been one of the study topics in his research group. A number of *Phaeocystis*-orientated theses has been published under his care. Winfried Gieskes has been a faithful participant of the various *Phaeocystis* meetings that have been organised throughout Europe since 1989. These developments culminated in his chairing the SCOR *Phaeocystis* Working Group. The SCOR Working Group, and the entire oceanographic community, would like to thank him most warmly for his contribution to science.

This symposium could not have been organised without the support of staff members of the Department of Marine Biology, the RuG Facilitair Bedrijf and SCOR. This document is based on work partially supported by the U.S. National Science Foundation to SCOR under Grants No. OCE- 0003700 and OCE- 0326301. Financial support was furthermore provided by the Dutch Royal Academy of Sciences (KNAW), the University of Groningen, RuG-GUF, the Directorate-General of Public Works and Water Management, and the University's research school CEES (Centre for Evolutionary and Environmental Studies). We would like to thank the editors of *Biogeochemistry* for agreeing to publish the proceedings of the *Phaeocystis* meeting and for practical support. We are indebted to all those colleagues who helped review all the manuscripts.

Biogeochemistry (2007) 83:3–18
DOI 10.1007/s10533-007-9087-1

REVIEW PAPER

A taxonomic review of the genus *Phaeocystis*

Linda Medlin · Adriana Zingone

Received: 20 October 2005 / Accepted: 22 August 2006 / Published online: 15 March 2007

Abstract *Phaeocystis* is recognized both as a nuisance and as an ecologically important phytoplankton species. Its polymorphic life cycle with both colonial and flagellated cells causes many taxonomic problems. Sequence variation among 22 isolates representing a global distribution of the genus has been compared using three molecular markers. The ribulose-1,5-bisphosphate carboxylase/oxygenase (RUBISCO) spacer is too conserved to resolve species. The most conserved 18S ribosomal deoxyribonucleic acid (rDNA) analysis suggests that an undescribed unicellular *Phaeocystis* sp. (isolate PLY559) is a sister taxon to the Mediterranean unicellular *Phaeocystis jahnii;* this clade branched prior to the divergence of all other *Phaeocystis* species, including the colonial ones. The internal transcribed spacer (ITS) region shows sufficient variation that some spatial population structure can be recovered, at least in *P. antarctica*. *P. globosa* and *P. pouchetii* have multiple different ITS copies, suggestive of cryptic species that are still able to hybridize. A molecular clock has been constructed that estimates the divergence of the cold water colonial forms from the warm-water colonial forms to be about 30 Ma and the divergence of *P. antarctica* and *P. pouchetii* to be about 15 Ma. A short description of the colonial stage and the flagellated stage for each formally recognized species is provided. Morphological information is also provided on a number of undescribed species. These include the strain Ply 559, consisting of non-colonial cells with peculiar tubular extrusomes, a second non-colonial species from the north western Mediterranean Sea producing a lot of mucus, and a colonial species with scale-less flagellates found in Italian waters. In addition, three flagellated morphotypes with scales different from those of *P. antarctica* were reported in the literature from Antarctic waters. The picture emerging from both molecular and morphological data is that the number of species in the genus is still underestimated and that cryptic or pseudocryptic diversity requires a sound assessment in future research of this genus. Based on all published observations, an emended description of the genus is provided.

Keywords Molecular clock · *Phaeocystis antarctica* · *P. cordata* · *P. globosa* · *P. jahnii* · *P. pouchetii* · *P. scrobiculata* · rDNA analysis

L. Medlin (✉)
Alfred Wegener Institute for Polar and Marine Research, Am Handelshafen 12, Bremerhaven, 27570, Germany
e-mail: lkmedlin@awi-bremerhaven.de

A. Zingone
Stazione Zoologica Anton Dohrn, Villa Comunale, 80121 Naples, Italy

Introduction

Phaeocystis Lagerheim is a cosmopolitan bloom-forming alga that is often recognized both as a nuisance alga and an ecologically important member of the phytoplankton (Davidson 1985; Lancelot et al. 1987; Smith et al. 1991; Davidson and Marchant 1992; Baumann et al. 1994; Schoemann et al. 2005; Veldhuis and Wassmann 2005). Its various life forms can make large-scale blooms that are often avoided by fish (Chang 1983) and appear detrimental to the growth and reproduction of shellfish and macrozooplankton (Davidson and Marchant 1992) or are ichthyotoxic (Shen et al. 2004). Massive areas of pollution are created when dissolved organic compounds released by *Phaeocystis* during declining bloom conditions accumulate, foam and then wash onshore (Lancelot et al. 1987). *Phaeocystis* is a major contributor to the global sulphur budget by releasing substantial quantities of dimethylsulfide propionate (DMSP) (Keller et al. 1989; Baumann et al. 1993), which is metabolized to dimethylsulfide (DMS) as the cells are grazed or infected and lysed by viruses. It may play yet another important ecological role with its production of ultraviolet B (UV-B)-absorbing compounds (Marchant et al. 1991; Davidson and Marchant 1992).

Phaeocystis has a polymorphic life cycle with both colonial and flagellated cells (Kornmann 1955; Whipple et al. 2005). The colonial stage, with cells very loosely interconnected and enclosed in a thin skin (Hamm et al. 1999), is most easily recognized, although some species may form mucilaginous colonies or do not seem to have a colonial stage. Thousands of cells can occur in a colony that may reach 2 cm in diameter (Jahnke and Baumann 1987; Verity et al. 1988; Rousseau et al. 1994; Davidson and Marchant 1992). Colony sizes of 3 cm or more have been reported in blooms from China (Shen et al. 2004). The difficulty in assigning a specific name to the colonial stage has caused much taxonomic confusion. Flagellated cells have two parietal chloroplasts and two flagella, which may be equal or unequal in length and heterodynamic. A short haptonema is present between the two flagella, which may or may not have a swollen end. The flagellated cells may be naked or have two layers of different shaped organic scales. Some flagellated cells also produce groups of filaments, which are extruded from the cell and assume a characteristic pattern.

The genus was erected by Lagerheim in 1893 to accommodate the colonial stage of an alga described originally as *Tetraspora poucheti* by Hariot in Pouchet (1892). *Phaeocystis pouchetii* (its correct orthography) occurs in cold waters and in its globular, lobed colonies, cells are arranged in packets of four (see Jahnke and Baumann 1987 for illustrations). *Phaeocystis globosa* was described by Scherffel (1900) from temperate waters and forms spherical colonies with cells arranged homogeneously within the colony (Jahnke and Baumann 1987), whereas older stages can assume distorted pear shapes (Bätje and Michaelis 1986). Early workers separated *P. pouchetii* and *P. globosa* based on different distributions and colonial morphologies until Kornmann (1955) doubted the differentiation between the two species. From his life-cycle studies, he considered that *P. globosa* cell types appeared to be juvenile forms of *P. pouchetii*. Since that report, colony morphology has been judged an unreliable specific character.

Sournia (1988) reviewed the diagnostic features of *Phaeocystis*, and discussed the reliability of the nine valid species published since the last century. He discarded two species from the genus, *P. fuscescens* (Braun) De Toni and *P. giraudyi* (Derbès and Solier) Hamel, because they did not fit the genus and probably not even the class characteristics. The descriptions of four species, two from cold waters, *P. antarctica* Karsten and *P. brucei* Mangin, and two from temperate waters, *P. amoeboidea* Büttner and *P. sphaeroidea* Büttner, were all judged as very superficial. The poor illustrations and unlikely features, including one chloroplast per cell and no haptonema (Büttner 1911), were probably the reasons why the two temperate species have never been mentioned again in the literature. For similar reasons the two Antarctic species were reported rarely and not studied again. As for the two most frequently recorded species, *P. pouchetii* (Hariot in Pouchet) Lagerheim and *P. globosa* Scherffel, they had been studied in more detail yet no element was available to keep them separate. Therefore Sournia suggested that the name *P. pouchetii* or, better, *P.* cf. *pouchetii* should have been used for the colonial *Phaeocystis* species pending new information. The only other reliable species was *Phaeocystis scrobiculata* Moestrup, described with modern methods but known only from the flagellated state (Moestrup 1979). Without convincing information on the differentiation between *P. pouchetii* and *P. globosa* (see paragraph below), most marine ecologists until the mid 1990s followed Sournia's advice and reported *Phaeocystis* colonies as *P. pouchetii* (the older name) or as *Phaeocystis* sp. to avoid confusion.

Baumann and Jahnke (1986), Jahnke and Baumann (1986, 1987) and Jahnke (1989) regarded this as over-simplification. Their observations on colony shape maintenance in long term culture showed that both juvenile and older stages of *P. globosa* and *P. pouchetii* were distinct and these data supported the recognition of the two entities as separate species. Also, detailed studies of the temperature and light tolerances suggested separation at the species level. *P. globosa* was a temperate species and *P. pouchetii* was a cold-water form. A third, unnamed colonial species from Antarctic waters was recognized by Baumann et al. (1993), which had a combination of features of *P. globosa* and *pouchetii*, as suggested earlier by Moestrup and Larsen (1992). The colonies resembled those of *P. globosa* (Larsen and Moestrup 1989), whereas temperature tolerance was similar to that of *P. pouchetii*. Notably, strains from the Antarctic had different pigment spectra (Buma et al. 1991; Vaulot et al. 1994) and DNA content (Vaulot et al. 1994). First indications that further undescribed species could exist were provided by Pienaar (1991, 1996) who illustrated a *Phaeocystis* flagellate with cup-shaped scales from South African waters suggesting it was a new species but not publishing a formal description.

To resolve the species issue in *Phaeocystis*, a molecular analysis of various clones was begun (Lange 1997). Medlin et al. (1994) were the first to propose that *P. globosa* and *P. pouchetii* were separate species based on genetic evidence. They also showed that colonial *Phaeocystis* from the Antarctic was genetically distinct from the other two species. They resurrected the species *P. antarctica* described by Karsten (1905) for the colonial isolates previously termed *P. globosa* or *pouchetii* from these waters. Zingone et al. (1999) added two more species to the genus, *P. cordata* and *P. jahnii*, but these were basically unicellular species, although the latter species could make simple clusters of cells that could be termed colonial. In this review we summarize the molecular and morphological information available to date for *Phaeocystis* species and provide evidence that a high diversity is still hidden in the genus.

Molecular analysis

A global distribution of the prymnesiophyte genus *Phaeocystis* was compared using nuclear-encoded 18S rDNA genes and two non-coding regions, the ribosomal DNA internal transcribed spacer 1 (ITS1) separating the 18S rDNA and 5.8S rDNA genes and the plastid ribulose-1,5-bisphosphate carboxylase/oxygenase (RUBISCO) spacer flanked by short stretches of the adjacent large and small subunits (*rbcL* and *rbcS*) (Lange et al., 2002). The RUBISCO spacer regions were highly conserved and generally uninformative among all *Phaeocystis* strains (Lange et al. 2002). The 18S rDNA analysis suggests that an undescribed unicellular *Phaeocystis* sp. (isolate PLY559) is a sister taxon to the occasionally colonial Mediterranean *P. jahnii* with which it forms a clade that diverges at the same time from a clade with all other *Phaeocystis* species, including those forming typical colonies wrapped in a skin (Fig. 1). In this latter clade, the unicellular *P. cordata* diverges before the colonial ones, which can be divided into a cold-water complex (*P. pouchetii* from the Arctic and *P. antarctica* from the Antarctic) and a warm-water complex consisting of *P. globosa*. Thus, all of the variation seen earlier using morphological and physiological criteria had a strong genetic basis and separation at the species level was warranted for all colonial species originating from major climatic regions, i.e., the Arctic, the Antarctic and temperate/tropical regions.

In contrast, ITS1 exhibited substantial inter- and intra-specific sequence divergence and showed more resolution among the strains (Fig. 2). Markedly different copies of the ITS1

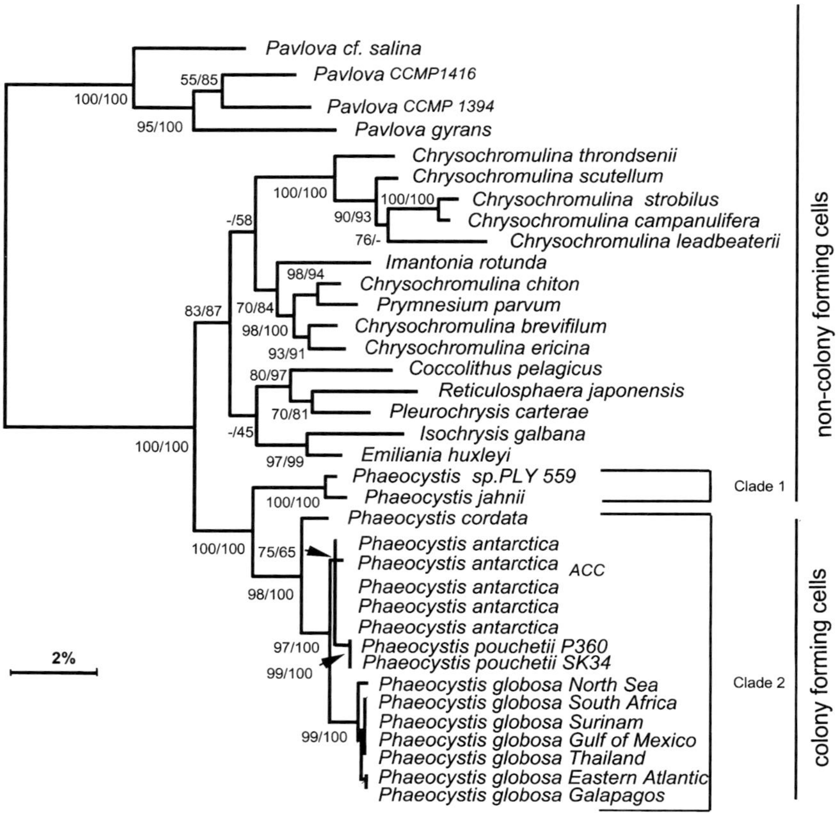

Fig. 1 Maximum-likelihood phylogeny (fastDNAml) of 17 *Phaeocystis* species/strains and other prymnesiophytes inferred from 18S rDNA. The class Pavlovophyceae was used as outgroup. Bootstrap values are placed on the nodes that are identical from ML/NJ/MP analyses. The scale bar corresponds to two base changes per 100 nucleotides. Redrawn from Lange et al. (2002)

region were found in *P. globosa* even among cloned DNA from a single strain, suggesting that it is a species complex composed of at least three species. This observation was also supported by different DNA content among different clones of *Phaeocystis* (Vaulot et al. 1994). Multiple copies of the ITS1 were also found in a single strain of *P. pouchetii*, suggesting that this is also a species complex. These species complexes would appear still to be able to interbreed with one another because multiple different haplotypes from different clades can be found within a single strain. Similar findings of multiple haplotypes being present in known hybrids of flowering plant species are quite common (Chase et al 2003) and it is known that in some hybrids of flowering plants, the hybrid can loose one of the parental haplotypes in as few as 12 generations (Bateman, pers. com.) However, among nine *P. antarctica* strains, only one type of ITS1 haplotype was found per strain, although it was variable among the strains. Using the branching order in the ITS1 tree (Fig. 3) we have attempted to trace the biogeographic history of the dispersal of strains in Antarctic coastal waters. We see that the first divergence among the Antarctic clade is strain SK 22, which was isolated from the Antarctic circumpolar current (ACC). The second divergence is a strain from Prydz Bay (T4-2) and this is then followed by a nearly simultaneous divergence of the remaining strains. Among these divergences is a cluster of strains from Prydz Bay (DE2, A1-3 and T9-1). The other divergence includes strain CCMP 1374 from the Ross Sea and two *Phaeocystis* strains from the Weddell Sea (SK 20, SK 23) (Fig. 3) and finally strain D4-5 from Prydz Bay, which shares a last common ancestor with the Weddell Sea strain SK 23, as the final cluster to diverge among the Antarctic strains. Populations of *P. antarctica* within the continental boundary water masses appear to be well-mixed because currents move around the Antarctic continent rather quickly and may effectively act as a barrier to significant population structure. Strain SK 22 isolated within the ACC, however, is clearly different. An earlier hypothesis, proposed

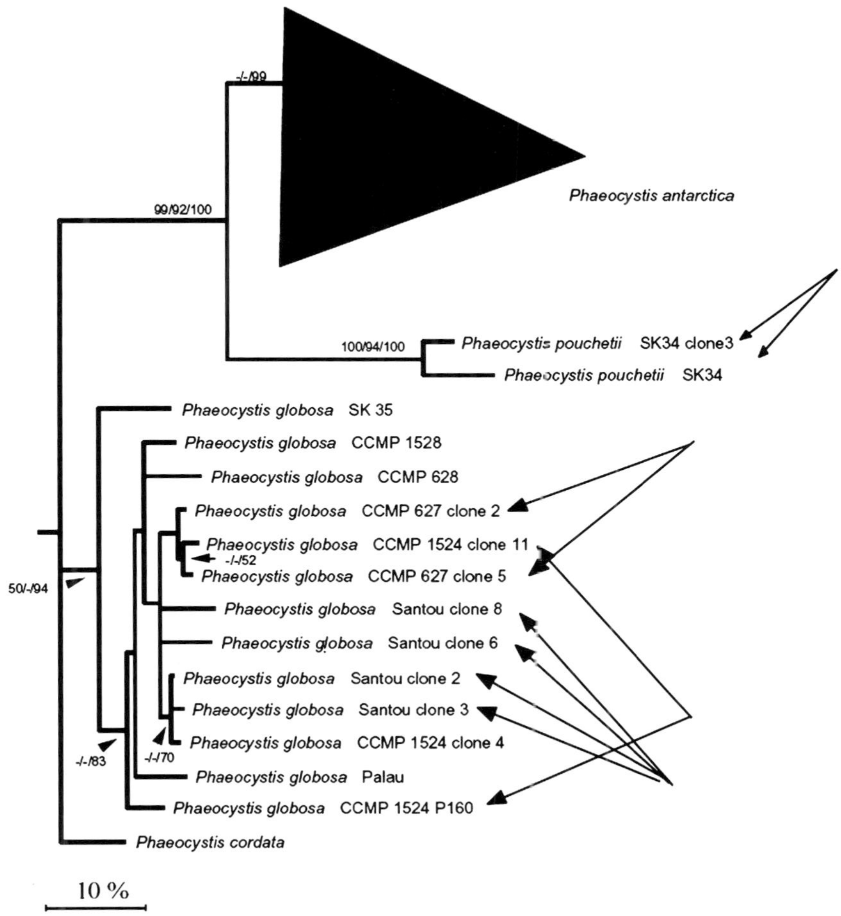

Fig. 2 Maximum-likelihood phylogeny of the ITS regions showing the multiple sequences from a single strain of *P. globosa* and *P. pouchetii*. Each different sequence comes from a different bacterial vector clone. All haplotypes from one strain of *Phaeocystis* are connected by the arrows. *P. antarctica* exhibited a single sequence per strain and these are collapsed into a triangle. Redrawn from Lange et al. (2002)

from rDNA data (Medlin et al. 1994), that ancestral populations in the Antarctic gave rise to present day *P. antarctica* and *P. pouchetii* populations appears to be supported by ITS1 analysis of the cold-water *Phaeocystis* strains. *P. antarctica* and *P. pouchetii*, both polar, are more closely related to one another than either is to the cold and warm temperate to tropical populations of present-day *P. globosa*. This suggests that dispersal did not occur from present-day warm-water populations into present-day cold-water populations but that gene flow has occurred from pole to pole across tropical oceans. Arctic *P. pouchetii* populations thus probably arose by a dispersal event from the south to the north during colder climate periods that allowed populations to survive the crossing of equatorial waters, as has been documented for other organisms (Crame 1993; Darling et al. 2000, 2004; Montresor et al. 2003). A subsequent warming event will then isolate the two polar populations. Evidence for this can be found in a study of Antarctic surface-water temperatures since the Cretaceous (Crame 1993, Fig. 4c).

If we follow the branching order in Fig. 1, we hypothesize the following scenario: *Phaeocystis* likely originated as a warm-water genus because first divergences in our tree are warm-water species. Ancestral populations in the Antarctic were derived from ancestors of the present-day warm-water species, after being isolated in Antarctic waters. The opening of the Drake passage and the formation of the Antarctic circumpolar current (ACC) are the most likely geological events that could have isolated populations in the Antarctic to separate them from warm-water ancestors. It can be inferred from Fig. 3 that presumed descendants of these warm-water ancestors were first entrained in the ACC because these are the first

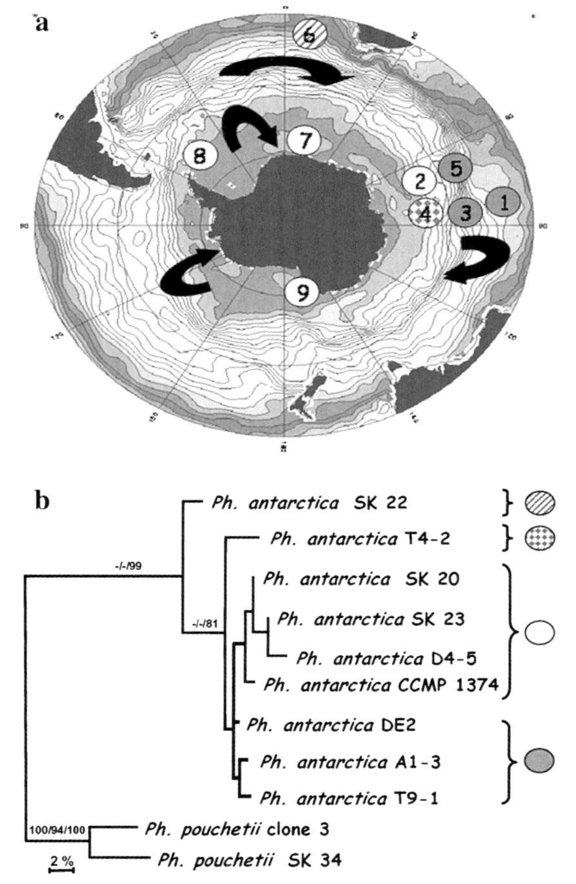

Fig. 3 (**a**) Locations of the strains of *P. antarctica* used in Lange et al. (2002). The location of different clades is indicated by the different patterns in the large circles and correspond to those clades in (**b**). Prydz Bay locations in E. Antarctica are slightly displaced for visual clarity. (**b**) Maximum-likelihood tree inferred from ITS1 sequences from *P. antarctica* with *P. pouchetii* as outgroup. Bootstrap values are placed at the nodes from a maximum likelihood analysis (100 replicates) a neighbor-joining analysis (500 replicates) and a maximum parsimony analysis (500 replicates). The scale bar corresponds to two changes per 100 nucleotide positions. Redrawn from Lange et al. (2002). Map of Antarctica redrawn from Olbers et al. (1962)

divergences. Some of these ancestral populations must have been transported northward and across the Equator shortly after the Drake passage opened because the *P. pouchetii* populations are sister to the *P. antarctica* populations. The ACC today encircles the Antarctic continent every 1–2 years. Water is entrained from this current into the major gyres of the continental water masses (Treshnikov 1964). Using the branching order in Fig. 3 we can trace the dispersal of the clones from the ACC, although the bootstrap support for the branching order is weak to strong among the clades. The first entrainment with a bootstrap support of 99% appears to be into Prydz Bay because strain T4–2 isolated from this bay is the first divergence in our tree. These populations then established themselves in the Eastern Antarctic in Prydz Bay. Subsequent divergences in the tree indicate that populations were then entrained into the Ross Sea and almost simultaneously they were entrained into the Weddell Sea (bootstrap support 54%). Both isolates from the Weddell Sea were the last to diverge before the populations were again entrained back into Prydz Bay from populations in the Weddell Sea because isolates from this bay are some of the last divergences in the tree (bootstrap 54%). The distribution of these isolates in this fashion follows the predominant current patterns of surface waters in the Antarctic today. What we do not know is how different the surface-water circulation was 30 Ma before the ACC was established.

Other studies have also shown the effect of mixing on the homogenization of the genetic structure of Antarctic populations. Krill species within the Antarctic continental water masses are very similar as documented by both mtDNA (Patarnello et al. unpubl.) and isozyme analysis (Fevolden and Schneppenheim 1989). The mtDNA study also suggested that the formation of the ACC effectively isolated krill species in Antarctic water masses from those north of the ACC. Calculation of the time of divergence between species groups found either side of the ACC coincided with the timing of the ACC, approximately 30 Ma. Thus, the molecular data is consistent with our hypothesized historical biogeographic reconstruction of the distribution of *Phaeocystis* based on the circulation patterns developed with the formation of the ACC.

Molecular clock

A molecular clock has been constructed from our 18S rDNA tree and calibrated with fossil dates from the haptophyte coccolithophorid species (Fig. 4). Our molecular clock calculations indicate

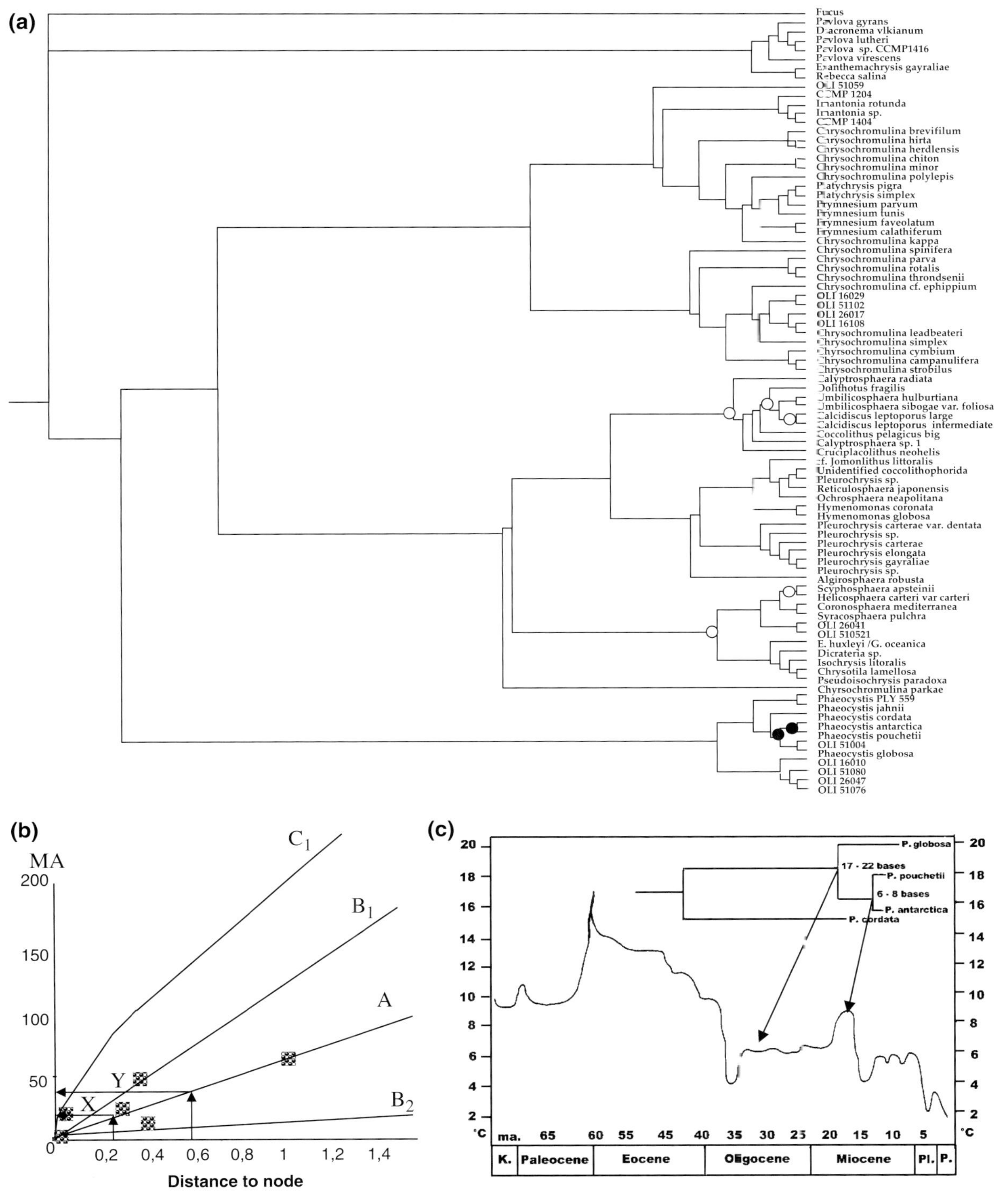

Fig. 4 Calculation of a molecular clock. (**a**) The ml tree shown in Edvardsen et al. (2000) has been linearized (Kooistra and Medlin 1996) so that all rates of evolution are the same. Fossil dates from coccolithophore taxa are placed on nodes where these taxa have their first appearance in the fossil record (open circles on tree). (**b**) A regression of branch lengths against fossil dates has been performed according to Kooistra and Medlin (1996). From this regression line we have extrapolated the divergence of the warm- and cold-water *Phaeocystis* (Y) and the divergence of *P. pouchetii* from *P. antarctica* (×) (solid circles on tree). (**c**) Temperature of Antarctic surface waters since the Cretaceous with the molecular tree plotted proportional to the time and the temperature, redrawn from Crame (1993)

that the warm-water *Phaeocystis* species diverged from the cold-water species approximately 30 Ma, which coincides with the time that the Drake passage opened and the ACC system was formed. This would have effectively isolated ancestral populations in the Antarctic sufficiently to allow them to speciate from their warm-water ancestors. The separation of *P. pouchetii* from *P. antarctica* is approximately 15 Ma, which coincides with a major warming event in the world's oceans at this time (Fig. 4). Before this time populations must have been able to cross the equator from the south to the north because water temperatures were cool enough to allow survival, but this warming event separated the two polar populations to allow them to diverge into the two species we have today at the poles. Similar results have been found for foraminfera (Darling et al. 2000, 2004).

Thus molecular data have defined our species well and suggest which species are likely to be composed of cryptic species. We detail below basic descriptions of each of the formally described species and provide some indication as to other undescribed species where this information is known.

Formally described species

An overview of the validly published taxa that have been re-examined in recent years is presented in the following. These include species listed in the genus in its most recent review (Sournia 1988) and new species described after that date, with the exception of three taxa that have not been studied since their description, *P. amoeboidea* Büttner, *P. sphaeroidea* Büttner and *P. brucei* Mangin. The first two of these do not have features characteristic of the genus, much the division Haptophyta, so it is likely that only the latter species may still be a valid species of *Phaeocystis*. The main distinctive characters of the species included in this section are summarized in a table in Jacobsen (2002).

P. pouchetii (Hariot in Pouchet) Lagerheim (Pouchet 1892) forms cloud-like colonies with cells in packets of four (Fig. 5a). Molecular data suggest that this is a species complex but here very few strains have been examined with molecular techniques so this is only a very preliminary suggestion.

Flagellated stages of *P. pouchetii* (Fig. 6a) were the subject of recent morphological studies (Jacobsen 2000, 2002). Cells are rounded, with an average diameter of 5 μm. The flagella are equal in length, ca 11 μm, and heterodynamic. The haptonema is extremely short, 1–2 μm, with a slight swelling, and is not easily seen with light microscopy. Body scales are of two types: almost circular flat plates, 0.24 × 0.25 μm, with raised rims, forming an external layer and smaller oval plates with slightly inflexed rims, 0.19 × 0.15 μm, underneath the larger scales. Both types of scales show thin radiating ridges. Filaments (up to 30 μm) arranged in groups of five with the typical pentagonal proximal structure are seen outside the cells, or are coiled up in vesicles under the cell surface. The presence of silica is reported in these filaments (Jacobsen 2002). The ultrastructure is similar to that of *P. globosa* (Parke et al. 1971) and of the other *Phaeocystis* for which this information is available (Zingone et al. 1999), with the nucleus located posteriorly, the two chloroplasts with the embedded pyrenoids, and the Golgi body between them. This cell stage can be infected by viruses (Jacobsen et al. 1996).

P. globosa Scherffel (Scherffel 1900) forms globular colonies with the cells evenly distributed throughout the colony (Fig. 5b). Molecular data and DNA content suggest that this is a complex of up to three or four cryptic species, but to date no morphological investigations exist to support this.

Flagellated stages of *P. globosa* (Fig. 6b, c) were described for the first time by Parke et al. (1971) under the name of *P. pouchetii*, at the time when these two species were considered as stages within the life cycle of the same species. Cells are 3–6 μm, more frequently between 3 and 4.5 μm. The two flagella and the haptonema emerge from a depression in the cell body. Flagella are equal in length, 1.5× the cell length, and heterodynamic. The haptonema is a quarter to a third the length of the flagella. It is stiff and has a clear distal swelling. The haptonema is easily seen in live cells, where it is directed forward while cells move. Body scales show all radiating ridges on both surfaces and are of two types: almost circular flat plates, 0.18 × 0.19 μm, with raised rims, forming

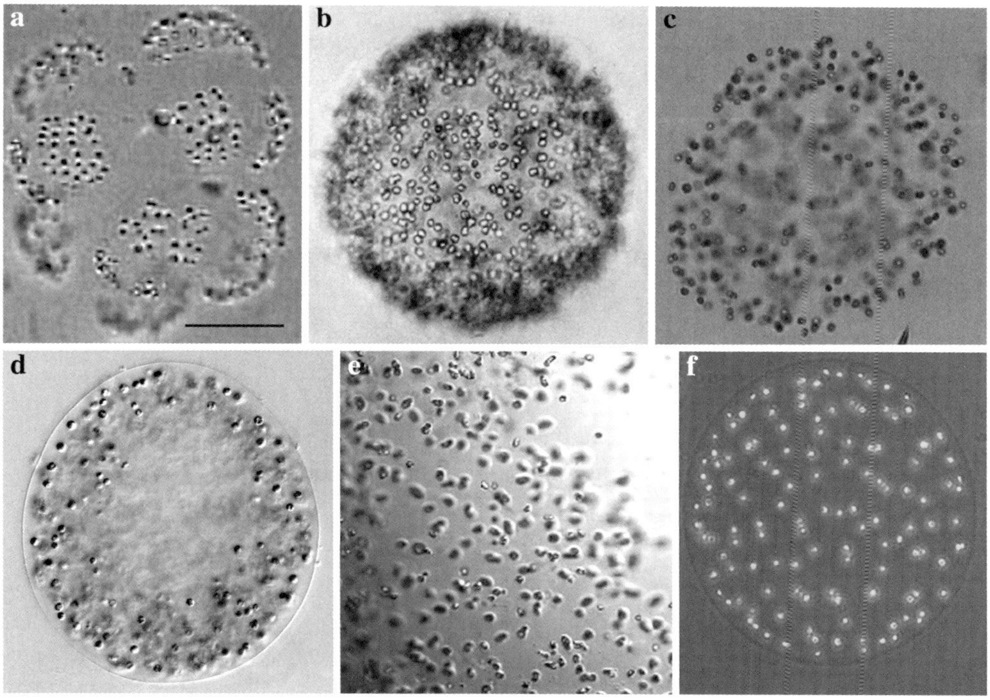

Fig. 5 Light microscopic micrographs of colony stages of *Phaeocystis*. (**a**) *P. pouchetii*, (**b**) *P. globosa*, (**c**) *P. antarctica*, young colony, (**d**) *P. antarctica*, older colony, (**e**) *P. jahnii*, (**f**) *Phaeocystis* sp.2 (**a**, **b**) taken from http://www.jochemnet.de/fiu/OCB3043_21.html Scale bar = 100 μm

an external layer and smaller oval plates, 0.10 × 0.13 μm, with strongly inflexed rims, underneath the larger scales. The ultrastructure is typical for the genus, with two golden-brown chloroplasts, with internal fusiform pyrenoids, and some refringent vesicles probably including storage products. The nucleus is posterior, whereas the Golgi body, with a high number of stacked cysternae, is seen in the space between the chloroplasts and the nucleus. The flagellar bases have typical distal and proximal plates in the transition zone. One of the two clones observed by Parke et al. (1971) (clone 147) forms long filaments (up to 20 μm) in groups of five when discharged outside, with the proximal ends arranged in a very typical pentagonal structure, surrounded by a faint vesicle with a pore in the centre. In the cells, the undischarged threads are found within a vesicle under the cell surface. Chrétiennot-Dinet et al. (1997) showed that these filaments contain alpha-chitin.

P. antarctica Karsten (Karsten 1905) is the least known *Phaeocystis* from the morphological point of view. It also forms globular colonies with cells randomly distributed under the colony surface (Fig. 5c, d). These colonies can become quite distorted and elongated with age.

Flagellated stages of *P. antarctica* (Fig. 6d) have received very little study. Only one illustration of scales from an Antarctic *Phaeocystis* was available (Larsen and Moestrup 1989), showing oval scales of two different sizes (0.27 × 0.19 μm and 0.18 × 0.14 μm, respectively). Recently, three different morphs have been illustrated from field material from the Antarctic having scales that are different in size as compared to those shown by Larsen and Moestrup and having a haptonema without a bulge on its tip (Marchant et al. 2005, Figs. 6f, 7d–f). Scales of still different sizes have been detected in flagellated stages obtained from the isolation of colonies, which are phylogenetically close to SK 22 (Zingone and Montresor, unpublished). The molecular data published to date are derived only from colonial stages or from flagellate stages that were originally colonial. There is only a single type of ITS sequence pres-

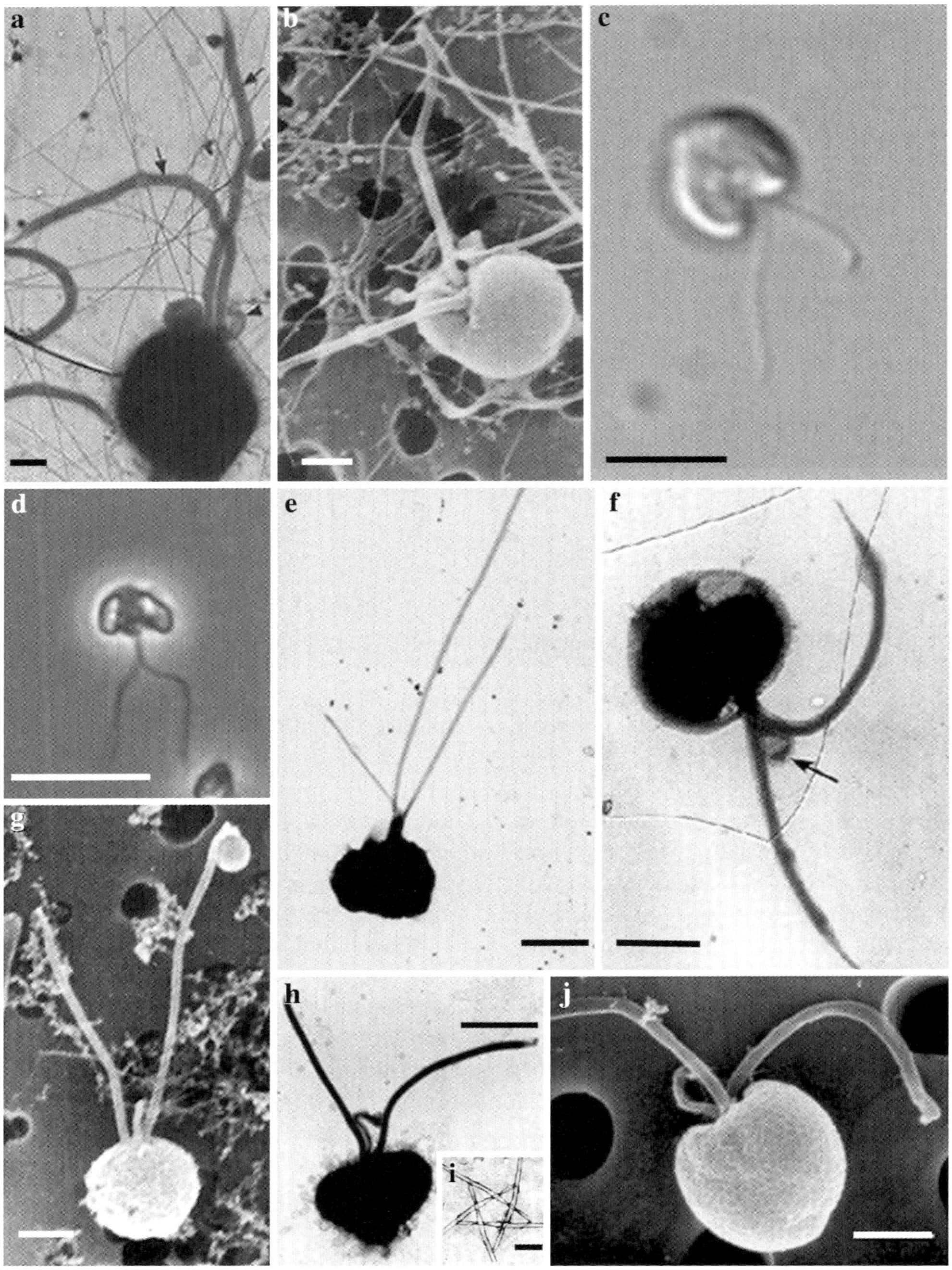

Fig. 6 Light microscopy (LM), transmission electron microscopy (TEM) and scanning electron microscopy (SEM) micrographs of flagellated stages of *Phaeocystis*. (**a**) *P. pouchetii*, TEM, (**b**) *P. globosa*, SEM, (**c**) *P. globosa*, LM, (**d**) *P. antarctica*, LM, (**e**) *P. jahnii*, TEM, (**f**) *Phaeocystis* sp. from Antarctic waters *TEM*, (**g**) *Phaeocystis* sp. 3, SEM, (**h**) *P. cordata TEM*, (**i**) *P. cordata*, star-like pattern in the center of the five-filament structure, TEM, (**j**) *P. cordata*, SEM. (**a**) from Jacobsen (2002), (**c**) taken by Anna Noordeloos, (**d**) taken by Dr. P. Assmy, (**f**) from Scott and Marchant (2005), (**e**, **h**, **i**) from Zingone et al. (1999), (**b**) and (**g**) from Vaulot et al. (1994). Scale bar = 1 μm on (**a**, **b**, **g**) Scale bar = 10 μm on (**c**, **d**) Scale bar = 2 μm on (**e**–**j**) Scale bar = 0.3 μm on (**i**)

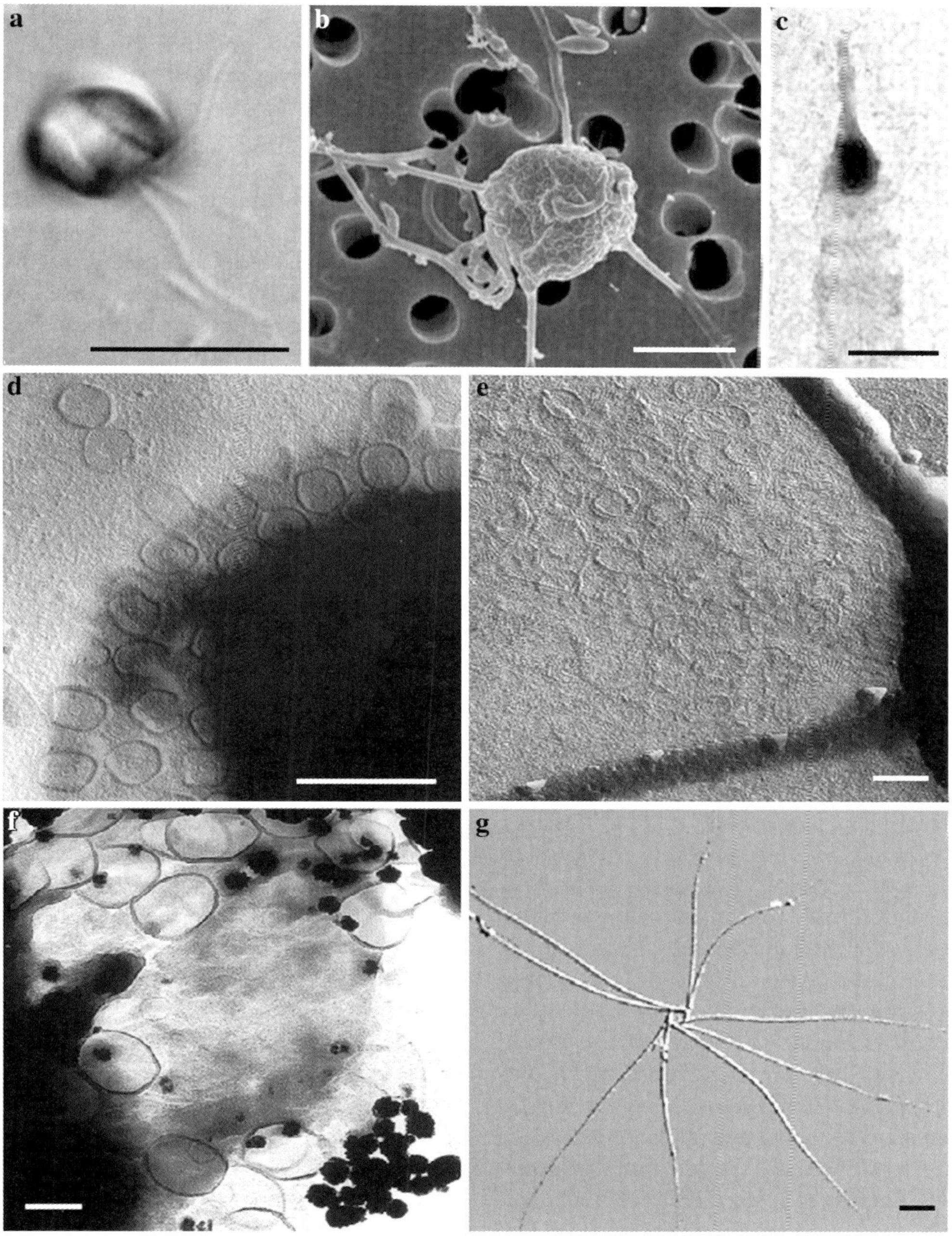

Fig. 7 LM, TEM and SEM micrographs of flagellated stages of *Phaeocystis*. (**a**) *Phaeocystis* sp. 1 (PML 559), LM, (**b**) *Phaeocystis* sp. 1 (PML 559), SEM, (**c**) Tip of the tube-like structure ejected from *Phaeocystis* sp. 1 (PML 559), TEM. (**d–f**) Morphs 1, 2, 3 of *P. antarctica*, (**g**) filaments of *P. scrobiculata*, (**b**, **c**) taken by Gandi Forlani, (**d–f**) from Scott and Marchant (2005). (**g**) From www.marbot.gu.se/SSS/others/Phaeocystis_scrobiculata.GIF. Scale bar = 10 μm on (**a**) Scale bar = 2 μm on (**b**), (**g**) Scale bar = 0.5 μm on (**c–f**)

ent in each strain in contrast to the multiple ITS variants in *P. pouchetii* and *P. globosa*, which strongly suggests that colonial *Phaeocystis* from the Antarctic are not a species complex. The similarity among ITS sequences from Antarctic strains indicates that they could be all a single species, whereas these new unicellular morphs in field samples may belong to species as yet uncultivated

(Fig. 7d–f). Alternatively, different flagellate morphotypes could belong to different sub-clades in the *P. antarctica*-clade. All this information indicates a high morphological diversity and suggests that there might be more than one species present in *Phaeocystis* from the Antarctic. Parallel morphological investigations are warranted on strains belonging to distinct *P. antarctica* sub-clades.

P. jahnii Zingone (Fig. 5e) forms colonies very different from all other *Phaeocystis* colonies (Zingone et al. 1999). These are loose aggregates of non-motile cells embedded in a sticky mucilaginous matrix probably of polysaccharide nature, with no external layer nor a definite shape. In culture material the colonies may form wide sheets with margins at times sticking to the cell tube. Colonial cells range from 6 to 8.5 μm and have 2–4 chloroplasts.

Flagellated cells of *P. jahnii* (Fig. 6e) are rounded, 3.5–5 μm diameter, with flagella of markedly unequal length (8.5–12 μm and 5.5–6.5 μm, respectively). The haptonema is relatively long (3–4.5 μm) and without a marked bulge at the end. As compared to the other *Phaeocystis* species, scales are thinner and more delicate, with a very faint radiating pattern lacking in the central part of the scale. The larger scales (0.35 × 0.28 μm) do not have an upraised rim, whereas the smaller underlying scales (0.18 × 0.14 μm) have the typical inflexed rim. A refringent yellow-orange body is often seen in the live cells in the space between the chloroplasts. Filaments have not been observed in this species.

P. cordata Zingone et Chrétiennot-Dinet (Fig. 6h–j) occurs only as single cells which are typically triangular, heart-shaped or oval, somewhat flattened, with a deep flagellar depression and more or less pointed antapical end (Zingone et al. 1999). The average size is 3–3.5 μm long, 3–4 μm wide, and ca. 2.5 μm thick. The two flagella are slightly subequal, 5.5–7.5 and 4.5–6 μm length, respectively. The haptonema is very short (2.2–2.5 μm) and hardly visible in light microscopy, with a bulging end observed in the electron microscope. Cells generally swim with the flagellar pole directed backwards, and the two flagella straight, completely hiding the haptonema. Cells rotate around their longitudinal axis while moving. Rarely cells are seen moving with the flagellar pole forward. Both larger and smaller scales are oval, 0.25 × 0.18 μm and 0.18 × 0.13 μm, respectively. The larger scales have upraised rims and a slight central knob, and form the external cell investment. The smaller scales have inflexed rims and form an inner layer adjacent to the plasmalemma. The filaments are seen in disk-like vesicles underneath the cell surface (up to three in a cell) or discharged, with the typical five-ray star pattern (Fig 6i). Ultrathin sections show the two flagella and the haptonema inserted along a line that is transversal to the plane crossing the plastids. Comparable information is not available for other species yet. The internal microanatomy is similar to that of the other species of the genus.

P. scrobiculata Moestrup (Fig. 7g) was described from field material collected in New Zealand waters (Moestrup 1979) as a unicell. There is no evidence that it makes colonies nor any molecular work has been done on it. Its cells are 8 μm in diameter with two types of scales, 0.6 × 0.45 μm and 0.19 × 0.21 μm in size. Both types of scales are structureless on the dorsal side, but with ridges radiating from a plain centre on the ventral side. Its flagella and haptonema are twice the length found in *P. globosa* and the scales are about two times larger. Another distinguishing feature is the filaments that it produces, which are in groups of nine (eight pairs and one single), in contrast to the production of five single filaments from the other species that produce filaments (Fig. 7g). The centre pattern of the filaments is rather irregular and does not form the characteristic star shape in the middle. Filaments arranged with the same pattern have also been found in Australian waters (Hallegraeff 1983) and in the Mediterranean Sea (Zingone et al. 1999). However, scales were smaller in both Australian and Mediterranean specimens, which suggests a possible higher diversity to be explored within this taxon as well.

Undescribed species

In addition to the species described in the literature, a number of taxa ascribed to *Phaeocystis* that are currently under morphological and molecular investigation are presented in the following.

Phaeocystis sp. 1 (PML 559) (Fig. 7a–c) seems only to be present in single cells. The two flagella are 8.5–12 μm and 5.5–6.5 μm, respectively. The haptonema is 3–4.5 μm long, without a bulge at the end. The larger scales are similar to those of *P. cordata*, though larger (ca. 0.35 × 0.22 μm), with thick upraised rim and a central knob. Smaller scales (0.25 × 0.17 μm) have inflexed rims. An unusual feature of this species is that it produces tube-like structures with peculiar ends that are ejected from the cell. These bodies may be present in number of 5–7 per cell (Fig. 7b) and leave a large depression once extruded. Many benthic stages are formed as the culture ages and it is likely that the tube-like bodies help to attach the cells to the substrate.

Phaeocystis sp. 2 (Fig. 5f) is the only *Phaeocystis* so far cultivated from the Mediterranean Sea that has been shown to form typical colonies of spherical shape (Zingone, Borra, Forlani and Procaccini, in preparation). The flagellates have an irregular shape, with pronounced shoulders at the flagellar pole. The flagella are markedly unequal in length, the haptonema has no bulging end. No scales were ever detected on the cell surface, nor any kinds of filaments. 18S analyses demonstrate that this taxon belongs to the *P. globosa* clade, although it differs by nine base pairs, a difference that is comparable to that between *P. pouchetii* and *P. antarctica*. ITS sequence is unalignable with those of those available for the other *P. globosa* strains.

Phaeocystis sp. 3 (Fig. 6g) includes strains isolated from the North-Western Mediterranean Sea. Being single-celled and similar to *P. cordata* in scale morphology, it was preliminarily attributed to the latter species (as strains MEDNS2 and MEDNS3 in Zingone et al. 1999), but it has morphological differences that were initially unappreciated. As compared to *P. cordata*, *Phaeocystis* sp. 3 is somewhat larger, has a rounded body, shorter flagella and the larger body scales are almost circular rather than oval. Preliminary molecular analysis has placed it within the *P. globosa* complex.

Clearly, morphological details of the species encountered in recent years fall outside of the original description of the genus *Phaeocystis*, therefore we feel it necessary to emend the genus description as follows:

Phaeocystis Lagerheim 1893, Zingone and Medlin emended.

Motile cells with two more or less equal flagella and a shorter non-coiling haptonema; 1–4 parietal chloroplasts; cell body often covered with flat scales of two different sizes. Ejectile organelles known for several species. Complex life cycles involving the formation of non-motile stages, not known for all species. Non-motile cells usually without appendages and scales, either single or arranged in spherical, lobed, sheathed or irregular gelatinous colonies; if appendages present, usually shorter or incomplete.

Outlook

From the observations we have to date, including field and cultured material and molecular data, it is clear that we have come a long way from just 10 years ago, when we had only one species of *Phaeocystis*: *P. globosa* with a cosmopolitan distribution. We now have a much clearer picture of the species in the genus and their distribution. However *Phaeocystis* still holds many mysteries. Clearly, there are more species of *Phaeocystis* than presently formally recognized. Some of these are morphologically distinct, whereas others require further research to assess whether they are cryptic species or, rather, they are morphologically distinct at least in some stages of their life cycle. New avenues of molecular and morphological investigation concern the taxa known only from field material, such as the three morphotypes of *P. antarctica*, and the as yet uncultured and rare *P. scrobiculata*, or the flagellate with cup-shaped plates from South African waters (Pienaar 1991, 1996). As flagellate stages appear to be more widespread and diverse as compared to colonial stages, material to study should be gathered through specific cultivation techniques (e.g., serial dilution techniques). Whereas the function of the thin filaments has not been fully clarified, the role of the peculiar extrusomes found in PLY 559 is even more difficult to understand. Presumably these are different attachment mechanisms, but they could also be involved in potential overwintering stage formation (Gäbbler et al. unpublished observations). The molecular tools that have been used so far have significantly contributed to delineate *Phaeocystis* species. What remains to be clarified is the genetic diversity within the major species and how this diversity changes in time and space, which will require the set up of new high-resolution methods (see Gäbbler et al. (2007)) for the latest developments in population genetic analysis of *Phaeocystis*). This information, coupled with a better circumscription of species, is the prerequisite for significant advancements in the understanding of the ecology of one of the key players of the world ocean's plankton.

Acknowledgements Dr. Philipp Assmy kindly provided photographs of *P. antarctica*. Gandi Forlani provided photographs of *Phaeocystis* sp. 1 (PML 559). Photographs taken from Fig. 5.2 from Scott and Marchant (2005) were reproduced with permission from F.J. Scott and H.J. Marchant (Eds), *Antarctic Marine Protists* 258, (2005), Copyright Australian Biological Resources Study, Australian Antarctic Division and Andrew Davidson. Figures reproduced from 'Morphology, relative DNA content and hypothetical life cycle of *Phaeocystis pouchetii* (Prymnesiophyceae); with special emphasis on the flagellated cell type' by Jacobsen (2002) from Sarsia, www.tandf.no/sarsia, 2002, 87: 338–349, by permission of Taylor and Francis AS. Figures 2a, g in Vaulot et al. (1994) and Fig. 6, 9, 32 in Zingone et al. (1999) were reproduced with permission of the Phycological Society of America. This review falls within the scopes of the EU Network of Excellence MARBEF (Marine Biodiversity and Ecosystem Functioning).

References

Bätje M, Michaelis H (1986) *Phaeocystis pouchetii* blooms in the east Frisian coastal waters (German Bight, North Sea). Mar Biol 93:21–27

Baumann MEM, Jahnke J (1986) Marine Planktonalgen der Arktis. I. Die Haptophycee *Phaeocystis pouchetii*. Mikrokosmos 75:262–265

Baumann MEM, Brandini FP, Staubes R (1993) The influence of light and temperature on carbon specific DMS-release by cultures of *Phaeocystis antarctica* and three Antarctic diatoms. Mar Chem 45:56–78

Baumann MEM, Lancelot C, Brandini FP, Sakshaug E, John DM (1994) The taxonomic identity of the cosmopolitan prymnesiophyte *Phaeocystis*: a morphological and ecophysiological approach. J Mar Syst 5:23–39

Buma AGJ, Bano N, Veldhuis MJW, Kraay GW (1991) Comparison of the pigmentation of two strains of the prymnesiophyte *Phaeocystis* sp. Neth J Sea Res 27:173–182

Büttner J (1911) Die farbigen Flagellaten des Kieler Hafens. Wiss Untersuch, NF Abt Kiel 12:119–133

Chang FH (1983) The mucilage producing *Phaeocystis pouchetii* (Prymnesiophyceae) cultured from the 1981 'Tasman Bay slime'. N Z J Mar Freshw Res 17:165–168

Chrétiennot-Dinet MJ, Giraud-Guille M-M, Vaulot D, Putaux J-L, Saito Z, Chanzy H (1997) The chitinous nature of the filaments ejected by *Phaeocystis* (Prymnesiophyceae). J Phycol 33:666–672

Crame JA (1993) Latitudinal range fluctuations in the marine realm through geological times. Trends Ecol Evol 8:162–166

Chase MW, Knapp S, Cox AV, Clarkson JJ, Butsko Y, Joseph J, Savolainen V, Parokonny AS 2003. Molecular systematics, GISH and the origin of hybrid taxa in *Nicotiana* (Solanaceae). Ann Bot 92:107–127

Darling KF, Wade CM, Stewart IA, Kroon D, Dingle R, Brown AJL (2000) Molecular evidence for genetic mixing of Arctic and Antarctic subpolar populations of plankton foraminifers. Nature 405:43–47

Darling KF, Kucera M, Pudsey CJ, Wade CM (2004) Molecular evidence links cryptic diversification in polar planktonic protists to Quaternary climate dynamics. PNAS 101:7657–7662

Davidson AT (1985) Aspects of the biology of *Phaeocystis pouchetii* (Prymnesiophyceae) (Hons. Thesis). University of Tasmania

Davidson AT, Marchant H (1992) The biology and ecology of *Phaeocystis* (Prymnesiophyceae). In: Round FE, Chapman DJ (eds) Progress in phycological research, vol. 8. Biopress, Bristol, pp 1–45

Edvardsen B, Eikrem W, Green JC, Andersen RA, Moon-Van Der Staay SY, Medlin LK (2000). Phylogenetic reconstructions of the Haptophyta inferred from rRNA sequences and available morphological data. Phycologia 39:19–35

Fevolden SE, Schneppenheim R (1989) Genetic homogeneity of krill (*Euphausia superba* Dana) in the Southern Ocean. Polar Biol 9:533–539

Gäbbler S, Hayes PK, Medlin LK (2007) Methods used to reveal genetic diversity in the colony forming prymnesiophytes *Phaeocystis antarctica, P. globosa* and *P. pouchetii*—preliminary results. Biogeochemistry (this volume) doi 10.1007/s10533-007-9084-4

Hallegraeff GM (1983) Scale-bearing and loricate nanoplankton from the East Australian Current. Bot Mar 36:493–515

Hamm C, Simson DA, Merkel R, Smetacek V (1999) Colonies of *Phaeocystis globosa* are protected by a thin but tough skin. Mar Ecol Prog Ser 187:101–111

Jahnke J (1989) The light and temperature dependence of growth rate and elemental composition of *Phaeocystis globosa* Scherffel and *P. pouchetii* (Har.) Lagerh. in batch cultures. Neth. J Sea Res 23:15–21

Jahnke J, Baumann MEM (1986) Die marine Planktonalge *Phaeocystis globosa*: eine Massenform unserer Küstengewässer. Mikrokosmos 75:357–359

Jahnke J, Baumann M (1987) Differentiation between *Phaeocystis pouchetii* (Har.) Lagerheim and *Phaeocystis globosa* Scherffel. I. Colony shapes and temperature tolerances. Hydrobiol Bull 21:141–147

Jacobsen A (2000) New aspects of bloom dynamics of *Phaeocystis pouchetii* (Haptophyta) in Norwegian Waters. PhD Thesis. University of Bergen, Norway, 138 pp

Jacobsen A (2002) Morphology, relative DNA content and hypothetical life cycle of *Phaeocystis pouchetii* (Prymnesiophyceae); with special emphasis on the flagellated cell type. Sarsia 87:338–349

Jacobsen A, Bratbak G, Heldal M (1996) Isolation and characterization of a virus infecting *Phaeocystis pouchetii* (Prymnesiophyceae). J Phycol 32:923–927

Karsten G (1905) Das Phytoplankton des Antarktischen Meeres nach dem Material der Deutschen Tiefsee-Expedition 1898–1899. Wiss. Ergeb. Deutch. Tief. Exp. Valdivia 1898–1899 Band I. , Teil 2, 136 pp

Keller MD, Ellows WKB, Guillard RL (1989) Dimethyl sulfide production in marine phytoplankton. In: Saltzmann E, Cooper W (eds) Biogenic Sulfur in the Environment. American Chemical Society, Washington DC, pp 167–182

Kooistra WHCF, Medlin LK (1996) Evolution of the diatoms (Bacillariophyta): IV. A reconstruction of their age from small subunit rRNA coding regions and the fossil record. Mol Phyl Evol 6:391–407

Kornmann P (1955) Beobachtungen an *Phaeocystis*-Kulturen. Helgol Wiss Meeres 5:218–233

Lagerheim G (1893) *Phaeocystis* nov. gen. grundadt på *Tetraspora poucheti* Har. Bot Not 1:32–33

Lancelot C, Billen G, Sournia A, Weisse T, Colijn F, Veldhuis MJW, Davies A, Wassmann P (1987) *Phaeocystis* blooms and nutrient enrichment in the continental coastal zones of the North Sea. Ambio 16:38–46

Lange M (1997) Molecular genetic investigation within the genus *Phaeocystis* (Prymnesiophyceae). P.D. dissertation, University of Bremen, Germany, 170 pp

Lange M, Chen Y-Q, Medlin LK (2002) Molecular genetic delineation of *Phaeocystis* species (Prymnesiophyceae) using coding and non-coding regions of nuclear and plastid genomes. Eur J Phycol 37:77–92

Larsen J, Moestrup Ø (1989) Guide to toxic and potentially toxic marine algae. Fish Inspection Service, Minister of Fisheries, Copenhagen

Medlin LK, Lange M, Baumann MEM (1994) Genetic differentiation among three colony-forming species of *Phaeocystis*: further evidence for the phylogeny of the Prymnesiophyta. Phycologia 33:199–212

Marchant HJ, Davison AT, Kelly GY (1991) UV-B protecting compounds in the marine alga *Phaeocystis pouchetii* from Antarctica. Mar Biol Berlin 109:391–395

Marchant HJ, Scott FJ, Davidson ST (2005) Haptophytes: Order Prymnesiales. In: Scott FJ, Marchant HJ (eds) Antarctic marine protists. Australian Biological Resources Study, Canberra, pp 255–275

Moestrup Ø (1979) Identification by electron microscopy of marine nanoplankton from New Zealand including the description of four new species. N Z J Bot 17:61–95

Moestrup Ø, Larsen J (1992) Potentially toxic phytoplankton 1. Haptophyceae (Prymnesiophyceae). In: Lindley S (ed) ICES Identification leaflets for plankton, leaflet No. 179. Natural Environmental Research Council, Plymouth, pp 1–11

Montresor M, Lovejoy C, Orsini L, Procaccini G (2003) Bipolar distribution of the cyst-forming dinoflagellate *Polarella glacialis*. Polar Biol 26:186–194

Olbers D, Gouretski V, Seiss G, Shröter J (1962) Hydrographic atlas of the Southern Ocean. Druckhaus Nord: Bremerhaven. 82 plates

Parke M, Green JC, Manton I (1971) Observations on the fine structure of zoids of the genus *Phaeocystis* (Haptophyceae). J Mar Biol Assoc UK 51:927–941

Pienaar RN (1991) Thread formation in the motile cells of *Phaeocystis*. Electr. Microsc Soc S Africa 21:135–137

Pienaar RN (1996) Observations on the disc bearing phase of *Phaeocystis* in South African Waters. In: Borg M, Semesi A, Pederson M, Bergman B (eds) Current trends in marine botanical research in the East African Region (77–98) Sida, Marine Science Program, Department Of Research Co-Operation, SAREC. ISBN> 91-630-4907-4

Pouchet G (1892) Sur une algue pélagique nouvelle. Compte Rendues séance à 16 Janvier 44:34–36

Rousseau V, Vaulot D, Casotti R, Cariou V, Lenz J, Gunkel J, Baumann M (1994) The life cycle of *Phaeocystis* (Prymnesiophyceae): evidence and hypotheses. J Mar Syst 5:23–39

Scherffel A (1900) *Phaeocystis globosa* nov. spec. nebst einigen Betrachtungen über die Phylogenie niederer, insbesondere brauner Organismsen. Wiss Meer Abt Helgo 4:1–28

Schoemann V, Becquevort S, Stefels J, Rousseau V, Lancelot C (2005) *Phaeocystis* blooms in the global ocean and their controlling mechanisms: a review. J Sea Res 53:43–66

Scott FJ, Marchant HJ (2005) Antarctic marine protists. Australian Biological Resources Study, Canberra

Shen P, van Rijssel M, Wang Y, Songhui L, Jufang C, Qi Y (2004) Toxic *Phaeocystis globosa* strain from China grow at remarkably high temperatures. In: Steidinger KA, Landsberg JH, Tomas CR, Vargo GA (eds) Harmful Algae 2002. Florida Fish and Wildlife Conservation Commission, Florida Institute of Oceanography and Intergovernmental Oceanographic Commission of UNESCO, St. Petersburg, pp 396–398

Smith WO, Codispoti LA, Nelson DM, Manley T, Buskey EJ, Niebauer HJ, Cota GF (1991) Importance of *Phaeocystis* blooms in the high-latitude ocean carbon cycle. Nature 352:514–516

Sournia A (1988) *Phaeocystis* (Prymnesiophyceae): How many species? Nova Hedwigia 47:211–217

Treshnikov AF (1964) Surface water circulation in the Antarctic Ocean. In: Russian. Sovet. Antarkticheskaia Eksped., Inform. biull., 45:5–8. Eng. transl. (1965) In: Soviet Antarctic Expedition, Information Bulletin, 5:81–83

Vaulot D, Birrien J-L, Marie D, Casotti R, Veldhuis MJW, Kraay GW, Chrétiennot-Dinet M-J (1994) Morphology, ploidy, pigment composition and genome size of cultured strains of *Phaeocystis* (Prymnesiophyceae). J Phycol 30:1022–1035

Veldhuis MJW, Wassmann P (eds) (2005) Bloom dynamics and biological control of *Phaeocystis*: a HAB species in European coastal waters. Harmful Algae 4:805–964

Verity PG, Villareal TA, Smayda TJ (1988) Ecological investigations of blooms of colonial *Phaeocystis pouchetii*. II. The role of life-cycle phenomena in bloom termination. J Plankton Res 10:749–766

Whipple SJ, Patten BC, Verity PG (2005) Life cycle of the marine alga *Phaeocystis*: A conceptual model to summarize literature and guide research. J Mar Syst 57:83–110

Zingone A, Chrétiennot-Dinet MJ, Lange M, Medlin LK (1999) Morphological and genetic characterization of *Phaeocystis cordata* and *P. jahnii* (Prymnesiophyceae), two new species from the Mediterranean Sea. J Phycol 39:1322–1337

Biogeochemistry (2007) 83:19–27
DOI 10.1007/s10533-007-9084-4

SHORT COMMUNICATION

Methods used to reveal genetic diversity in the colony-forming prymnesiophytes *Phaeocystis antarctica*, *P. globosa* and *P. pouchetii*—preliminary results

Steffi Gaebler · Paul K. Hayes · Linda K. Medlin

Received: 12 October 2005 / Accepted: 17 March 2006 / Published online: 12 April 2007

Abstract Previous work on the genetic diversity of *Phaeocystis* used ribosomal DNA and internal transcribed spacer (ITS) sequence analyses to show that there is substantial inter- and intraspecific variation within the genus. First attempts to trace the biogeographical history of strains in Antarctic coastal waters were based on a comparison of ITS sequences. To gain deeper insights into the population structure and bloom dynamics of this microalga it is necessary to quantify the genetic diversity within populations of *P. antarctica* from different locations (i.e., each of the three major gyres in the Antarctic continental waters) and to calculate the gene flow between them. Here we describe methods to quantify genetic diversity and our preliminary results for *P. antarctica* in comparison to two other colonial species: *P. globosa* and *P. pouchetii*. For this study of genetic diversity, two fingerprinting techniques were used. First, amplified fragment-length polymorphisms (AFLPs) were established as a pre-screening tool to assess clone diversity and to select divergent clones prior to physiological investigations. Second, the more-powerful microsatellite markers were established to assess population structure and biogeography more accurately. Results show differences in the AFLP patterns between isolates of *P. antarctica* from different regions, and that a wide variety of microsatellite motifs could be obtained from the three *Phaeocystis* species.

Keywords AFLP Microsatellite marker · *Phaeocystis* · *P. antarctica* · *P. globosa* · *P. pouchetii*

S. Gaebler (✉) · L. K. Medlin
Alfred Wegener Institute for Polar and Marine Research, Am Handelshafen 12, 27570 Bremerhaven, Germany
e-mail: sgaebler@awi-bremerhaven.de

P. K. Hayes
School of Biological Sciences, University of Bristol, Woodland Road, Bristol, BS8 1UG, UK

Introduction

The genus *Phaeocystis* was erected by Lagerheim (1893/1896) to accommodate the colonial alga *Tetraspora pouchetii* described by Hariot in Pouchet (1892). The newly combined species, *Phaeocystis pouchetii*, can be found in Arctic waters. Two other colonial species were described soon after that: *P. globosa* by Scherffel (1899, 1900) from temperate waters and *P. antarctica* by Karsten (1905) from the Antarctic. Kornmann (1955) expressed doubt that these species were separate and lumped up all colonial species into a single taxon *Phaeocystis pouchetii*. Despite physiological studies (Baumann and Jahnke 1986; Jahnke and Baumann 1986, 1987; Jahnke 1989) that confirmed the separation of these taxa, it took a molecular study (Medlin et al. 1994) to end the controversy over the validity of the three colonial species. Since that time further molecular studies using other gene loci have confirmed the separation of these taxa (Lange et al. 2002) and other species have also been included in the genus (Zingone et al. 1999).

Phaeocystis is known to play an important role in ecology and biogeochemistry because it is distributed worldwide and forms massive blooms. Its blooms can fix a high amount CO_2 and produce a substantial amount of dimethylsulfoniopropionate (DMSP), which is the biological precursor of the climatically important trace gas of the atmosphere dimethylsulfide (DMS) (Smith et al. 1991; Stefels 1997; Arrigo 1999; Verity and Smetacek 1996). *P. antarctica* is widely distributed in the Southern Ocean where it is among the most abundant primary producers and is thus a major contributor to organic matter vertical fluxes. It is known from physiological studies from many phytoplankton species that there is a high variability among strains for every trait examined (Wood and Leatham 1992). Thus, a study of the genetic diversity and gene flow among *Phaeocystis* strains around the Antarctic is both timely and necessary. To pursue this we chose two techniques to assess the genetic diversity within and among *Phaeocystis* spp. One technique, amplified fragment-length polymorphisms, (AFLPs) provides a rapid means to screen the entire genome for polymorphic genetic loci as a pre-screening tool to select most divergent clones for physiological investigations, whereas analysis of microsatellite loci provides a more rigorous method by which population genetic statistics can be applied to assess the genetic diversity in populations and gene flow between them. Both of these techniques will be applied to each of the three colonial species, but we report here preliminary data to establish these techniques for *P. antarctica*.

Amplified fragment length polymorphism

The AFLP technique developed by Vos et al. (1995) was used as described in John et al. (2004). Genomic DNA of 48 *Phaeocystis* strains previously extracted in earlier studies (Medlin et al. 1994; Lange et al. 2002) was digested over night at 37°C with two restriction enzymes (*Eco*RI and *Mse*I, New England BioLabs, Frankfurt a. Main, Germany). This enzyme combination consists of a rare (*Eco*RI, six-base-pair recognition sequence) and a frequent (*Mse*I, four-base-pair recognition sequence) cutter. Subsequent to digestion, site-specific adapters were ligated to the ends of the restriction fragments (see Fig. 1).

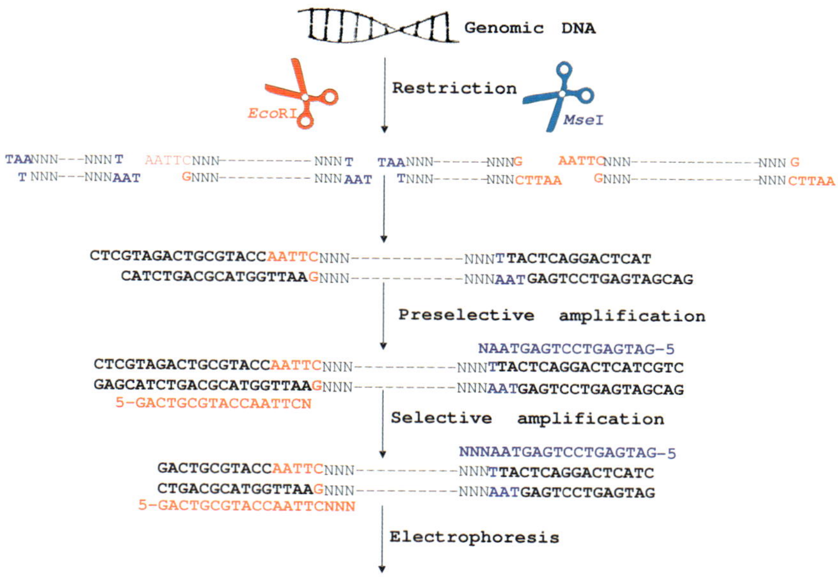

Fig. 1 Schematic diagram showing the construction of the AFLP fragments (from Müller 2005)

To amplify a subset of DNA restriction fragments, one to six additional nucleotides can be added to the 3′-end of primers complementary to the adapter and restriction site sequences (Vos et al. 1995). With two additional polymerase chain reaction (PCR) steps, the preselective and the selective amplification, only a specific subset of fragments is amplified, so that the number of the resulting PCR products provides sufficient resolution among the strains. Too many fragments make analysis difficult, whereas too few do not provide enough resolution. A primer set with one selective nucleotide per primer for the preselective amplification and different numbers of selective nucleotides per primer for the selective amplification were compared (Table 1). To be able to visualize the amplification products one primer was fluorescently labelled for the second amplification (*Eco*RI*6Fam + NNN, Applied Biosystems, Germany). After the second amplification, the samples were electrophoresed using a capillary sequencer (ABI PRISM 3100 AVANT, Applied Biosystems, Germany). For this part of the AFLP optimization, 10 primer sets for the selective amplification were tested using only four out of the 48 different *P. antarctica* isolates (Table 3). In Fig. 2 eight of these primer sets are compared for the isolate *P. antarctica* A1-3 because two of them produced no products. For all four isolates tested it was found that fewer selective nucleotides used in the second amplification step provided better resolution. Primer set *Mse*I + **CT**/*Eco*RI + **AT** seemed to be the best choice because it not only gave greater resolution of the fragment pattern from zero to 500 bp (base pairs), but also more-intense fragments (Fig. 2).

Table 1 Selective nucleotides (SN) used for selective amplification. X = these primer sets have been used

Primer		*Mse*I + SN		
		C	CC	CT
*Eco*R1 + SN	AA			
	AC			X
	AG			X
	AT			X
	AAG	X	X	X
	ATG	X	X	X

Fragment patterns for the *Phaeocystis* isolates were compared for this primer set (Fig. 3) and for primer set *Mse*I + **CT**/*Eco*RI + **AA** (Fig. 4). Differences and similarities among the isolates were easily noted by the presence of bands shared between the isolates. Differences in the genetic fingerprints could result from their biogeographical distance and the supposed gene flow around the Antarctic waters. Comparing Figs. 3 and 4 the band patterns of all isolates *are* distinct from one another. These differences are more clearly shown with the primer set *Mse*I + **CT**/*Eco*RI + **AT** (Fig. 3). From the fragment patterns of the *P. antarctica* isolates obtained with this primer set, a distance matrix (fragment presence/absence) was manually calculated for amplification fragments of between 100 bp and 500 bp only to illustrate the similarity and differences among these four strains. Fragment lengths from 100 bp to 500 bp were arbitrarily chosen because this size range appeared to present those lengths that could be most easily scored. The presence/absence of 41 bands was scored between these lengths. Peak heights below 200 fluorescence units (see Figs. 2, 3 and 4) were ignored: fragment lengths <100 bp were not clearly resolved. Maximum parsimony analysis of the AFLP data set (Fig. 5: PAUP, Swofford 2002) shows that genetic diversity of *P. antarctica* in the Antarctic region is highly variable. The isolate *P. antarctica* SK22 from the Antarctic circumpolar current (ACC) (Fig. 6) was defined as the outgroup based on internal transcribed spacer (ITS) sequence data (Lange et al. 2002). Isolate SK21 (Weddell Sea) and D5 (Prydz Bay) seem to be more closely related to each other than to A1-3 (Prydz Bay), which, to judge by its branch length, contains more unique fragment lengths. The divergence of A1-3 (Prydz Bay) before the divergence of SK21 (Weddell Sea) and D5 (Prydz Bay) was also recovered in the ITS analysis of Lange et al. (2002). The next step is to increase the resolution of the amplification patterns and to optimize primer sets for all *Phaeocystis* strains. Reproducibility of the AFLP technique must be assessed by performing replicate reactions using more than one DNA isolation from the same strain; however we have, with our limited strain analysis, recovered the same tree topology as the

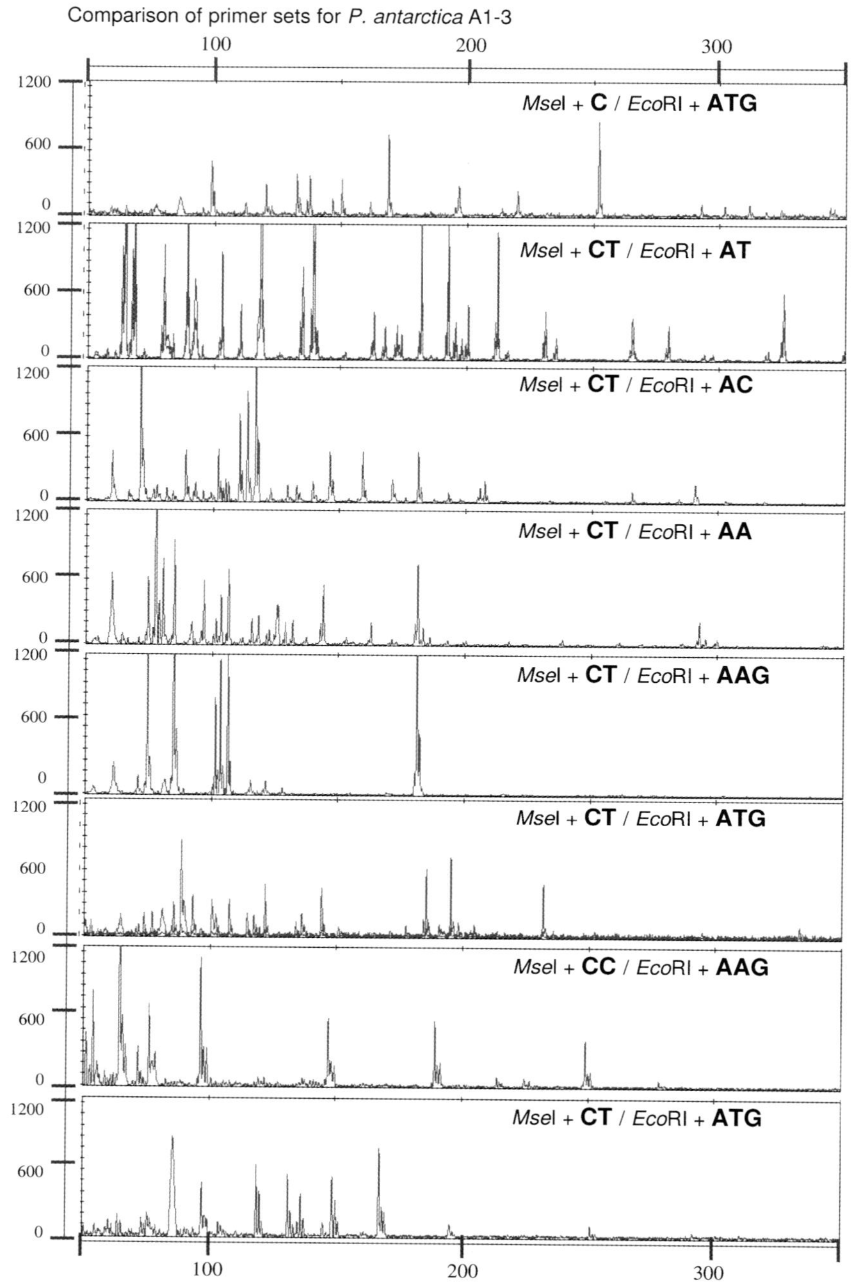

Fig. 2 AFLP electropherograms (resolution from 50–350 base pairs) of eight different primer sets for *P. antarctica* A1-3 (Prydz Bay, Fig. 6)

ITS analysis with more strains (Lange et al. 2002).

Microsatellite markers

Total DNA was isolated from *P. globosa* (CCMP1528), *P. pouchetii* (SK 34) and *P. antarctica* (SK 23) (Table 2) using a modified hexadecyltrimethylammonium bromide (CTAB) extraction protocol (Doyle and Doyle 1990; Lange 1997). Nuclear DNA was purified by ultracentrifugation through a caesium chloride—ethidium bromide density gradient (Lange 1997). This DNA was used to create enriched microsatellite libraries

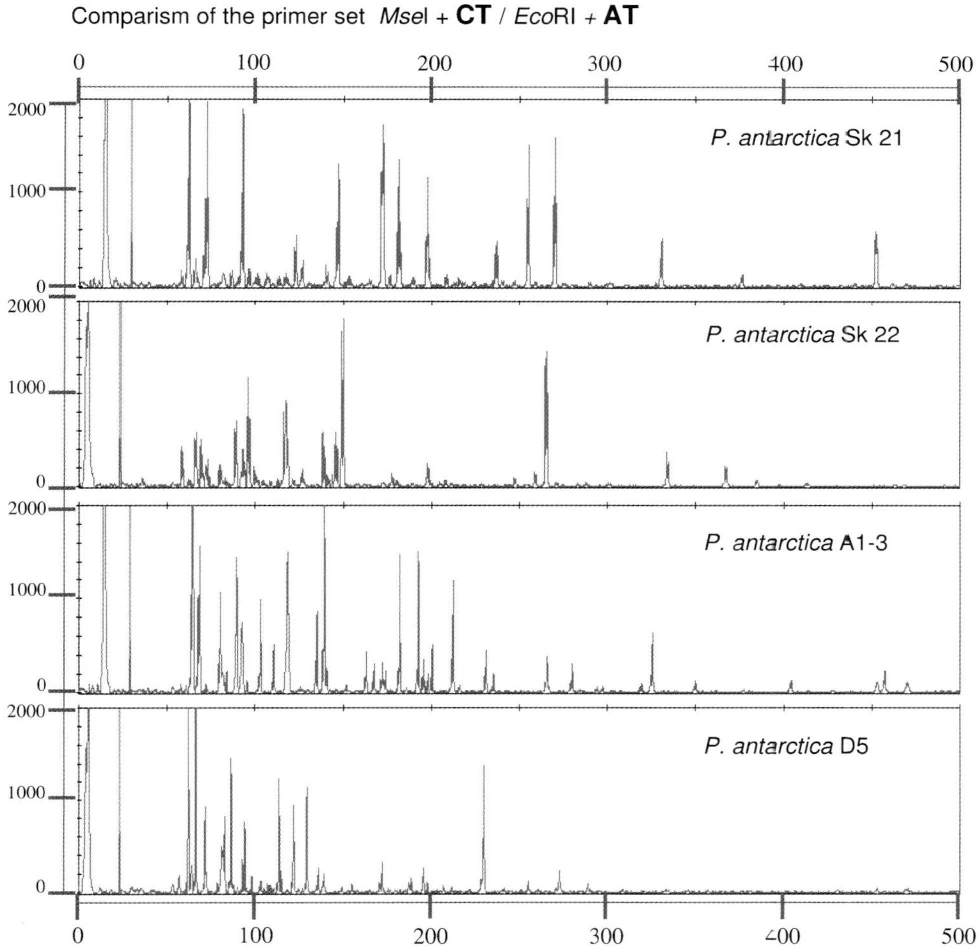

Fig. 3 AFLP electropherograms (resolution from 0–500 base pairs) of the primer set *Mse*I + **CT**/*Eco*RI + **AT** for the *P. antarctica* isolates (Table 2)

for the three colony-forming species. The protocol for the construction of these libraries was modified slightly from that of Evans and Hayes (2004), which was based on Edwards et al. (1996). Two biotinylated microsatellite oligonucleotides Bio-GA and Bio-GT (Thermo Electron GmbH, Germany) were immobilized on magnetic beads (Dynabeads, Dynal Biotech Invitrogen, Germany). These oligonucleotides were used to bind and capture CT and CA motifs in restriction fragments to which oligonucleotide adapters had been ligated. A second round of enrichment was achieved by recapturing post-amplification Dynabead products and repeating all subsequent steps in the protocol (Evans and Hayes 2004). Enrichment PCR fragments were cloned into TOPO vector (pCR®2.1-Topo, Invitrogen, Germany) and plasmid DNA was isolated from *Escherichia coli* using the mini-preparation-scale Wizard™ Minipreps DNA Purification System (Serva, Heidelberg, Germany). The resultant Miniprep DNA was washed with 70% EtOH and resuspended in H_2O. Cloned fragments were directly sequenced using the BigDye® Terminator v3.1 Cycle sequencing Kit (Applied Biosystems, Germany) and the vector primer set: M13 HedgeF—5′-GTTTTCCCA GTCACGACGTTG 3′; M13 HedgeR—5′-TGA GCGGATAACAATTTCACACAG 3′ (Operon, Germany).

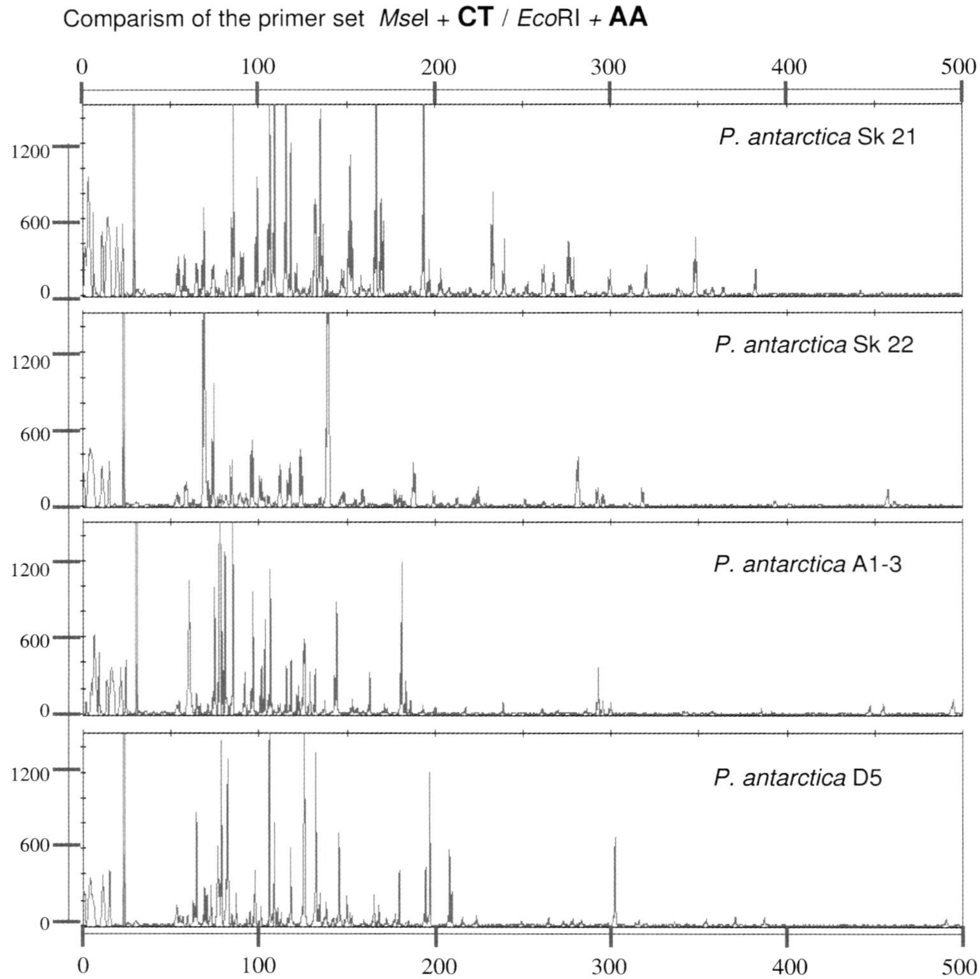

Fig. 4 AFLP electropherograms (resolution from 0–500 base pairs) of the primer set *Mse*I + **CT**/*Eco*RI + **AA** for the *P. antarctica* isolates (Table 2)

Eighteen cloned fragments from *P. antarctica* and 17 cloned fragments from both *P. globosa* and *P. pouchetii* were sequenced. Of these, 14 cloned fragments from *P. antarctica* and nine cloned fragments from *P. globosa* and *P. pouchetii* contained repeating sequence motifs. Preliminary results indicate a very high variability in the microsatellite motifs (Table 3). At least 50 further clones will be sequenced from each species to allow the characterization of additional microsatellite repeats and their flanking sequences. Primer pairs that anneal within the flanking sequences either side of the microsatellite repeats will be designed with the program PRIMER 3.0 (http://www.frodo.wi.mit.edu/cgi-bin/primer3/primer3_www.cgi). Of the microsatellite motifs shown in Table 3, only the simple repeats will be investigated further: complicated nested motifs are too difficult to analyse.

These initial results suggest that *Phaeocystis* is genetically diverse and that this diversity can be quantified and interpreted based on the use of AFLPs and MS. With the microsatellites, we will be able to calculate gene flow around the Antarctic for *P. antarctica*.

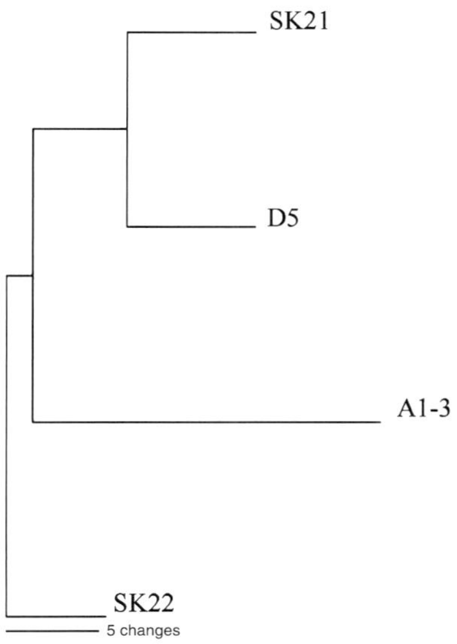

Fig. 5 Most parsimonious phylogram of ± presence of fragment lengths of the *P. antarctica* isolates obtained from the primer set *Mse*I + **CT**/*Eco*RI + **AT**

Table 2 Algal species used to test best primer set and to create microsatellite-enriched libraries

Strains	Culture number, Culture facility[b,c,d]	Geographic origin
P. antarctica	SK 21[c]	Antarctica, 65°15′ S, 39°22′ W
P. antarctica	SK 22[a,c]	Antarctica, 54°20′ S, 3°20′ W
P. antarctica	A1-3[a,b]	Antarctica, 63°11.5′ S, 85°45.3′ E
P. antarctica	D5[a,b]	Antarctica, 68°47.5′ S, 73°30.2′ E
P. globosa[e]	CCMP1528[d]	Galapagos
P. pouchetii[e]	SK 34[a,b]	Greenland Sea, East Greenland current
P. antarctica[e]	SK 23[a,b]	Antarctica, 63°15′ S, 58°20′ W

[a] Culture no longer available

[b] CSIRO Division of Fisheries, Hobart, Tasmania, Australia

[c] Alfred-Wegener-Institute for Polar and Marine Research, Am Handelshafen 12, Bremerhaven, Germany

[d] Provasoli-Guillard Culture Center for Marine Phytoplankton, Bigelow Laboratory for Ocean Sciences, West Boothbay Harbour, Maine, USA

[e] Used for microsatellite-enriched library

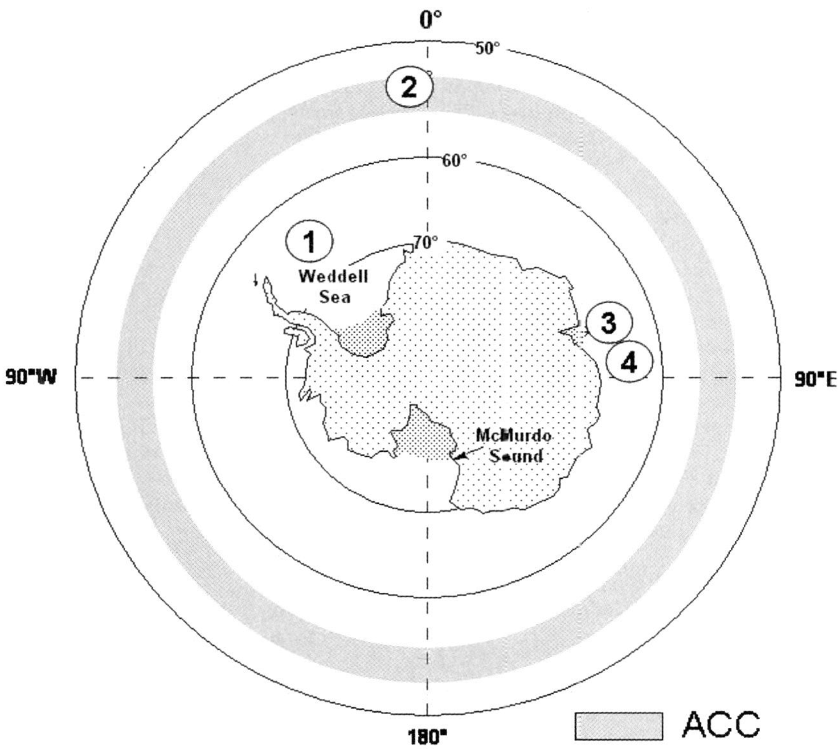

Fig. 6 Geographic origin of the four *P. antarctica* strains (Table 2). To indicate the position of the Antarctic circumpolar current (ACC) encircling the Antarctic a schematic presentation was chosen (Lange et al. 2002; Olbers et al. 1962). [1: SK21, 2: SK22, 3: D5, 4: A1-3]

Table 3 Microsatellite (MS) motifs of the investigated *Phaeocystis* strains. Some clones contained multiple motifs

Microsatellite motifs	*P. globosa*	*P. pouchetii*	*P. antarctica*
$(TG)_n$	4	1	6
$(CA)_n$	1	3	3
$(GA)_n$		2	1
$(TC)_n$		1	2
$(CAA)_n$			1
$(TTAC)_n$	1		
$(GCTC)_n$		1	
$(TATG)_n$			1
$[(TG)_n(AG)_n]_n$			2
$[(CA)_n(TA)_n]_n$			1
$(TTAGGG)_n$	1	1	
$[(TG)_nCG(TG)_n]_n$	1	5	
$[TTT(GT)_nGC]_n$	3		
	17 clones 9 with MS	17 clones 9 with MS	18 clones 14 with MS

Acknowledgement This research was funded by the German Science Foundation (DFG) through a postgraduate research fellowship (ME 1480/2).

References

Arrigo KR, Robinson DH, Worthen DL, Dunbar RB, DiTullio GR, van Woert M, Lizotte MP (1999) Phytoplankton community structure and the drawdown of nutrients and CO_2 in the Southern Ocean. Science 283:365–367

Baumann MEM, Jahnke J (1986) Marine Planktonalgen der Arktis. I. Die Haptophycee *Phaeocystis pouchetii*. Mikrokosmos 75:262–265

Doyle JJ, Doyle JL (1990) Isolation of plant DNA from fresh tissue. Focus 12:13–15

Edwards KJ, Barker JHA, Daly A, Jones C, Karp A (1996) Microsatellite libraries enriched for several microsatellite sequences in plants. Biotechniques 20:758–760

Evans KM, Hayes PK (2004) Microsatellite markers for the cosmopolitan marine diatom *Pseudo-nitzschia pungens*. Mol Ecol Note 4:125–126

Jahnke J, Baumann MEM (1986) Die marine Planktonalge *Phaeocystis globosa*: eine Massenform unserer Küstengewässer. Mikrokosmos 75:357–359

Jahnke J, Baumann MEM (1987) Differentiation between *Phaeocystis pouchetii* (Har.) Lagerheim and *Phaeocystis globosa* Scherffel I. Colony shapes and temperature tolerances. Hydrobiol Bull 21:141–147

Jahnke J (1989) The light and temperature dependence of growth rate and elemental composition of *Phaeocystis globosa* Scherffel and *P. pouchetii* (Har.) Lagerh. in batch cultures. Neth J Sea Res 23:15–21

John U, Groben R, Beszteri B, Medlin LK (2004) Utility of Amplified Fragment Length Polymorphisms (AFLP) to analyse genetic structures within the *Alexandrium tamarense* species complex. Protist 155:169–179

Karsten G (1905) Das Phytoplankton des Antarktischen Meeres nach dem Material der deutschen Tiefsee-Expedition 1898–1899. In: Chun C (ed) Wissenschaftliche Ergebnisse der deutschen Tiefsee-Expedition auf dem Dampfer "Valdivia" 1898–1899. Gustaf Fischer Verlag, Jena

Kornmann P (1955) Beobachtungen an *Phaeocystis*-Kulturen. *Helgoländer* wiss. Meeresunters 5:218–233

Lagerheim G (1893) *Phaeocystis*, nov. gen., grundadt pa *Tetraspora Poucheti* Har. Botaniska Notiser. 32–33

Lagerheim G (1896) Über Phaeocystis *Poucheti* (Har.) Lagerh., eine Plankton-Flagellate. Öfvers. af Vet. Akad Förhandl 53:277–288

Lange M (1997) Phylogeny and Taxonomy of the genus *Phaeocystis* (Prymnesiophyceae). Dissertation, University of Bremen, Germany, 170 pp

Lange M, Chen Y, Medlin LK (2002) Molecular genetic delineation of *Phaeocystis* species (Prymnesiophyceae) using coding and non-coding regions of nuclear and plastid genomes. Eur J Phycol 37:77–92

Medlin LK, Lange M, Baumann MEM (1994) Genetic differentiation among three colony-forming species of *Phaeocystis*. Phycologia 33:199–212

Müller J (2005) Genetic fingerprints of microalgal culture strains: amplified fragment length polymorphism (AFLP) for investigations below the species level. Dissertation, University of Goettingen, Germany, 100 pp

Olbers D, Gouretski V, Seiss G, Shröter J (1962) Hydrographic Atlas of the Southern Ocean. *Druckhaus Nord*, Bremerhaven, 82 plates

Pouchet MG (1892) Sur une algue pélagique nouvelle. Compt. Rend. Hebd. Séances Mém. Soc Biol 44:34–36

Scherffel A (1899) *Phaeocystis globosa* n.sp. (Vorläufige Mitteilung). Berichte der Deutschen Botanischen Gesellschaft 17:317–318

Scherffel A (1900) *Phaeocystis globosa* nov. spec. nebst einigen Betrachtungen, Über die Phylogenie niederer, insbesondere brauner Organismen. Wissenschaftliche Meeresuntersuchungen Abteilung Helgoland N.F. Bd. 4:1–29

Smith WO, Codispoti LA, Nelson DM, Manley T, Buskey EJ, Niebauer HJ, Cota GF (1991) Importance of *Phaeocystis* blooms in the high-latitude ocean carbon cycle. Nature 352:514–516

Stefels J (1997) The smell of the Sea. Production of dimethylsulphoniopropionate and its conversion into dimethylsulphide by the marine phytoplankton genus *Phaeocystis*. Dissertation, University of Groningen

Swofford DL (2002) PAUP*: Phylogenetic Analysis Using Parsimony (and Other Methods) 4.0 Beta, Florida State University, CD-ROM

Verity PG, Smetacek V (1996) Organism life cycles, predation, and the structure of marine pelagic ecosystems. Mar Ecol Prog Ser 130:277–293

Vos P, Hogers R, Bleeker M, Reijans M, Vandelee T, Hornes M, Frijters A, Pot J, Peleman J, Kuiper M, Zabeau M (1995) AFLP—a new technique for DNA fingerprinting. Nucleic Acids Res 23:4407–4414

Wood AM, Leatham T (1992) The species concept in phytoplankton ecology. J Phycol 28:723–729

Zingone A, Crétiennot-Dinet MJ, Lange M, Medlin LK (1999) Morphological and genetic characterisation of *Phaeocystis cordata* and *P. jahnii* (Prymnesiophyceae), two new species from the Mediterranean Sea. J Phycol 35:1322–1337

Biogeochemistry (2007) 83:29–47
DOI 10.1007/s10533-007-9085-3

REVIEW PAPER

The life cycle of *Phaeocystis*: state of knowledge and presumptive role in ecology

Véronique Rousseau · Marie-Josèphe Chrétiennot-Dinet · Anita Jacobsen · Peter Verity · Stuart Whipple

Received: 9 March 2006 / Accepted: 19 May 2006 / Published online: 13 April 2007

Abstract Despite numerous investigations, the number and role of morphotypes involved in the life cycle of *Phaeocystis* species remain under debate. This is partly due to the application of different methodologies such as light, transmission, scanning electron microscopy and flow cytometry on specific samples. This heterogeneity of approaches results in the incomplete morphometric description of the different cell types existing within one species according to relevant criteria and the indetermination of the ploidy level of each observed stage. We review here the different morphotypes observed within each of the six *Phaeocystis* species recognized up to now. Four different cell types have been observed. In common to all six species is the occurrence of a scaly flagellate producing star-forming filaments (all species except *P. jahnii*) or not (*P. globosa* and *P. jahnii*). In three colony-forming species, *P. globosa*, *P. pouchetii* and *P. antarctica*, three morphotypes are observed: a flagellate with scales and filaments, a colonial cell, and a flagellate devoid of scales and filaments. In the non-colony-forming species, *P. scrobiculata* and *P. cordata*, only flagellates with scales and filaments have been observed. While suspected in *P. pouchetii* and *P. antarctica*, a haploid–diploid life cycle has only been evidenced for *P. globosa*. The two main prominent features of this cycle are that sexuality is prevalent in colony bloom formation and termination and that two types of vegetative reproduction exist. The ecological relevance of alternating haploid and diploid stages is not clearly apparent on the basis of existing ecological studies.

Keywords Ecological niche · Haploid–diploid · Life cycle stages · Morphotype · *Phaeocystis* species · Sexual processes

V. Rousseau (✉)
Ecologie des Systèmes Aquatiques, ULB, CP 221, Boulevard du Triomphe, 1050 Brussels, Belgium
e-mail: vrousso@ulb.ac.be

M.-J. Chrétiennot-Dinet
Observatoire Biologique de Banyuls, UMR 7621, Lab. Arago, BP 44, 66651 Banyuls-sur-Mer, France

A. Jacobsen
Department of Biology, University of Bergen, PO Box 7800, 5020 Bergen, Norway

P. Verity · S. Whipple
Skidaway Institute of Oceanography, 10 Ocean Science Circle, Savannah, GA, 31411, USA

Introduction

The genus *Phaeocystis* is a worldwide colony-blooming species that has a significant role in biogeochemical cycles (Schoemann et al. 2005) including the global sulphur cycle (Liss et al. 1994). Despite numerous investigations devoted to its ecophysiology and the role and impact of colonies in ecosystem processes, knowledge of some major biological features of the genus is still limited. Such is the case for the *Phaeocystis* life cycle and its controlling mechanisms that, 50 years after Kornmann's (1955) classic paper, remain under debate. Problems of taxonomic confusion, lack of fine morphometric description of the different cell types within one species, and inadequacy of cell nomenclature have precluded a complete understanding of the *Phaeocystis* life cycle.

Since early description of the genus *Phaeocystis* by Lagerheim in 1893, the number of inclusive species has long been a matter of discussion (e.g. Kornmann 1955; Kashkin 1963; Parke et al. 1971; Sournia 1988; Medlin et al. 1994). This is mainly because the criteria used to distinguish *Phaeocystis* species were based on phenotypic characters such as the morphometry of the colonial stage, and/or physiological and biochemical properties (Jahnke and Baumann 1987; Baumann et al. 1994; Vaulot et al. 1994). Six species are now recognized based on small subunit (SSU) rDNA sequence analysis and morphological characterization. These are: *P. antarctica* Karsten, *P. globosa* Scherffel, *P. pouchetii* (Hariot) Lagerheim, *P. jahnii* Zingone, *P. scrobiculata* Moestrup and *P. cordata* Zingone et Chrétiennot-Dinet (Moestrup 1979; Medlin et al. 1994; Zingone et al. 1999; Edvardsen et al. 2000; Lange et al. 2002). Colonial forms have been reported for the first four species. It is now considered that probably more than six *Phaeocystis* species exist (Lange et al. 2002; Medlin and Zingone this issue).

Comparative descriptions of cell types existing within one species using morphometric criteria, i.e., presence/absence of body scales, flagella, haptonema and star-forming filaments, and ploidy levels have been made (e.g. Zingone et al. 1999; Peperzak et al. 2000a). However, a complete study of all morphotypes occurring within one species is still missing (Lancelot and Rousseau 2002). Our current knowledge of *Phaeocystis* cell types relies on composite independent investigations combining light (LM), transmission (TEM) and scanning electron microscopy (SEM) as well as flow cytometry. Each of these methodologies provides part of the information needed for a complete identification of the morphotype. LM is useful for observations of cell shape, size, number, presence of flagella, and swimming activity. SEM and TEM with higher resolution and magnification are needed for morphological and ultrastructural details of the cell covering, appendages and organelles. Flow cytometry is required for determining the ploidy levels of each cell type. In addition, it is essential that sample preservation and fixation procedures be fully described because such procedures may lead to methodological biases. Use of fixatives can indeed cause cell shrinkage, loss of appendages (Peperzak et al. 2000a; Wassmann et al. 2005) or colony disintegration, releasing colonial cells into the medium, and therefore lead to possible misinterpretation (Wassmann et al. 2005). This mixed approach results in a confuse nomenclature of the various cell types, i.e. solitary flagellates and nonflagellates, free-living single cells, colonial flagellates, motile free-living cells, swarmers, zoids, microflagellates and microzoospores. These terms are often used loosely, and this can lead to misinterpretation of life cycle events.

The number and role of cell types involved in the life cycle of the six *Phaeocystis* species, and whether these are the same within each species, are still among the main questions not yet resolved (Lancelot and Rousseau 2002). Of particular interest is the identification of the stage persisting between two colony bloom events, as well as the nature of colony-forming cells. The persistence of *Phaeocystis* as a flagellate between two colony blooms has been suggested (Kornmann 1955; Parke et al. 1971; Veldhuis et al. 1986; Verity et al. 1988b), but the type of flagellate was never described from field observations due to its low cell density and possible confusion with other nanoplanktonic species. On the other hand, senescent colonies or aggregates have also been proposed as over-wintering forms of *P. globosa* (Cadée 1991). Still unknown are factors responsible for the transition between life stages. The ecological significance of the different life cycle stages, flagellates and colonies, has recently been discussed by Verity and Medlin (2003). Further investigation is however needed to discriminate between the different flagellates that have been identified within some species.

In this paper, we review and synthesize the available information gained from field and culture observations on the different morphotypes occurring within each of the six *Phaeocystis* species, focusing on cell morphology and ploidy level. We also present unpublished data on the morphology of *P. globosa* and *P. antarctica* cells. The role of the different morphotypes within the life cycle will be addressed based on field and culture observations on their occurrence. Finally, the ecological relevance of the free-living and colonial stages will be discussed based on knowledge of physiology and trophic significance of the various morphotypes. Because it is the best-known species, *P. globosa* will be used as a model.

Morphotypes among *Phaeocystis* species

This section reviews the different cell types reported for the six *Phaeocystis* species which are considered here according to their revised taxonomic status as recommended by Baumann et al. (1994), Medlin et al. (1994) and Vaulot et al. (1994). This is particularly relevant for the species *globosa*, which has long been referred to as *pouchetii* in the previous literature (e.g., Parke et al. 1971; Kayser 1970; Admiraal and Venekamp 1986; Veldhuis et al. 1991; Davidson and Marchant 1992a). The seasonal distribution of the different cell types in the natural environment will also be considered in order to assess their role in the life cycle.

Morphotypes of *P. globosa*

A careful analysis of the literature published since the first description of *P. globosa* cells by Scherffel (1899, 1900) suggests that four morphotypes exist: diploid colonial cells, diploid flagellates, and two types of haploid flagellates.

Colonial cells

Colonial cells have 2–4 parietal chloroplasts, are deprived of body scales, haptonema, and flagella and are embedded in a mucilaginous matrix (Scherffel 1899; Kornmann 1955). They possess on their flagellar pole two short appendages, the role and nature of which being presently unknown (Fig. 1; Rousseau et al. submitted). Reported size ranges for live colonial cells are 4.5–8.0 μm (Kornmann 1955) and 5.8–10.4 μm (Rousseau et al. submitted) while they are 5.6–8.3 μm (Peperzak et al. 2000a) and 4.6–7.8 μm (Rousseau et al. submitted) for Lugol and Lugol-glutaraldehyde fixed cells. These cells are diploid (Cariou et al. 1994; Vaulot et al. 1994). They are evenly distributed in the colony 15–20 μm beneath a thin skin, and are weakly interconnected with dilute gel (Kornmann 1955; van Rijssel et al. 1997; Hamm et al. 1999). The colony skin is strong and semipermeable with pore size 1.0–4.4 nm in diameter, and presents plastic and elastic properties (Hamm et al. 1999). Diameter of colonies typically ranges from 10 μm to 8–9 mm (Kornmann 1955; Jahnke and Baumann 1987; Rousseau et al. 1990), but may occasionally reach 20–30 mm (Kayser 1970; Gieskes and Kraay 1975; Chen et al. 2002). Colonies are originally spherical but deviate into nonspherical shapes when growing larger or when subjected to hydrodynamical stress (Kornmann 1955; Bätje and Michaelis 1986; Rousseau et al. 1994).

Non-motile free-living cells from colonial origin have also been reported in *P. globosa* cultures. These cells are morphologically similar to colonial cells: same size range, lack of flagella, haptonema (Kornmann 1955; Rousseau et al. 1990; 1994; Peperzak 1993; Peperzak et al. 2000a; Dutz and Koski 2006), body scales (Peperzak et al. 2000a), thread-like filaments and stars (Peperzak et al. 2000a; Dutz and Koski 2006) and were assumed to have the same ploidy level, i.e., diploidy (Peperzak et al. 2000a). On this basis, non-motile free-living and colonial cells should not be considered anymore as distinct morphotypes (Peperzak et al. 2000a; Dutz and Koski 2006).

Haploid flagellates

The fine structure of the flagellate stage of *P. globosa* was first described as *P. pouchetii* from TEM by Parke et al. (1971). These cells have been reported as swarmers (Scherffel 1900), microzoospores (Kornmann 1955), small and large zoids (Parke et al. 1971), microflagellates

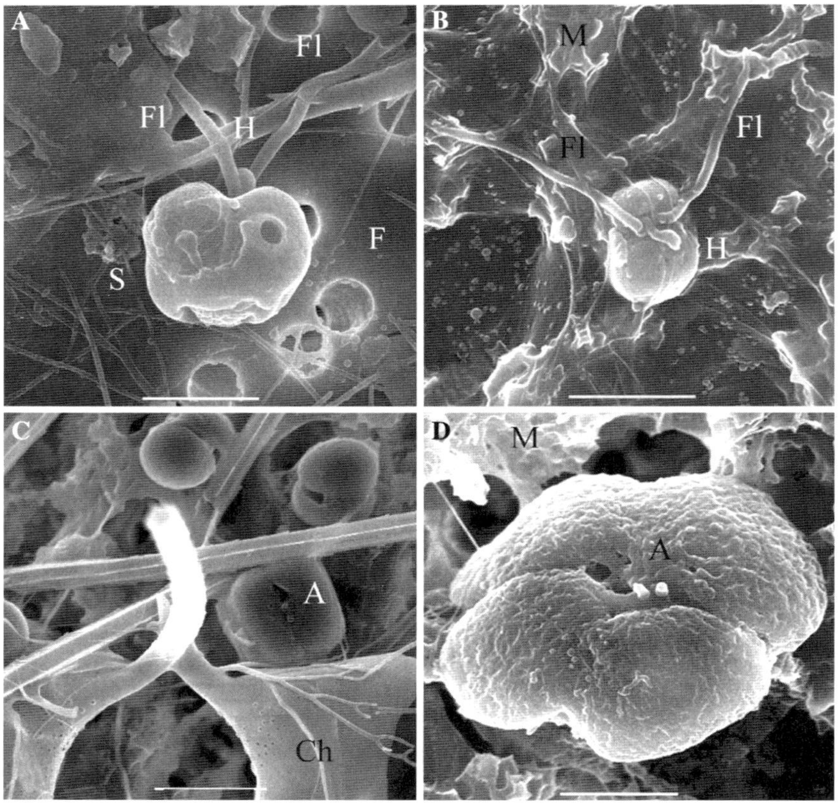

Fig. 1 SEM photographs of *P. globosa* cells observed in the Belgian coastal waters. Small-sized flagellates with stars and filaments but no scales observed (**A**) in February during the pre-bloom period; bar = 3 μm and (**B**) in May still embedded within the colony mucus at the end of the colonial stage; bar = 2 μm; colonial cells with the typical short appendages observed (**C**) in the early stage of the bloom within *Chaetoceros* setae; bar = 4 μm and (**D**) at the end of the bloom just before formation of haploid flagellates; bar = 2.4 μm. H: haptonema; Fl: flagella; S: star; F: filament; M: mucus; A: short appendage; Ch: Chaetoceros cell"

(Veldhuis et al. 1986) and micro- and mesoflagellates (Peperzak et al. 2000a). They have a rounded shape, and are smaller than colonial cells, with a diameter of 3–5 μm (Kornmann 1955); 3–6 μm (Parke et al. 1971) and 3.6–5.8 μm (Rousseau et al. submitted) for live cells observed under LM. Their small size has long been used for distinguishing them from other *P. globosa* cell types (e.g. Veldhuis et al. 1986) before Vaulot et al. (1994), using flow cytometry, demonstrated they were the haploid stage of *P. globosa*.

These flagellates are capable of rapid vegetative reproduction (Kornmann 1955; Parke et al. 1971; Rousseau et al. 1994; Vaulot et al. 1994) and swim very actively (Kornmann 1955; Parke et al. 1971). They possess two equal heterodynamic flagella, 10–15 μm in length, and a short haptonema (3–4 μm) characterized by a distal swelling. They present an anterior depression and two golden-brown plastids (Parke et al. 1971). The cell body is covered by two types of organic scales showing a pattern of radiating ridges, visible on both sides (Parke et al. 1971; Peperzak et al. 2000a). Two types of haploid flagellates were distinguished by Parke et al. (1971). One type, 3–6 μm in size, possesses two superficial, bright vesicles located on the body surface (Parke et al. 1971). These vesicles release filaments, 20 μm in length, 0.05 μm in diameter, and made of alpha-chitin (Chrétiennot-Dinet et al. 1997), which form a highly characteristic pentagonal star (Parke et al. 1971). The function of these filaments is unknown. It has been hypothesized that they could act as anchors for attachment to solid structures before colony initiation (Chrétiennot-Dinet 1999), or alternatively have a role in defense against grazers (Peperzak et al. 2000a; Dutz and Koski 2006). The other haploid flagellate, 3–5 μm in size, lacks the vesicles and filaments (Parke et al. 1971; Peperzak et al. 2000a). Such a difference in size and ability to produce filaments has also been observed in field samples for cells described as zoids (Manton and Leadbeater 1974 cited in Peperzak et al. 2000a) or meso- and microflagellates (Peperzak et al. 2000a).

These flagellates were observed in senescent cultures, swimming inside spherical colonies of various sizes, and were associated with colony disappearance (Kornmann 1955). In the natural environment and in mesocosms, large numbers of these flagellates were repeatedly reported, at the decline of *P. globosa* colony blooms, either inside spherical colonies (Scherffel 1899, 1900; Cadée 1991; Peperzak et al. 1998, 2000a), or released into the medium leaving ghost colonies or as free-living flagellates (Jones and Haq 1963; Admiraal and Venekamp 1986; Veldhuis et al. 1986; Verity et al. 1988b; Escaravage et al. 1995). These small-sized flagellates were observed invariably associated with the presence of chitinous filaments and stars, a feature specific to the haploid stage of *P. globosa* Scherffel (Vaulot et al. 1994; Zingone et al. 1999). Released from colonies at the end of the bloom, these flagellates were recorded at different periods of the year in the water column of the Channel and the Southern Bight of the North Sea, at cell densities varying from 80×10^3 cells L^{-1} to 220×10^3 cells L^{-1} (Fig. 1; Rousseau and Chrétiennot-Dinet, unpublished data). They apparently represent the life stage persisting between two blooms of colonial cells, as suggested by Kornmann (1955) and Parke et al. (1971).

Diploid flagellates

The third morphotype of *P. globosa* is a flagellate of the same size range as colonial cells when observed live, i.e., 4.5–8.0 μm (Kornmann 1955), 6–7 μm (Peperzak et al. 2000a) and 6.1–9.3 μm (Rousseau et al. submitted) containing two flagella and one haptonema but lacking scales, filaments and stars (Rousseau et al. submitted). This flagellate has been reported as an asexual swarmer by Kornmann (1955), as a large flagellate by Cariou et al. (1994) and Rousseau et al. (1994), and as a macroflagellate by Peperzak et al. (2000a). It typically appears in cultures within 24 h when colonial cells are released mechanically from the colony (Kornmann 1955; Rousseau et al. 1990; Cariou et al. 1994). This flagellate is diploid as no ploidy change was found during the transformation of colonial cells into flagellates (Cariou et al. 1994; Rousseau et al. 1994).

These flagellates are able to rapidly form new colonies within a day after adhesion to a surface (Kornmann 1955; Kayser 1970; Cariou et al. 1994; Rousseau et al. 1994). Inanimate particles (Rousseau et al. 1994), walls of culture vessels (Kayser 1970; Cariou et al. 1994), and diatoms (Weisse et al. 1986; Boalch 1987; Rousseau et al. 1994) have been observed as adhesion sites. This property of attachment to surfaces, specific to this life stage, led to the assumption that a benthic stage, acting as an overwintering form, exists in the natural environment (Kayser 1970). It is not clear if this flagellate is able to mitotically divide. Its rapid transformation into a colony suggests it is short-lived (Kornmann 1955; Rousseau et al. 1994), but vegetative multiplication was also reported by Kayser (1970). The short lifespan of this morphotype could well explain why it is observed only occasionally in the field (Kornmann 1955; Peperzak et al. 2000a; Rousseau et al. submitted).

Such flagellates were occasionally observed inside colonies in cultures by Kornmann (1955) who considered them as a distinct cell types, the macrozoospores, and in the natural environment (Peperzak et al. 2000a).

Morphotypes of *P. pouchetii*

At the present time, two *P. pouchetii* cell types have been confirmed based on EM observations and cytometric analysis: a diploid colonial cell and a diploid flagellate (Jacobsen 2002). However, other reports suggest that a larger flagellate could exist within the *P. pouchetii* life cycle (Sukhanova and Flint 2001; Wassmann et al. 2005).

Colonial cells

Diploid colonial cells are in the size range of 5–7 μm, have an anterior longitudinal groove and are deprived of filamentous appendages and scale coverings (Jacobsen 2002). Actively growing colonial cells are typically distributed in groups on lobes of cloud-like colonies (Jahnke and Baumann 1987; Gunkel 1988; Baumann et al. 1994; Rousseau et al. 1994; Jacobsen 2002). *Phaeocystis pouchetii* colonies, which are spherical up to 0.1 mm (Rousseau et al. 1994), have a maximum size of 1.5–2 mm (Jahnke and Baumann 1987). They are characterized by a delicate mucus which disrupts easily compared to the solid mucilage of *P. globosa* (van Rijssel et al. 1997; Jacobsen 2002; Wassmann et al. 2005). Non-motile free-living cells morphologically similar to colonial cells, i.e., deprived of flagella, haptonema, filamentous appendages, scales, can be found together with the colony stage due to colony disruption (Eilertsen 1989; Jacobsen 2002; Wassmann et al. 2005).

Flagellates

A filament-bearing flagellate has been described in detail by Jacobsen et al. (1996) and Jacobsen (2002) on the basis of LM and TEM observations. This flagellate, which originates from a colony when brought into culture, is round and has an average diameter of 5 μm when live. It has two golden-brown parietal chloroplasts, two heterodynamic equally long (11 μm) flagella, and a short non-coiling haptonema (1.5 μm). The cell body is covered by two types of scales, both with radiating ridges visible on both surfaces. The filaments are coiled inside one or two superficial vesicles, and form five-ray stars when unwound. Contrary to the filament-producing cell of *P. globosa*, this flagellate was assumed to be diploid (Jacobsen 2002). This flagellate was observed during winter, increasing in abundance prior to colonial development, and has been thought to be the precursor of the colonial stage (Jacobsen 2002; Jacobsen and Veldhuis 2005). Such flagellates were observed inside colonies at the end of spring blooms, being subsequently released in the water column (Jacobsen 2002).

From LM observation of field samples, two types of flagellates were reported within the *P. pouchetii* life cycle based on size criteria, i.e., large (6 μm; Sukhanova and Flint 2001; Wassmann et al. 2005) and small heart-shaped (3–4 μm; Wassmann et al. 2005) cells. The large flagellates co-occurred with colonies and were abundant, sometimes being dominant over colonial cells in terms of cell density (Sukhanova and Flint 2001; Wassmann et al. 2005). The small flagellates were found free-living (Wassmann et al. 2005) or within decaying colonies before their disappearance (Lagerheim 1896; Wassmann et al. 2005). They might well correspond to the flagellate described by Jacobsen (2002), with the size difference explained by cell shrinkage due to the use of Lugol-glutaraldehyde as sample fixative. Interestingly, small flagellate formation and their further release into the water column have been shown to originate from colonies assuming a spherical shape (Gunkel 1988; Whipple et al. this issue).

Morphotypes of *P. antarctica*

Three morphotypes have been observed in *P. antarctica*: colonial cells and two types of flagellates. Two ploidy levels have been recorded.

Colonial cells

The colonial cell of *P. antarctica* has no flagella, no haptonema, no scales, and no vesicles or star-forming filaments (Davidson 1985). This cell has an anterior depression and two short appendages similar to those observed on *P. globosa* colonial cells (Fig. 2; Chrétiennot-Dinet and Rousseau, unpublished). The size range reported for colonial cells of *P. antarctica* is quite large, depending on preservation and fixation procedures. Size range includes: 10 μm for live cells (Moestrup and Larsen 1992), and 3.2–7.9 μm (Mathot et al. 2000), 4.7–5.6 μm (Vaulot et al. 1994), 4–6 μm (this study) for Lugol-gluraldehyde fixed cells. Colonial cells of *P. antarctica*, assumed to be diploid (Vaulot et al. 1994), are evenly distributed along the periphery of colonies characterized by a solid mucus. Field *P. antarctica* colonies are typically spherical but can include numerous morphs with a maximum size of 9 mm (Baumann et al. 1994; Marchant and Thomsen 1994).

Flagellates

One *P. antarctica* flagellate bears scales and produces filaments and stars; the other is devoid of scales, filaments and stars. The scale-bearing flagellate was described in detail by Davidson (1985): it has an anterior depression, two chloroplasts with a large central pyrenoid, and bears two flagella and an haptonema. Cell size ranges between 3.5 μm and 7 μm when fixed with

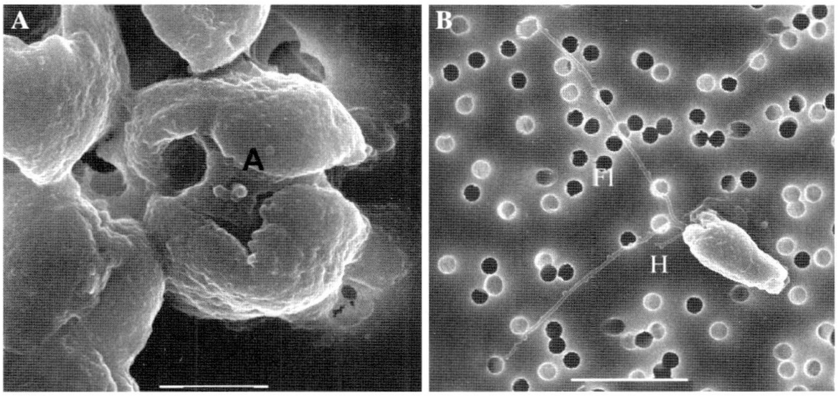

Fig. 2 SEM photographs of *P. antarctica* (strain CCMP 1871): (**A**) Colonial cells with four plastids and the two short appendages typical of the colonial cell on the flagellar pole; bar = 2 μm: (**B**) Flagellates co-occurring with colonies (diploid?) deprived of scales and filaments; bar = 6 μm. H: haptonema; Fl: flagella; A: short appendage

glutaraldehyde and observed with TEM. The two flagella are equally long (6–10 μm). The haptonema presents a bulbous tip and is 1.5–2 μm long. The body cell is covered by two types of scales with a pattern of radiating ridges visible on both sides. Thread-like material is regularly arranged within circular posterio-lateral vesicles. When deployed, the threads have a length of 25 μm and form a pentagonal star.

In the natural environment, such scale-bearing flagellates increase in number at the beginning of the colonial bloom, and decline during the bloom; they are present in large numbers after the bloom (Davidson and Marchant 1992b). Similar flagellates, that have a diameter of 5.2 μm when fixed with glutaraldehyde, two flagella of 14.3 μm in length, and filamentous appendages of 40 μm in length forming a pentagonal star, were found at the ice edge of the Weddell Sea during the austral summer (Buck and Garrison 1983). Flagellates with thread-like material were observed in the Bransfield Strait region during post-bloom period (Kang and Lee 1995). On the other hand, an increase in the abundance of small-sized free-living flagellates (2.4–5.5 μm when fixed with glutaraldehyde) was observed in the Ross Sea in early austral spring just before colony formation occurred (Mathot et al. 2000; Smith et al. 2003). Although no fine morphological description was provided, these small flagellates could well correspond to the flagellates described by Buck and Garrison (1983) and Davidson (1985), the size difference resulting from cell shrinkage and preparation for microscopic observations. Their relative abundance decreased to a minimum, while colonial cells dominated in late spring before they were observed again inside colonies (Smith et al. 2003) or as free-living cells (Putt et al. 1994) at the end of the bloom.

The second *P. antarctica* flagellate type has the same size as the colonial cell, ranging from 6.5 μm (Garrison and Thomsen 1993) to 7.5 μm (Fryxell 1989). This flagellate bears two flagella and a haptonema but lacks scales, filaments and stars (Fig. 2, Garrison and Thomsen 1993; Marchant and Thomsen 1994; Chrétiennot-Dinet and Rousseau, unpublished). This flagellate was formed inside and released from spherical or elongated colonies at the ice edge during austral spring (Fryxell 1989; Garrison and Thomsen 1993; Marchant and Thomsen 1994). Some 5–6 h after formation, it was found attached to the spines and processes of large diatoms (Garrison and Thomsen 1993; Marchant and Thomsen 1994) where it subsequently formed new colonies (Fryxell 1989).

Morphotypes of *P. jahnii*

Two cell types have been described for *P. jahnii*: colonial cells and flagellates (Zingone et al. 1999).

Colonial cells

Phaeocystis jahnii colonial cells possess the two short appendages typical of *P. globosa* and *P. antarctica* colonial cells, have a size of 6–8.5 μm when live and are irregularly distributed in loose, irregular colonies. Non-motile free-living cells of the same size have also been reported (Zingone et al. 1999).

Table 1 Different morphotypes reported for the six *Phaeocystis* species recognized to date

	Flagellates with scales, threads and stars	Flagellates with scales	Flagellates without scales, threads and stars Same size range as colonial cells	Colonial cells
P. globosa	X (n)	X (n)	X (2n)	X (2n)
P. pouchetii	X (2n)	–	X	X(2n)
P. antarctica	X	–	X	X(2n)
P. jahnii	–	X	–	X
P. cordata	X (n)	–	–	–
P. scrobiculata	X	–	–	–

Measured or assumed ploidy levels are indicated
— not observed

Flagellates

The *P. jahnii* flagellate is round, 3.5–5.0 μm in diameter when observed live, and has two to four golden-brown parietal chloroplasts. It bears two unequally long flagella (8.5–12 μm and 5.5–6.5 μm), a non-coiling haptonema 3.0–4.5 μm in length, and two scale types (Zingone et al. 1999). Filaments were not observed.

Morphotype of *P. cordata*

Phaeocystis cordata has only been described as a flagellated cell (Zingone et al. 1999). Live cells are 3–3.5 μm long and 3–4 μm wide. They have two unequally long flagella of 5.5–7.5 μm and 4.5–6 μm in length, and a non-coiling haptonema 2.2–2.5 μm in length. They possess two scale types, both with faint radiating ridges. These cells produce filaments forming pentagonal figures.

Morphotype of *P. scrobiculata*

Phaeocystis scrobiculata, which has been observed occasionally in field samples and never in cultures, was only reported as a flagellate (Moestrup 1979; Hallegraeff 1983; Estep et al. 1984; Chrétiennot-Dinet unpublished data). The cell, 8 μm in length when fixed with osmium tetroxide, bears two equal flagella and a non-coiling haptonema. It is covered by a periplast of two types of scales, and produces filaments longer than 50 μm which form a nine-ray figure rather than the five-ray star observed in the other *Phaeocystis* species (Moestrup 1979; Hallegraeff 1983). Large differences have been reported in the length of the flagella and haptonema, and in body scale sizes (Moestrup 1979; Hallegraeff 1983; Estep et al. 1984; Hoepffner and Haas 1990) depending on the preparation procedures.

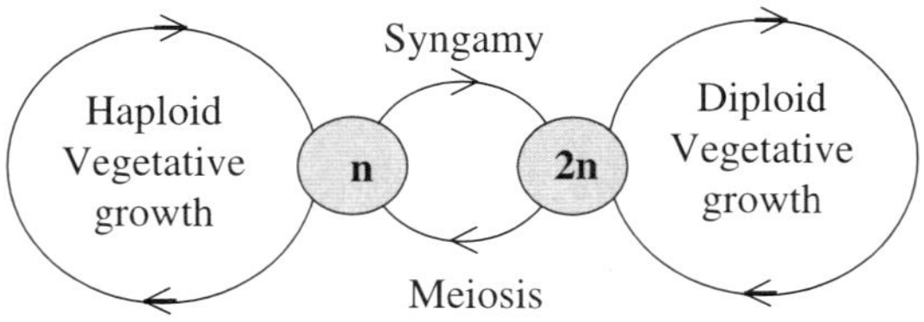

Fig. 3 Schematic representation of a haploid–diploid life cycle (from Valero et al. 1992)

Synthesis of the observed morphotypes

A careful examination of the literature suggests that four different cell types exist within *Phaeocystis* species (Table 1). In common to all six species, whether colonial or not, is the occurrence of scaly flagellates (Table 1) which are of two types. One produces star-forming filaments and has been reported for all species except *P. jahnii*. The other, deprived of filaments and stars, has been found in *P. globosa* (Parke et al. 1971; Peperzak et al. 2000a) and *P. jahnii* (Zingone et al. 1999). These two cell types are particularly important as they provide relevant taxonomic criteria to compare and distinguish the different species of the genus. These criteria include difference in size, structure and arrangement of scales and star-forming threads (Moestrup 1979; Moestrup and Larsen 1992; Baumann et al. 1994; Medlin et al. 1994; Zingone et al. 1999; Jacobsen 2002; Lange et al. 2002). These cells have been shown to be haploid in *P. globosa* (Vaulot et al. 1994) but diploid in *P. pouchetii* (Jacobsen 2002).

In common to three colony-forming species, *P. globosa*, *P. pouchetii* and *P. antarctica*, is the occurrence of three cell types: a flagellate with

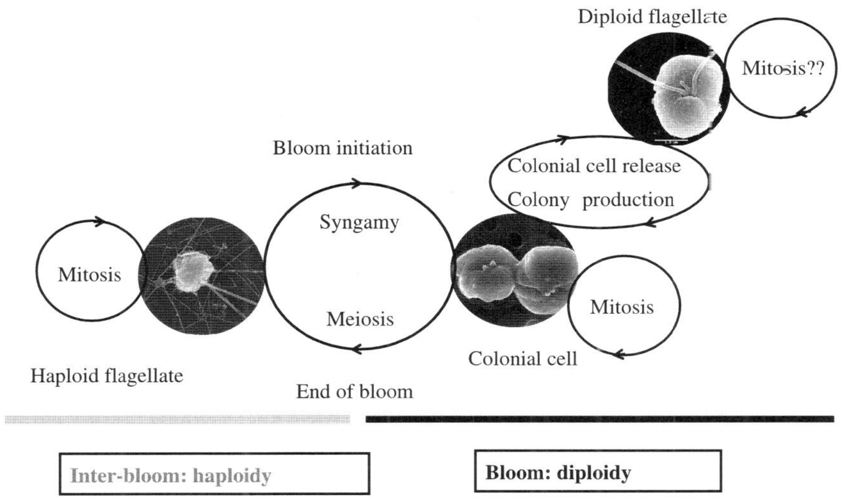

Fig. 4 The haploid–diploid life cycle of *P. globosa*. The haploid flagellates are characterized by stars, filaments, scales and have a size in the range 3.6–5.8 μm when live. Colonial cells, in the size range 5.8–10.4 μm when live, present two short appendages on their apical side, are deprived of haptonema, stars, filaments and scales. Diploid flagellates, of the same size range than colonial cells, have two flagella, a haptonema and lack the stars, filaments and scales

scales and filaments, a colonial cell, and a flagellate devoid of scales and filaments (Table 1). The flagellates with scales and filaments have been observed to increase in number before colonial cell blooms, to disappear when colonial cells are present, and are massively formed inside colonies at the end of the colonial stage. The presence of larger flagellates is usually found restricted to that of colonial cells. In species that do not form colonies, *P. scrobiculata* and *P. cordata*, only flagellates with scales and filaments have been observed (Table 1).

The *Phaeocystis* life cycle

The haploid–diploid life cycle of *P. globosa*

The existence of different morphotypes, two ploidy levels related to phase changes, and the ability of both haploid and diploid stages to divide mitotically (Kornmann 1955; Rousseau et al. 1994; Vaulot et al. 1994), support the existence of a haploid–diploid life cycle in *P. globosa*. In such life cycles, both haploid and diploid stages are related by sexual processes, meiosis and syngamy, and both are capable of mitotic division (Fig. 3; Valero et al. 1992; Houdan et al. 2004). Based on available information from cultures and field studies presented in the previous section, the *P. globosa* haploid–diploid life cycle has been reconstructed (Fig. 4). The two main prominent features of this cycle are that sexuality is prevalent in colony bloom formation and termination, and that two types of vegetative reproduction exist.

Phaeocystis globosa colony blooms result from sexual processes

The occurrence of haploid flagellates in the water column between two blooms of diploid colonial cells as observed in the southern North Sea (Rousseau and Chrétiennot-Dinet unpublished data) provides evidence that *P. globosa* colony bloom initiation and termination involve sexual processes, with the length of the diploid phase being restricted to the colony blooms. The formation of a diploid non-motile colonial cell from haploid flagellates implies that syngamy (fusion of the cytoplasm and nuclei of two haploids and subsequent zygote production) must occur at the time of colony bloom initiation. The observation of two morphologically distinguishable haploid flagellates (Parke et al. 1971; Peperzak et al.

2000a) suggests that anisogamy occurs in *P. globosa*, but this still awaits proof. Filaments characteristic of haploid flagellates could possibly play a role in mating. These structures have indeed been suggested to be related to sexuality (Vaulot et al. 1994) or to play a role in attachment (Chrétiennot-Dinet 1999). The inability to regenerate diploid colonial cells from clones of haploid flagellates (Parke et al. 1971; Vaulot et al. 1994) suggests that different mating types, i.e., compatible gametes able to form zygotes, exist within *P. globosa* (Vaulot et al. 1994). It is not known if homothallism (self-fertile colonies) or heterothallism (self-sterile colonies) is the rule in *P. globosa*.

Conversely, meiosis must happen to form haploid flagellates from diploid colonial cells. This may well occur during the massive production of haploid flagellates within colonies often reported at the end of the *P. globosa* bloom, before disappearance of the colonial stage (Rousseau et al. 1994; Peperzak et al. 2000a). Haploid flagellates are however produced in a restricted number of colonies. Most of them senesce and subsequently aggregate, being progressively invaded by various heterotrophic organisms that develop complex microbial networks (Lancelot and Rousseau 1994; Lancelot et al. 2002).

Vegetative reproduction in P. globosa

The vegetative reproduction of the diploid stage occurs through two distinct pathways involving colonial cells and diploid flagellates (Fig. 4). One consists of mitotic division of colonial cells within the colony, i.e., colony growth (Kornmann 1955; Rousseau et al. 1994; Veldhuis et al. 2005). This process can lead to colony division and budding as observed in mesocosms (Verity et al. 1988a) and in the field (Rousseau et al. 1994) but seems to be of minor importance. The second pathway involves the transition through short-lived diploid flagellates that are released from colonies and are able to reinitialize the colonial stage within a day (Kornmann 1955; Cariou et al. 1994; Rousseau et al. 1994). Diploid flagellates therefore co-occur with colonial cells and propagate the colonial stage, a pathway commonly used to produce colony cultures in the laboratory. However, its significance in the natural environment is highly questionable and is probably reduced due to the short life span of diploid flagellates (Kornmann 1955; Cariou et al. 1994). The natural occurrence of diploid flagellates in the field is difficult to estimate. It results from disruption of large colonies when the sea is stormy and under turbulent conditions (Kornmann 1955, Peperzak et al. 2000a). Occurrence of these cells, in association with non-motile free-living cells, could however also result from sample manipulation during collection, size fractionation, and incubation. The minor role of the diploid flagellate in the natural environment is also suggested by the observation of massive and synchronous generation of small colonies in the early phase of blooms (Rousseau et al. 1990). A significant vegetative reproduction would instead result in more-continuous colony production. However, this pathway could not be excluded and has been suggested to provide the inoculum for colony blooms (Cadée 1991).

Factors inducing phase changes within the P. globosa life cycle

Several factors have been hypothesized to play a role in transitions between *P. globosa* morphotypes. The formation of colonies from free-living cells has been related to phosphate depletion (Veldhuis and Admiraal 1987; Cariou et al. 1994), light intensity (Kornmann 1955; Peperzak 1993; Peperzak et al. 1998), chemical substances produced by vernal diatoms, especially some *Chaetoceros* species (Weisse et al. 1986; Boalch 1987; Rousseau et al. 1994), and turbulence (Schapira 2005; Shapira et al. 2006). The requirement of a solid substrate for cell attachment has also been suggested from the observation of small colonies attached to *Chaetoceros* setae at the early stage of the bloom (Fig. 1; Boalch 1987; Rousseau et al. 1994; Chrétiennot-Dinet et al. 1997).

However, a careful examination of literature shows that most studies on *P. globosa* colony formation have been performed in laboratory cultures using the vegetative reproduction pathway, i.e., with an inoculum of diploid flagellates originating from colony disruption (Table 2). These studies show that solid substrate, turbulence, and phosphate are factors promoting the vegetative generation of colonies (Table 2). From field

Table 2 Factors involved in the transition from free-living cell to colonial stage in *P. globosa*

References	Factors investigated	Exp. cond.	Free-living cell origin	Terms used and available cell characterization
Kornmann (1955)	Light	C	Natural seawater	Flagellates
Kayser (1970)	Solid substrate	C	Colonial cells released from colony disruption	Non-motile free-living cells
Boalch (1987)	Solid substrate Diatom exsudate	F	Natural pre-bloom populations	Motile stage, biflagellated cells
Veldhuis and Admiraal (1987)	Phosphate	C	Fractionation (<20 μm) of a culture (colony + cells)	Single cells Free-living motile cells (7 μm diameter)
Peperzak (1993)	Light	C	Fractionation (<20 μm) of culture (colony + cells) Clone Ph91	>100 W h m^{-2} day^{-1}: mesoflagellates (4.1 μm) < 100 W h m^{-2} day^{-1}: microflagellates (3.1 μm)
Rousseau et al. (1994)	Solid substrate	C	Colonial cells released from colony disruption	Free-living cells
Cariou et al. (1994)	Phosphate Solid substrate	C	Colonial cells released from colony disruption	Non flagellated, flagellated cells, diploid cells
Schapira (2005)	Turbulence	C	Fractionation (<5 μm) of culture (colony + cells)	Flagellated and non-flagellated cells

Experimental conditions are reported as C: cultures; F: field; M: mesocosms

observations, it appears that the vernal growth of diatoms and light level would play a key role in colony generation from haploid flagellates (Table 2). The role of light in colony formation from haploid flagellate cultures has been confirmed in laboratory studies (Kornmann 1955; Peperzak 1993). The formation of haploid flagellates in colonies at the end of a *P. globosa* bloom has been observed in mesocosms and related to nutrient depletion (Verity et al. 1988b; Escaravage et al. 1995) or rapid decrease in temperature (Verity et al. 1988b). Light limitation has also been suspected to play a role both in the laboratory and in the natural environment associated after sinking of healthy colonies in low ambient light conditions (Peperzak et al. 1988).

In future studies, the *P. globosa* life stage transition should be considered in terms of induction of sexual processes, i.e., syngamy and meiosis. The possible interplay of endogenous (biological clock) and environmental factors should be considered, as this has been shown relevant for other haptophytes such as coccolithophorids (Houdan et al. 2004). Some factors, in particular light, nutrient status, and production of gamone (pheromone for attraction, recognition of compatible cells and mating), should be investigated. Knowledge of these factors coupled to high-resolution microscopy and flow cytometry techniques would allow observation of such short-lived processes as syngamy and meiosis in cultures and in the field.

The life cycle of the other *Phaeocystis* species

The existence of a haploid–diploid life cycle is well supported for *P. globosa*. Such haploid–diploid life cycles could also possibly exist within the three other *Phaeocystis* colonial species, as already hypothesized for *P. pouchetii* (Eilertsen 1989; Jacobsen 2002) and *P. antarctica* (Medlin and Zingone this issue). Besides two ploidy levels, similar events to those occurring in the *P. globosa* life cycle were indeed observed in the bloom dynamics of these two species. Filament-bearing scaly flagellates, thought to be colony precursors (Jacobsen 2002), were observed to increase in abundance before *P. pouchetii* colony development. These cells and young colonies with 2–40 cells were later found attached to diatoms (mostly *Chaetoceros* spp; Eilersten 1989; Sukhanova and Flint 2001; Jacobsen 2002), suggesting that diatoms could be a key factor for colony initiation. At the end of the colonial stage, small-sized flagellates assumed to be scaly flagellates, are massively produced intracolonially or are found free-living (Wassmann et al. 2005). In the same way, small flagellates of *P. antarctica* were shown to be mostly present before and at the end of colony blooms (Putt et al. 1994; Smith et al. 2003).

Some intraspecific differences may exist however in the significance of the vegetative pathway for colony production and the occurrence of large solitary cells. As only one ploidy level (diploidy) has been identified up to now (Jacobsen, 2002), the vegetative pathway of colony formation, either for bloom initiation and propagation, seems to be predominant in *P. pouchetii*. Colony propagation from cells released from colonies could be particularly important. In northern studies, the contribution of colonies and flagellates was indeed shown to be highly variable depending on the site and the year investigated (Wassmann et al. 2005). Although sample preservation and handling procedures could strongly affect the observed proportion of free-living and colonial cells due to colony disintegration, large flagellates of *P. pouchetii* can at times exceed the abundance of colonial cells (Wassmann et al. 2005). Such high number of flagellates may reflect the more delicate nature of the *P. pouchetii* colony matrix compared to that of *P. globosa*, allowing easier colony disruption and subsequent vegetative reproduction. In addition, colony budding might be an important process for *P. pouchetii* colony proliferation and bloom development (Whipple et al. 2005).

The observation of only one morphotype in *P. scrobiculata* and *P. cordata* suggests either that the whole cycle has not been yet observed, or these species lose the ability to form colonies. The assumption of Vaulot et al. (1994) that filaments forming five- or nine-rayed stars are associated with haploid cells suggests that *Phaeocystis* would occur as haploid flagellates in oceanic and oligotrophic environments.

The haploid–diploid life cycle of Prymnesiophytes

With morphologically different types of flagellates and one non-motile cell type, *P. globosa* shares the features of most Haptophytes (except the Pavlovophyceae), i.e., a life cycle involving alternation of morphologically different haploid and diploid generations, each capable of independent asexual reproduction (Billard 1994; Edvardsen 2002; Billard and Inouye 2004; Houdan et al. 2004). Houdan et al. (2004) suggest from the phylogenetic distribution that this is the primitive state across the Prymnesiophyceae.

These haploid and diploid phases present typically distinct patterns of scale ornamentation characteristic of each ploidy stage. The morphological difference between life stages may be minor, as is the case for *Chrysochromulina polylepis* (Paasche et al. 1990; Edvardsen and Paasche 1992; Edvardsen and Vaulot 1996) and *Prymnesium parvum/patelliferum* (Larsen and Edvardsen 1998; Larsen 1999), in which the two motile cell types are distinguished only by minor details in body scale morphology. Heteromorphic life stages of coccolithophores are more clearly differentiated and show remarkable diversity (Billard and Inouye 2004) alternating diploid heterococcolithophores and, depending on the family, five different types of haploid stages (noncalcifying motile stage, holococcoliths, aragonitic coccoliths, nannoliths and noncalcifying benthic stages; Billard and Inouye 2004). Observation of syngamy and meiosis, providing direct evidence for the existence of sexuality is, however, restricted to very few coccolithophorids species such as *Pleurochrysis pseudoroscoffensis* (Gayral and Fresnel 1983) and more recently *Coccolithus pelagicus* (Houdan et al. 2004).

The ecological relevance of the haploid–diploid life cycle of *P. globosa*

The advantage of haploid–diploid life cycles

Haploid–diploid life cycles are widespread among diverse sexual eukaryotic organisms such as red algae, most brown algae, many green algae, some fungi, foraminiferans, mosses, and ferns. This suggests that such life cycles result from an adaptive evolution providing selective advantages to organisms (Valero et al. 1992; Mable and Otto 1998). Based on theoretical and empirical considerations, haploid–diploid life cycles have been shown to combine advantages from sexual reproduction and those from being haploid and diploid (Kondrashov and Crow 1991; Mable and Otto 1998). The advantage of sexual reproduction is a homogenizing effect on the species gene pool, resulting in high genetic plasticity. Because there are two copies of each gene, diploidy masks deleterious recessive mutations within the genome but harbors twice as many genes capable of beneficial mutations. The larger pool of mutant alleles provides a source of genetic variability that allows a higher rate of adaptation to a changing environment (Mable and Otto 1998). By contrast, haploids express each mutation within their genome. Deleterious mutations are more efficiently eliminated and haploid populations tend to carry fewer mutations (Mable and Otto 1998).

At the ecological level, the diploid stage has been suggested to represent a *r*-selected strategy with high growth rates, use of inorganic nutrients, and resistance to turbulence, while the haploid stage would have a *K*-selected strategy, being better adapted to nutrient-limited conditions with low growth rates, motility and mixotrophy (Valero et al. 1992). The nutritional advantage of haploid over diploid cells under nutrient-limited conditions is due to their smaller size (Lewis 1985). Surface-to-volume ratios calculated for two cells of 4 μm and 8 μm in diameter, typical of live haploid and colonial cells of *P. globosa*, respectively, vary from 1.5 to 0.75 respectively. This suggests that haploid cells increase by a factor two their morphological potential to acquire nutrients when they are present at low concentrations. In addition, haploid cells may also halve the energetic cost of DNA replication during cell division (Lewis 1985). Maintenance of a haploid–diploid heteromorphic life cycle is generally considered as an evolutionary adaptation to an environment that is seasonally variable or contains two different niches (Stebbins and Hill 1980 cited in Valero et al. 1992).

The ecology of *P. globosa* morphotypes

Many tentative hypotheses have been proposed to understand the dominance, alternation and succession of *P. globosa* life forms, i.e., nanoplanktonic cells and large gelatinous colonies (Lancelot and Rousseau 1994; Lancelot et al. 2002; Verity and Medlin 2003). In general, these studies conclude an advantageous status accrues to the colonial stage over free-living cells due to higher growth characteristics and resistance to losses.

The success of *P. globosa* colonies during blooms has been related to the peculiar physiology and ecology of the gelatinous colonies (Lancelot and Rousseau 1994; Lancelot et al. 2002). The colony matrix was shown to act as an energy (Lancelot and Mathot 1985; Veldhuis and Admiraal 1985; Lancelot and Rousseau 1994) and nutrient (Fe and Mn; Schoemann et al. 2001) reservoir. This provides a competitive advantage to *P. globosa* colonies when resources are limiting (Lancelot and Rousseau 1994; Lancelot et al. 2002), and when nitrate is the nitrogen source (Lancelot et al. 1998), allowing higher growth rates than free-living cells (Peperzak et al. 2000b; Veldhuis et al. 2005). On the other hand, some advantages of being a free-living cell are related to a greater ability to compete under ammonium- and phosphate-limited conditions (Veldhuis et al. 1991; Riegman et al. 1992), reduced nutrient uptake (Ploug et al. 1999), and alternative nutritional modes such as phagotrophy (Verity and Medlin 2003).

Large *P. globosa* colonies are generally not, or insignificantly, grazed by mesozooplankton, in particular the small neritic copepods (Hansen and van Boekel 1991, Hansen et al. 1993; Gasparini et al. 2000; Verity 2000; Koski et al. 2005). This has been related to a prey-predator size mismatch, chemical deterrence, physical inhibition and/or other growth-state-related mechanisms (Verity 2000; Turner et al. 2002; Dutz et al. 2005; Long and Hay 2006). The presence of grazers has been shown to induce colony size increase (Jakobsen and Tang 2002) through the release of infochemicals (Tang 2003). The colonial skin also provides a protection to colonial cells from fast growing grazers, virus and pathogen infection, and bacterial colonization (Hamm 2000). In mesocosms, the resistance of *P. globosa* colonies to viral infection has been demonstrated, while free-living cells undergo severe infection and mortality (Brussaard et al. 2005). Besides the physical barrier of the colonial skin, some antimicrobial effects might also be due to the high acrylate accumulation inside the *P. globosa* matrix (Noordkamp et al. 2000). Contrary to the colonial form, *P. globosa* free-living cells may be heavily grazed by at least some microzooplankton and some stages of mesozooplankton (Admiraal and Venekamp 1986; Verity and Smayda 1989; Weisse and Scheffel-Möser 1990b; Weisse et al. 1994; Verity 2000; Tang et al. 2001; Turner et al. 2002). However, as recently shown by Long and Hay (2006), the grazing response for different copepods may vary more between different growth states of *P. globosa*, than compared to other prey types. Thus, further studies are needed to quantify the importance of different zooplankton grazing on this species.

A review of the *P. globosa* free-living cells used in physiology, grazing, and lysis experiments in field and laboratory studies shows that the morphological characterization of the cells, referred mostly as solitary, single or free-living cells, is very poor (Table 3). A careful analysis of culture and sampling procedures suggests that *P. globosa* free-living cells are either colonial (non-motile free-living cells) or diploid flagellates co-occurring with the colonial stage. Field studies were indeed mostly performed during the spring bloom, i.e. when diploid flagellates and colonial cells released from colonies are present (Table 3). In laboratory experiments, free-living cells were often separated from colonies using size fractionation or the vegetative pathway involving colony formation from diploid cells (Table 3). In a recent study, Dutz and Koski (2006) demonstrated that grazing losses and trophic transfer efficiency are highly dependent on the *P. globosa* morphotypes used. The sensitivity of flagellates bearing threads and forming stars, i.e., the haploid flagellates to protozooplankton grazing has been shown lower than that of colonial cells lacking the filamentous appendages (Dutz and Koski 2006). However, the evaluation of grazing pressure on the different *Phaeocystis* life forms is a complex issue that has been critically reviewed (Nejstgaard et al., this issue).

Based on the literature analysis, there is presently no experimental evidence to demonstrate a different bottom-up growth control for the *P. globosa* morphotypes. However, there is some evidence that haploid and diploid morphotypes belong to different trophic groups. Haploid flagellates are indeed inedible autotrophic nanoplankton while diploid flagellates and free-living colonial cells are preyed by microzooplankton.

Conclusions

Knowledge of the life stages of the different *Phaeocystis* species is still incomplete and is very variable. The identification of all life stages, through morphological characterization and determination of ploidy levels, are still missing for most of *Phaeocystis* species. Complete understanding of the life cycle events requires combining flow cytometry, electron microscopy, genetic techniques, and light microscopy on the same material, in either field samples or cultured strains. To avoid confusion, standardized nomenclature should be developed to refer to different morphotypes within different species. This should be done by scientist groups analyzing together the same material but containing different species, using both sophisticated and routine techniques. While suspected in *P. pouchetii* and *P. antarctica*, a haploid–diploid life cycle has only been confirmed for *P. globosa*. Further research is however needed to substantiate this life cycle, by direct observation of short-lived syngamy and meiosis, the characterisation of the mating system and capability (isogamy versus anisogamy, homothallism versus heterothallism), and the knowledge of the factors inducing sexuality. The study of genetic regulation of sexual processes will be an important step in the characterization of life cycles (Guillou and Biegala 2002).

At this stage, comparative studies of the autoecology and grazing sensitivity of *Phaeocystis* haploid and diploid cells are insufficient to demonstrate an ecological differentiation between stages and to provide support for ecological niche separation. Understanding the ecological significance of blooming as diploid cells but persisting as haploid stage throughout the year should be appraised through eco-physiological characterization of pure cultures of haploid and diploid stages combined with cell characterization (morphology and ploidy).

Table 3 Types of *P. globosa* free-living cells used in physiology, grazing and viral lysis experiments

References	Experiments	Exp. cond.	Free-living cells origin	Terms used and available cell characterization
Growth regulation				
Veldhuis and Admiraal (1987)	Phosphorus metabolism	C	Fractionation (<20 µm) of a culture (colonies + cells)	Single cells (7 µm diameter)
Weisse and Sheffel-Möser (1990)	Growth rates	F	Zooplankton net fractionation (<200 µm) of natural populations during the bloom	Single cells (5.3 µm; 9.5 pgC cell^{-1})
Van Boekel and Veldhuis, (1990)	Phosphatase activity	C	Fractionation (<20 µm) of culture (colonies + cells)	Single cells
Lancelot et al. (1991)	Growth rates Photosynthesis	C F	Fractionation (<20 µm) of culture (colonies + cells) and natural populations during the bloom	Free-living cells/Solitary cells
Veldhuis et al. (1991)	Phosphorus uptake	C	Fractionation (<20 µm) of a culture (colonies + cells)	Single cells Free-living motile cells (7 µm diameter)
Riegman et al. (1992)	Nutrient competition	C	Isolated from field as a colony	Non-motile free-living vegetative cells Non colonial flagellate cell types
Peperzak et al. (2000b)	Growth rates	C	Culture of 2 types of solitary cells (Clone Ph91)	Diploid solitary non-flagellate cells (6.1 µm) Haploid flagellates (4.2 µm)
Veldhuis et al. (2005)	Growth rates	M	Culture (colonies + flagellates)	Single cells/Flagellated single cells (4–6 µm)
		F	Natural populations	Single cells slightly smaller than colonial cells
Grazing and lysis experiments				
Admiraal and Venekamp (1986)	Grazing	F	Removal of colonies from natural populations during the bloom	Remaining single cells
Verity and Smayda (1989)	Grazing	C	Frequent transfers of solitary cells into f/2 medium	Solitary cells (3–5 µm; 20 pgC cell^{-1})
Claustre et al. (1990)	Grazing	F	Filtration (<22 µm) of natural populations during thebloom	Free-living cells
Weisse and Sheffel-Möser (1990)	Grazing	F	Zooplankton net fractionation (<200 µm) of natural populations during the bloom	Single cells (5.3 µm; 9.5 pgC cell^{-1})
Hansen et al. (1993)	Grazing	C	Culture without colony formation	Single cells or Flagellates
Verity (2000)	Grazing	C	Fractionation (<20 µm) of culture (colonies + cells)	Solitary cells
Tang et al. (2001)	Grazing	C	Gravity filtration (<11 µm) of culture (colonies + cells) of strain CCMP 1528	Single cells (4.4 µm; 9.3 pgC cell^{-1})
Jakobsen and Tang (2002)	Grazing	C	Gravity filtration (<11 µm) of culture (colonies + cells) of strain CCMP 1528	Solitary cells
Turner et al. (2002)	Grazing	C	Cell-dominated mixed culture (colonies + cells)	Solitary cells (5–6 µm length)
Tang (2003)	Grazing	C	Gravity filtration (<11 µm) of culture (colonies + cells) of strain CCMP 1528	Solitary cells
Koski et al. (2005)	Grazing	M	Cultured populations (colonies + cells)	Single cells
Brussaard et al. (2005)	Viral infection	M	Cultured populations (colonies + cells)) inoculated from a culture of colonies and flagellated cells	Single cells/Solitary cells—no distinction between flagellates and non-flagellates; correspond to the largest cell types (Brussaard, pers. comm)
Dutz and Koski (2006)	Grazing	C	5 strains from Southern North Sea	Three flagellates with threads and stars (3.6–4.2 µm; haploid) Two solitary non flagellate (5.6–5.7 µm; colonial cells)

Experimental conditions are reported as C: cultures; F: field; M: mesocosms

Acknowledgments The authors are indebted to François Lantoine from the laboratory of Banyuls (France) for providing the electron microscopy photographs of the diploid flagellate of *Phaeocystis globosa*. We also thank J. Nejstgaard for his helpful comments on grazing issues, and C. Lancelot for fruitful discussion on life cycle. We thank B. Patten for detailed recommendations of an early draft and three anonymous reviewers for their constructive comments. The present review is a contribution to the SCOR WG 120 "*Phaeocystis*, major link in the biochemical cycling of climate-relevant elements." V. Rousseau has been supported by the advanced modelling and research on eutrophication (AMORE) project of the Belgian Programme "Scientific Support Plan for a Sustainable Development Policy—Sustainable Management of the North Sea" funded by the Belgian Science Policy under contract EV/11/19. P. Verity and S. Whipple have been supported by U.S. National Science Foundation grant OPP-00-83381.

References

Admiraal W, Venekamp LAH (1986) Significance of tintinnid grazing during blooms of *Phaeocystis pouchetii* (Haptophyceae) in Dutch coastal waters. Neth J Sea Res 20:61–66

Bätje M, Michaelis H (1986) *Phaeocystis pouchetii* blooms in the East Frisian coastal waters (German Bight, North Sea). Mar Biol 93:21–27

Baumann MEM, Lancelot C, Brandini FP, Sakshaug E, John DM (1994) The taxonomic identity of the cosmopolitan prymnesiophyte *Phaeocystis*: a morphological and ecophysiological approach. In: Lancelot C, Wassmann P (eds) Ecology of *Phaeocystis*-dominated ecosystems. J Mar Syst 5:5–22

Billard C (1994) Life cycles. In Green JC, Leadbeater BSC (Eds) The haptophyte algae. The systematics association special, vol 51. Clarendon Press, Oxford, pp 167–186

Billard C, Inouye I (2004) What is new in coccolithophore biology? In: Thierstein H, Young J (Eds). Coccolithophores – from molecular processes to global impact. Springer-Verlag, Berlin Heidelberg, pp 1–30

Boalch GT (1987) Recent blooms in the Western English Channel. Rapp PV Réun Cons Int Explor Mer.187:94–97

Brussaard CPD, Kuipers B, Veldhuis MJW (2005) A mesocosm study of *Phaeocystis globosa* population dynamics. I. Regulatory role of viruses in bloom control. In: Veldhuis MJW, Wassmann P (eds) Harmful algae, vol 4. pp 859–874

Buck KR, Garrison DL (1983) Protists from the ice-edge region of the Weddell Sea. Deep Sea Res 30:1261–1277

Cadée GC (1991) *Phaeocystis* colonies wintering in the water column? Neth. J Sea Res 28:227–230

Cariou V, Casotti R, Birrien J-L, Vaulot D (1994) The initiation of *Phaeocystis* colonies. J Plankton Res 16:457–470

Chen Y-Q, Wang N, Zhang P, Zhou H, Qu L-H (2002) Molecular evidence identifies bloom-forming *Phaeocystis* (Prymnesiophyta) from coastal waters of southeast China as *Phaeocystis globosa*. Biochem Syst Ecol 30:15–22

Chrétiennot-Dinet M-J, Giraud-Guille M-M, Vaulot D, Puteaux J-L, Saito H, Chanzy Y (1997) The chitinous nature of filaments ejected by *Phaeocystis* (Prymnesiophyceae). J Phycol 33:666–672

Chrétiennot-Dinet M-J (1999) An enigma in marine nanoplankton. The role of star-like structures produced by *Phaeocystis*. In: Seckbach J (ed) Enigmatic microorganisms and life in extreme environments. Kluwer Academic Publishers, The Netherlands, pp 205–213

Claustre H, Poulet SA, Williams R, Marty J-C, Coombs S, Ben Mlih F, Hapette AM, Martin- Jezequel V (1990) A biochemical investigation of a *Phaeocystis* sp. bloom in the Irish sea. J Mar biol Ass UK. 70:197–207

Davidson AT (1985) Aspects of the biology of *Phaeocystis pouchetii* (Prymnesiophyceae). B Sc Thesis, University of Tasmania, 212 pp

Davidson AT, Marchant HJ (1992a) The biology and ecology of *Phaeocystis* (Prymnesiophyceae). In: Round FE, Chapman DJ (Eds) Progress in phycological research, vol 8. Biopress, Bristol, pp 1–45

Davidson AT, Marchant HJ (1992b) Protist abundance and carbon concentration during a *Phaeocystis*-dominated bloom at an Antarctic coastal site. Polar Biol 12:387–395

Dutz J, Koski M (2006) Trophic significance of solitary cells of the prymnesiophyte *Phaeocystis globosa* depends on cell type. Limnol Oceanogr 51(3):1230–1238

Dutz J, Klein Breteler WCM, Kramer G (2005) Inhibition of copepod feeding by exsudates and transparent exopolymer particles (TEP) derived from a *Phaeocystis globosa* dominated phytoplankton community. In: Veldhuis MJW, Wassmann P (eds) Harmful algae, vol 4. pp 915–940

Edvardsen B (2002). Life cycle strategies in the haptophyte genera *Chrysochromulina* and *Prymnesium*. In: Garcés E, Zingone A, Montresor M, Reguera B, Dale B (eds) Proceedings of the LIFEHAB workshop: life histories of microalgal species causing harmful blooms, vol 12. Calvia, Majorca, Spain, October 2001. Research in Enclosed Seas Series, pp 67–70

Edvardsen B, Paasche E (1992) Two motile stages of *Chrysochromulina polylepis* (Prymnesiophyceae): morphology, growth and toxicity. J Phycol 28:104–114

Edvardsen B, Vaulot D (1996) Ploidy analysis of the two motile forms of *Chrysochromulina polylepis* (Prymnesiophyceae). J Phycol 32:94–102

Edvardsen B, Eikrem W, Green JC, Andersen RA, Moon-van der Staay SY, Medlin LK (2000) Phylogenetic reconstructions of the Haptophyta inferred from 18S ribosomal DNA sequences and available morphological data. Phycologia 39(1):19–35

Eilertsen HC (1989) *Phaeocystis pouchetii* (Hariot) Lagerheim, a key species in Arctic marine ecosystems: life history and physiology. Rapp PV Réun Cons Int Explor Mer 188:131

Escaravage V, Peperzak L, Prins TC, Peeters JCH, Joordens JCA (1995) The development of a *Phaeocystis* sp. bloom in a mesocosm experiment in relation to nutrients, irradiance and coexisting algae. Ophelia 42:55–74

Estep KW, Davis PG, Hargraves PE, Sieburth JM (1984) Chloroplast containing microflagellates in natural populations of north Atlantic nanoplankton, their identification and distribution: including a description of five new species of *Chrysochromulina* (Prymnesiophyceae). Protistol 20:613–634

Fryxell GA (1989) Marine phytoplankton at the Weddell Sea Ice Edge: Seasonal changes at the specific level. Polar Biol 10:1–18

Gasparini S, Daro M-H, Antajan E, Tackx M, Rousseau V, Parent J-Y, Lancelot C (2000) Mesozooplankton grazing during the *Phaeocystis globosa* bloom in the Southern Bight of the North Sea. J Sea Res 43:345–356

Garrison DL, Thomsen HA (1993) Ecology and biology of ice biota. Berichte zur Polarforschung 121:68–74

Gayral P, Fresnel J (1983) Description, sexualité et cycle de développement d'une nouvelle Coccolithophoracée (Prymnesiophyceae): *Pleurochrysis pseudoroscoffensis* sp. Nov Protistologica 19:245–261

Gieskes WWC, Kraay GW (1975) The phytoplankton spring bloom in Dutch coastal waters of the North Sea. Neth J Sea Res 9:166–196

Guillou L, Biegala I (2002) Different molecular techniques to examine life cycles of phytoplankton or HAB species. In: Garcés E, Zingone A, Montresor M, Reguera B, Dale B (eds) Proceedings of the LIFEHAB workshop: life histories of microalgal species causing harmful blooms, vol 12. Calvia, Majorca, Spain, October 2001. Research in Enclosed Seas Series, pp 99–101

Gunkel J (1988) Zür Verbreitung von *Phaeocystis pouchetii* in Phytoplankton der Framstrasse ünter besonderer Berücksichtigung der Kolonie-bildung. PhD thesis, Kiel University, 118 pp

Hallegraeff GM (1983) Scale-bearing and loricate nanoplankton from the east Australian current. Bot Mar 26:493–515

Hamm CE, Simson DA, Merkel R, Smetacek V (1999) Colonies of *Phaeocystis globosa* are protected by a thin but tough skin. Mar Ecol Progr Ser 187:101–111

Hamm FC (2000) Architecture, ecology and biogeochemistry of *Phaeocystis* colonies. J Sea Res 43:307–315

Hansen FC, van Boekel WHM (1991) Grazing pressure of the calanoid copepod *Temora longicornis* on a *Phaeocystis* dominated spring bloom in a Dutch tidal inlet. Mar Ecol Progr Ser 78:123–129

Hansen FCR, Reckermann M, Klein Breteler WCM, Riegman R (1993) *Phaeocystis* blooming enhanced by copepod predation on protozoa: evidence from incubation experiments. Mar Ecol Progr Ser 102:51–57

Hoepffner N, Haas LW (1990) Electron microscopy of nanoplankton from the North Pacific Central Gyre. J Phycol 26:421–439

Houdan A, Billard C, Marie D, Not F, Saez AG, Young JR, Probert J (2004) Holococcolithophore-heterococcolithophore (Haptophyta) life cycles: flow cytometric analysis of relative ploidy levels. Systemat Biodiv 4:453–465

Jacobsen A (2002) Morphology, relative DNA content and hypothetical life cycle of *Phaeocystis pouchetii* (Prymnesiophyceae) with special emphasis on the flagellated cell type. Sarsia 87:338–349

Jacobsen A, Bratbak G, Heldal M (1996) Isolation and characterization of virus infecting *Phaeocystis pouchetii* (Prymnesiophyceae). J Phycol 32:923–927

Jacobsen A, Veldhuis MJW (2005) Growth characteristics of flagellated cells of *Phaeocystis pouchetii* revealed by diel changes in cellular DNA content. In: Veldhuis MJW, Wassmann P (eds) Harmful Algae, vol 4. pp 811–821

Jahnke J, Baumann M (1987) Differentiation between *Phaeocystis pouchetii* (Har.) Lagerheim and *Phaeocystis globosa* Scherffel. Colony shapes and temperature tolerances. Hydrobiol Bull 21:141–147

Jakobsen HH, Tang KW (2002) Effects of protozoan grazing on colony formation in *Phaeocystis globosa* (Prymnesiophyceae) and the potential costs and benefits. Aquat Microb Ecol 27:261–273

Jones PGW, Haq SM (1963) The distribution of *Phaeocystis* in the eastern Irish sea. J Cons Intern Expl Mer 28:8–20

Kang SH, Lee S (1995) Antarctic phytoplankton assemblage in the western Bransfield Strait region, February 1993: composition, biomass, and mesoscale distributions. Mar Ecol Progr Ser 129:253–267

Kashkin NI (1963). Materials on the ecology of *Phaeocystis pouchetii* (Hariot) Lagerheim, 1893 (Chrysophyceae) II. Habitat and specifications of biogeographical characteristics. Okeanologiya, Moscow 3:697–705

Kayser H (1970) Experimental-ecological investigations on *Phaeocystis pouchetii* (Haptophyceae): cultivation and waste water test. Helgoländer Wiss. Meeresunt 20:195–212

Kondrashov AS, Crow JF (1991) Haploidy or diploidy: which is better? Nature 351:314–317

Kornmann P (1955) Beobachtungen an *Phaeocystis*-Kulturen. Helgoländer Wiss Meeresunt 5:218–233

Koski M, Dutz J, Klein Breteler WCM (2005). Selective grazing of *Temora longicornis* in different stages of a *Phaeocystis globosa* bloom – a mesocosm study. In: Veldhuis MJW, Wassmann P (eds) Harmful Algae 4:915–927

Lagerheim G (1893) *Phæocystis*, nov. gen., grundadt på *Tetraspora Poucheti* Har. Botaniska Notiser, pp 32–33

Lagerheim G (1895). Ueber *Phæocystis Poucheti* (Har.) Lagerh., eine Plankton-Flagellate. Öfversigt af K. Vetensk.-Akad. Förhanlingar, Stockholm 4:277–288

Lancelot C, Mathot S (1985) Biochemical fractionation of primary production by phytoplakton in Belgian coastal waters during short- and long-term incubations with ^{14}C-bicarbonate. Mar Biol 86:219–226

Lancelot C, Rousseau V (1994) Ecology of *Phaeocystis*: the key role of colony forms. In: Green JC, Leadbeater BSC (eds) The Haptophyte Algae, vol 51. Clarenton Press, Oxford, pp 229–245

Lancelot C, Rousseau V (2002) Life cycle in Haptophyta. In: Garcés E, Zingone A, Montresor M., Reguera B, Dale B (eds) Proceedings of the LIFEHAB workshop: life histories of microalgal species causing harmful blooms, vol 12. Calvia, Majorca, Spain, October 2001, Research in Enclosed Seas Series, pp 88–90

Lancelot et al. (1991) Dynamics of *Phaeocystis* blooms in nutrient enriched coastal zones. In: Lancelot C, Billen G, Barth H (eds) Proceedings of the third workshop Plymouth, March 6–7, 1990, pp 106. *Water Pollution Research* report 23

Lancelot C, Keller MD, Rousseau V, Smith WO, Mathot S (1998) Autecology of the marine Haptophyte *Phaeocystis* sp. In: Anderson DM, Cembella AD, Hallagraeff GM (Eds) Physiological ecology of harmful algal blooms, vol. G 41. Springer-Verlag, Berlin, pp 209–224

Lancelot C, Rousseau V, Schoemann V, Becquevort S (2002). On the ecological role of the different life forms of *Phaeocystis*. In: Garcés E, Zingone A, Montresor M., Reguera B, Dale B (eds) Proceedings of the workshop LIFEHAB: Life histories of microalgal species causing harmful blooms, vol 12. Calvia, Majorca, Spain, octobre 2001. Research in Enclosed Seas series, pp 71–75

Lange M, Chen Y-Q, Medlin LK (2002) Molecular genetic delineation of *Phaeocystis* species (Prymnesiophyceae) using coding and non-coding regions of nuclear and plastid genomes. Eur J Phycol 37:77–92

Larsen A (1999) *Prymnesium parvum* and *P. patelliferum* (Haptophyta) – one species. Phycologia 38:541–543

Larsen A, Edvardsen B (1998) Relative ploidy levels in *Prymnesium parvum* and *P. patelliferum* (Haptophyta) analyzed by flow cytometry. Phycologia 37(6):412–424

Lewis WM (1985). Nutrient scarcity as an evolutionary cause of haploidy. Am Nat 125:692–701

Liss PS, Malin G, Turner SM, Holligan PM (1994) Dimethyl sulphide and *Phaeocystis*: a review. In: Lancelot C, Wassmann P (eds) Ecology of *Phaeocystis*-dominated ecosystems. J Mar Syst 5:41–53

Long JD, Hay ME (2006) When intraspecific exceeds interspecific variance: Effects of phytoplankton morphology and growth phase on copepod feeding and fitness. Limnol Oceanogr 51:988–996

Mable BK, Otto SP (1998) The evolution of life cycles with haploid and diploid phases. BioEssays 20:453–462

Marchant H, Thomsen HA (1994) Haptophytes in polar waters. In: Green JC, Leadbeater BSC (Eds). The haptophyte algae, vol 51. Clarenton Press, Oxford, pp 209–228

Mathot S, Smith WOJ, Carlson CA, Garrison DL, Gowing MM, Vickers CL (2000) Carbon partitioning within *Phaeocystis antarctica* (prymnesiophyceae) colonies in the Ross Sea, Antarctica. J Phycol 36:1049–1056

Medlin LK, Lange M, Baumann MEM (1994) Genetic differentiation among three colony-forming species of *Phaeocystis*: further evidence for the phylogeny of the Prymnesiophyta. Phycologia 33:199–212

Medlin LK and Zingone A (this issue) A taxonomic review of the genus Phaeocystis. Biogeochemistry

Moestrup Ø (1979) Identification by electron microscopy of marine nanoplankton from New Zealand, including the description of four new species. N Z J Bot 17:61–95

Moestrup Ø, Larsen J (1992) Potentially toxic phytoplankton. 1. Haptophyceae (Prymnesiophyceae). In: Lindley JS (ed) ICES identification leaflets for plankton No 179. International Council for the Exploration of the Sea, Copenhagen, pp 11

Nejstgaard JC, Tang KW, Steinke M, Dutz J, Koski M, Antajan E, Long JD (this issue) Zooplankton grazing on *Phaeocystis*: a quantitative status and future challenges. Biogeochemistry. this volume

Noordkamp JB, Gieskes WWC, Gottschal JC, Forney LJ, Van Rijssel M (2000) Acrylate in *Phaeocystis* colonies does not affect the surrounding bacteria. J Sea Res 43:287–296

Paasche E, Edvardsen B, Eikrem W (1990) A possible alternate stage in the life cycle of *Chrysochromulina polylepis* Manton et Parke (Prymnesiophyceae). Nova Hedwigia Beihefte 100:91–99

Parke M, Green JC, Manton I (1971) Observations on the fine structure of zoids of the genus *Phaeocystis* (Haptophyceae). J Mar Biol Assoc UK 51:927–941

Peperzak L (1993) Daily irradiance governs growth rate and colony formation of *Phaeocystis* (Prymnesiophyceae). J Plankton Res 15:809–821

Peperzak L, Colijn F, Gieskes WWC, Peeters JCH (1998) Development of the diatom-*Phaeocystis* spring bloom in the Dutch coastal zone of the North Sea: the silicon depletion versus the daily irradiance threshold hypothesis. J Plankton Res 20:517–537

Peperzak L, Colijn F, Vrieling EG, Gieskes WWC, Peeters JCH (2000a) Observations of flagellates in colonies of *Phaeocystis globosa* (Prymnesiophyceae): a hypothesis for their position in the life cycle. J Plankton Res 22:2181–2203

Peperzak L, Duin RMN, Colijn F, Gieskes WWC (2000b) Growth and mortality of flagellates and non-flagellate cells of *Phaeocystis globosa* (Prymnesiophyceae). J Plankton Res 22:107–119

Ploug H, Stolte W, Jørgensen BB (1999) Diffusive boundary layers of the colony-forming plankton alga, *Phaeocystis* sp. – implications for nutrient uptake and cellular growth. Limnol Oceanogr 44:1959–1967

Putt M, Miceli G, Stoecker DK (1994) Association of bacteria with *Phaeocystis* sp. in McMurdo Sound, Antarctica. Mar Ecol Progr Ser 105:179–189

Riegman R, Noordeloos A, Cadée GC (1992) *Phaeocystis* blooms and eutrophication of the continental coastal zones of the North Sea. Mar Biol 112:479–484

Rousseau V, Mathot S, Lancelot C (1990) Calculating carbon biomass of *Phaeocystis* sp. from microscopic observations. Mar Biol 107:305–314

Rousseau V, Vaulot D, Casotti R, Cariou V, Lenz J, Gunkel J, Baumann M (1994) The life cycle of *Phaeocystis* (Prymnesiophyceae): evidence and hypotheses. In: Lancelot C, Wassmann P (eds) Ecology of *Phaeocystis*-dominated ecosystems. J Mar Syst 5:23–40

Schapira M (2005) Dynamique spatio-temporelle de *Phaeocystis globosa* en Manche Orientale: effets de la turbulence et des apports sporadiques en sels nutritifs. PhD Thesis. Université de Lille, France, 316 pp

Schapira M, Seuront L, Gentilhomme V (2006) Effects of small-scale turbulence on *Phaeocystis globosa* (Prymnesiophyceae) growth and life cycle. J Exp Mar Biol Ecol 335(1):27–38

Scherffel A (1899) *Phaeocystis globosa* n. sp. (Vorläufige Mittheilung). Berichte der Deutschen Botanischen Gesellschaft 17:317–318

Scherffel A (1900) *Phaeocystis globosa* nov. spec. nebst einigen Betrachtungen über die Phylogenie niederer, insbesondere brauner Organismen. Wissenschaftliche Meeresuntersuchungen Abteilung Helgoland NF Bd 4:1–29

Schoemann V, Wollast R, Chou L, Lancelot C (2001) Effects of photosynthesis on the accumulation of Mn and Fe by *Phaeocystis* colonies. Limnol Oceanogr 46:1065–1076

Schoemann V, Becquevort S, Stefels J, Rousseau V, Lancelot C (2005). *Phaeocystis* blooms in the global ocean and their controlling mechanisms: a review. J Sea Res 53:43–66

Smith WOJr, Dennett MR, Mathot S, Caron D (2003) The temporal dynamics of the flagellated and colonial stages of *Phaeocystis antarctica* in the Ross Sea. Deep-Sea Res II 50:605–617

Sournia A (1988) *Phaeocystis* (Prymnesiophyceae): How many species? Nova Hedwigia 47:211–217

Sukhanova IN, Flint MV (2001) *Phaeocystis pouchetii* at the Eastern Bering Sea Shelf. Oceanology 41:75–85

Tang KW (2003). Grazing and colony size development in *Phaeocystis globosa* (Prymnesiophyceae): the role of a chemical signal. J Plankt Res 25(7):831–842

Tang KW, Jacobsen HH, Visser AW (2001) *Phaeocystis globosa* (Prymnesiophyceae) and the planktonic food web: feeding, growth, and trophic intercations among grazers. Limnol Oceanogr 46(8):1860–1870

Turner JT, Ianora A, Esposito F, Carotenuto Y, Miralto A (2002). Zooplankton feeding ecology: does a diet of *Phaeocystis* support good copepod grazing, survival, egg production and egg hatching success. J Plankton Res 24(11):1185–1195

Valero M, Richerd S, Perrot V, Destombe C (1992) Evolution of alternation of haploid and diploid phases in life cycles. Trends Ecol Evol 7(1):25–29

van Boekel WHM, Veldhuis MJW (1990) Regulation of alkaline phosphatase synthesis in *Phaeocystis* sp. Mar Ecol Progr Ser 61:281–289

van Rijssel M, Hamm CE, Gieskes WWC (1997) *Phaeocystis globosa* (Prymnesiophyceae) colonies: hollow structures built with small amounts of polysaccharides. Eur J Phycol 32:185–192

Vaulot D, Birrien J-L, Marie D, Casotti R, Veldhuis MJ, Kraay GW, Chrétiennot-Dinet M-J (1994) Morphology, ploidy, pigment composition and genome size of cultured strains of *Phaeocystis* (Prymnesiophyceae). J Phycol 30:1022–1035

Veldhuis MJW, Admiraal W (1985) Transfer of photosynthetic products in gelatinous colonies of *Phaeocystis pouchetii* (Haptophyceae) and its effect on the measurement of excretion rate. Mar Ecol Progr Ser 26:301–304

Veldhuis MJW, Admiraal W (1987) Influence of phosphate depletion on the growth and colony formation of *Phaeocystis pouchetii*. Mar Biol 95:47–54

Veldhuis MJW, Colijn F, Venekamp LAH (1986) The spring bloom of *Phaeocystis pouchetii* (Haptophyceae) in Dutch coastal waters. Neth J Sea Res 20:37–48

Veldhuis MJW, Colijn F, Admiraal W (1991) Phosphate utilization in *Phaeocystis pouchetii* (Haptophyceae). Mar Ecol 12:53–62

Veldhuis MJW, Brussaard CPD, Noordeloos AAM (2005) Living in a *Phaeocystis* colony: a way to be successful algal species. In: Veldhuis MJW, Wassmann P (eds) Harmful Algae 4:841–858

Verity PG (2000). Grazing experiments and model simulations of the role of mesozooplankton in *Phaeocystis* food webs. J Sea Res 43:317–344

Verity PG, Smayda TJ (1989) Nutritional value of *Phaeocystis pouchetii* (Prymnesiophyceae) and other phytoplankton for *Acartia* spp. (Copepoda): Ingestion, egg production, and growth of nauplii. Mar Biol 100:161–171

Verity PG, Medlin LK (2003) Observations on colony formation by the cosmopolitan phytoplankton genus *Phaeocystis*. J Mar Syst 423:153–164

Verity PG, Villareal TA, Smayda TJ (1988a) Ecological investigations of blooms of colonial *Phaeocystis pouchetii*. I. Abundance, biochemical composition and metabolic rates. J Plankton Res 10:219–248

Verity PG, Villareal TA, Smayda TJ (1988b) Ecological investigations of blooms of colonial *Phaeocystis pouchetii*. II. The role of life cycle phenomena in bloom termination. J Plankton Res 10:749–766

Wassmann P, Ratkova T, Reigstad M (2005) The contribution of single and colonial cells of *Phaeocystis pouchetii* to spring and summer blooms in the north-eastern North Atlantic. In: Veldhuis MJW, Wassmann P (eds) Harmful algae, vol 4. pp 823–840

Weisse T, Scheffel-Möser U (1990). Growth and grazing loss rates in single-celled *Phaeocystis* sp. (Prymnesiophyceae). Mar Biol 106:153–158

Weisse T, Grimm N, Hickel W, Martens P (1986) Dynamics of *Phaeocystis pouchetii* blooms in the Wadden Sea of Sylt (German Bight, North Sea). Est Coast Shelf Sc 23:171–182

Weisse T, Tande K, Verity P, Hansen F, Gieskes W (1994) The trophic significance of *Phaeocystis* blooms. In: Lancelot C, Wassmann P (eds) Ecology of *Phaeocystis*-dominated ecosystems. J Mar Syst 5:67–79

Whipple SJ, Patten BC, Verity PG (2005) Colony growth and evidence for colony multiplication in *Phaeocystis pouchetii* (Prymnesiophyceae) isolated from mesocosm blooms. J Plank Res 27:495–501

Whipple SJ, Patten BC, Verity PG, Nejstgaard JC, Long JD, Anderson JT, Jacobsen A, Larsen A, Martinez-Martinez A, and Borrett, SR (this issue) Gaining integrated understanding of Phaeocystis spp. (Prymnesiophyceae) life cyle through model-driven laboratory and mesocosm studies Biogeochemistry

Zingone A, Chrétiennot-Dinet M-J, Lange M, Medlin L (1999) Morphological and genetic characterization of *Phaeocystis cordata* and *P. Jahnii* (Pymnesiophyceae), two new species from the Mediterranean Sea. J Phycol 35:1322–1337

Biogeochemistry (2007) 83:49–60
DOI 10.1007/s10533-007-9082-6

ORIGINAL PAPER

Phaeocystis colony distribution in the North Atlantic Ocean since 1948, and interpretation of long-term changes in the *Phaeocystis* hotspot in the North Sea

W. W. C. Gieskes · S. C. Leterme · H. Peletier · M. Edwards · P. C. Reid

Received: 13 October 2005 / Accepted: 12 June 2006 / Published online: 10 March 2007

Abstract Monitoring of *Phaeocystis* since 1948 during the Continuous Plankton Recorder survey indicates that over the last 5.5 decades the distribution of its colonies in the North Atlantic Ocean was not restricted to neritic waters: occurrence was also recorded in the open Atlantic regions sampled, most frequently in the spring. Apparently, environmental conditions in open ocean waters, also those far offshore, are suitable for complete lifecycle development of colonies (the only stage recorded in the survey).

In the North Sea the frequency of occurrence was also highest in spring. Its southeastern part was the *Phaeocystis* abundance hotspot of the whole area covered by the survey. Frequency was especially high before the 1960s and after the 1980s, i.e., in the periods when anthropogenic nutrient enrichment was relatively low. Changes in eutrophication have obviously not been a major cause of long-term *Phaeocystis* variation in the southeastern North Sea, where *total* phytoplankton biomass was related significantly to river discharge. Evidence is presented for the suggestion that *Phaeocystis* abundance in the southern North Sea is to a large extent determined by the amount of Atlantic Ocean water flushed in through the Dover Strait.

Since *Phaeocystis* plays a key role in element fluxes relevant to climate the results presented here have implications for biogeochemical models of cycling of carbon and sulphur. Sea-to-air exchange of CO_2 and dimethyl sulphide (DMS) has been calculated on the basis of measurements during single-year cruises. The considerable annual variation in phytoplankton and in its *Phaeocystis* component reported here does not warrant extrapolation of such figures.

Keywords Annual variation · North Sea hotspot · North Atlantic-wide · *Phaeocystis*

W. W. C. Gieskes
Department of Ocean Ecosystems,
University of Groningen, Haren, The Netherlands
e-mail: gieskess@xs4all.nl

S. C. Leterme
M.B.E.R.C., School University of Plymouth,
Plymouth, PL4 8AA, UK

H. Peletier
R.I.K.Z.-Rijkswaterstaat, Haren,
The Netherlands

S. C. Leterme (✉) · M. Edwards · P. C. Reid
S.A.H.F.O.S., Citadel Hill,
Plymouth, PL1 2PB, UK
e-mail: sophie.leterme@plymouth.ac.uk

Introduction

In the North Atlantic and the North Sea seasonal variations in marine phytoplankton composition are well known, but annual variations in abundance have hardly been described, with only a few exceptions. Monitoring has been maintained for several decades close to British (Russell et al. 1971), Dutch (Cadée and Hegeman 1991), German (Hickel et al. 1996) research institutes, off Belgium between 1988 and 2000 (Breton et al. 2006). Observed changes in plankton have been related to shifts in ocean current patterns (Russell 1935, 1973; Lindley et al. 1990; Taylor et al. 1998; Reid et al. 2003) that are assumed to be governed by large-scale meteorological phenomena (Beaugrand et al. 2002; Edwards et al. 2002; Reid and Edwards 2001; Drinkwater et al. 2003; Leterme et al. 2005; Breton et al. 2006), to climate (Radach 1984; Seuront and Souissi 2001; Reid et al. 1998), or to increased coastal eutrophication (Richardson 1989, 1997; Greve et al. 1996), but the response of plankton to anthropogenic interference (Cadée and Hegeman 2002; Beaugrand 2004) has not always been demonstrated in a convincing way.

The Continuous Plankton Recorder survey is the longest-running monitoring programme in the North Atlantic and the North Sea, with a remarkably wide coverage (Fig. 1a), based on one standard sampling method (Glover 1967; Colebrook 1975; Warner and Hays 1995). Since the time series spans several decades, it has been tempting to interpret striking changes in abundance of the plankton caught on the nets ('silks') of Hardy's recorders as the consequence of long-term environmental change because variations often took place simultaneously over very large areas (Colebrook and Robinson 1964; Edwards et al. 2002). Actually, it has often been taken for granted that hydroclimatic forcing in relation to global warming must have controlled the long- and short-term plankton variability. The northward shift in subtropical and temperate zooplankton groups and the gradual disappearance of cold-adapted copepod species in the eastern North Atlantic (Beaugrand et al. 2002) has indeed been quite conspicuous. However, trends in ocean water temperature, in the Atlantic multidecadal oscillation, or in the state of the North Atlantic oscillation and related hydrographical features (e.g., vertical stratification: Martin and Hall 1975; Fromentin and Planque 1996, Reid et al. 2001) could often only be related to the abundance of plankton components and transitions in taxonomic composition after sophisticated statistical treatment (Beaugrand et al. 2000, 2003; Beaugrand 2004) or on assumptions involving differential lags in the response of ecosystem components (Drinkwater et al. 2003; Kane 2005).

The Continuous Plankton Recorder survey has provided a tremendous data set on the occurrence of *Phaeocystis* sp. *Phaeocystis* was already known to vary as much as any other group of plankton in the North Atlantic (Owens et al. 1989) and the North Sea, in a way apparently independent of the phytoplankton in general or of other microalgal species groups such as dinoflagellates and diatoms (Gieskes and Kraay 1977a), for reasons not known at the time. We present here an overview of the occurrence of *Phaeocystis* throughout the North Atlantic. In our discussion we will focus on its presence in the North Sea, where it is most abundant. We offer an attempt to link occurrence to the pattern of long-term change in hydrographic events in the eastern North Atlantic. We end the presentation of our analysis by highlighting the biogeochemical implications of the long-term changes in the abundance of this species, which is known to affect profoundly the cycling and sea-to-air exchange of climate-relevant elements, especially carbon and sulphur, components in the greenhouse gases CO_2 and dimethyl sulphide (DMS; see review by Schoemann et al. 2005).

Materials and methods

The Continuous Plankton Recorder

The Continuous Plankton Recorder (CPR) survey consists of a dense network of transects across the North Atlantic and the North Sea. Plankton recorders are towed on a monthly basis at a depth of 8–10 m from ships-of-oppportunity that travel at 10–18 knots. The sampling mechanism inside the recorders consists of a narrow band of filtering silk (mesh 270 μm) that is driven by an impeller at the rear of the recorder at a speed adjusted according to the speed of the ship. The silk catches particles entering the 12 mm^2 aperture while it passes (at a rate of 10 cm per 10 nautical miles, 18.5 km) through the end of a wide tunnel behind the narrow opening in front. About 3 m^3

of seawater is filtered every 18.5 km. Clogging of silks is an exception, but it happens.

Phaeocystis colonies are recorded visually as present or absent on lengths of 10 cm of silk (a sample). From this information a time series from all the CPR samples for a given region can be used to create maps of the frequency of occurrence (the percentage of samples with *Phaeocystis* presence). 0% means that for a given month not a single sample contained *Phaeocystis*, 100% that it was present (no matter how much) in every single sample taken in that month. Each section covered by the CPR constitutes a variable number of samples. Total coverage of a section is determined as an average of all percentages.

Details of the sampling procedure have been described by Warner and Hays (1995). Total phytoplankton biomas' is assessed in four categories by comparing the colouration of the silks (mostly caused by algal pigments) with a colour chart. The colour values [Phytoplankton Colour index (PCI)] are related to chlorophyll concentration on the silks and in the field (Gieskes and Kraay 1977a; Batten et al. 2003).

Standard regions of the CPR survey are shown in Fig. 1a; each consists of smaller rectangles; rectangle 'Ow', located in the southeastern part of the Southern Bight of the North Sea just off The Netherlands where major rivers of the continent (Rhine, Scheldt) discharge, has received particular attention in the present study.

Rijkswaterstaat-RIKZ survey

The chlorophyll concentrations of the monitoring programme run by Rijkswaterstaat-RIKZ off The

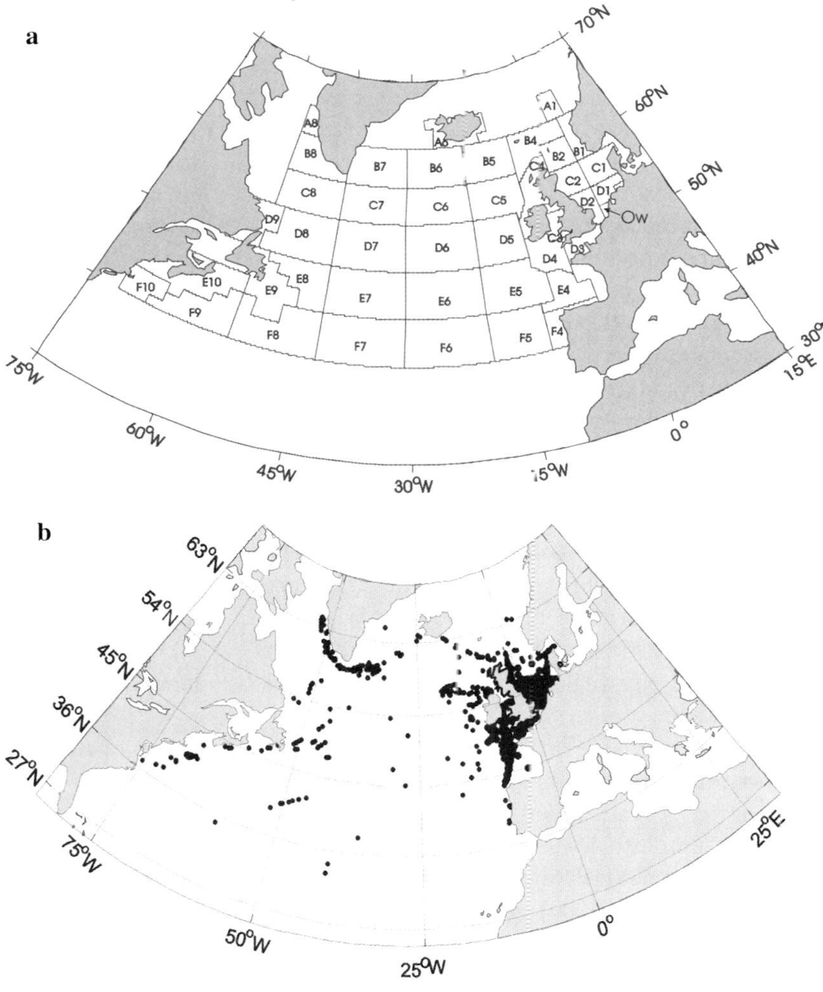

Fig. 1 (**a**) Standard areas of the continuous plankton recorder survey. (**b**)Distribution of *Phaeocystis* colonies over the North Atlantic and the North Sea over the years 1948–2003. Annual variations: see Fig. 2

Netherlands since the early 1970s and analysed here (Fig. 4) have been measured by fluorometry and later by high-performance liquid chromatography (HPLC). These data, also those of Rhine river discharge, are available free of charge to the public on the internet because Rijkswaterstaat is a government agency funded by taxes; *waterbase* is the keyword for access. Briefly, samples are taken near the surface; they are considered to be representative for the whole water column because normally Dutch coastal waters are mixed from the surface to bottom by tides and wind. Information on the frequency of sampling, the station network, numbers of samples taken, and other details of the sampling procedure are implicit in the data set that can be extracted by using *waterbase*. We have chosen section Noordwijk (N) of this programme because it starts just north of the Rhine outflow and samples taken there are likely to be influenced most by this river. The numbers (N2, N10, etc.) refer to the number of kilometers from the coast.

Data presentation

The monthly mean phytoplankton and *Phaeocystis* plots presented here (months versus years, 1948–2003) are so-called Hovmöller diagrams, two-dimensional (2D) plots are often used to display large amounts of data in a readily understandable way (Hovmöller 1949); the computing and statistical analysis has been presented by Leterme et al. (2005). In order to reveal anomalies in the trends in the North Sea more clearly, the Phytoplankton Colour index (PCI) (chlorophyll) and *Phaeocystis* colony percent frequency-of-occurrence data were also standardized to zero mean and unit variance in the North Sea survey regions D1, D2, C1 and C2 together (refered to herein as 'the North Sea'), the cumulative sums method (Ibañez et al. 1993). The calculation consists of subtracting a reference value (here the mean of the series) from the data. The residuals are then successively added, forming a cumulative function. Cumulative sum plots have been introduced into the SAHFOS data set interpretation (Beaugrand et al. 2000) to summarise major changes and identify transitional periods in phytoplankton time series. The cumulative function results in the smoothing of high-frequency interannual variability and highlights changes in mean values along the time series.

Results

The distribution of *Phaeocystis* colonies over the North Sea and the North Atlantic (Fig. 1b) seems to support the notion of an ocean-wide species with peak abundance in spring everywhere (Fig. 2). The centre of abundance over the study period, spanning five decades, was restricted nearly entirely to the North Sea, with the southeastern part clearly being a hotspot. In the open and central Atlantic regions the long record suggests low (cf. Owens et al. 1989) but persistent occurrence (Fig. 2). Abundance was also low (but equally persistent) in the northern North Sea (regions C1 and C2) where Atlantic water flows in from the northwest and dominates the water budget. Low colony frequencies were seen (only until 1985) in the front between the North Atlantic current in the Irminger Sea and the East Greenland current, south of Greenland (Figs. 1b and 2); the season of highest occurrence was spring also there.

Annual variation of *Phaeocystis* colony abundance did not at all resemble the variation in phytoplankton biomass [the chlorophyll-related Phytoplankton Colour index (PCI)] anywhere, certainly not in the bloom period, i.e. the spring (see Fig. 3). Year–month plots of true chlorophyll concentrations, derived from the Dutch monitoring programme, in CPR rectangle 'Ow' (see Fig. 1b) are presented in Fig. 4. Notice the gradually earlier chlorophyll spring bloom over the monitoring period at all stations (2, 10, 20 and 70 km off the coast of Noordwijk, The Netherlands).

The Phytoplankton Colour (PCI) cumulative sum plot showed a (single) shift (Fig. 5) in the mid 1980s. This was very different from the two shifts in the *Phaeocystis* plot (mid 1960s and early 1990s, see Fig. 5), which shows again that *Phaeocystis* does not reflect the variation observed in phytoplankton biomass. Between 1970 and 1985, low phytoplankton abundance (PCI) in the North Sea coincided with a late spring bloom. This

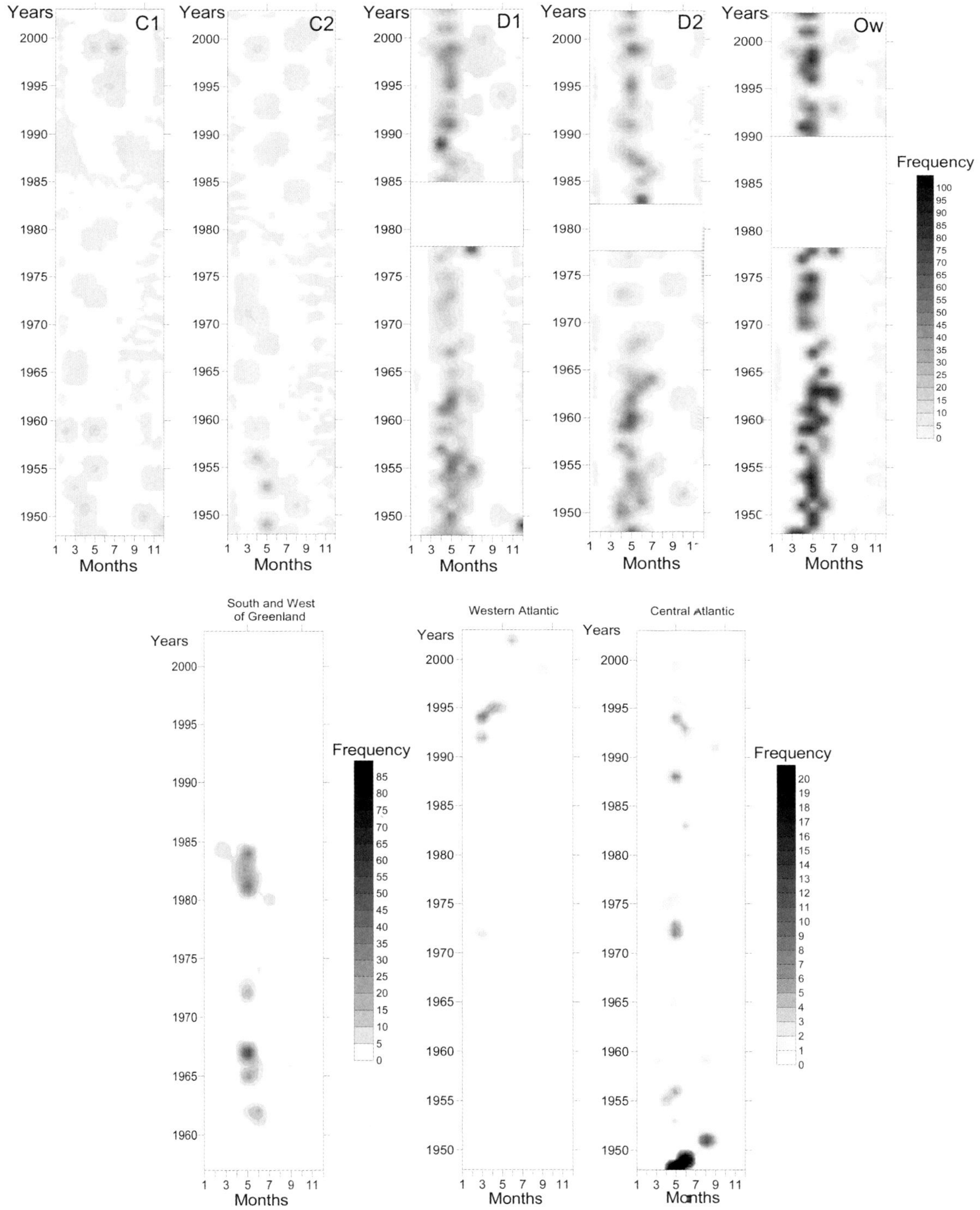

Fig. 2 Hovmöller diagrams (years 1948–2003 versus months) of *Phaeocystis* frequency in different regions of the North Sea and the Atlantic Ocean (see Fig. 1b for the location of the regions). Rectangle 'Ow' is the southeastern part of the Southern Bight of the North Sea, directly off the Netherlands. South and West Greenland corresponds to the regions A8, B7 and B8 (see Fig. 1). Western Atlantic regrouped regions E9, E10 and F10. Central Atlantic corresponds to regions E5 to E8 and D5 to D8

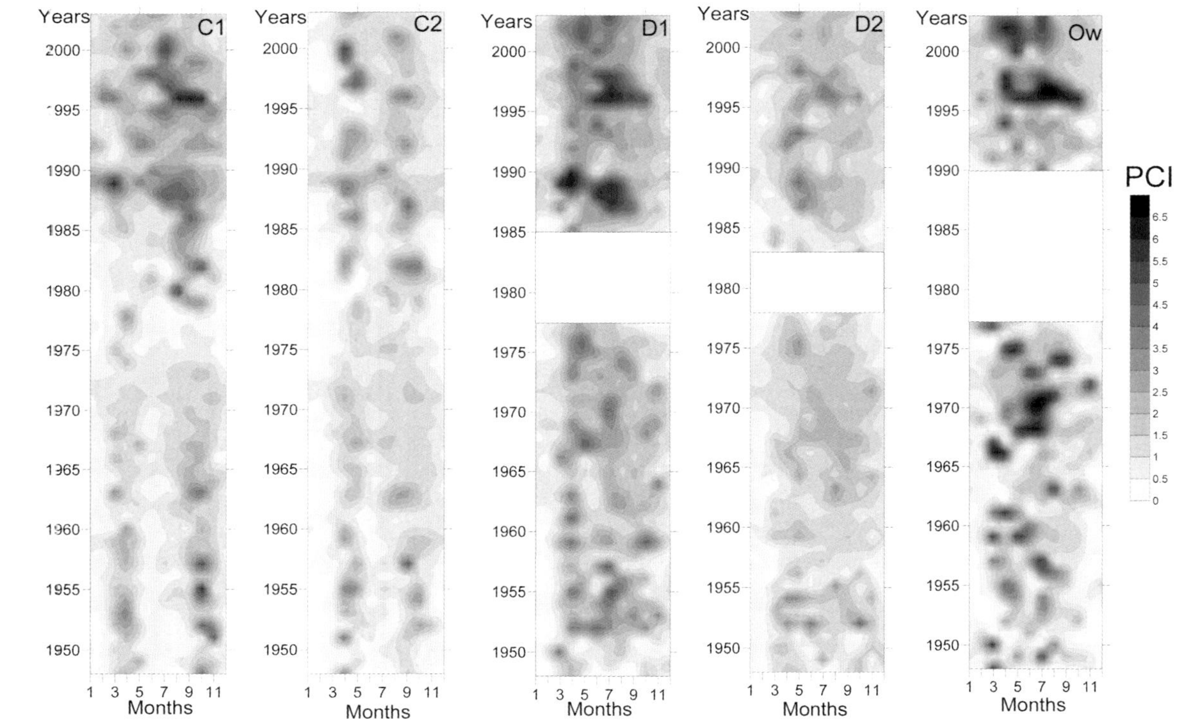

Fig. 3 Hovmöller diagrams (years 1948–2003 versus months) of the Phytoplankton Colour index (PCI) in the North Sea. For region location see Fig. 1b. Rectangle 'Ow' is the southeastern part of the Southern Bight of the North Sea, directly off The Netherlands

pattern is in contrast with a high frequency of occurrence of *Phaeocystis* and its long season of presence before 1955. After 1995 the *Phaeocystis* season was again longer, an observation confirmed by Breton et al. (2006) who described the *Phaeocystis* record since 1988 off the Belgian coast, and by the *Phaeocystis* record of Philippart, Cadeé and van Iperen (not published) that covers the period 1975–2006.

Discussion

The distribution map of *Phaeocystis* presented here can hardly be improved, not even by ocean colour remote sensing: the frequent cloud cover over the North Atlantic hinders satellite-borne observations, but not the CPR survey. Also, the pigment signature of *Phaeocystis* is not very specific (Antajan et al. 2004) so the absorption spectrum and therewith the light reflectance hardly differs from that of other fucoxanthin- and fucoxanthin-derivative-containing phytoplankton groups, frustrating remote distinction of this particular species group. Without the survey it would not have been possible to construct the detailed year–month diagrams spanning many decades; those of low but persistent *Phaeocystis* occurrence in spring far offshore in the mid-Atlantic regions (Fig. 2) and in the Atlantic water of the northern North Sea, in areas C1 and C2 (Fig. 2) are probably the most unexpected. Not only in neritic water but apparently also in the open Atlantic, far away from anthropogenic nutrient sources (eutrophication), environmental conditions are clearly quite suitable for lifecycle development from single cells to colonies (the only stage recorded during the CPR survey).

Even so, changes in eutrophication have often been invoked to explain variation in *Phaeocystis* abundance and length of the season of blooming and occurrence, but this interpretation has not been satisfactory in view of the restricted number of stations, often limited to just one (Hickel et al. 1996; Cadée and Hegeman 2002). Breton et al. (2006), who also used observations at a single

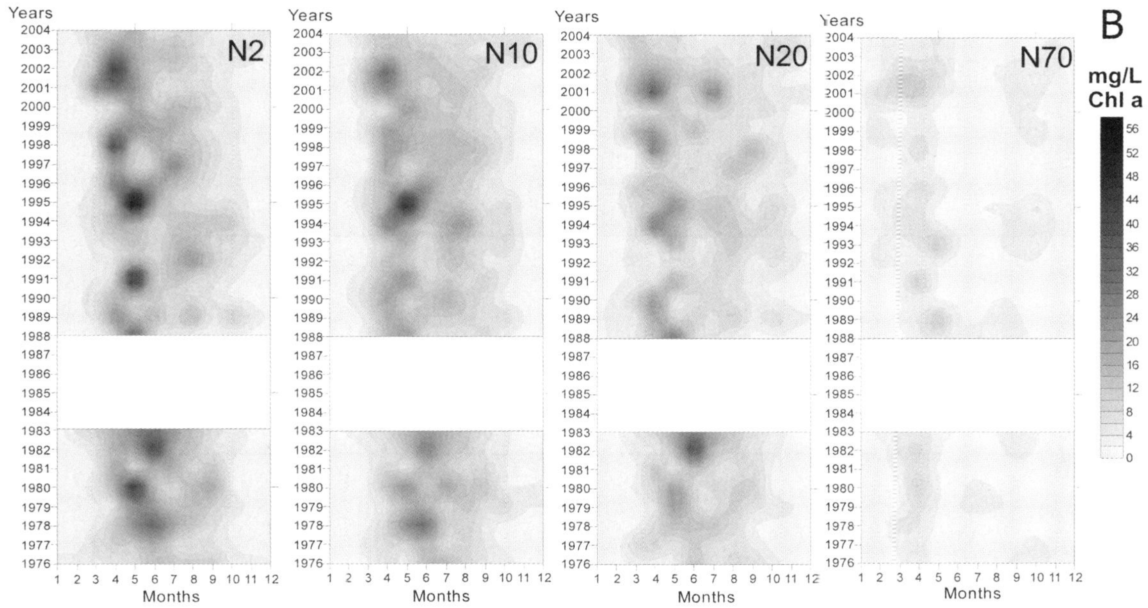

Fig. 4 Hovmöller diagrams of chlorophyll-a concentration on a section perpendicular to the Dutch coast starting from the city of Noordwijk (just north of the river Rhine outflow), at 2, 10, 20 and 70 km from the coast. Notice the gradual shift of the start of the vegetative season to an earlier date at all four stations

station (off Belgium), drew attention to the influence of larger-scale phenomena, namely the North Atlantic Oscillation that governs climate. Much earlier, an analysis of CPR survey results obtained in the same area had revealed trends of *Phaeocystis* (Gieskes and Kraay 1977a) that did not simply follow the eutrophication trend (Philippart et al. 2000): a decrease of *Phaeocystis* frequency of occurrence in the 1960s and early 1970s, a period when anthropogenic nutrient input in the southeastern North Sea increased.

The higher frequency of occurrence and longer growing season of *Phaeocystis* before 1965, when eutrophication was still quite low (Philippart et al. 2000), is particularly striking (Figs. 2, 5). Actually, colony abundance was already reported to be extensive more than 80 years ago (Savage 1930). This also suggests that *Phaeocystis* abundance is not determined by anthropogenic nutrient input only. On the other hand, phytoplankton biomass in general has always suggested a direct relation with eutrophication (Cadée and Hegeman 2002). Also in our study a relationship can be established between chlorophyll concentrations measured off the Dutch coast (Fig. 4) and river run-off (not shown), which confirms previous findings by Schaub and Gieskes (1991); recently Breton et al. (2006) have also pointed out this link. Increased nutrient availability of North Sea coastal waters after high freshwater fluxes following periods of heavy rainfall (a response to the North Atlantic Oscillation, cf. Breton et al. 2006) in the catchment area of the continent's rivers clearly sets the stage for phytoplankton abundance in the southeastern North Sea. Such a batch culture-like response to nutrient enrichment (light is apparently not limiting primary production here) can be observed to the present day (Fig. 4). It can readily be explained by nutrient flushed out from the agricultural lands along the continent's rivers, and probably also resuspension of nutrient-rich bottom sediment when river flow is rapid after heavy downpours upstream; Schaub and Gieskes (1991) stated that phytoplankton biomass near the Dutch coast can be predicted on the basis of meteorological data on precipitation, which has now also been suggested for Belgian coastal waters by Breton et al. (2006). There is a suggestion of a gradual shift of the start of the growing season to an earlier date in the southern North

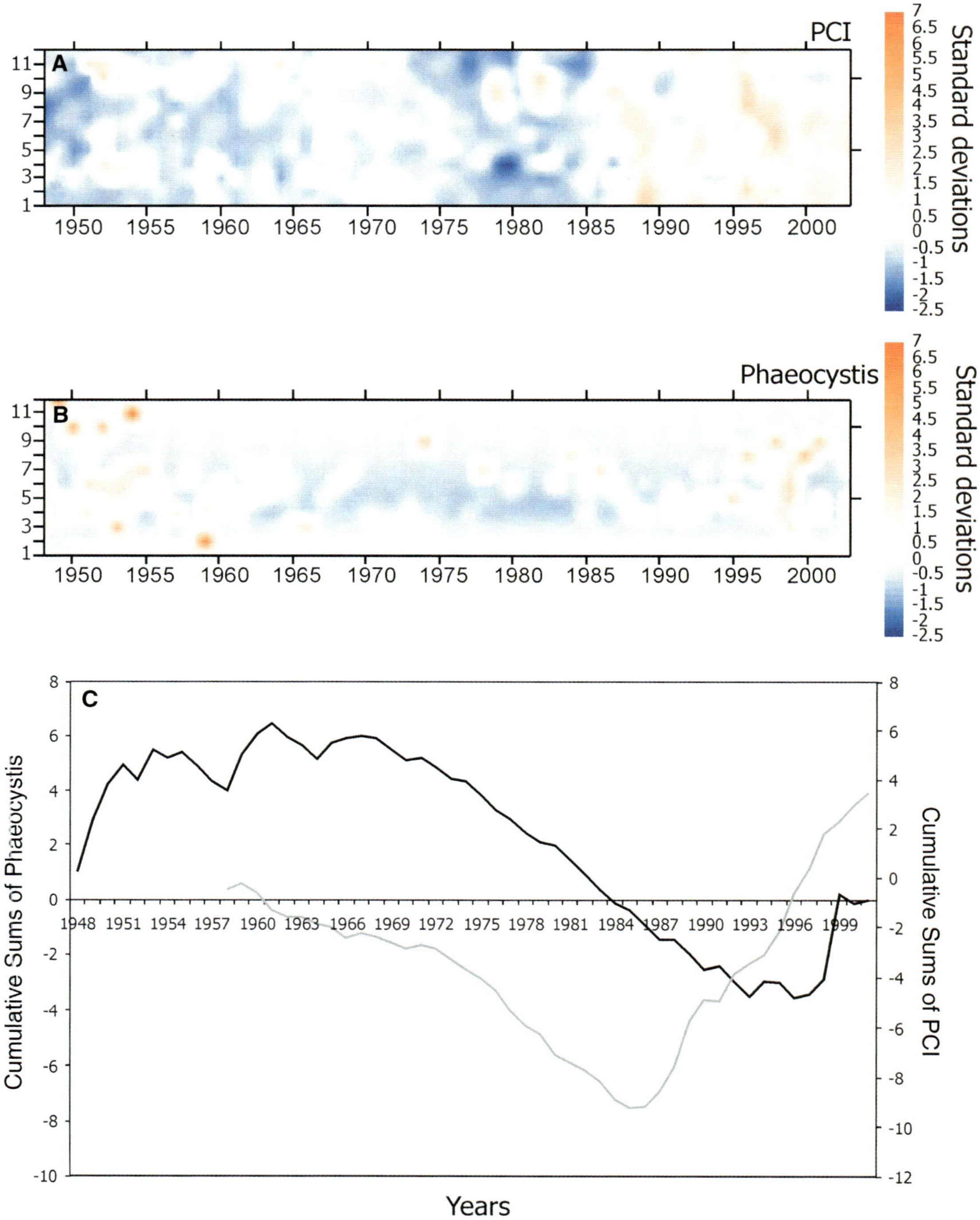

Fig. 5 Hovmöller diagrams, years 1948–2003 versus months of the year of: (A) standard deviations of the phytoplankton colour index (PCI), (B) standard deviations of *Phaeocystis* occurrence frequency, data averaged over North Sea regions C1, C2, D1 and D2 (cf. Fig. 1a). and (C) cumulative sum (see "Materials and methods" section) plots of mean phytoplankton colour index (PCI, in gray) and *Phaeocystis* frequency of occurrence (black) for the same region

Sea (Fig. 4). Whether or not this is the consequence of a slight but gradual warming of southern North Sea waters remains to be shown.

The *Phaeocystis* component of the phytoplankton is not necessarily related to river-induced eutrophication: long-term changes of *Phaeocystis* are quite different from those of the phytoplankton (as Phytoplankton Colour index PCI, i.e. chlorophyl, in the "Materials and methods" section) in general (Fig. 5). As we have seen above, environmental conditions in open ocean water, not necessarily those of coastal waters, promote colony development from single cells. Atlantic water inflow into the North Sea may therefore well be an important factor for the development of *Phaeocystis* in the southern North Sea, a suggestion made earlier for phytoplankton in general by De Jonge et al. (1996) and De Jonge (1997). The inflow of Atlantic water into the North Sea through the Dover Straits has unfortunately never been monitored over longer periods; it has usually been no more than deduced, e.g. from the potential induced in submarine cables by the water flow (Prandle 1978), most often by constructing hydrodynamic models that simulate currents on the basis of tides, wind stress over the North Sea and freshwater input (Salomon et al. 1993; Laane et al. 1996; Skogen and Søiland 1998; Siegismund 2001). Rather sudden changes in climate–ocean interactions that influenced regime shifts in North Sea ecosystem components took place in the late 1970s, late 1980s and late 1990s (Weijerman et al. 2005). The long-term changes in *Phaeocystis* described here (Fig. 5) do not correspond well with these regime shift periods, probably because both variability in the mix of nutrient sources in the southern North Sea and in the sampling route of the plankton recorders do not allow corresponding biological and hydrographical observations.

Phaeocystis has been reported to accumulate preferentially far offshore, not on the coast, in the southern North Sea in the 1920s (Savage 1930). Later, Gieskes and Kraay (1977b) also described preferential abundance offshore. According to Brunet et al. (1996) accumulation on French beaches of the British Channel is the effect of eastward transport of offshore *Phaeocystis* blooms, not of in situ growth. The line of increased abundance off the southern coasts of Greenland, in the region of the front between the extension of the North Atlantic current in the Irminger Sea and the East Greenland current (Fig. 1b), may be explained in these terms. In view of the current system the *Phaeocystis* distribution here suggests advection from the Denmark Straits or even from the west coast of Iceland (regions not covered by the CPR survey), where *Phaeocystis* has been reported to be abundant (Thordardottir and Astthorsson 1986; Stefansson and Olafsson 1991). The total absence of colonies since 1985 can be explained simply (albeit speculatively) by assuming a shift in the front to an area outside the plankton recorder routes.

Offshore accumulation of *Phaeocystis* along frontal zones as suggested here may well be a more general phenomenon; in the southern North Sea it would explain an apparent dependence on the inflow of Channel/Atlantic water through the Dover Straits, a source of nutrients in this region that can be more important than local river sources of nutrients (Radach and Lenhart 1995; Laane et al. 1996; De Jonge et al. 1996; De Jonge 1997). The high *Phaeocystis* abundance far off the Netherlands, all the way between Holland and England in 1925 and 1926 (Savage 1930), can be understood in this light. The Atlantic Ocean water mass entering the North Sea from the southeast forms a front with the coastal water off the coasts of Belgium and the Netherlands. Interestingly, part of the decrease of *Phaeocystis* abundance in the CPR record in the 1960s and 1970s may be explained by a change of the position of the front, in line with the suggested front shift south of Greenland that may have brought *Phaeocystis* out of the reach of the CPR survey in that region. In the southeastern North Sea the sampling route of the CPR survey has remained the same geographically while the coastal water/ocean water front has shifted up to 20 km eastward to shallow Dutch coastal waters since the gradual man-made closure of the estuaries in the south-estern Netherlands following the catastrophic storm surge of 1953 (J. de Kok and L. Villerius, Rijkswaterstaat-RIKZ, pers. comm.).

Wind forcing may of course broaden the Rhine plume considerably, especially when winds from the east persist (Gieskes 1974). Wind forcing, later often suggested to be a dominant factor in setting the stage for water mass distribution and flow in the North Sea (Salomon et al. 1993; Laane et al. 1996; Smith et al. 1996; Pingree 2005), is a parameter with a long-term measurement record at many meteorological stations along the coasts. Exploitation of such data to hindcast and predict large-scale current patterns would enhance and improve existing models to estimate inflow of Atlantic Water through the Dover Straits. These models currently produce contradictory results, and no trend has so far been revealed (Otto et al. 1990). Low phytoplankton abundance in the North Sea and a late spring bloom and changes in other ecosystem components between the mid 1970s and the late 1980s have been ascribed to a cold-water period considered to be the consequence of low Atlantic water inflow into the North Sea (Corten and van de Kamp 1992, 1996; Lindeboom et al. 1995; Taylor et al. 1998; Edwards et al. 2002; Weijerman et al. 2005). It is well known that biological responses may magnify changes in the physical or chemical environment, and the annual variation of *Phaeocystis* colony occurrence in the southeastern North Sea may well be an example of this concept.

Phaeocystis plays a key role in element fluxes relevant to climate (reviewed by Schoemann et al. 2005) and therefore the results presented here have implications for biogeochemical models of cycling of carbon and sulphur. Sea-to-air exchange of carbon dioxide (CO_2) and dimethyl sulphide (DMS) has been calculated on the basis of measurements made during cruises in a single year. Manizzi et al. (2005) already issued a warning to prevent this practice. Thomas et al. (2004) presented CO_2 fluxes in the southern North Sea based on measurements from August 2001 to May 2002 only. They concluded that coastal seas would take up 0.4 Pg C per year. In view of the large annual variation both in phytoplankton and in *Phaeocystis* abundance presented in this paper such extrapolations cannot be correct. Discrepancies noted by Thomas et al. (2004) with results obtained in the same area by Borges and Frankignoulle (2003) may well have been the consequence of differences in plankton activity between the years in which the cruises were carried out.

Acknowledgements Financial support from the Dutch Royal Academy of Sciences (KNAW) allowed W.W.C.G. to spend a sabbatical at the Sir Alister Hardy Foundation of Ocean Sciences in Plymouth (UK); the SAHFOS directory is gratefully acknowledged for generous hospitality. We thank the Dutch National Institute for Coastal and Marine Management (RIKZ) for access to their monitoring data through *waterbase*, and a group of students of Groningen University coached by Rob Middag for the analyses. Comments by two unknown referees and editing by Dr M.A. van Leeuwe have improved the manuscript considerably. We acknowledge the assistance of the reviewers, and of colleagues who inspected earlier versions of the manuscript.

References

Antajan E, Chrétiennot-Dinet M-J, Leblanc C, Daro M-H, Lancelot C (2004) 19′Hexanoyloxyfucoxanthin may not be the appropriate pigment to trace occurrence and fate of *Phaeocystis*: the case of *P. globosa* in Belgian coastal waters. J Sea Res 52:165–177

Batten SD, Walne AW, Edwards M, Groom SB (2003) Phytoplankton biomass from continuous plankton recorder data: an assessment of the phytoplankton colour index. Plankton Res 25:697–702

Beaugrand G (2004) The North Sea regime shift: evidence, causes, mechanisms and consequences. Progr Oceanogr 60:245–262

Beaugrand G, Ibañez F, Lindley JA (2003) An overview of statistical methods applied to CPR data. Progr Oceanogr 58:253–262

Beaugrand G, Ibañez F, Reid PC (2000) Spatial, seasonal and long-term fluctuations of plankton in relation to hydrodynamic features in the English Channel, Celtic Sea and Bay of Biscay. Mar Ecol Progr Ser 200:93–102

Beaugrand G, Reid PC, Ibañez F, Lindley JA, Edwards M (2002) Reorganization of North Atlantic marine copepod biodiversity and climate. Science 296:1692–1694

Borges AV, Frankignoulle M (2003) Distribution of surface carbon dioxide and air-sea exchange in the English Channel and adjacent areas. J Geophys Res 108(C5):3140 (doi:10. 1029/2000JC 000571)

Breton E, Rousseau V, Parent J-Y, Ozer J, Lancelot C (2006) Hydroclimatic modulation of diatom/*Phaeocystis* blooms in nutrient-enriched Belgian coastal waters (North Sea). Limnol Oceanogr 51:1401–1409

Brunet C, Brylinski JM, Bodineau L, Thommelin G, Bentley D, Helde D (1996) Phytoplankton dynamics during the spring bloom in the south-eastern Channel. Estuar Coast Shelf Sci 43:469–483

Cadée GC, Hegeman J (1991) Historical phytoplankton data of the Marsdiep. Hydrobiol Bull 24:111–118

Cadée GC, Hegeman J (2002) Phytoplankton in the Marsdiep at the end of the 20th century; 30 years monitoring biomass, primary production, and *Phaeocystis* blooms. J Sea Res 48:97–110

Colebrook JM (1975) The continuous plankton recorder survey: automatic data processing methods. Bull Mar Ecol 8:123–142

Colebrook JM, Robinson GA (1964) Continuous plankton records: annual variability of plankton, 1948–1960. Bull Mar Ecol 6:52–69

Corten A, van de Kamp G (1992) Natural changes in pelagic fish stocks of the North Sea in the 1980s. ICESMar Sci Symp 195:402–417

Corten A, van de Kamp G (1996) Variation in the abundance of southern fish species in the southern North Sea in relation to hydrography and wind. ICES J Mar Sci 53:1113–1119

De Jonge VN (1997) High remaining productivity in the Dutch Western Wadden Sea despite decreasing nutrient inputs from riverine sources. Mar Poll Bull 34:427–436

De Jonge VN, Bakker JF, van Stralen M (1996) Recent changes in the contributions of river Rhine and North Sea to the eutrophication of the western Dutch Wadden Sea. Neth J Aq Ecol 30:27–39

Drinkwater KF, Belgrano A, Borja A, Conversi A, Edwards M, Greene CH, Ottersoen G, Pershing AJ, Walker H (2003) The response of marine ecosystems to climate variability associated with the North Atlantic Oscillation: climatic significance and environmental impact. Geophys Monogr 134:211–234 (Amer. Geophys. Union)

Edwards M, Beaugrand G, Reid PC, Rowden AA, Jones MB (2002) Ocean climate anomalies in the ecology of the North Sea. Mar Ecol Progr Ser 239:1–10

Fromentin JM, Planque B (1996) *Calanus* and environment in the eastern North Atlantic. II. Influence of the North Atlantic Oscillation on *C. finmarchicus* and *C. helgolandicus*. Mar Ecol Progr Ser 134:111–118

Gieskes WWC (1974) Phytoplankton and primary productivity studies in the Southern Bight of the North Sea, eastern part, in 1972. Ann Biol 29:54–61

Gieskes WWC, Kraay GW (1977a) Continuous plankton records: changes in the plankton of the North Sea and its eutrophic Southern Bight from 1948 to 1975. Neth J Sea Res 11:334–364

Gieskes WWC, Kraay GW (1977b) Primary production and consumption of organic matter in the southern North Sea during the spring bloom of 1975. Neth J Sea Res 11:146–167

Glover RS (1967) The continuous plankton recorder survey of the North Atlantic. Symp Zool Soc Lond 19:189–210

Greve W, Reimers F, Nast J (1996) Biocoenotic changes of the zooplankton in the German Bight: the possible effects of eutrophication and climate. ICES J Mar Sci 53:951–956

Hickel W, Eickenhoff M. Spindler H, Berg J, Raabe T, Mueller R (1996) Auswertungen von Langzeituntersuchungen von Naehrstoffen und Phytoplankton in der Deutschen Bucht. Berlin (UBA) 213 pp

Hovmöller E (1949) The trough and ridge diagram. Tellus 1:62–66

Ibañez F, Fromentin JM, Castel J (1993) Application de la méthode des sommes cumulées à l'analyse des séries chronologiques océanographiques. CR Acad Sci Paris, Sciences de la vie/Life sciences 316:745–748

Kane J (2005) The demography of *Calanus finmarchicus* (copepoda, calanoida) in the middle Atlantic Bight, USA, 1977–2001. J Plankton Res 27:401–414

Laane RWPM, Groeneveld G, de Vries A, van Bennekom AJ, Sydow JS (1996) Nutrients (P, N, Si) in the Channel and the Dover Strait: seasonal and year-to-year variation and fluxes to the North Sea. Oceanol Acta 16:607–616

Leterme SC, Edwards M, Seuront L, Attrill MJ, Reid PC, John AWG (2005) Decadal basin-scale changes in diatoms, dinoflagellates, and phytoplankton color across the North Atlantic. Limnol Oceanogr 50:1244–1253

Lindeboom H, Raaphorst W, Beukema J, Cadée GC, Swennen C (1995) (Sudden) changes in the North Sea and Wadden Sea: oceanic influences underestimated? Dt Hydrogr Z Suppl 2:87–100

Lindley JA, Roskell J, Warner AJ, Halliday N, Hunt HG, John AWG, Jonas TD (1990) Doliolids in the German Bight in 1989:evidence for exceptional inflow into the North Sea. J Mar Biol Ass UK 70:679–682

Manizzi M, Metzl N, Le Quéré C (2005) Interannual variability of the ocean carbon cycle. SOLAS News 1:10–11

Martin JHA, Hall WB (1975) Changes in physical, chemical and phytoplankton parameters in the northern North Sea, 1961–1970. ICES, CM 1975/C:33:1–4

Otto L, Zimmermann JTF, Furnes GK, Mork M, Saetre R, Becker G (1990) Review of physical oceanography of the North Sea. Neth J Sea Res 26:161–238

Owens NJP, Cook D, Colebrook M, Hunt H, Reid PC (1989) Long-term trends in the occurrence of *Phaeocystis* sp. In the North-East Atlantic Ocean. J Mar Biol Ass UK 69:813–821

Philippart CJM, Cadée GC, van Raaphorst W, Riegman R (2000) Long-term phytoplankton–nutrient interactions in a shallow coastal sea: algal community structure, nutrient budgets, and denitrification potential. Limnol Oceanogr 45:131–144

Pingree RD (2005) North Atlantic and North Sea Climate Change: curl up, shut down, NAO and Ocean Colour. J Mar Biol Assoc UK 85:1301–1315

Prandle D (1978) Monthly mean residual flow through the Dover Strait, 1949–1972. J Mar Biol Ass UK 58:965–973

Radach G (1984) Variations in the plankton in relation to the climate. Rapp P-v Reun Cons Int Explor Mer 185:234–254

Radach G, Lenhart HJ (1995) Nutrient dynamics in the North Sea: fluxes and budgets in the water column derived from ERSEM. Neth J Sea Res 33:301–335

Reid PC, Borges MF, Svendsen E (2001) A regime shift in the North Sea circa 1990. Fish Res 50:163–171

Reid PC, Edwards M (2001) Long-term changes in the pelagos, benthos and fisheries of the North Sea. Senckenb Marit 31:107–115

Reid PC, Edwards M, Beaugrand G, Skogen M, Stevens D (2003) Periodic changes in the zooplankton of the North Sea during the 20th century linked to oceanic inflow. Fish Oceanogr 12:260–269

Reid PC, Planque B, Edwards M (1998) Is observed variability in the long-term results of the continuous plankton recorder survey a response to climate change? Fish Oceanogr 7:282–288

Richardson K (1989) Algal blooms in the North Sea: the good, the bad and the ugly. Dana 8:83–89

Richardson K (1997) Harmful or exceptional blooms in the marine ecosystem. Adv Mar Biol 31:301–385

Russell FS (1935) On the value of certain plankton animals as indicators of water movements in the English Channel and North Sea. J Mar Biol Ass UK 20:309–331

Russell FS (1973) A summary of the observations on the occurrence of plankton stages of fish off Plymouth, 1924–1972. J Mar Biol Ass UK 53:347–355

Russell FS, Southward AJ, Boalch GT, Butler EI (1971) Changes in biological conditions in the English Channel off Plymouth during the last half-century. Nature (Lond) 234:468–470

Salomon JC, Breton M, Guéguénat P (1993) Computed residual flow through the Dover Strait. Oceanol Acta 16:449–455

Savage RE (1930) The influence of *Phaeocystis* on the migration of the herring. Fishery Invest II, XII(2):5–13

Schaub BEM, Gieskes WWC (1991) Eutrophication of the North Sea: the relation between Rhine river discharge and chlorophyll-a concentration in Dutch coastal waters. In: Estuaries and coasts: spatial and temporal intercomparisons. ECSA 19 Symposium. Publ. Olsen and Olsen, pp 85–90

Schoemann V, Becquevort S, Stefels J, Rousseau V, Lancelot C (2005) *Phaeocystis* blooms in the global ocean and their controlling mechanisms: a review. J Sea Res 53:43–66

Seuront L, Souissi S (2001) Climatic control of *Phaeocystis* spring bloom in the eastern English Channel (1991–2000). La Mer 40:41–51

Siegismund F (2001) Long-term changes in the flushing times of the ICES-boxes. Senckenb Marit 31:151–167

Skogen M, Søiland H (1998) A user's guide to NORWECOM v2.0. The Norwegian Ecological Model system. Tech. Rep. Fisken og Havet 18/98. Institute of marine Research, Pb.1870, N-5024 Bergen. 42 pp

Smith JA, Damm P, Skogen MD, Flather RA, Paetsch J (1996) An investigation into the variability of circulation and transport on the North-west European shelf using three hydrodynamic models. German J Hydr 8:325–348

Stefansson U, Olafsson J (1991) Nutrient and fertility of Icelandic waters. Rit Fiskideildar 12:1–56

Taylor AH, Jordan MB, Stephens JA (1998) Northward shifts of the Gulf Stream following ENSO events. Nature 393:638

Thomas H, Bozec Y, Elkelay K, de Baar HJW (2004) Enhanced open ocean storage of CO_2 from shelf sea pumping. Science 304:1005–1008

Thordardottir T, Astthorsson OS (1986) Chlorophyll a and zooplankton distribution in Icelandic waters in spring 1982, 1983 and 1984. Ann Biol 41:63–66

Warner A, Hays GC (1995) Sampling by the continuous plankton recorder survey. Prog Oceanogr 34:237–256

Weijerman M, Lindeboom H, Zuur AF (2005) Regime shifts in marine ecosystems of the North Sea and the Wadden Sea. Mar Ecol Prog Ser 298:21–39

Biogeochemistry (2007) 83:61–70
DOI 10.1007/s10533-007-9083-5

ORIGINAL PAPER

Photosynthetic responses in *Phaeocystis antarctica* towards varying light and iron conditions

M. A. van Leeuwe · J. Stefels

Received: 23 March 2006 / Accepted: 10 August 2006 / Published online: 23 March 2007

Abstract The effects of iron limitation on photoacclimation to a dynamic light regime were studied in *Phaeocystis antarctica*. Batch cultures were grown under a sinusoidal light regime, mimicking vertical mixing, under both iron-sufficient and -limiting conditions. Iron-replete cells responded to changes in light intensity by rapid xanthophyll cycling. Maximum irradiance coincided with maximum ratios of diatoxanthin/diadinoxanthin (dt/dd). The maximum quantum yield of photosynthesis (F_v/F_m) was negatively related to both irradiance and dt/dd. Full recovery of F_v/F_m by the end of the light period suggested successful photoacclimation. Iron-limited cells displayed characteristics of high light acclimation. The ratio of xanthophyll pigments to chlorophyll *a* was three times higher compared to iron-replete cells. Down-regulation of photosynthetic activity was moderated. It is argued that under iron limitation cells maintain a permanent state of high energy quenching to avoid photoinhibition during exposure to high irradiance. Iron-limited cells could maintain a high growth potential due to an increased absorption capacity as recorded by in vivo absorption, which balanced a decrease in F_v/F_m. The increase in the chlorophyll *a*-specific absorption cross section was related to an increase in carotenoid pigments and a reduction in the package effect. These experiments show that *P. antarctica* can acclimate successfully to conditions as they prevail in the Antarctic ocean, which may explain the success of this species.

Keywords Fluorescence · Iron · Light · *Phaeocystis* · Pigments · Xanthophyll cycling

M. A. van Leeuwe
Department of Marine Biology, Biological Centre, University of Groningen, Haren, The Netherlands

J. Stefels (✉)
Laboratory for Plant Physiology, Biological Centre, University of Groningen, P. O. Box 14, Haren 9750 AA, The Netherlands
e-mail: j.stefels@rug.nl

Introduction

In the Southern Ocean, the factors light and iron exert synergistic growth control on microalgae (De Baar et al. 2005). In vast areas, iron concentrations are often too low to support growth (Martin et al. 1990; van Leeuwe et al. 1997). In addition, frequent wave action induces deep vertical mixing of the water column. Consequently, its planktonic inhabitants are subject to a highly dynamic light regime (Mitchell and Brody 1991). Under these conditions, the ability to photoacclimate determines the growth success of algae. Since iron exerts control on various photosynthetic processes, its low ambient concentration may impede photoacclimation (Greene et al. 1991; van Leeuwe and De Baar 2000). Yet, the impact of iron limitation on photoacclimation has so far received little attention.

Photoacclimation to a dynamic light regime involves optimisation of the photosystems in such a way that energy generation and utilisation are in balance (Kana et al. 1997). Conditions of rapid vertical mixing require that cells can acclimate directly to changes in light intensity that range from almost darkness to excessively high irradiance levels. Dim light conditions deep in the water column necessitate an extensive light-harvesting complex, whereas high irradiance levels at the surface require small complexes that are resistant to photodamage. When more energy is generated than can be utilised for growth, photodamage can occur (Demmig-Adams and Adams 1996). Under variable light conditions, periods between low- and high-light conditions are usually too short for acclimation by de novo synthesis of light-harvesting or photoprotective pigment-protein complexes, respectively. Moreover, de novo synthesis is a costly process. Cells that are subjected to variable light conditions will therefore benefit from flexible photosystems (see Muller et al. 2001 for review). Processes involved in photoacclimation on short time scales (minutes) are energy-dependent quenching (qE) and state-transition quenching (qT). These processes vary with species (e.g. van Leeuwe et al. 2005). Green flagellates possess adaptable photosystems that support the rapid down-regulation of photosynthetic activity by xanthophyll pigment cycling and state transitions. In these organisms, alternative quenching processes that involve mobility of the photosynthetic membranes (Anderson et al. 1995) may also play a role, but are not well described. In contrast, rapid photoacclimation in diatoms is limited to xanthophyll cycling. This appears to be related to a more-homogeneous, and therefore less-flexible, organisation of the thylakoid membranes of diatoms (Larkum et al. 2003).

The make-up of the photosynthetic apparatus and consequently to the capacity of photoacclimation depends on the availability of iron. Previously it has been found that iron-limited cells contain less chlorophyll *a*, but relatively more photoprotective pigments (Geider et al. 1993; van Leeuwe and Stefels 1998; van Leeuwe and De Baar 2000). In addition, various protein-pigment complexes that constitute the photosystems require iron for synthesis (cytochromes and FeS proteins). Under iron limitation, impairment of those protein complexes results in a decrease in the efficiency of electron transport (Greene et al. 1991). Experiments that were carried out under low, stable light conditions showed that while iron limitation resulted in a reduction in photosynthetic efficiency, the absorption efficiency had increased due to a reduction in the package effect (Greene et al. 1991; van Leeuwe and De Baar 2000). Accordingly, iron-limited cells were acclimated to low irradiance to a large extent. However, nothing is known about the effects of iron limitation on the photosynthetic response towards variable light conditions. A disorder in the architecture of the photosynthetic membranes may have negative consequences for the flexibility of the photosystems through constraining effects on the coupling and decoupling of pigment-protein complexes that are involved in xanthophyll cycling (Muller et al. 2001). Even so, alternative quenching processes that depend on conformational changes of membrane proteins are likely to be disturbed.

So far, few studies addressed photoacclimation to rapidly changing light conditions. The data that are available show that diatoms acclimate to the average irradiance levels they encounter. Under variable light conditions, this strategy has negative consequences for their growth success as it makes the algae more susceptible to photoinhibition when exposed to high irradiance levels (Flameling and Kromkamp 1995; Flameling 1998, van Leeuwe et al 2005). No studies are currently available on the effects of iron limitation on photoacclimation towards a variable light regime. Here, we present the first data on photosynthetic responses by the marine microalga *Phaeocystis antarctica* towards varying conditions of light and iron availability. *P. antarctica* is abundant in the Southern Ocean, and it plays a prominent role in various biogeochemical cycles (Smith et al. 1998; Arrigo et al. 1999; Vaillancourt et al. 2003). The alga may form massive blooms that can potentially sequester CO_2 from the atmosphere. In addition, *P. antarctica* is known for its production of dimethylsulphide (DMS), a semivolatile sulphur compound that once it has escaped into the atmosphere induces cloud formation thus affecting the global climate system (Crocker et al. 1995; Stefels and van Leeuwe 1998). The success of *Phaeocystis* as a cosmopolitan species is still not well understood. Previously, rapid photophysiological acclimation was found in *P. antarctica* (Moisan et al. 1998; van Leeuwe and Stefels 1998; Moisan and Mitchell 1999). These studies were performed under stable light conditions and they only partly explain its success in the open ocean areas that are characterised by highly dynamic light conditions (Mitchell and Brody 1991). Here we will show that *P. antarctica* can be so abundant in the Southern Ocean because of its high capacity to acclimate its photosystems to ambient irradiance levels, even under conditions of iron limitation. Photosynthetic efficiency was recorded by means of dual-modulation fluorescence analysis. Analysis of the pigment composition and recording of in vivo absorption spectra were used to determine responses of the antennae complex.

Methods

Culture conditions

Phaeocystis antarctica (CCMP 1871; non-axenic) was cultured under dynamic light conditions simulating vertical mixing through the water column. The light regime was enforced by a Venetian Blind system that was mounted between the cultures and the light source (Philips MHN-TD). The system consisted of horizontal slats the position of which could be altered by a step motor. Two independent units were used to enforce variable light conditions; the diurnal intensity of solar irradiance was simulated by a sinusoidal light course. Vertical mixing was simulated by a second oscillation that was superimposed on the solar pattern (see Fig. 1). The maximum intensity of irradiance was 400 μmol photons m^{-2} s^{-1}. The light inside the culture flasks was measured with a spherical sensor (Biosperical Instruments Inc. QSL-100). The cultures were grown in 2L polycarbonate erlenmeyer flasks at 4°C and kept in suspension by daily shaking. Medium was prepared as described by Morel et al. (1979). Chelexed, nutrient-poor seawater originating from the Atlantic sector of the Southern Ocean was enriched with 140 μM NO_3, 7.5 μM PO_4 and 60 μM Si. Solutions of $FeCl_3$ and ethylenediaminetetraacetic acid (EDTA) were added separately at final concentrations of 10^{-6} M EDTA for all cultures and 10^{-6} M iron for iron-rich cultures. Initial iron concentrations (measured by solvent extraction followed by atomic absorption spectrophotometry; Nolting and de Jong 1994) were 15 nM. Before starting the experiments, cultures were stressed by iron deficiency and adapted to light conditions for several weeks. During that period, the main growth form of the cells had gradually changed: the iron-limited cultures consisted mainly of single cells; the iron-replete cultures consisted mainly of colonies.

Photosynthetic parameters

Chlorophyll *a* fluorescence was measured with a dual-modulated fluorometer (Photosystem Instruments, Trtílek et al. 1997). Minimum (F_o) and maximum fluorescence (F_m) was recorded after dark adaptation (5 min at 4°C, sufficient to attain stabilization of the fluorescence signal for all light regimes). The maximum quantum yield of photosynthesis F_v/F_m was calculated as $(F_m–F_o)/F_m$ (Krause and Weis 1991).

Samples for pigment analysis were filtered gently (<15 KPa) over Whatman GF/F filters and subsequently snap-frozen in liquid nitrogen and stored at –80°C until analysis. Before extraction in 90% acetone filters were freeze-dried during 48 h. Pigments were analysed by high-performance liquid chromatography on a Waters system equipped with a photodiode array (van Leeuwe et al. 2006). A Waters DeltaPak reversed-phase column (C18, fully end-capped) was used. Pigment standards were obtained from the DHI Water Quality Institute (Horsholm, Denmark).

In vivo light absorption spectra (400–700 nm) were recorded with a Varian-Cary spectrophotometer equipped with an integrating sphere. Samples were taken to an approximate optical density of 0.1. Data were corrected for scattering by subtracting a baseline value as measured at 725 nm. The spectrally averaged, chlorophyll *a* specific absorption cross section ($\bar{a}_{ph}$, m^2 mg Chl a^{-1}) was calculated as follows:

$$\bar{a}_{ph} = \frac{\int_{400}^{700} a_{chl}(\lambda) * Q(\lambda)\delta(\lambda)}{\int_{400}^{700} Q(\lambda)\delta(\lambda)}$$

where $a_{chl}(\lambda)$ is the wavelength specific absorption (m^2 mg Chl a^{-1}), $Q(\lambda)$ is spectral irradiance of the light source (µmol photons m^{-2} s^{-1}) and $\delta(\lambda)$ is the wavelength interval of the spectral band (2 nm).

Sampling strategy and statistical analyses

All experiments were performed in duplicate. Samples for photosynthetic parameters were taken over the course of one day only (13 data points). A day was chosen during which cells were still growing exponentially while cell densities were high enough to allow reliable in vivo fluorescence measurements. Growth rates were established by cell counts in the days preceding the 24 h experiments. Cells were counted with an inverse microscope after settling for 24 h in a counting chamber. Effects of iron limitation on photosynthetic parameters (Table 2) were tested for significance by use of a two-factor analyses of variance. The daily dynamics of the photosynthetic parameters (Figs. 1–3) were tested for significance by multiple regression analyses. Differences were considered significant at $P < 0.05$.

Results

The main pigments of *Phaeocystis antarctica* were chlorophyll *a*, chlorophyll c_{1+2}, fucoxanthins, diadinoxanthin and diatoxanthin. The ratio of the photoprotective xanthophyll pigments diadino- and diatoxanthin to chlorophyll *a* was 0.229 in iron-limited cells and only 0.077 in iron-replete cells (Table 1). The ratio of 19′-hexanoyloxyfucoxanthin to chlorophyll *a* was also higher under iron limitation: 0.734 versus 0.344 for iron-replete cells. Iron limitation resulted in 10% growth reduction (Table 2). The reduction in growth was associated to a 20% reduction in the maximum quantum yield of photosynthesis (F_v/F_m) and

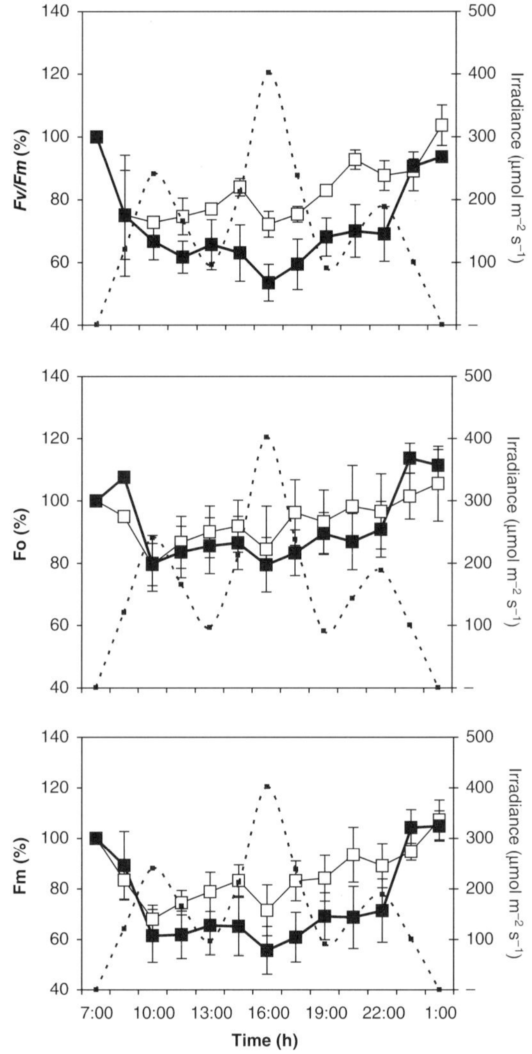

Fig. 1 Fluorescence characteristics of photosynthesis determined over the day for iron-limited (*open symbols*) and iron-replete cells (*closed symbols*), plotted alongside irradiance (photons m^{-2} s^{-1}; dashed line). Fluorescence dynamics are expressed in percentages, taking the maximum morning values as reference (i.e. 100%; $n = 3$). (**A**) Maximum quantum yield of fluorescence (F_v/F_m). (**B**) Minimum fluorescence (F_o). (**C**) Maximum fluorescence (F_m)

Table 1 The pigment composition of *Phaeocystis antarctica* for iron-limited and iron-replete cells, expressed as ratios to chlorophyll *a* (n = 3) (average values ± s.d.)

Ratio to Chl *a*	Fe-limited	± s.d.	Fe-replete	± s.d.
Chlorophyll *c3*	0.116	0.012	0.091	0.009
Chlorophyll *c2*	0.197	0.017	0.202	0.013
19′-Butanoyloxyfucoxanthin	0.013	0.003	0.009	0.003
Fucoxanthin	0.011	0.004	0.017	0.009
19′-Hexanoyloxyfucoxanthin	0.734	0.021	0.344	0.120
Diadinoxanthin	0.214	0.037	0.057	0.048
Diatoxanthin	0.017	0.012	0.025	0.024
Zeaxanthin	0.026	0.002	0.009	0.003
Beta-carotene	0.016	0.004	0.014	0.007
Total fucos*	0.760	0.025	0.381	0.138
Diadino-plus diatoxanthin	0.229	0.044	0.077	0.065

* 19′-Butanoyloxyfucoxanthin + Fucoxanthin + 19′- Hexanoyloxyfucoxanthin

Table 2 Photosynthetic parameters of *Phaeocystis antarctica* for iron-limited and iron-replete cells (n = 3) (average values ± s.d.)

	μ (d^{-1})	± s.d.	F_v/F_m^a	± s.d.	Chlorophyll *a* (pg cell^{-1})	± s.d.	$\bar{a}_{ph}$ (m^2 mg Chl^{-1})	± s.d.	POC[b] (pg cell^{-1})	rgc [c]
Fe-limited	0.49	0.02	0.40	0.04	0.08	0.02	0.057	0.023	14	0.13
Fe-replete	0.55	0.07	0.50	0.02	0.15	0.03	0.030	0.022	19	0.12

Differences between the treatments were significant [analysis of variations (ANOVA) $P < 0.05$]) for all parameters

[a] Morning values

[b] Data from Stefels and van Leeuwe (1998)

[c] Relative growth capacity, see discussion for details

almost 50% decrease in cellular chlorophyll *a* content (Table 2, Fig. 4).

The F_v/F_m of 0.5 recorded for iron-replete cells was close to the maximum values measured in our lab, though they are lower than the values reported for natural populations that were dominated by *Phaeocystis antarctica* (Vaillancourt et al. 2003). Whereas our cultures were in good condition, the experimental conditions provided were certainly not the most optimal for growth *P. antarctica*. Maximum values for F_v/F_m were in fact achieved at more stable light conditions (pers. obs.).

The maximum quantum yield of photosynthesis (F_v/F_m) varied with daily dynamics in light intensity in both the iron-limited and iron-replete cells (Fig. 1A). Maximum values were recorded in the morning, followed by decreasing values towards the middle of the light period. In replete cells, the midday depression was almost 50% of the morning values; F_v/F_m decreased by about 30% under iron limitation. The F_v/F_m recovered towards the end of the light period (significant relationship between irradiance and F_v/F_m; ANOVA, $P < 0.05$). Depressions in irradiance coincided with moderate increases in the fluorescence parameters. Minimum (F_o) and maximum (F_m) fluorescence values varied along with F_v/F_m (Fig. 1B, C). The dynamics were most pronounced for the maximum fluorescence signal in iron-replete cells.

In concordance with the diel dynamics in F_v/F_m, active xanthophyll cycling was observed over the day (Fig. 2A, B). The ratio of diatoxanthin (dt) to diadinoxanthin (dd) was closely linked to irradiance, with maximum conversion of dd into dt recorded during maximum irradiance. Whereas the total pool of dd + dt was highest in iron-limited cells (Table 1), the dt/dd ratio was significantly higher in iron-replete cells: 0.9 versus 0.15 for Fe-limited cells (Fig. 2).

In iron-deplete cells, the pool of fucoxanthins was not stable over the day. The ratio of 19′-hexanoyloxyfucoxanthin to fucoxanthin

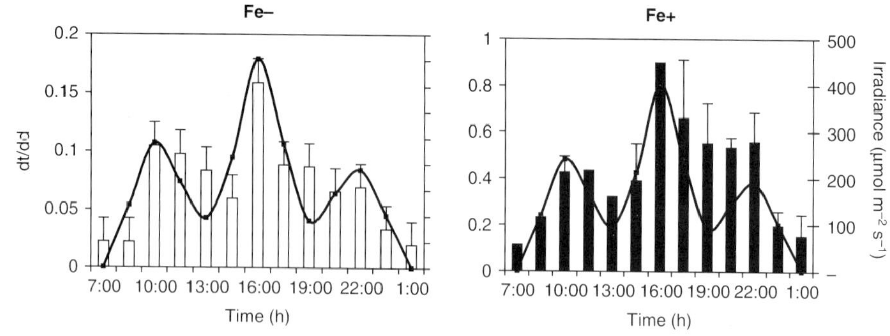

Fig. 2 Daily dynamics of the xanthophylls cycle (ratio of diadinoxanthin to diatoxanthin; dt:dd) plotted alongside irradiance (photons m^{-2} s^{-1}; solid line). (**A**) Iron-limited cells. (**B**) Iron-replete cells

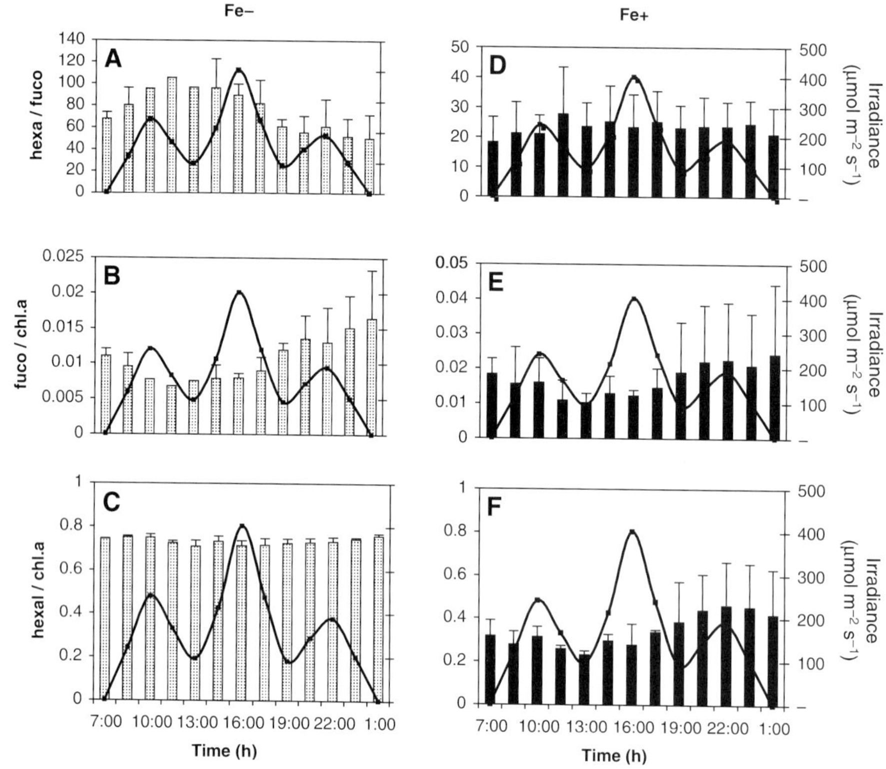

Fig. 3 Daily dynamics of the fucoxanthin pigments plotted alongside irradiance (photons m^{-2} s^{-1}; solid line) in iron-limited (Fe–) and iron-replete (Fe+) cells. (**A, D**) Ratio of 19′-hexanoyloxyfucoxanthin to fucoxanthin (hexa/fuco). (**B, E**) Ratio of fucoxanthin to chlorophyll *a*. (**C, F**) Ratio of 19′hexanoyloxyfucoxanthin to chlorophyll *a*

(HEXA/FUCO) decreased towards the end of the light period (Fig. 3A; $P < 0.05$), which was associated with a 50% increase in FUCO/Chl *a* (Fig. 3B), whereas HEXA/Chl *a* remained invariable at a high level (Fig. 3C). In iron-replete cells, no significant daily dynamics in HEXA and FUCO were observed (Fig. 3D–F).

In both treatments, the in vivo light absorption characteristics were invariable over the day (data not shown). Maximum values in the chlorophyll-specific absorption coefficient were observed near 675 nm (Chl *a*), 470 nm (carotenoids) and 430 nm (Chl *a*) (Fig. 4). The absorption cross section was 0.057 m^2 mg Chl a^{-1} under iron limitation and

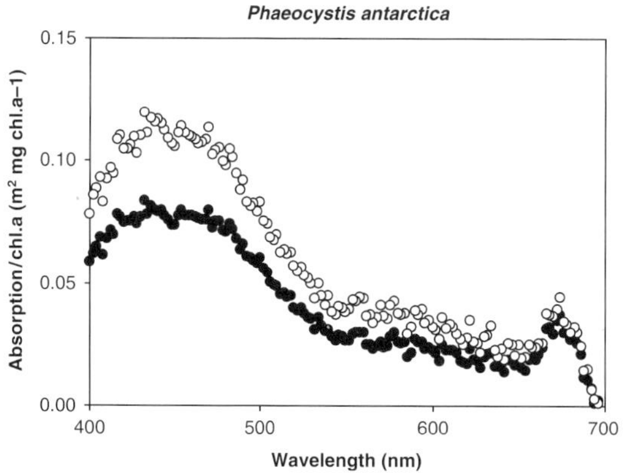

Fig. 4 In vivo absorption spectra expressed per chlorophyll *a* for iron-limited (open symbols) and iron-replete cells (closed symbols)

almost two times lower in iron-replete cells (Table 2). Elevated values in the Soret region for iron-limited versus iron-replete cells reflected the relative increase in fucoxanthins and photoprotective pigments.

Discussion

Cells that are exposed to varying irradiance levels need to tune their photosynthetic apparatus in a way that allows optimal use of irradiance during exposure to the lower intensities and simultaneously find a way to avoid photoinhibition when subjected to high irradiance levels. On short time scales, xanthophyll pigment cycling plays an important role in photoacclimation to high irradiance (Olaizola et al. 1994). De-epoxidation of the light-harvesting pigment diadinoxanthin into the heat dissipating diatoxanthin requires only seconds to minutes, depending on the species. Until now, physiological studies on xanthophyll cycling mostly involved short-term experiments during which abrupt light shifts were applied. A role for xanthophyll cycling under conditions of continuous variations in light intensity only had been recorded in an Antarctic diatom and a green flagellate (van Leeuwe et al. 2005). Here we show that also in *Phaeocystis antarctica*, photoacclimation to variable light conditions is promoted by the xanthophyll pigments. A significant relationship was recorded between the de-epoxidation state of the xanthophyll pigments and the maximum quantum yield of photosynthesis (F_v/F_m). Minimum values for F_v/F_m, coinciding with increasing irradiance, were mirrored by enhanced ratios of diato- to diadinoxanthin. The close relationship between the two parameters unequivocally establishes the role of the xanthophyll pigments in the down-regulation of photosynthetic activity. Daily dynamics in fluorescence signals have also been observed in experiments with cultures of *P. globosa* (Flameling and Kromkamp 1995). The authors could, however, not demonstrate a role for xanthophyll pigments. Possibly, in those experiments the levels in diatoxanthin were too low to determine accurately. A role for xanthophyll pigments in photoacclimation by *Phaeocystis* sp. has been shown for stable light conditions (Moisan et al. 1998; van Leeuwe and Stefels 1998; Meyer et al. 2000). Our study confirms these findings, and adds an active role in the regulation of photosynthesis under conditions of a—ecologically more relevant—dynamic light climate.

Since we worked in batch cultures and iron concentrations in the unamended media were at nanomolar levels, initial growth rates were only slightly reduced in the iron-limited cultures (Table 2). The addition of EDTA did not set this right; it appears as though *Phaeocystis*, amongst other algal species, might access EDTA-complexed iron (see Sedwick et al. 2006). Iron limitation was achieved in the course of time, as most clearly reflected in the photosynthetic parameters that were recorded during the final stages of exponential growth. A modest albeit significant reduction in F_v/F_m was observed in iron-limited cells. The impact of nutrient limitation on this fluorescence parameter per se is under debate (see Kruskopf and Flynn 2006 and references therein). Yet, the daily dynamics provided the more important information. Down-regulation of F_v/F_m was suppressed under iron limitation. The moderation in regulation was related to a weaker response in F_m (Fig. 1). Iron limitation exerted a minor effect on F_0. These effects are most likely related to a reduction in xanthophyll pigment cycling. Although the patterns in de-epoxidation state were similar for both iron conditions, the conversion of dd into dt was relatively minor in iron-limited cells. The enlarged xanthophyll pool in iron-limited cells most probably compensated for the lower dt/dd ratio, thereby maintaining a high level of heat dissipation throughout the day, with minor variations over the light period. This pattern characterises high light acclimated cells, which confirms a high susceptibility towards photoinhibition known for iron-limited cells (Geider and LaRoche 1994; Morales et al. 1998; van Oijen et al. 2004). In addition, it can be hypothesised that the single cell shape will have rendered iron-limited cells more sensitive to high irradiance compared to the colonial form that dominated under iron-replete conditions because of an increase in specific absorption (Morel and Bricaud 1981; Fujiki and Tagushi 2002).

Similar differences in pigment composition between iron-limited and iron-replete cells have previously been recorded for *P. antarctica*, albeit not under variable light conditions (van Leeuwe and Stefels 1998). When subject to a dynamic light regime, more energy has to be invested in an increased xanthophyll pool if the photosystems are not flexible enough to respond to rapid increases in irradiance level. We hypothesise that a loss in flexibility is related to a role of iron in the integrity and fluidity of photosynthetic membranes. Photochemical quenching by xanthophyll cycling (qE) involves conformational changes of the photosynthetic membranes (Mohanty et al. 1995; Pascal et al. 2005). These membranes are made up of various protein complexes in many of which iron plays an essential role. Under iron limitation, it appears that the conformation of the complexes themselves and of the membrane as a whole is impaired. As a result, membranes become less flexible and the necessary changes during xanthophyll cycling may be hampered.

The almost full recovery of F_v/F_m under both iron conditions indicated successful photoacclimation. Damage to the photosystems can, however, not be excluded. Minor degradations will go by unnoticed when the rate of repair is sufficiently fast to sustain high rates of photochemistry (Morales et al. 1998; Han et al. 2000; Kana et al. 2002; van Leeuwe et al. 2005). The increased xanthophyll pool of iron-limited cells indicates that these cells experience an enhanced risk of photodamage, which could result in increased energy costs for repair. This exerts additional stress on iron-limited cells. Whereas iron-sufficient cells invest in larger photosystems (increased Chl *a* per cell) in order to efficiently use the periods of low irradiance, iron-limited cells have to invest in a mechanism to prevent photodamage and at the same time maximise energy production. This is attained by a relative reduction of (iron-expensive) Chl *a* per cell, which results in a reduced packaging effect (Greene et al. 1991; van Leeuwe and De Baar 2000) and a relative increase in light-harvesting pigments (increased ratio of total carotenoid pigments to Chl *a*). As a result, the Chl *a*-specific absorption cross section $\bar{a}_{ph}$ significantly increases under iron limitation. As discussed, the single-cell form that dominated under iron-deplete conditions may further contribute to an increase in $\bar{a}_{ph}$. The increase in absorption capacity appears to balance the reduction in photosynthetic efficiency to a large extent. If we consider growth as a function of these photosynthetic parameters we may estimate the relative carbon-normalised growth capacity (Table 2: rgc) by applying a simple relationship: rgc $\approx F_v/F_m * \bar{a}_{ph} *$Chl *a* cell^{-1}/POC cell^{-1}. This exercise shows that the capacity of iron-limited and iron-replete cells is almost equal: 0.13 and 0.12, respectively. It should be noted that this only is an approach of *gross* growth capacity, since maintenance costs are not included in this concept. *Net* growth rates as determined on the basis of cell divisions are significantly higher in iron-replete cells.

The ratio HEXA/Chl *a* was found to be twice as high in iron-limited cells compared to iron-replete cells. In previous experiments, similar high HEXA/Chl *a* ratios were recorded (van Leeuwe and Stefels 1998). Concomitant high ratios of HEXA/FUCO under high irradiance gave rise to a hypothesis on an alternative photoprotective pigment cycle. It was suggested that in response to exposure to high irradiance the efficient light-harvesting pigment FUCO would be converted into its structural relative HEXA. The efficiency of energy transfer by HEXA has not yet been described, which leaves room for a role as energy quencher. This hypothesis could not be confirmed in our short-term experiments. The ratio of HEXA/FUCO was again higher for iron-limited cells, but did not vary with fluctuating irradiance levels. As for the well-established xanthophyll pigment cycle, such a relationship should have been apparent in case of an active role in photochemical quenching. The role of HEXA is still enigmatic, but an influence on energy dissipation cannot be excluded.

This study has shown that *Phaeocystis antarctica* can readily respond to fast changes in light climate. Iron limitation exerted stress on the cells, but due to changes in pigment composition these cells were able to acclimate and apparently did not suffer from photoinhibition. Altogether, these characteristics are vital to maintain growth in the Antarctic Ocean, where deep wind-mixed layers predominate and where low iron concentrations suppress algal growth. Under such conditions *P. antarctica* appears a successful competitor. The alga may support carbon flow within so-called retention systems. Alternatively, *Phaeocystis* can contribute to large carbon fluxes from the surface waters into the oceans interior. So far, very few data are available on photosynthetic responses towards variable light conditions. With this paper we hope to improve our understanding of biogeochemical cycles in the Antarctic ecosystem.

Acknowledgements We would like to thank Ronald Visser for pigment analysis and technical support. We thank Walker Smith for a critical review of the manuscript. MAvL and JS were financially supported by the Dutch Polar Programme of the Research Council for Earth and Life Sciences (ALW) of the Netherlands Organisation for Scientific Research (NWO).

References

Anderson JM, Chow WS, Park Y-I (1995) The grand design of photosynthesis: acclimation of the photosynthetic apparatus to environmental cues. Photosynth Res 46:129–139

Arrigo KR, Robinson DH, Worthen DL, Dunbar RB, DiTullio GR, VanWoert M, Lizotte MP (1999) Phytoplankton community structure and the drawdown of nutrients and CO_2 in the Southern Ocean. Science 283:365–367

Crocker KM, Ondrusek ME, Petty RL, Smith RC (1995) Dimethylsulfide, algal pigment and light in an Antarctic *Phaeocystis* sp. bloom. Mar Biol 124:355–340

DeBaar HJW, et al (2005) Synthesis of iron fertilisation experiments: from the iron age in the age of enlightment. J Geophys Res 110:C09S16

Demmig-Adams B, Adams III WW (1996) Photoprotection and other responses of plants to high light stress. Annu Rev Plant Physiol Plant Mol Biol 43:599–626

Flameling IA (1998) Growth and photosynthesis of eukaryotic microalgae in fluctuating light conditions, induced by vertical mixing. University of Nijmegen. Thesis/Dissertation, 135 pp

Flameling I, Kromkamp J (1995) Regulation of photosynthesis of *Phaeocystis globosa* in fluctuating light conditions. In: Mathis P (ed) Photosynthesis: from light to biosphere. Kluwer Academic Publishers, pp 797–800

Fujiki T, Tagushi S (2002) Variability in chlorophyll *a* specific absorption coefficient in marine phytoplankton as a function of cell size and irradiance. J Plankton Res 24(9):859–874

Geider RJ, LaRoche J (1994) The role of iron in phytoplankton photosynthesis, and the potential role of iron limitation of primary productivity in the sea. Photosynth Res 39:275–301

Geider RJ, LaRoche J, Greene RM, Olaizola M (1993) Response of the photosynthetic apparatus of *Phaeodactylum tricornutum* (Bacillariophyceae) to nitrate, phosphate, or iron starvation. J Phycol 29:755–766

Greene RM, Geider RJ, Falkowski PG (1991) Effect of iron limitation on photosynthesis in a marine diatom. Limnol Oceanogr 36(8):1772–1782

Han B-P, Virtanen M, Koponen J, Straskraba M (2000) Effect of photoinhibition on algal photosynthesis: a dynamic model. J Plankton Res 22:865–885

Kana TD, Geider RJ, Critchley C (1997) Regulation of photosynthetic pigments in micro-algae by multiple environmental factors: a dynamic balance hypothesis. New Phytol 137:629–638

Kana R, Lazar D, Prasil O, Naus J (2002) Experimental and theoretical studies on the excess capacity of photosystem II. Photosynth Res 72:271–284

Krause GH, Weis E (1991) Chlorophyll fluorescence and photosynthesis: the basics. Annu Rev Plant Physiol Plant Mol Biol 42:313–349

Kruskopf M, Flynn KJ (2006) Chlorophyll content and fluorescence responses cannot be used to gauge reliable phytoplankton biomass, nutrient status or growth rate. New Phytol 169:525–537

Larkum AWD, Douglas SE, Raven JA (eds) (2003) Photosynthesis in algae. Kluwer Academic Publishers, Dordrecht, 480 pp

Martin JH, Gordon RM, Fitzwater SE (1990) Iron in Antarctic waters. Nature 345:156–158

Meyer AA, Tackx M, Daro N (2000) Xanthophyll cycling in *Phaeocystis globosa* and *Thalassiosira* sp.: a possible mechanism for species succession. J Sea Res 43:373–384

Mitchell BG, Brody EA (1991) Light limitation of phytoplankton biomass and macronutrient utilization in the Southern Ocean. Limnol Oceanogr 36:1662–1677

Mohanty N, Gilmore AM, Yamamoto HY (1995) Mechanism of non-photochemical chlorophyll fluorescence quenching. II. Resolution of rapidly reversible absorbance changes at 530 nm and fluorescence quenching by the effects of antimycin, dibucaine and cation exchanger, A23187. Aust J Plant Physiol 22:239–247

Moisan TA, Mitchell BG (1999) Photopysiological acclimation of *Phaeocystis antarctica* Karsten under light limitation. Limnol Oceanogr 44(2):247–288

Moisan TA, Olaizola M, Mitchell BG (1998) Xanthophyll cycling in *Phaeocystis antarctica*: changes in cellular fluorescence. Mar Ecol Prog Ser 169:113–121

Morales F, Abadia A, Abadia J (1998) Photosynthesis, quenching of chlorophyll fluorescence and thermal energy dissipation in iron-deficient sugar beet leaves. Aust J Plant Physiol 25:403–412

Morel A, Bricaud A (1981) Theoretical results concerning light absorption in a discrete medium, and application to specific absorption of phytoplankton. Deep-Sea Res 28A(11): 1375–1393

Morel FMM, Rueter JG, Anderson DM, Guillard RRL (1979) AQUIL: a chemically defined phytoplankton culture medium for trace metals. J Phycol 15:135–141

Muller P, Li X-P, Niyogi KK (2001) Non-photochemical quenching. A response to excess light energy. Plant Physiol 125:1558–1566

Nolting RF, de Jong JTM (1994) Sampling and analytical methods for the determination of trace metals in surface seawater. Intern J Environ Anal Chem 57:189–196

Olaizola M, La Roche J, Kolber Z, Falkowski PG (1994) Non-photochemical fluorescence quenching and the diadinoxanthin cycle in a marine diatom. Photosynth Res 41:357–370

Pascal A et al (2005) Molecular basis of photoprotection and control of photosynthetic light-harvesting. Nature 436:134–137

Sedwick PN, Garcia NS, Riseman SF, Marsay CM, DiTullio GR (2006) Evidence for high iron requirements of colonial *Phaeocystis antarctica* at low irradiance Biogeochemistry, this volume

Smith WO, Carlson CA, Ducklow HW, Hansell DA (1998) Growth dynamics of *Phaeocystis antarctica*-dominated plankton assemblages from the Ross Sea. Mar Ecol Prog Ser 168:229–244

Stefels J, van Leeuwe MA (1998) Effects of iron and light stress on the biochemical composition of Antarctic *Phaeocystis* sp. (Prymnesiophyceae). I. Intracellular DMSP concentrations. J Phycol 34:486–495

Trtílek M, Kramer DM, Koblízek M, Nedbal L (1997) Dual-modulation LED kinetic fluorometer. J Lum 72–74:597–599

Vaillancourt RD, Sambrotto RN, Green S, Matsuda A (2003) Phytoplankotn biomass and photosynthetic competency in the summertime Metz Glacier Region of East Antarctica. Deep-Sea Res II 50:1415–1440

van Leeuwe MA, De Baar HJW (2000) Photoacclimation by the Antarctic flagellate *Pyramimonas* sp. (Prasinophyceae) in response to iron limitation. Eur J Phyol 35:295–303

van Leeuwe MA, Scharek R, de Baar HJW, de Jong JTM, Goeyens L (1997) Iron enrichment experiments in the Southern Ocean: physiological responses of plankton communities. Deep-Sea Res II 44:189–207

van Leeuwe MA, Stefels J (1998) Effects of iron and light stress on the biochemical composition of Antarctic *Phaeocystis* sp. (Prymnesiophyceae). II. Pigment composition. J Phycol 34:496–503

van Leeuwe MA, van Sikkelerus B, Gieskes WWC, Stefels J (2005) Taxon-specific differences in photoacclimation to fluctuating irradiances in an Antarctic diatom and a green flagellate. Mar Ecol Prog Ser 288:9–19

van Leeuwe MA, Villerius LA, Roggeveld J, Visser RJW, Stefels J (2006) An optimized method for automated analysis of algal pigments by HPLC. Mar Chem 102:267–275

van Oijen T, van Leeuwe MA, Gieskes WWC, de Baar HJW (2004) Effect of iron limitation on photosynthesis and carbohydrate metabolism in the Antarctic diatom *Chaetoceros brevis* (Bacillariophyceae). Eur J Phycol 39:161–171

Biogeochemistry (2007) 83:71–81
DOI 10.1007/s10533-007-9080-8

ORIGINAL PAPER

Effects of iron concentration on pigment composition in *Phaeocystis antarctica* grown at low irradiance

G. R. DiTullio · N. Garcia · S. F. Riseman · P. N. Sedwick

Received: 2 January 2006 / Accepted: 10 August 2006 / Published online: 17 April 2007

Abstract Interpretation of photosynthetic pigment data using iterative programs such as CHEMTAX are widely used to examine algal community structure in the surface ocean. The accuracy of such programs relies on understanding the effects of environmental parameters on the pigment composition of taxonomically diverse algal groups. *Phaeocystis antarctica* is an important contributor to total autotrophic production and the biogeochemical cycling of carbon and sulfur in the Southern Ocean. Here we report the results of a laboratory culture experiment in which we examined the effects of ambient dissolved iron concentration on the pigment composition of colonial *P. antarctica*, using a new *P. antarctica* strain isolated from the southern Ross Sea in December 2003. Low-iron (<0.2 nM dissolved Fe) filtered Ross Sea seawater was used to prepare the growth media, thus allowing subnanomolar iron additions without the use of EDTA to control dissolved iron concentrations. The experiment was conducted at relatively low irradiance (~20 μE m^{-2} s^{-1}), with *P. antarctica* primarily present in the colonial form—conditions that are typical of the southern Ross Sea during austral spring. Relative to the iron-limited control treatments (0.22 nM dissolved Fe), iron addition mediated a decrease in the ratio of 19′-hexanoyloxyfucoxanthin to chlorophyll *a*, and an increase in the ratio of fucoxanthin to chlorophyll *a*. Our results also suggest that the ratio of 19′-hexanoyloxyfucoxanthin to chlorophyll c_3 (Hex:Chl c_3 ratio) may be a characteristic physiological indicator for the iron-nutritional status of colonial *P. antarctica*, with higher Hex:Chl c_3 ratios (>3) indicative of Fe stress. We also observed that the ratio of fucoxanthin to 19′-hexanoyloxyfucoxanthin (Fuco:Hex ratio) was highly correlated ($r^2 = 0.82$) with initial dissolved Fe concentration, with Fuco:Hex ratios <0.05 measured under iron-limited conditions (dissolved Fe <0.45 nM). Our results corroborate and extend the results of previous experimental studies, and, combined with pigment measurements from the southern Ross Sea, are consistent with the hypothesis that the interconversion of fucoxanthin and 19′-hexanoyloxyfucoxanthin by colonial *P. antarctica* is used as a photo-protective or light-harvesting mechanism, according to the availability of dissolved iron.

Keywords Iron · *Phaeocystis antarctica* · Pigments · Ross Sea

G. R. DiTullio (✉) · N. Garcia · S. F. Riseman
Hollings Marine Laboratory, College of Charleston, 331 Fort Johnson, Charleston, SC 29412, USA
e-mail: ditullioj@cofc.edu

P. N. Sedwick
Bermuda Biological Station for Research, Ferry Reach, St. George's, Bermuda GE01, Bermuda

Introduction

Phaeocystis antarctica is an important phytoplankton species in the Southern Ocean, although the relative contribution of *Phaeocystis* to total autotrophic biomass and primary production in this vast oceanic region is poorly constrained, for a number of reasons. First, *P. antarctica* has a complex heteromorphic life cycle, alternating between gelatinous colonies and free-living flagellate cells, which hampers identification using simple microscopy (Rousseau et al. 1994). Second, blooms of *P. antarctica* display pronounced temporal and spatial variability, especially near the Antarctic continental margins (El-Sayed et al. 1983; Palmisano et al. 1986). And finally, although numerous studies have focused on this species in near-surface waters, *P. antarctica* may also dominate algal biomass in the lower euphotic zone in the Southern Ocean (DiTullio et al. 2003a, b). In addition, the role of *P. antarctica* in the biogeochemical cycling of carbon, sulfur and nutrient elements in the Southern Ocean remains to be elucidated. Studies on the Antarctic continental shelf suggest that *P. antarctica* plays a significant role in the regional cycling of carbon (Schoemann et al. 2005; DiTullio et al. 2000) and sulfur (Gibson et al. 1990; DiTullio and Smith 1995), as well as impacting the stoichiometry of macronutrient cycling in Southern Ocean waters (Arrigo et al. 1999; Arrigo 2005). Thus there are compelling reasons to improve our understanding of the abundance, distribution and growth requirements of *P. antarctica*.

One of the primary methods for determining algal community structure in polar waters is based on the analysis of photosynthetic pigments by high-performance liquid chromatography (HPLC). Chemotaxonomic interpretations of HPLC data using programs such as CHEMTAX (Mackey et al. 1996) rely on the input of environmentally representative initial pigment ratios for taxonomically diverse phytoplankton groups. To apply such chemotaxonomic methods to *P. antarctica* in the Southern Ocean thus requires the use of laboratory culture studies in order to establish representative pigment ratios for this species under various growth conditions. Previous studies (e.g., Buma et al. 1991) have demonstrated that *P. antarctica* contains the two accessory pigments 19′-hexanoyloxyfucoxanthin (Hex) and fucoxanthin (Fuco), although diatoms, also common in the Southern Ocean, are also known to contain Fuco. The presence of Hex in significant concentrations relative to other photosynthetic pigments has been used to identify *P. antarctica* blooms in Antarctic waters (e.g., DiTullio and Smith 1995; Crocker et al. 1995), and a strong correlation has been observed between Hex concentrations and *P. antarctica* cell number estimated by microscopy in the southern Ross Sea (DiTullio et al. 2003b), which is dominated by *P. antarctica* blooms during the austral spring (November–December; El-Sayed et al. 1983; Arrigo et al. 1998; Smith et al. 2003).

Previous laboratory culture experiments have shown that relative concentrations of Fuco and Hex in *P. antarctica* vary as a function of ambient dissolved Fe concentration in the growth media (van Leeuwe and Stefels 1998). However, these experiments were performed under relatively high dissolved Fe concentrations (nM–μM), compared to open Antarctic surface ocean waters, and, as in most other culture studies, the synthetic organic chelating compound EDTA was used to buffer free dissolved Fe concentrations. Gerringa et al. (2000) have presented a strong argument for the need to conduct such culture experiments using realistically low dissolved iron concentrations without the use of synthetic chelating compounds such as EDTA. Following these suggestions, we have attempted to examine the effect of iron availability on the growth and pigment composition of *P. antarctica*, using culture experiments conducted at low (i.e., sub-nanonolar) dissolved iron concentrations and without using EDTA (or other chelators) to control dissolved iron levels in the growth media. These experiments were performed using a new strain of *P. antarctica*, which was isolated from native phytoplankton collected in the southern Ross Sea during December 2003. Here we report the results of an experiment that examines the effect of sub-nanomolar iron additions on the pigment composition of colonial *P. antarctica* under relatively low irradiance—conditions that are typical of the southern Ross Sea during early austral spring—and discuss the ecological implications of our results within the context of field observations from this region.

Methods and materials

Isolation of *Phaeocystis antarctica*

Algal cultures of *P. antarctica* were isolated from native phytoplankton assemblages collected in the Ross Sea, Antarctica, during cruise NBP03-05A of the RV/IB *Nathaniel B. Palmer* in December 2003. For our experiments, we used a strain of *P. antarctica* isolated from phytoplankton collected in the polynya region of the southern Ross Sea at ca. 76°05′S, 170°08′E. The mixed field assemblage was maintained at 0°C in filtered Ross Sea seawater amended with L1 growth medium (Guillard and Hargraves 1993), and then transported to the Hollings Marine Laboratory, where a uni-algal culture of *P. antarctica* was subsequently isolated. This *P. antarctica* strain was then maintained in semicontinuous culture at 0°C, in primarily colonial form, using L1-amended filtered Ross Sea seawater as a growth medium. In addition, a large volume of 0.2 μm-filtered, low-iron (<0.2 nM dissolved Fe) seawater was collected from the same area of the southern Ross Sea in December 2003, using a trace-metal clean protocol (Sedwick et al. this issue), and then transported to the Hollings Marine Laboratory for use in our culture experiments.

Laboratory iron-addition dose-response experiment

As described by Sedwick et al. (this issue), a semicontinuous culture of the Ross Sea *P. antarctica* strain was successively diluted with low-iron (<0.2 nM dissolved Fe) filtered Ross Sea seawater, over a period of several weeks, using a rigorous trace-metal-clean protocol. Using this process, we obtained an 'inoculum culture' of colonial *P. antarctica* in L1 medium diluted 9,140-fold with filtered, low-iron (0.17 nM dissolved Fe) Ross Sea seawater, for which we calculate dissolved Fe and EDTA concentrations of 1.42 nM and 1.25 nM, respectively. Prior to commencing the experiment described here, this inoculum culture was acclimated to a constant irradiance of ~20 μE m^{-2} s^{-1} for 11 days at 0°C. The experiment was initiated by diluting 770 ml of this inoculum culture with 37.8 l of filtered, low-iron (0.17 nM dissolved Fe) Ross Sea seawater that had been chilled overnight to 0°C, and this diluted inoculum culture ('starting seawater') was then dispensed into multiple acid-cleaned 1.2 l polycarbonate bottles, using a rigorous trace-metal clean protocol. As detailed by Sedwick et al. (this issue), bottles were then amended with iron in four different treatments: control (no added Fe), +0.2 nM Fe, + 0.6 nM Fe, and + 1.8 nM Fe. These treatments correspond to initial dissolved iron concentrations of 0.22 nM (measured in the starting seawater), 0.42 nM, 0.82 nM and 2.02 nM, respectively. All bottles were then incubated at a temperature of 0°C and a constant irradiance of ~20 μE m^{-2} s^{-1}, using a refrigerated incubator fitted with cool white fluorescent lights.

For each different iron treatment, bottles were terminally sampled in duplicate or triplicate after incubation periods of 16, 25 and 31 days. Visual observations indicated that all treatments were dominated by colonial (rather than solitary) *P. antarctica* cells during the entire course of the experiment. Measurements of dissolved nitrate + nitrite, particulate chlorophyll *a* (Chl *a*) and photosynthetic efficiency in subsamples from the incubation bottles (Sedwick et al. this issue; Garcia et al. manuscript in preparation) suggest that exponential growth phase commenced at around day 16, whereas days 25 and 31 correspond to the early and late stages of the exponential growth phase. For measurements of photosynthetic pigments in the incubation cultures, duplicate 40 ml subsamples were passed through GF/F filters, which were then stored at –80°C until analysis by HPLC (see below). The low biomass observed after 16 days incubation (chlorophyll *a* concentration ~0.2 μg l^{-1}; Sedwick et al. this issue) precluded analysis of accessory pigments in the day–16 incubation bottles. Owing to low biomass in the starting seawater, initial concentrations of photosynthetic pigments were calculated from the analysis of material filtered from 40 ml subsamples of the inoculum culture immediately prior to dilution with the low-Fe Ross Sea seawater.

Ross Sea pigment samples

Water-column samples for the analysis of phytoplankton pigments were collected during three cruises in the Ross Sea as part of the the Research on Ocean-Atmosphere Variability and Ecosystem Response in the Ross Sea (ROAVERRS) program. In this paper, we focus on samples collected aboard the RV/IB *Nathaniel B. Palmer* during the ROAVERRS III expedition in mid spring-early summer (November 11–December 11, 1998), when colonial *P. antarctica* blooms typically dominate the southern Ross Sea polynya (Arrigo et al. 1998, 1999; Smith et al. 2000, 2003). Water samples were collected using 10-l Bullister bottles attached to a conductivity-temperature-depth (CTD) rosette. From these water samples, 0.2–2.2 l aliquots were filtered under low-light conditions onto GF/F filters, quick-frozen in liquid nitrogen, and then stored at –80°C for on-shore processing and subsequent HPLC analysis (see below). The pigment measurements discussed in this paper are mean values for water-column samples collected within the 30–50 m depth range (within or below the base of the surface mixed layer), where mean in-situ irradiance is expected to be comparable to that (20 μE m^{-2} s^{-1}) used in our laboratotry experiment (Smith and Van Hilst 2003).

HPLC pigment analyses

Prior to analysis, filters were homogenized with 1.5 mL 100% acetone and extracted for 2–4 h at –20°C. Extracts were then centrifuged at 0°C, filtered, and then injected into a Hewlett Packard Series 1050 liquid chromatograph using an autosampler. The system was equipped with HP 1050 photodiode array and HP 1046A fluorescence detectors. The pigment samples from the laboratory iron-addition experiment were analyzed using a modification of the Zapata et al. (2000) method, as described in DiTullio and Geesey (2002), whereas the Ross Sea pigment samples were measured using a modification of the ammonium acetate pairing method (Wright et al. 1991) as described by DiTullio et al. (1993). HPLC pigment separation employed a Phenomenex ODS II Spherisorb C18 column for the ammonium acetate gradient method and a Waters C8 Symmetry column for the Zapata et al. (2000) gradient elution method. Purified pigment standards were isolated from uni-algal cultures. The coefficient of variation on replicate standard injections was typically <3%. Pigments were identified from relative peak retention times. In addition, visible-light absorption spectra from each eluted peak were compared to stored library spectra of purified pigment standards, to verify the identity and relative purity of eluted pigments. Pigment ratios were calculated by weight. CHEMTAX analyses were performed on the ROAVERRS III data using initial pigment ratios optimized for the Ross Sea. Those initial pigment ratios for the Ross Sea are described elsewhere (DiTullio et al. 2003b). During this austral spring cruise the phytoplankton Chl *a* biomass was dominated (>95%) by *P. antarctica* (DiTullio et al. 2003b).

Results and discussion

Hex:Chl *a* ratios

Our experimental results show that sub-nanomolar additions of dissolved Fe had a significant effect on the pigment ratios of our Ross Sea *P. antarctica* culture. The ratio of Hex to chlorophyll *a* (Hex:Chl *a* ratio) decreased from approximately 1.10 ± 0.20 under low-iron (0.22 nM dissolved Fe) conditions to 0.56 ± 0.02 under high-iron (2.02 nM dissolved Fe) conditions (Fig. 1a). The range of Hex:Chl *a* ratios observed in this experiement are consistent with the overall mean Hex:Chl *a* ratio measured in water samples collected in the Ross Sea during all three ROAVVERS expeditions (DiTullio et al. 2003b). In addition, the trend of decreasing Hex:Chl *a* ratios with increasing dissolved Fe concentration observed in our laboratory experiment is qualitatively consistent with the data trend for the ROAVERRS field samples: during the three ROAVERRS cruises, the highest Hex:Chl *a* ratio (1.08) was observed during the

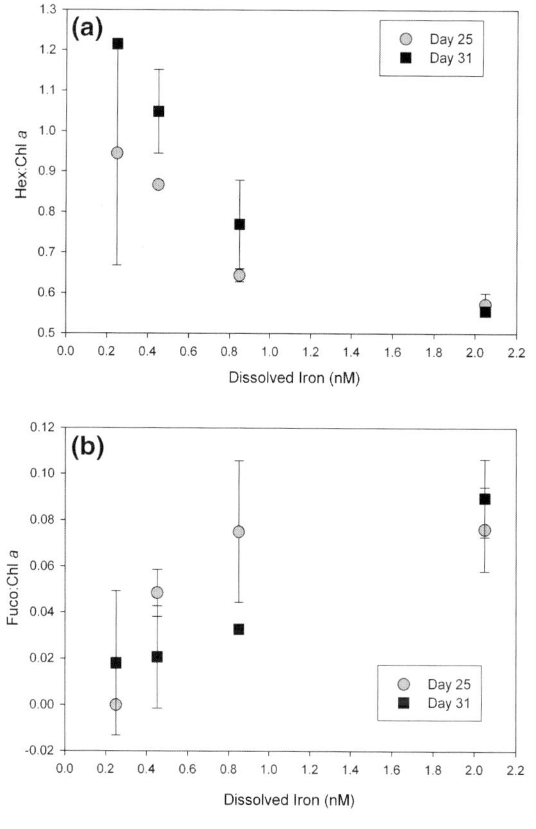

Fig. 1 (**a**) Hex:Chl *a* and (**b**) Fuco:Chl *a* ratios of Ross Sea colonial *P. antarctica* culture as a function of initial dissolved iron concentration, after incubation for 25 days (duplicate bottles sampled) and 31 days (triplicate bottles sampled) at an irradiance of 20 μE m^{-2} s^{-1}. The data points show mean values ± one standard deviation

austral summer period (DiTullio et al. 2003b), when dissolved Fe concentrations in the upper water column of the southern Ross Sea polynya are typically low (<0.2 nM; Fitzwater et al. 2000; Sedwick et al. 2000; Coale et al. 2005); whereas the lowest Hex:Chl *a* ratio (0.62) was observed in the southern Ross Sea during the early spring (DiTullio et al. 2003b), when dissolved Fe concentrations in the upper water column are typically higher than the summer values (Fitzwater et al. 2000; Sedwick et al. 2000; Coale et al. 2005). Although these field samples were collected in regions dominated by *P. antarctica* (as verified by microscopy; DiTullio et al. 2003b), the likely co-occurrence of diatoms, which will contribute to Chl *a* but not Hex to the field pigment samples, means that these field ratios represent lower limits for the Hex:Chl *a* ratio in native *P. antarctica*.

Fuco:Chl *a* ratios

The concentrations of Fuco in our Ross Sea *P. antarctica* isolate remained relatively low for all iron treatments used in our laboratory iron-addition experiment. However, the ratio of Fuco to chlorophyll *a* (Fuco:Chl *a* ratio) increased eightfold in response to iron addition, from approximately 0.01 ± 0.02 under iron-limited conditions (0.22 nM dissolved Fe) to 0.08 ± 0.02 under high-iron (2.02 nM dissolved Fe) conditions (Fig. 1b). In comparison, the mean Fuco:Chl *a* ratio in the *P. antarctica*-dominated region of the Ross Sea during the ROAVERRS expeditions was somewhat higher than our experimental values, at 0.12 ± 0.05 (DiTullio et al. 2003b). In bulk field samples, however, this pigment ratio is sensitive to the presence of diatoms, since these organisms contain Fuco but not Hex. Whereas Hex in the Ross Sea is diagnostic for the occurrence of *P. antarctica*, the presence of Fuco is not necessarily highly correlated with the presence of diatoms, since *P. antarctica* are known to produce Fuco.

Chl c_3:Chl *a* ratios

In our laboratory experiment, iron addition had mediated a decrease in the ratio of chlorophyll c_3 to chlorophyll *a* (Chl c_3:Chl *a* ratio) at day 25 (from ~0.27 at low iron to ~0.13 at high iron). This pigment ratio, however, had reached a relatively constant value of ~0.25 in all iron treatments during the late exponential growth phase (Fig. 2a). In comparison, analyses of the Chl c_3:Chl *a* ratio in *P. antarctica*-dominated regions in the Ross Sea during the three ROAVERRS cruises showed an average Chl c_3:Chl *a* ratio of 0.24 ± 0.11 (DiTullio et al. 2003b). The data trend for our laboratory experiment at day 25 are qualitatively consistent with the seasonal trend in the ROAVERRS field data: the highest mean Chl c_3:Chl *a* ratio (0.34) was observed during the summer, and the lowest mean Chl c_3:Chl *a* ratio (0.12) was observed during the spring (DiTullio et al. 2003b), when dissolved Fe concentrations

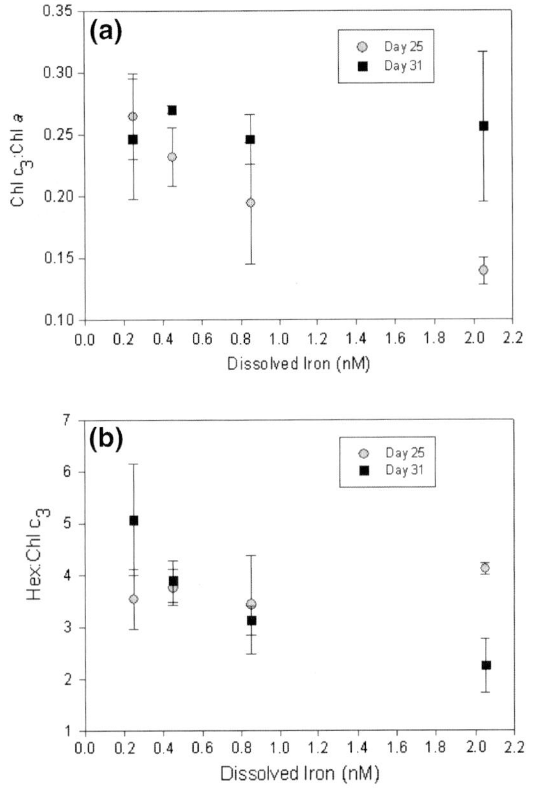

Fig. 2 (**a**) Chl c_3:Chl *a* and (**b**) Hex:Chl c_3 ratios of Ross Sea colonial *P. antarctica* culture as a function of initial dissolved iron concentration, after incubation for 25 days (duplicate bottles sampled) and 31 days (triplicate bottles sampled) at an irradiance of 20 μE m^{-2} s^{-1}. The data points show mean values ± one standard deviation

are typically low (<0.2 nM) and high (>0.2 nM), respectively (Fitzwater et al. 2000; Sedwick et al. 2000; Coale et al. 2005). Thus, in the southern Ross Sea, there may be a general seasonal increase in the Chl c_3:Chl *a* ratio of colonial *P. antarctica* that is associated with a corresponding seasonal decrease in dissolved iron availability, although these trends are not entirely consistent with our experimental results.

Hex:Chl c_3 ratios

Since diatoms and *P. antarctica* dominate the algal community in the southern Ross Sea polynya (DiTullio and Smith 1996; Arrigo et al. 1999), a potentially unique pigment signature for *P. antarctica* populations in this region is the ratio of Hex to chlorophyll c_3 (Hex:Chl c_3 ratio), because diatoms produce neither of these accessory pigments. In our laboratory iron-addition experiment, during the early exponential growth phase (i.e., day 25), the Hex:Chl c_3 ratio was relatively constant (~3.5) with respect to dissolved Fe concentration (Fig. 2b). However, by day 31, as Chl *a* biomass accumulated during the later exponential growth phase, lower Hex:Chl c_3 ratios were observed at higher dissolved Fe concentrations (Fig. 2b). Van Leeuwe and Stefels (1998) have reported a similar relationship between the Hex:Chl c_3 ratio and dissolved Fe concentration for a Weddell Sea *P. antarctica* culture grown at both high (110 μE m^{-2} s^{-1}) and low (25 μE m^{-2} s^{-1}) irradiance. Although the iron concentrations used in that study were significantly higher than those employed in our experiment, the results of van Leeuwe and Stefels (1998) are consistent with our experimental data, and there are no clear differences that might be attributed to the different *P. antarctica* strains (Weddell Sea versus Ross Sea) used in the two experimental studies.

The results of both studies indicate that Hex:Chl c_3 ratios are greater than ~3 for ambient dissolved Fe concentrations below ~1 nM, an iron concentration at which the growth rate of colonial *P. antarctica* appears to be significantly reduced at an irradiance of 20 μE m^{-2} s^{-1} (Sedwick et al. this issue). The fact that Hex:Chl c_3 ratios in the lower euphotic zone of the southern Ross Sea were ~3–5 during the ROAVERRS III spring-early summer cruise (Fig. 4d) is suggestive of the importance of iron availability in limiting the growth and biomass of colonial *P. antarctica*, as discussed by Sedwick et al. (this issue). However, we note that the experiments of van Leeuwe and Stefels (1998) also demonstrate a significant effect of irradiance on the Hex:Chl c_3 ratio of colonial *P. antarctica*, with growth under iron-replete conditions, producing Hex:Chl c_3 ratios of 0.6 at high irradiance (110 μE m^{-2} s^{-1}) and 0.06 at low irradiance (25 μE m^{-2} s^{-1}). These values, however, were approximately an order of magnitude lower compared to the high Hex:Chl c_3 ratios (3–5) observed under iron-limited conditions. Thus, it is clear that iron limitation results in significantly higher Hex:Chl c_3 ratios in *P. antarctica* compared to iron-sufficient conditions. Because these two pigments are diagnostic pigments for *P. antarctica* in the Ross Sea, the Hex:Chl c_3 ratio may serve as a physiological indicator of iron stress for *P. antarctica* in this region. Further work comparing this ratio with dissolved iron concentrations in the region should provide a good test for this hypothesis.

Fuco:Hex ratios

The results of our laboratory experiment show that sub-nanomolar iron additions to our low-iron (0.22 nM dissolved Fe) growth medium mediated an increase in the Fuco:Chl *a* ratio and a decrease in the Hex:Chl *a* ratio of colonial *P. antarctica* at an irradiance of 20 μE m^{-2} s^{-1}. Not surprisingly, our data reveal a strong correlation ($r^2 = 0.82$) between the ratio of Fuco to Hex (Fuco:Hex ratio) after 25 and 31 days incubation and the initial dissolved Fe concentration of our experimental treatments (Fig. 3a). This observed increase in the Fuco:Hex ratio with dissolved iron availability appears to extend well above the range of dissolved iron concentrations used in our laboratory experiment. For instance, when grown under iron-replete conditions (~500 nM dissolved Fe, in a 20-fold seawater dilution of L1 medium) at an irradiance of 20 μE m^{-2} s^{-1}, the same strain of colonial *P. antarctica* exhibited a Fuco:Hex ratio of ~0.45 (Fig. 3b), which was around 40 times higher than the Fuco:Hex ratio (ca. 0.01) in our experimental control treatments (0.22 nM dissolved Fe). In addition, two older *P. antarctica* isolates that had been maintained under iron-replete conditions for more than 15 years displayed Fuco:Hex ratios > 1 (Fig. 3b), which is consistent with other reported data for *P. antarctica* laboratory cultures (Buma et al. 1991).

Our experimental results also corroborate the findings of van Leeuwe and Stefels (1998), who have reported an increase in the Fuco:Hex ratio of a Weddell Sea *P. antarctica* isolate with increasing iron availability, based on experiments using growth media with EDTA-buffered iron concentrations on the order of 1 nM ('iron-deplete' conditions) and 1 μM ('iron-replete' conditions). However, in contrast to this earlier work, our experimental results may be directly

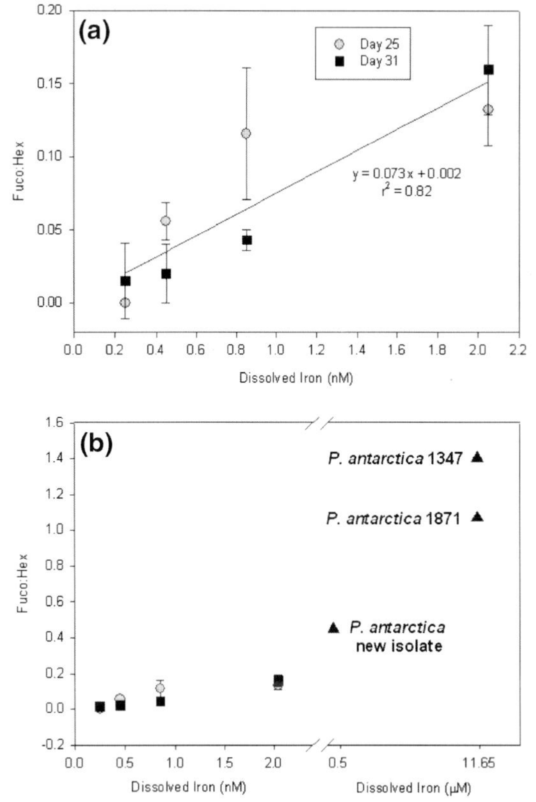

Fig. 3 (**a**) Fuco:Hex ratios of Ross Sea colonial *P. antarctica* culture as a function of initial dissolved iron concentration, after incubation for 25 days (duplicate bottles sampled) and 31 days (triplicate bottles sampled) at an irradiance of 20 μE m^{-2} s^{-1}. The data points show mean values ± one standard deviation. A linear regression using all data points yields the following relationship: $y = 0.073x + 0.002$ ($r^2 = 0.82$), where y is the Fuco:Hex ratio and x is the initial dissolved Fe concentration in nM. (**b**) Same data as figure (**a**), as well as Fuco:Hex ratios measured for two strains (1347 and 1871) of *P. antarctica* grown under iron-replete conditions (obtained from the Bigelow culture collection) and our Ross Sea *P. antarctica* culture (new strain) grown at a dissolved Fe concentration of 0.5 μM and an irradiance of 20 μE m^{-2} s^{-1}. Note the scale break on the x-axis

compared with field measurements of pigments in Antarctic surface waters. Specifically, we have used relatively low dissolved iron concentrations (~0.2–2 nM) that are typical of Antarctic surface waters during spring and summer, without using EDTA to control dissolved iron concentrations, thereby allowing iron speciation to be controlled by the natural iron-binding ligands present in Antarctic surface seawater (see discussion by

Gerringa et al. 2000). In addition, our experiment was conducted using a recently isolated strain of *P. antarctica*, in an effort to avoid any genetic and/or physiological biases that might arise when algal clones are maintained over a period of years under unnatural conditions (e.g., relatively high dissolved iron concentrations).

Our experimental estimates of growth rate as a function of iron concentration yielded a half-saturation constant for growth of 0.45 nM dissolved Fe for colonial *P. antarctica* at an irradiance of ~20 μE m^{-2} s^{-1} (Sedwick et al., this issue). These results imply that at this relatively low irradiance, colonial *P. antarctica* was iron-limited (sensu Morel et al. 1991) at ambient dissolved Fe concentrations < 0.45 nM (Fuco:Hex < 0.035), and was still significantly iron-stressed (sensu Morel et al. 1991) at dissolved iron concentrations of 1–2 nM (Fuco:Hex ~0.075–0.15). Thus, the mean Fuco:Hex ratio of 0.076 measured in subsurface, *P. antarctica*-dominated waters of the Ross Sea during mid spring and early summer 1998 (DiTullio et al. 2003b; Figs. 4a, b) would suggest that ambient dissolved Fe concentrations were limiting growth. Although no iron measurements were made during the ROAVERRS III cruise, the results of other studies in the southern Ross Sea suggest that dissolved Fe concentrations in the upper water column are typically subnanomolar during mid spring-early summer (Fitzwater et al. 2000; Sedwick et al. 2000; Coale et al. 2005).

When comparing our experiental data with field observations, an important caveat that must be considered is the potential effect of irradiance on the pigment composition of *P. antarctica* in relation to iron availability. The experiments of van Leeuwe and Stefels (1998) revealed relatively low Fuco:Hex ratios (ca. 0.3 and 0.7) for *P. antarctica* grown under low iron conditions at high irradiance (110 μE m^{-2} s^{-1}) and low irradiance (25 μE m^{-2} s^{-1}), respectively. Under

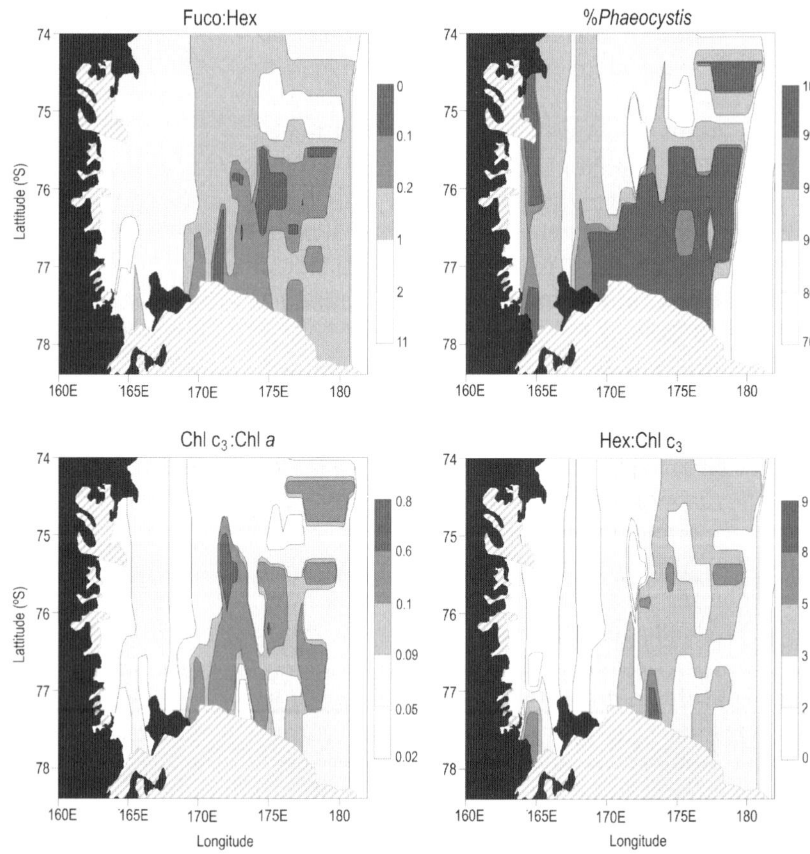

Fig. 4 Pigment ratios and *P. antarctica* abundance for 30–50 m depth in the Ross Sea during the ROAVERRS III expedition in November–December 1998: (**a**) Fuco:Hex ratios; (**b**) % *P. antarctica* as estimated by CHEMTAX analyses and corroborated by microscopy (see DiTullio et al. 2003b for details); (**c**) Chl c_3:Chl *a* ratios; and (**d**) Hex:Chl c_3 ratios

iron-replete conditions, the Fuco:Hex ratio in that study averaged 5 and 41 under high- and low-light conditions, respectively. Hence, similar to the trend in the Hex:Chl c_3 ratios, the effects of irradiance on the Fuco:Hex ratios were relatively small compared to the observed changes due to iron availability. Similarly, the results of a shipboard bioassay experiment in the Ross Sea using a *P. antarctica*-dominated assemblage suggest that the Fuco:Hex ratio of colonial *Phaeocystis* is more sensitive to changes in iron availability than to changes in irradiance (Sedwick et al., this issue). However, the relationship between iron availability and the pigment composition of *P. antarctica* may be complicated by the interrelated influence of iron and light on phytoplankton growth, whereby algal iron requirements are expected to decrease under increased irradiance (Raven 1990; Sunda and Huntsman 1997). Indeed, the results of recent culture experiments in our laboratory indicate that the half-saturation constant for the growth of colonial *P. antarctica* with respect to iron is significantly less than 0.45 nM at irradiances greater than 20 μE m^{-2} s^{-1} (Garcia et al. manuscript in preparation). This finding has potentially important implications for the pigment composition of *P. antarctica* in the southern Ross Sea, where there are seasonal decreases in dissolved iron concentration and increases in mean irradiance between the early spring and mid summer (see discussion by Sedwick et al., this issue).

Thus there is a need for further experimental work to assess the relationship between the pigment composition of *P. antarctica* and dissolved iron concentrations as a function of irradiance. To this end, we are currently undertaking the analysis of pigment samples from iron-addition culture experiments conducted at irradiances > 20 μE m^{-2} s^{-1}. In addition, there are other factors that may limit the utility of pigment ratios such as Fuco:Hex in assessing the physiological status and/or ambient growth conditions of *P. antarctica* in Antarctic waters. These factors include growth limitation/co-limitation by other resources, such as zinc (Coale et al. 2003) or vitamin B_{12} (Bertrand et al. in press); luxury uptake of iron; integrated effects of prior changes in availability of iron and/or light; physiological differences between genetically distinct *P. antarctica* ecotypes; and life-stage of *P. antarctica* (i.e., colonial versus solitary cells, each of which may have quite different growth requirements).

Physiological and ecological implications

The use of algal pigment measurements in field samples as a chemotaxonomic tool begs a mechanistic understanding of the environmental factors that impact the pigment composition of relevant phytoplankton species. With regard to understanding the distribution and ecology of *P. antarctica* in the Ross Sea and the wider Southern Ocean, it is of interest to consider the physiological basis behind the significant changes in the Hex:Chl *a*, Fuco:Chl *a*, Hex:Chl c_3 and Fuco:Hex ratios in response to sub-nanomolar iron additions that were observed in our culture experiment. These experimental trends are qualitatively consistent with the results of previous work, which has demonstrated decreases in both the cellular chlorophyll content and the Fuco:Hex ratio of colonial *P. antarctica* in response to decreased iron availability (van Leeuwe and Stefels 1998; Schoemann et al. 2005). This decrease in cellular chorophyll is readily explained by the involvement of iron in the synthesis of chlorophyll (Greene et al. 1992; Schoemann et al. 2005). With regard to the decrease in the Fuco:Hex ratio, van Leeuwe and Stefels (1998) have proposed that this reflects the conversion of Fuco to Hex, which acts as a photoprotective mechanism under conditions of low iron availability, when iron-deficient cells are more susceptible to photo-damage. Field measurements from the ROAVERRS cruises provide some support for this hypothesis, in that the Hex:Chl *a* ratio in *P. antarctica*-dominated waters was highest during the summer, when dissolved Fe levels are typically low and mean mixed-layer irradiance is relatively high (see discussion by Sedwick et al. this issue). Conversely, under conditions of high iron availability, Hex may be converted to the more-efficient light-harvesting pigment Fuco, in order to maximize the capture of light energy by the iron-replete cells (van Leeuwe and Stefels 1998). There is evidence for this in the observed increase in the Fuco:Hex ratio of our *P. antarctica* cultures with increasing dissolved iron concentration, with the trend extending to dissolved Fe concentrations as high as 0.5 μM (Fig. 3b).

At present, it is not clear whether the Ross Sea *P. antarctica* strain used in our experiments is genetically distinct from other *P. antarctica* strains, particularly those isolated from open-ocean surface waters of the Antarctic circumpolar current (ACC), where dissolved iron levels are likely to be low (<0.5 nM) year-round (Martin et al. 1990; Coale et al. 2005), and mean summer mixed-layer depths are typically greater (hence mean irradiance lower) than those over the Antarctic continental shelf (Trull et al. 2001). It may be that our experimental data are readily applicable to *P. antarctica* in these ACC waters, given that our experiment was conducted using sub-nanomolar iron additions and relatively low irradiance. On the other hand, long-term adaptation of open-ocean *P. antarctica* strains to the iron and light regimes of the ACC, which differ significantly from those in the southern Ross Sea, may limit the application of our experimental results to *P. antarctica* ecotypes from the Ross Sea/Antarctic shelf region. Ongoing molecular studies of relevant *P. antarctica* isolates, as well as further experimental work using open-ocean *P. antarctica* strains, are expected to provide answers to these questions.

Acknowledgements We are grateful to Mark Geesey, who performed the CHEMTAX and HPLC measurements for the ROAVERRS III samples, and Christopher Marsay, who performed the analyses of dissolved iron in the Ross Sea seawater and our experimental media. We also acknowledge the outstanding support of the officers, crew and support personnel aboard RV/IB *Nathaniel B. Palmer*. The primary funding for this research was provided by the United States National Science Foundation grants OPP–0230513 (to GRD), OPP-0230559 (to PNS) and to DGE-0139313 (to NSG through M. VanSickle and G. Tempel).

References

Arrigo KR, Weiss AM, Smith WO (1998) Physical forcing of phytoplankton dynamics in the southwestern Ross Sea. J Geophys Res 103(C1):1007–1022

Arrigo KR, Robinson DH, Worthen DL, Dunbar RB, DiTullio GR, van Woert M, Lizotte MP (1999) Phytoplankton community structure and the drawdown of nutrients and CO_2 in the Southern Ocean. Science 283:365–367

Arrigo KR (2005) Marine microorganisms and global nutrient cycles. Nature 437(7057):349–355

Bertrand EM, Saito M, Rose JM, Riesselman C, Lohan MC, Noble AE, Lee P, DiTullio GR (in press) Vitamin B12 and iron co-limitation of phytoplankton growth in the Ross Sea. Limnol Oceanogr

Buma AG, Bano JN, Veldhuis MJW, Kraay GW (1991) Comparison of the pigmentation of two strains of the prymnesiophyte *Phaeocystis* sp. Neth J Sea Res 27:173–182

Coale KH, Wang X, Tanner SJ, Johnson KS (2003) Phytoplankton growth and biological response to iron and zinc addition in the Ross Sea and Antarctic circumpolar current along 170°W. Deep-Sea Res II 50:635–653

Coale KH, Gordon RM, Wang X (2005) The distribution and behavior of dissolved and particulate iron and zinc in the Ross Sea and Antarctic circumpolar current along 170°W. Deep-Sea Res I 52:295–318

Crocker KM, Ondrusek ME, Petty RL, Smith RC (1995) Dimethylsulfide, algal pigments and light in an Antarctic *Phaeocystis* sp. bloom. Mar Biol 124:335–340

DiTullio GR, Hutchins DA, Bruland KW (1993) Interaction of iron and major nutrients controls phytoplankton growth and species composition in the tropical North Pacific Ocean. Limnol Oceanogr 38:495–508

DiTullio GR, Smith WO Jr (1995) Relationship between dimethylsulfide and phytoplankton pigment concentrations in the Ross Sea, Antarctica. Deep-Sea Res I 42:873–892

DiTullio GR, Smith WO Jr (1996) Spatial patterns in phytoplankton biomass and pigment distributions in the Ross Sea. J Geophys Res 101:18467–18477

DiTullio GR, Grebmeier JM, Arrigo KR, Lizotte MP, Robinson DH, Leventer AR, Barry JP, van Woert M, Dunbar RB (2000) Rapid and early export of *Phaeocystis antarctica* blooms in the Ross Sea, Antarctica. Nature 404:595–598

DiTullio GR, Geesey ME (2002) Photosynthetic pigments in marine algae and bacteria. In: G Bitton (ed) The Encyclopedia of environmental microbiology. John Wiley & Sons, Inc, New York, NY, pp 2453–2470

DiTullio GR, Geesey ME, Jones DR, Daly KL, Campbell L, Smith WO Jr (2003a) Phytoplankton assemblage structure and primary productivity along 170° W in the South Pacific Ocean. Mar Ecol Prog Ser 255:55–80

DiTullio GR, Geesey ME, Leventer AR, Lizotte MP (2003b) Algal pigment ratios in the Ross Sea: Implications for CHEMTAX analysis of Southern Ocean Data. In: DiTullio GR, Dunbar RB (eds) Biogeochemistry of the Ross Sea. AGU *Antarctic Research Series* 78, Washington, DC, pp 35–52

El-Sayed SZ, Biggs D, Holm-Hansen O (1983) Phytoplankton standing crop, primary productivity, and near-surface nitrogenous nutrient fields in the Ross Sea, Antarctica. Deep-Sea Res I 30:871–886

Fitzwater SE, Johnson KS, Gordon RM, Coale KH, Smith WO (2000) Trace metal concentrations in the Ross

Sea and their relationship with nutrients and growth. Deep-Sea Res II 47:3159–3179

Gerringa LJA, de Baar HJW, Timmermans KR (2000) A comparison of iron limitation of phytoplankton in natural oceanic waters and laboratory media conditioned with EDTA. Mar Chem 68:335–346

Gibson JA, Garrick ERC, Burton HR, McTaggart AR (1990) Dimethylsulfide and the alga *Phaeocystis pouchetti* in Antarctic coastal waters. Mar Biol 104:339–346

Greene RM, Geider RJ, Kolber Z, Falkowski PG (1992) Iron-induced changes in light harvesting and photochemical energy conversion processes in eukaryotic marine algae. Plant Physiol 100(2):565–575

Guillard RRL, Hargraves PE (1993) *Stichochrysis immobilis* is a diatom, not a chrysophyte. Phycologia 32: 234–236

Mackey MD, Mackey DJ, Higgins HW, Wright SW (1996) CHEMTAX—a program for estimating class abundances from chemical markers: application to HPLC measurements of phytoplankton. Mar Ecol Prog Ser 144:265–283

Martin JH, Fitzwater SE, Gordon RM (1990) Iron deficiency limits phytoplankton growth in Antarctic waters. Global Biogeochem Cycles 4:5–12. *Research Series*, 78:209–220, Washington, DC

Morel FMM, Hudson RJM, Price NM (1991) Limitation of productivity by trace metals in the sea. Limnol Oceanogr 36:1742–1755

Palmisano AC, SooHoo JB, SooHoo SL, Kottmeier ST, Craft LL, Sullivan CW (1986) Photoadaptation in *Phaeocystis pouchetii* advected beneath annual sea ice in McMurdo Sound, Antarctica. J Plankton Res 8:891–906

Raven JA (1990) Predictions of Mn and Fe use efficiencies of phototrophic growth as a function of light availability for growth and C assimilation pathway. New Phytol 116:1–17

Rousseau V, Vaulot D, Casotti R, Cariou V, Lenz J, Gunkel J, Baumann M (1994) The life cycle of *Phaeocystis* (Prymnesiophyceae): evidence and hypotheses. J Mar Syst 5(1):23–39

Schoemann V, Becquevort S, Stefels J, Rousseau V, Lancelot C (2005) *Phaeocystis* blooms in the global ocean and their controlling mechanisms: a review. J Sea Res 53:43–66

Sedwick PN, DiTullio GR, Mackey DJ (2000) Iron and manganese in the Ross Sea, Antarctica: seasonal iron limitation in Antarctic shelf waters. J Geophys Res 105(C5):11, 321–11, 336

Sedwick PN, Garcia NS, Riseman SF, DiTullio GR (2007) Evidence for high iron requirements for colonial *Phaeocystis antarctica* at low irradiance. Biogeochemistry (this issue)

Smith WO, Marra J, Hiscock MR, Barber RT (2000) The seasonal cycle of phytoplankton biomass and primary productivity in the Ross Sea, Antarctica. Deep Sea Res II 47:3119–3140

Smith WO Jr, Dennett MR, Mathot S, Caron DA (2003) The temporal dynamics of the flagellated and colonial stages of *Phaeocystis antarctica* in the Ross Sea. Deep Sea Res II 50:605–617

Smith WO Jr, Van Hilst CM (2003) Effects of assemblage composition on the temporal dynamics of carbon and nitrogen uptake. In: DiTullio GR, Dunbar RB (eds) Biogeochemistry of the Ross Sea, American Geophysical Union, Washington DC, pp197–208

Sunda WG, Huntsman SA (1997) Interrelated influence of iron, light and cell size on marine phytoplankton growth. Nature 390:389–392

Trull TS, Rintoul R, Hadfield M, Abraham ER (2001) Circulation and seasonal evolution of polar waters south of Australia: implications for iron fertilization of the Southern Ocean. Deep-Sea Res II 48:2439–2466

Van Leeuwe MA, Stefels J (1998) Effects of iron and light stress on the biochemical composition of Antarctic *Phaeocystis* sp. (Prymnesiophyceae). II. Pigment composition. J Phycol 34:496–503

Wright SW, Jeffrey SW, Mantoura RFC, Llewellyn CA, Bjørnland T, Repeta D, Welschmeyer NA (1991) Improved HPLC method for the analysis of chlorophylls and carotenoids from marine phytoplankton. Mar Ecol Prog Ser 77:183–196

Zapata M, Rodriguez F, Garrido JL (2000) Separation of chlorophylls and carotenoids from marine phytoplankton:a new HPLC method using a reversed phase C_8 column and pyridine-containing mobile phases. Mar Ecol Prog Ser 195:29–45

Biogeochemistry (2007) 83:83–97
DOI 10.1007/s10533-007-9081-7

ORIGINAL PAPER

Evidence for high iron requirements of colonial *Phaeocystis antarctica* at low irradiance

P. N. Sedwick · N. S. Garcia · S. F. Riseman · C. M. Marsay · G. R. DiTullio

Received: 1 December 2005 / Accepted: 12 June 2006 / Published online: 13 March 2007

Abstract We have carried out field and laboratory experiments to examine the iron requirements of colonial *Phaeocystis antarctica* in the Ross Sea. In December 2003, we performed an iron/light-manipulation bioassay experiment in the Ross Sea polynya, using an algal assemblage dominated by colonial *Phaeocystis antarctica*, collected from surface waters with an ambient dissolved Fe concentration of ~0.4 nM. Results from this experiment suggest that *P. antarctica* growth rates were enhanced at high irradiance (~50% of incident surface irradiance) but were unaffected by iron addition, and that elevated irradiance mediated a significant decrease in cellular chlorophyll *a* content. We also conducted a laboratory iron dose–response bioassay experiment using a unialgal, non-axenic strain of colonial *P. antarctica* and low-iron (<0.2 nM) filtered seawater, both collected from the Ross Sea polynya in December 2003. By using rigorous trace-metal clean techniques, we performed this dose–response iron-addition experiment at ~0°C without using organic chelating reagents to control dissolved iron levels. At the relatively low irradiance of this experiment (~20 μE m^{-2} s^{-1}), estimated nitrate-specific growth rate as a function of dissolved iron concentration can be described by a Monod relationship, yielding a half-saturation constant with respect to growth of 0.45 nM dissolved iron. This value is relatively high compared to reported estimates for other Antarctic phytoplankton. Our results suggest that seasonal changes in the availability of both iron and light play critical roles in limiting the growth and biomass of colonial *Phaeocystis antarctica* in the Ross Sea polynya.

Keywords Iron · Light · *Phaeocystis antarctica* · Ross Sea

P. N. Sedwick (✉) · C. M. Marsay
Bermuda Biological Station for Research, Ferry Reach, St. George's, GE01 Bermuda
e-mail: psedwick@bbsr.edu

N. S. Garcia · S. F. Riseman · G. R. DiTullio
College of Charleston, Charleston, SC 29412, USA

Introduction

The colonial prymnesiophyte *Phaeocystis antarctica* is a keystone species on the Antarctic continental shelf, where it plays an important role in the biogeochemical cycling of carbon, sulfur and nutrient elements (Gibson et al. 1990; Smith et al. 1991; DiTullio and Smith 1995; Arrigo et al. 1999; DiTullio et al. 2000; Sweeney et al. 2000; Schoemann et al. 2005). In the polynya region of the southern Ross Sea, extensive blooms of colonial *P. antarctica* typically form during spring and early summer, then wane during the mid to late summer (Smith and

Gordon 1997; Arrigo et al. 1998; Smith et al. 2000; Mathot et al. 2001; Arrigo and van Dijken 2004). Field observations have documented a general decrease in the depth of the surface mixed layer during the growing season, from more than 100 m in early spring to less than 25 m in summer (Arrigo et al. 1998; Gordon et al. 2000; Sweeney 2003; Worthen and Arrigo 2003; Smith and Van Hilst 2003). The results of several studies suggest that this seasonal change in stratification is accompanied by a decrease in the dissolved iron concentration of surface waters, from values of around 0.5 nM or more in spring to less than 0.1 nM in late summer (Sedwick and DiTullio 1997; Fitzwater et al. 2000; Sedwick et al. 2000; Coale et al. 2005). These dissolved iron levels are low enough to limit algal community growth rates in the Ross Sea during the summer (Martin et al. 1990; Sedwick and DiTullio 1997; Sedwick et al. 2000; Coale et al. 2003). Hence it has been argued that seasonal changes in irradiance and/or iron availability—bottom-up controls—regulate the timing and location of colonial *Phaeocystis antarctica* blooms in the Ross Sea (Arrigo et al. 1998; Boyd 2002; Smith et al. 2003; Worthen and Arrigo 2003; Arrigo and Tagliabue 2005; Tagliabue and Arrigo 2005). As yet, there are few published data to support this assertion.

With regard to irradiance, the results of field and laboratory studies indicate that *P. antarctica* can readily adapt to a wide range of ambient light conditions (Palmisano et al. 1986; Matrai et al. 1995; Hong et al. 1997; Moisan and Mitchell 1999; Robinson et al. 2003; Schoemann et al. 2005). However, the field and laboratory measurements discussed by Van Hilst and Smith (2002) and Smith and Van Hilst (2003) suggest that there is little difference between the photosynthesis-irradiance characteristics of colonial *Phaeocystis antarctica* and diatoms in the Ross Sea, implying that other factors, such as iron availability and grazing, must control the spatial and temporal distribution of the two main phytoplankton taxa in this highly productive shelf region.

So far there has been only limited research addressing the iron requirements of colonial *P. antarctica*. Sedwick et al. (2000) performed a shipboard iron-addition experiment with a colonial *Phaeocystis*-dominated algal assemblage collected from the southern Ross Sea in spring 1994, with their results suggesting that the ambient dissolved iron concentrations of ~0.8 nM were sufficient to meet the growth requirements of the native *P. antarctica* colonies. Evidence for iron limitation of *Phaeocystis antarctica* was reported by Olson et al. (2000), who measured the photosynthetic competency of individual *P. antarctica* cells in the shipboard iron-addition experiments that Coale et al. (2003) performed in the Ross Sea during late summer 1997. In these experiments, Coale et al. (2003) observed an increase in the abundance of prymnesiophytes, mainly *Phaeocystis*, in response to iron addition. The dose–response experiments performed by Coale et al. (2003) in the Ross Sea allowed them to estimate half-saturation constants (with respect to growth) of 0.005–0.043 nM dissolved Fe for prymnesiophytes, suggesting that the iron requirements of *Phaeocystis* in the Ross Sea may be quite low, although their experimental results suggest similarly low values for diatoms. However, the results described by Olson et al. (2000) and Coale et al. (2003) pertain to bottle incubations performed at full surface irradiance (i.e., light levels that were probably much higher than mean surface mixed-layer irradiance), and presumably apply to solitary *Phaeocystis antarctica* cells; therefore, their results may not be representative of the colonial *Phaeocystis antarctica* that bloom in the southern Ross Sea during spring and early summer.

Here we report the results of two experiments that provide further information on the iron requirements of colonial *Phaeocystis antarctica* in the Ross Sea. The first is a shipboard iron- and light-manipulation experiment, which was performed using a native algal assemblage dominated by colonial *P. antarctica*, collected in the northern Ross Sea polynya during December 2003. The second is a laboratory dose–response iron-addition experiment in which we used a pure culture of colonial *P. antarctica* and low-iron (<0.2 nM) filtered seawater, both collected from the southern Ross Sea polynya in December 2003. To our knowledge, this experiment represents the first time a dose–response iron-addition experiment has been carried out with colonial *Phaeocystis antarctica* using realistic, sub-nanomolar iron concentrations, and without using iron-complexing ligands such as Ethylene Diamine Tetra Acetic Acid (EDTA) to control dissolved iron concentrations in the growth medium.

Methods

Field collections

In late December 2003, we collected seawater and resident plankton from several sites in the Ross Sea polynya (Fig. 1) aboard the research vessel *Nathaniel B. Palmer*. At station A (ca. 73°24′ S, 173°14 E), in ice-free waters at the northern edge of the polynya, we used a Zodiac inflatable boat to collect surface seawater and resident plankton in an acid-cleaned 50-l polyethylene carboy. The seawater and plankton were collected by submerging an acid-cleaned 10-l polyethylene carboy from the bow of the work boat while slowly underway, then transferring the seawater from the 10-l carboy into the 50-l carboy on the work boat, using a protocol designed to avoid contamination of the seawater during the transfer process. The 50-l carboy of seawater and resident plankton were then stored for ~2 h in a shipboard cold room maintained at 2°C, prior to subsampling and processing for the shipboard iron-light manipulation experiment described below.

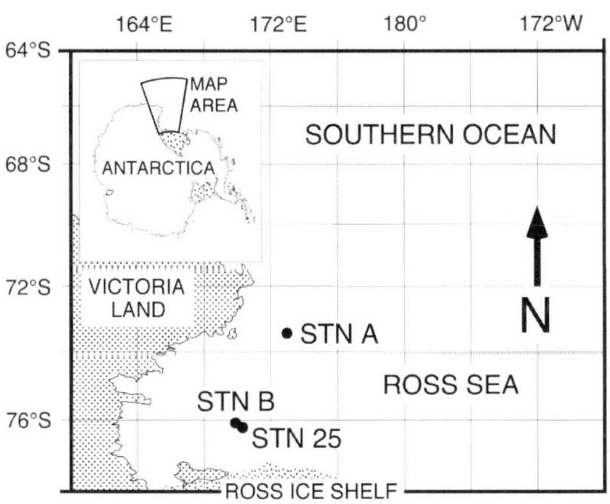

Fig. 1 Location of sampling sites in the Ross Sea, Antarctica, where seawater and plankton were collected for the experiments described in this paper. Seawater and resident plankton used in the shipboard iron-light experiment were collected at station A; uncontaminated seawater used in the laboratory iron experiment was collected at station B; the *P. antarctica* culture used in the laboratory iron experiment was isolated from water collected at station 25

Station 25 (ca. 76°05′ S, 170°08′ E), in the central polynya, was one of several stations where we collected near-surface seawater and resident plankton in 10-l Niskin bottles lowered on a rosette. Whole seawater samples from these stations were maintained in polystyrene culture flasks inside a shipboard incubator at ~0°C, prior to return to the Hollings Marine Laboratory (South Carolina, USA) for isolation of *Phaeocystis* cultures. In the nearby, ice-free waters at station B (ca. 76°02′ S, 169°53′ E), we collected ~4,000 l of seawater from the upper water column through acid-rinsed polyethylene tubing using an electrically operated Jabsco polypropylene-Santoprene double-diaphragm pump. The tubing inlet was lowered to ~10 m water depth from the side of the ship using a polyaramid line and epoxy-coated end weight, then seawater was pumped aboard while slowly underway at 1–2 knots. The pumped seawater was filtered through in-line, acid-rinsed 1-μm/0.2-μm high-capacity pleated-polypropylene cartridge filters (Cole-Parmer, Inc.), and then collected in acid-cleaned 125-l blue polyethylene barrels with airtight polyethylene screw closures. Subsequent measurements of iron in subsamples of this filtered seawater showed dissolved Fe concentrations of 0.07 ± 0.02 nM (n = 4). The barrels of low-iron filtered seawater (hereafter 'Station B seawater') were returned to the Hollings Marine Laboratory for use in the laboratory dose–response iron-addition experiment described below.

Shipboard iron-light manipulation experiment

Shipboard microscopy suggested that phytoplankton in seawater collected from Station A was dominated by healthy colonies of *P. antarctica*; this was later verified by high-performance liquid chromatography (HPLC) pigment analysis (see section "Laboratory dose–response iron-addition experiment"). Using the 50-l of cleanly collected seawater and resident plankton, we carried out a bioassay experiment that was designed to evaluate the effects of iron and light availability on the growth rate of colonial *Phaeocystis antarctica*. The seawater was gently mixed in the 50-l carboy, then transferred into acid-cleaned 1.2-l polycarbonate bottles, leaving ~200 ml headspace in each bottle, under class-100 filtered air using the rigorous trace-metal clean techniques described by Sedwick et al. (2000, 2002). Three iron treatments were used for these experimental incubation bottles: (1) 'high-Fe', to which 3 nM Fe was added as a 17.9 μM solution of ferric nitrate (the added nitrate was negligible relative to the ambient seawater nitrate concentration of ~20 μM); (2) 'control', to which there was no Fe added (initial ambient dissolved Fe concentration = 0.38 ± 0.03 nM); and (3) 'low-Fe', to which 3 nM of the iron-chelating ligand deferoxamine (DFOB or desferal) was added as a 38.4 μM solution of deferoxamine mesylate salt (Sigma Chemical).

After filling, the bottles were tightly sealed with polyvinyl chloride (PVC) tape, then set in circulating surface seawater at ~0°C in deckboard incubator baths shaded to two light levels using neutral-density screening: (1) 'low-light', shaded to ~15% of incident deckboard irradiance, and (2) 'high-light', shaded to ~50% of incident deckboard irradiance. The low-light treatment was chosen so as to fall within the range of mean in situ irradiance experienced by phytoplankton in the surface mixed layer, whereas the high-light treatment was intended to approximate the expected maximum value for mean in-situ irradiance (see discussion in section "Conclusions and directions for future research"). Our shipboard experiment thus incubated bottles containing seawater and resident plankton with six different treatments: (1) low-light, low-iron; (2) low-light, control-iron (this treatment was intended to approximate in situ irradiance and iron concentration); (3) low-light, high-iron; (4) high-light, low-iron; (5) high-light, control-iron; and (6) high-light, high-iron. Duplicate bottles for each treatment were harvested over an 80-h period, and subsampled for dissolved macronutrients, chlorophyll *a*, phytoplankton pigments, and other parameters.

Laboratory dose–response iron-addition experiment

At the Hollings Marine Laboratory, a pure culture of colonial *P. antarctica* was isolated from the native phytoplankton assemblage collected at Station 25, and subsequently maintained in nutrient-replete semicontinuous batch cultures, using Station B seawater amended with L1 medium (Guillard and Hargraves 1993). For the dose–response iron-addition experiment described here, we used a culture of this *P. antarctica* strain (predominantly colonial) that had been maintained for two months in L1 medium diluted 457-fold with station B seawater.

Eleven days before commencing the experiment, 50 ml of this culture was diluted to 1 l with 0.2 μm-filtered station B seawater in an acid-cleaned polycarbonate bottle, using a stringent trace-metal clean protocol. A peristaltic pump was used to transfer the Station B seawater from the 125-l polyethylene barrel (in which it had been stored since collection) through acid-cleaned silicone tubing and an acid-rinsed 0.2-μm CritiCap Supor capsule filter (Pall Corporation) into the 1-l polycarbonate bottle under class-100 filtered air. Subsamples of this 0.2-μm filtered station B seawater were collected for subsequent iron measurements, which confirmed that dissolved Fe concentrations were relatively low (0.17 ± 0.1 nM, $n = 4$). Based on the concentrations of $FeCl_3$ (11.65 μM) and EDTA (11.71 μM) in the L1 medium, the resultant 1-l inoculum of colonial *P. antarctica* had initial dissolved Fe and EDTA concentrations of approximately 1.42 and 1.25 nM, respectively.

After acclimating this *P. antarctica* inoculum to an irradiance of ~20 μE m^{-2} s^{-1} for 11 days at ~0°C, approximately 230 ml was removed for measurements of 'initial' pigment concentrations and qualitative observations, which showed the inoculum to be dominated by colonies rather than solitary cells. The remaining 770 ml of the *P. antarctica* inoculum was added to 37.8 l of station B seawater (~0.17 nM dissolved Fe) that had been filtered (0.2 μM Supor capsule filter) then chilled to ~0°C in an acid-cleaned 50-l polyethylene carboy. The calculated dissolved Fe concentration in the resulting large-volume *P. antarctica* inoculum was ~0.20 nM, which was subsequently confirmed by measurement of 0.22 ± 0.00 nM ($n = 2$) dissolved Fe in filtered subsamples, whereas the final concentration of EDTA in this solution was negligibly low

(~0.02 nM). This large-volume inoculum was gently mixed, and then used to fill 32 acid-cleaned 1.2-l polycarbonate incubation bottles, leaving 200 ml of headspace in each bottle. In addition, subsamples were taken from the 50-l carboy for measurements of initial chlorophyll *a* and dissolved macronutrient concentrations. In order to avoid inadvertent contamination with iron, the preparation of the large-volume *P. antarctica* inoculum and filling of the 1.2-l bottles were carried out using stringent trace-metal clean techniques, inside a plastic enclosure under positive pressure of class-100 filtered air; in addition, all materials contacting the station B filtered seawater and *P. antarctica* inoculums were rigorously cleaned using methods similar to those described by Sedwick et al. (2000).

Of the 32 incubation bottles, eight received no iron additions ('control treatments'), eight received an addition of 0.2 nM Fe ('+0.2 nM treatments'), eight received an addition of 0.6 nM Fe ('+0.6 nM treatments'), and eight received an addition of 1.8 nM Fe ('+1.8 nM treatments'). Iron was added as an aqueous 17.9 μM solution of ferric nitrate in 0.1% hydrochloric acid; the added nitrate and hydrochloric acid had negligible effect on the nitrate concentration and pH of the inoculum. The four iron treatments correspond to initial dissolved Fe concentrations of 0.22 nM ('control treatments'), 0.42 nM ('+0.2 nM treatments'), 0.82 nM ('+0.6 nM treatments'), and 2.02 nM ('+1.8 nM treatments'). The 32 incubation bottles were then placed in an incubator maintained at an irradiance of ~20 $\mu E\ m^{-2}\ s^{-1}$ and a temperature of ~0°C. For each different iron treatment, duplicate 1-l bottles were terminally sampled after incubation periods of 16, 25 and 31 days, with subsamples taken for measurements of dissolved macronutrients, chlorophyll *a*, and other chemical and biological parameters. These sampling times were chosen by following the concentration of particulate chlorophyll *a* in dedicated monitoring bottles (one of the eight bottles prepared for each iron treatment), which were repeatedly subsampled for chlorophyll measurements during the course of the experiment. Axenic conditions were not maintained in this experiment; however, in an effort to minimize bacterial contamination, the low-iron seawater used to prepare the experimental inoculums was filtered through the 0.2-μm CritiCap Supor capsule filter immediately prior to use, and the 50-l inoculum carboy and 1-l incubation bottles were thoroughly rinsed with this filtered seawater under class-100 air before filling.

Analytical methods

Dissolved inorganic nitrate + nitrite, phosphate and silicic acid were determined in samples filtered through 0.45-μm Acrodisc Supor syringe filters (Pall Corporation), using standard flow analysis methods by the Marine Science Institute Analytical Laboratory, University of California, Santa Barbara. Chlorophyll *a* was measured by fluorometry in a 90% acetone extract of particles collected under 1–2 psi vacuum on a Whatman GF/F filter, using a Turner Designs 10-AU fluorometer and standard Joint Global Ocean Flux Study (JGOFS) protocols (Knap et al. 1996). Water samples for the analysis of iron were collected and processed using the rigorous trace-metal clean protocols described by Sedwick et al. (1997, 2000, 2005), which are required to avoid inadvertent sample contamination. Dissolved iron was determined in 0.4 μm-filtered, acidified water samples by flow injection analysis, modified after the method of Measures et al. (1995), as described by Sedwick et al. (2005). Phytoplankton pigments were measured in particles collected on Whatman GF/F filters by high-performance liquid chromatography using the methods described by DiTullio and Geesey (2002).

Interpretation of experimental results

As in previous studies that have used bioassay growout experiments to examine resource limitation of phytoplankton growth (e.g. Martin et al. 1990; De Baar et al. 1990; DiTullio et al. 1993; Fitzwater et al. 1996; Hutchins et al. 2001; Sedwick et al. 2002), we infer relative net growth rates in our incubation bottles from the accumulation of algal biomass. Because changes in light intensity and iron availability are known to effect variations in the ratio of chlorophyll *a* to cell carbon for *Phaeocystis antarctica* (Van Leeuwe and Stefels 1998; Stefels and van Leeuwe 1998;

Schoemann et al. 2005), measurements of chlorophyll *a* may not provide a reliable estimate of relative changes in cell numbers (i.e., growth rate) in our incubation bottles. In addition, there are potential problems in using particulate organic carbon (POC) to estimate changes in the cellular biomass of colonial *Phaeocystis*, as a result of the POC that resides in the colonial mucilage rather than in the *Phaeocystis* cells (Schoemann et al. 2005). Thus, for the experiments reported here, we estimate relative net growth rates from net decreases (drawdown) in dissolved nitrate + nitrite, which are assumed to be proportional to net increases in cell carbon and cell number of colonial *Phaeocystis antarctica*, assuming balanced, exponential growth.

The experiments described here are principally diagnostic in nature: where cellular biomass was significantly enhanced in bottles after resource (iron or light) amendment, relative to control (or other) treatments, we infer that algal growth rates in the control (or other) treatments were limited by a deficiency in that resource. The statistical significance of differences between mean values of parameters measured in different treatments were assessed using a two-tailed *t*-test for comparisons between two treatments, or a one-way analysis of variance (ANOVA) for comparisons between three or more treatments, at a confidence level of 95% ($P = 0.05$).

Results and discussion

Shipboard iron-light manipulation experiment

The starting seawater collected for this experiment contained high concentrations of dissolved inorganic macronutrients that were typical of surface waters in the Ross Sea during early summer: ~20 μM nitrate + nitrite, ~1.5 μM phosphate, and ~55 μM silicic acid. Analysis of iron in 0.4-μM filtered subsamples of the starting seawater using methods described by Sedwick et al. (2005) revealed an initial dissolved Fe concentration of 0.38 ± 0.03 nM ($n = 2$). Thus, our additions of DFOB to the low-iron treatments were nearly an order of magnitude higher than the ambient dissolved iron concentration, on a molar basis. Initial phytoplankton biomass was relatively high, with ~4 μg l^{-1} chlorophyll *a*. Shipboard microscopy and subsequent HPLC measurements of the molar ratios of 19′-hexanoyloxyfucoxanthin to fucoxanthin (~3.8) and chlorophyll *c3* to chlorophyll *a* (~0.3) in the starting seawater indicate that algal biomass was overwhelmingly dominated by *Phaeocystis antarctica* (DiTullio and Smith 1996; Van Leeuwe and Stefels 1998; DiTullio et al., this issue).

The results from this shipboard incubation experiment are presented in Fig. 2, which shows the concentrations of dissolved nitrate + nitrite, particulate chlorophyll *a*, and particulate 19′-hexanoyloxyfucoxanthin:fucoxanthin in the incubation bottles during the course of the 80-h experiment. There was a significant decrease (drawdown) in dissolved nitrate + nitrite in all treatments during the experiment (Fig. 2a), relative to the starting seawater, suggesting net growth and accumulation of biomass in all treatments. There was no statistically significant difference between nitrate + nitrite drawdown among the different iron treatments at a given light level, suggesting that algal community growth rates were not increased by iron addition, nor decreased by DFOB addition. However, after 80 h incubation there was a significantly lower ($P < 0.05$) concentration of nitrate + nitrite (mean = 14.6 ± 0.6 μM) in the six high-light treatments than in the six low-light treatments (mean = 16.6 ± 0.2 μM), implying higher net growth rates in the bottles that were exposed to an irradiance higher than estimated in situ levels.

The chlorophyll *a* data (Fig. 2b) also show no significant difference between the various iron treatments for a given light level after 80 h incubation, although significantly lower ($P < 0.05$) levels of chlorophyll *a* (mean = 6.1 ± 1.6 μg l^{-1}) were measured in the high-light treatments relative to concentrations in the low-light incubations (mean = 9.5 ± 0.5 μg l^{-1}). If chlorophyll *a* were to be used as a proxy for cell biomass, this result would run contrary to our conclusions based on nitrate drawdown. A likely explanation for this discrepancy is that chlorophyll *a* concentrations in the high-light treatments reflect light-driven changes in cellular chlorophyll *a* content, with cells producing less chlorophyll *a* in

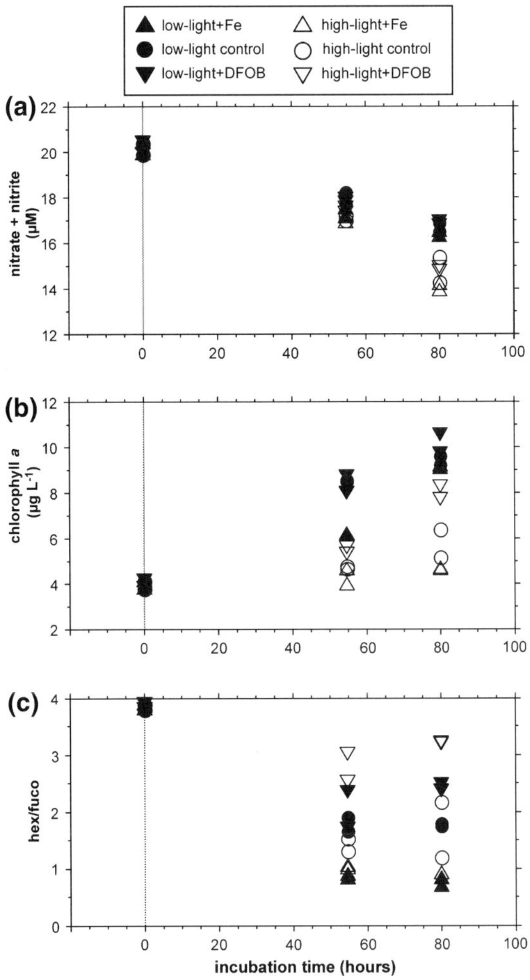

Fig. 2 Concentrations of (**a**) dissolved nitrate + nitrite, (**b**) particulate chlorophyll *a*, and (**c**) ratio particulate 19′-hexanoloxyfucoxanthin/fucoxanthin (hex/fuco) in the incubation bottles versus incubation time, from the shipboard iron-light manipulation experiment

the high-light treatments relative those incubated at low (approximately in-situ) irradiance, consistent with the experimental results of Van Leeuwe and Stefels (1998). This finding suggests that chlorophyll *a* is not a suitable proxy for cell biomass in such bioassay experiments with *P. antarctica*.

The ratio of 19′-hexanoloxyfucoxanthin (hex) to fucoxanthin (fuco) in the experimental bottles (Fig. 2c) showed a considerable range after 80 h incubation, from ~0.7 to ~3.2, with a general decrease relative to the ratio of ~3.8 in the starting seawater. In contrast to the data trends for nitrate + nitrite and chlorophyll *a*, there were no significant differences in the mean hex/fuco ratio of the low-light versus high-light treatments. However, there were statistically significant differences in the hex/fuco ratio between the high-iron, control-iron, and low-iron treatments after 80 h incubation (one-way ANOVA), with lower hex/fuco ratios observed in the Fe-amended bottles. These differences may reflect an increase in the relative abundance of diatoms following addition of iron (DiTullio and Smith 1996; Sedwick et al. 2000), and/or iron-driven changes in the hex/fuco ratio of *Phaeocystis* cells (Van Leeuwe and Stefels 1998; DiTullio et al., this issue). However, there was no significant drawdown of silicic acid in any of the treatments (data not shown), implying little growth by diatoms. This is consistent with our visual observations, and suggests that colonial *P. antarctica* dominated the algal biomass that accumulated in all incubation bottles. Thus the lower hex/fuco ratios in the Fe-amended bottles most likely reflect changes in the hex/fuco ratio of *Phaeocystis* cells, a conclusion corroborated by the results of pigment analyses from our laboratory dose–response iron-addition experiment (see next section and DiTullio et al., this issue).

Additions of nanomolar concentrations of the iron-chelating ligand DFOB to open-ocean algal assemblages have been shown to reduce the concentration of iron available to many types of phytoplankton (Wells et al. 1994; Wells 1999; Hutchins et al. 1999; Timmermans et al. 2001). However, there is evidence that some algal species can access DFOB-bound iron (Soria-Dengg and Horstman 1995; Hutchins et al. 1999; Maldonado and Price 1999; Eldridge et al. 2004; Maldonado et al. 2005). A potentially important observation from our shipboard experiment is that DFOB, when added to incubation bottles in significant excess (~10×) over dissolved iron, did not appear to limit the growth of colonial *Phaeocystis antarctica* over the course of the experiment. The majority of dissolved Fe in open-ocean surface waters is thought to be complexed by an excess of uncharacterized Fe(III)-binding organic ligands that have conditional stability constants $\left(K'_{\mathrm{FeL,Fe^{3+}}}\right)$ of around 10^{20}–10^{24} M^{-1} (Gledhill and van den Berg 1994; Rue and Bruland 1995,

1997; van den Berg, 1995; Wu and Luther 1995; Boye et al. 2005). A significant fraction of this organically complexed iron is thought to be available to phytoplankton (Rue and Bruland 1997; Hutchins et al. 1999; Maldonado and Price 1999, 2001; Boye et al. 2005; Maldonado et al. 2005). Assuming this to be the case for surface seawater at our Ross Sea station A, then the addition of 3 nM DFOB (which has a $K'_{FeL,Fe^{3+}}$ value similar to or higher than that of Fe-binding ligands in seawater; Rue and Bruland 1995; Maldonado et al. 2005) would have sequestered a significant fraction of the dissolved iron in the DFOB-Fe complex. Thus our experimental results imply that DFOB-bound iron, and potentially other organic-iron complexes, are, to some extent, biologically available to colonial *P. antarctica*. The mechanism by which *Phaeocystis* might access this DFOB-bound Fe remains unknown, but it may reflect the ability of haptophytes to produce specific Fe-binding ligands (C. Trick, personal communication, 2006).

Although the addition of DFOB in excess of dissolved iron had no clear effect on the growth rate of colonial *Phaeocystis antarctica*, the data shown in Fig. 2c suggest that DFOB addition did result in a higher hex/fuco ratio in *P. antarctica* relative to cells grown in the control treatments, presumably in response to complexation of dissolved Fe by the DFOB. This observation is consistent with experimental results reported by Van Leeuwe and Stefels (1998), and with the pigment data from our laboratory dose–response iron-addition experiment (DiTullio et al., this volume), which indicate that the hex/fuco ratio of *P. antarctica* cells may vary significantly in response to changes in the concentration and/or speciation of dissolved iron, even though dissolved iron concentrations are not low enough to limit algal growth rate. Van Leeuwe and Stefels (1998) suggest that *P. antarctica* may convert fuco to hex as a photoprotective mechanism under conditions of decreased iron availability. Thus the higher hex/fuco ratio observed in our DFOB-amended incubation bottles, relative to control samples, may reflect photoprotective adaptation to a decrease in the concentration of readily available dissolved Fe.

In summary, the results of our shipboard iron-light manipulation experiment suggest that: (1) the proximate control on the growth of the colonial *P. antarctica* at station A was irradiance, with higher growth rates observed in colonies exposed to irradiance higher than the range of estimated in-situ values; and (2) the ambient dissolved Fe levels of ~0.38 nM were sufficient to meet the growth requirements of colonial *Phaeocystis* in surface waters at this site. However, one caveat that must be considered here is the relatively short duration (~3 days) of this incubation experiment, which was imposed by our cruise schedule. Results of other shipboard incubation experiments in the Ross Sea (e.g., Martin et al. 1990; Sedwick et al. 2000; Coale et al. 2003) have shown time lags of 2 or more days between iron amendment and biological response, in terms of increases in algal biomass and drawdown in macronutrients. Thus we cannot say definitively that the growth rate of colonial *P. antarctica* collected at station A did not change in response to additions of Fe or DFOB.

Laboratory dose–response iron-addition experiment

As well as the results of the shipboard iron-light experiment described in the preceding section, a preliminary laboratory experiment that we performed using *Phaeocystis* isolated from Ross Sea station 25 provided further evidence that colonial *P. antarctica* are able to access iron from seawater in which DFOB is present in ~10× excess of dissolved iron concentration (Garcia et al., manuscript in preparation). This finding, and evidence for the ability of phytoplankton to access Fe bound by other organic ligands, including EDTA, raise concerns regarding the use of chelating compounds, such as EDTA and DFOB, to maintain low concentrations of biologically available iron in algal culture experiments (e.g., Anderson and Morel 1982; Brand et al. 1983; Sunda et al. 1991; Sunda and Huntsman 1995, 1997). Indeed, Gerringa et al. (2000) have eloquently argued against the use of EDTA or other added ligands to control dissolved inorganic Fe(III) in such experiments.

With this issue in mind, we designed the laboratory iron-addition experiment that is described in the methods section "Laboratory dose–response iron-addition experiment", whereby our starting culture of colonial *Phaeocystis antarctica* was successively diluted with low-iron (~0.17 nM dissolved Fe) station B seawater using stringent trace-metal clean techniques. We thus produced a unialgal culture of predominantly colonial *P. antarctica* in low-iron (~0.2 nM) seawater containing a negligible concentration of EDTA (~0.02 nM), which was then used in a dose–response iron-addition experiment (see the methods section "Laboratory dose–response iron-addition experiment"). To our knowledge, this represents the first successful application of an iron-clean culture experiment using colonial *Phaeocystis antarctica* in a land-based laboratory. A similar method has been applied to unialgal cultures of diatoms in shipboard experiments (Timmermans et al. 2001), and, more recently, in land-based laboratory experiments (Timmermans et al. 2004).

Visual observations indicated that *Phaeocystis* biomass was dominated by nearly spherical colonies in all incubation treatments during the course of our 31-day incubation experiment. Figure 3 shows the time course of dissolved nitrate + nitrite and particulate chlorophyll *a* concentrations in the incubation bottles. After 31 days incubation, the iron-amended bottles (except for the +1.8 nM Fe treatment) displayed a statistically significant decrease (drawdown) in dissolved nitrate + nitrite relative to the control treatments (Fig. 3a), with the +0.6 nM Fe treatments achieving the greatest mean drawdown in nitrate + nitrite (6.5 μM) relative to the starting seawater. Assuming that the drawdown in nitrate + nitrite reflects net accumulation of cellular biomass in the incubation bottles (see section "Interpretation of experimental results"), we conclude that the growth rate of colonial *P. antarctica* in the starting seawater, which contained 0.22 nM dissolved Fe, was limited by iron deficiency. The chlorophyll *a* data (Fig. 3b) also indicate Fe-driven increases in chlorophyll biomass, with the highest mean chlorophyll *a* concentration (5.3 μg l^{-1}) achieved in the +0.6 nM Fe treatment after 31 days. These chlorophyll *a* data are wholly consistent with the nitrate + nitrite results, implying that iron addition did not mediate large changes in cellular chlorophyll *a*. This result supports our assertion that changes in cellular chlorophyll *a* during the shipboard iron-light incubation experiment (see previous section) were largely driven by changes in irradiance.

Our experimental data indicate that the cellular biomass of colonial *Phaeocystis* increased significantly in response to added Fe, with the notable exception of the +1.8 nM Fe treatments, which displayed a lesser biological response than the +0.2 nM and +0.6 nM Fe treatments after 31 days incubation. This unexpected result for the +1.8 nM Fe treatment may reflect an effective solubility limit for dissolved Fe in such experiments, whereby bottles with an initial dissolved Fe concentration greater than ~1–2 nM (in the absence of EDTA) experience a significant loss of biologically available Fe during the incubation period, via precipitation of iron hydroxides (Kuma et al. 1996, 1998; Nakabayashi 2002). Following this reasoning, we suggest that the final concentration of dissolved Fe in the +1.8 nM Fe treatments was perhaps much less than the initial concentration of 2.02 nM. In support of this explanation, we note that the concentration of dissolved Fe rarely exceeds ~1–2 nM in open-ocean surface waters, even in the presence of much higher concentrations of particulate iron (e.g. see Sedwick and DiTullio 1997; Brown et al. 2005; Sedwick et al. 2005). This effective solubility limit for dissolved Fe in ocean surface waters is thought to be determined by the concentration of naturally occurring Fe(III)-binding organic ligands, which allow for dissolved Fe(III) concentrations that far exceed the solubility of iron hydroxides in organic-free seawater (Kuma et al. 1996, 1998; Wu et al. 2001; Liu and Millero 2002; Nakabayashi 2002).

We have estimated nitrate-specific net growth rates (μ_N) of *Phaeocystis* for each experimental treatment between days 25 and 31, using the mean values of nitrate + nitrite drawdown, relative to the starting seawater, at these timepoints. This calculation assumes that: (1) growth was exponential in all bottles between days 25 and 31, and (2) decreases in nitrate + nitrite were directly

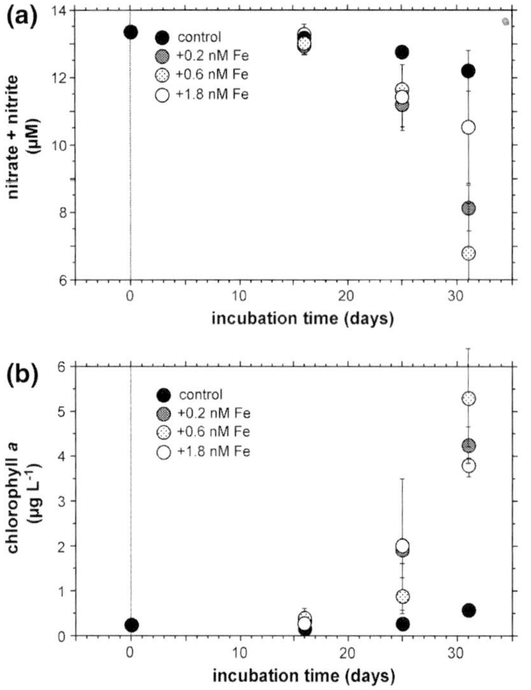

Fig. 3 Concentrations of (**a**) dissolved nitrate + nitrite, (**b**) particulate chlorophyll *a*, in the incubation bottles versus incubation time, from the laboratory dose–response iron-addition experiment. Data are plotted as the mean value of duplicate treatments, with error bars showing ± one standard deviation from the mean

proportional to net increases in the standing stock of colonial *Phaeocystis* cells (see the section "Interpretation of experimental results"). In Fig. 4, we plot these μ_N values against the initial concentration of dissolved Fe in the incubation bottles. In other studies that have used bioassay experiments to examine the relationship between algal growth rate and dissolved Fe concentration (e.g. Coale et al. 1996, 2003; Blain et al. 2002; Timmermans et al. 2001, 2004), the experimental data have been described by a Monod (Michaelis–Menten) saturation function of the form $\mu = \mu_{max}[dFe/(K_\mu + dFe)]$, where μ is the net growth rate, dFe is the dissolved Fe concentration in the experimental treatments, μ_{max} is maximum Fe-replete growth rate, and K_μ is the half-saturation constant for growth with respect to dissolved Fe. An important assumption of these studies is that dissolved Fe concentrations do not change significantly during the course of growout experiments; as discussed above, we suggest that this assumption is not valid for our +1.8 nM Fe treatments. If we exclude results from the +1.8 nM Fe treatments on this basis, then a Monod function can be fitted to the experimental data plotted in Fig. 4.

To do this we use the Eadie–Hofstee linear transformation of the Monod hyperbolic function, which is a relatively robust approach with regard to error-prone data such as our μ_N values, because it gives equal weight to all data points (Zivin and Waud 1982). The Eadie–Hofstee plot of μ_N versus μ_N/dFe yields a least-squares best fit with a slope ($=-K_\mu$) of –0.45 nM, a *y*-intercept ($=\mu_{max}$) of 0.33 d^{-1}, and an r^2 value of 0.85. This exercise yields a half-saturation constant for growth (K_μ) of 0.45 nM dissolved Fe for colonial *Phaeocystis antarctica* grown at an irradiance of ~20 μE m^{-2} s^{-1}. This K_μ estimate is within the range of experimental K_μ estimates that Timmermans et al. (2004) report for four species of Southern Ocean diatoms grown at an irradiance of 60 μE m^{-2} s^{-1} (K_μ = 0.19–1.14 nM Fe), but it is more than an order of magnitude higher than the K_μ values estimated for Ross Sea prymnesiophytes (presumably solitary *P. antarctica*) by Coale et al. (2003). It should be noted that the curve shown in Fig. 4 is constrained by only three data points, and that there are large uncertainties in the μ_N values calculated from the nitrate + nitrite data shown in Fig. 3a. As a result, there are large uncertainties associated with our estimates of K_μ and μ_{max}. Nonetheless, our experimental results clearly demonstrate that colonial *P. antarctica* require relatively high dissolved Fe concentrations—certainly well above 0.2 nM—to achieve maximum growth rates at the relatively low irradiance of ~20 μE m^{-2} s^{-1}.

Conclusions and directions for future research

Our experimental results must be considered in the context of seasonal changes in mean irradiance and dissolved iron concentrations for ocean surface waters in our study region. Smith and van Hilst (2003) estimate mean mixed-layer irradiances of 96 ± 58 μE m^{-2} s^{-1} and 180 ± 110 μE m^{-2} s^{-1} in the southern Ross Sea for the periods November

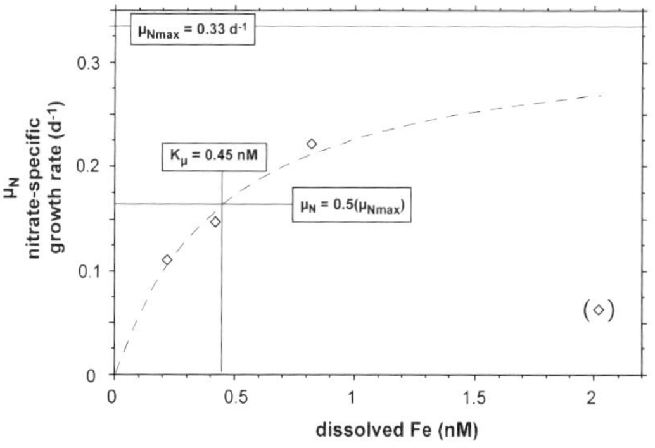

Fig. 4 Nitrate-specific growth rates (μ_N) of conlonial *Phaeocystis* in the incubation bottles between days 25 and 31, versus initial dissolved iron concentrations. The Monod hyperbola (dashed line) was fitted using an Eadie–Hofstee linear transformation ($r^2 = 0.85$), and excludes the +1.8 nM Fe datum (in parentheses). The half-saturation constant for growth (K_μ) and maximum nitrate-specific growth rate (μ_{Nmax}) are indicated

10–December 8, 1994 (mid spring–early summer) and December 20–January 15, 1995 (mid summer), respectively. Over these same periods, dissolved iron concentrations in the upper water column averaged 1.0 and 0.23 nM, respectively (Sedwick et al. 2000). It is also important to acknowledge the antagonistic relationship between the iron and light requirements of phytoplankton, whereby cellular iron requirements are expected to decrease as mean irradiance increases (Raven 1990; Sunda and Huntsman 1997). Thus, in the Ross Sea polynya, we would expect the cellular iron requirements of colonial *Phaeocystis* to decrease during the growing season, as the surface mixed-layer shoals and mean irradiance increases, and as dissolved iron concentrations decrease.

The results of our two experiments are consistent with these seasonal changes in the availability of light and iron, given the interrelated influences of irradiance and iron availability on phytoplankton growth rate. The data from our laboratory dose–response iron-addition experiment indicate a relatively high iron requirement for colonial *Phaeocystis* at an irradiance of ~20 μE m^{-2} s^{-1}, a value that is representative of the mean irradiance in the mixed layer of the southern Ross Sea during early spring (Smith et al. 2000; Smith and van Hilst 2003; Hiscock 2004). At that time, dissolved Fe concentrations are likely to exceed our estimated half-saturation constant of 0.45 nM dissolved Fe (Sedwick et al. 2000; Coale et al. 2005), thus the growth of colonial *P. antarctica* should not be strongly limited by iron availability. However, we would expect colonial *Phaeocystis* to have significantly lower iron requirements at the higher irradiance levels used in our shipboard iron-light manipulation experiment, which was conducted in late December. In the southern Ross Sea during mid summer, integrated clear-sky irradiance at the sea surface estimated from the model of Gregg and Carder (1990) is around 640 μE m^{-2} s^{-1} (Hiscock 2004), thus we estimate irradiance on the order of 100 μE m^{-2} s^{-1} and 300 μE m^{-2} s^{-1} for our low-light and high-light treatments, respectively. The expectation of lower iron requirements of *Phaeocystis* under such irradiance levels (compared with the much lower irradiance of 20 μE m^{-2} s^{-1} used in our laboratory dose–response experiment) are borne out by the results of our shipboard experiment, since we observed no evidence for growth limitation or iron stress at the ambient dissolved Fe concentration of 0.38 nM.

Boyd (2002) speculated that the growth of *Phaeocystis* in the Ross Sea may be limited by iron availability from spring through late summer, and by low irradiance from autumn through early spring, with the potential for colimitation by iron and light during spring and early fall. We propose a slightly different scenario, whereby the impact of decreasing Fe availability during the spring is mitigated by the increase in irradiance, which significantly lowers the cellular iron requirements of *Phaeocystis*, thus allowing blooms to develop through the spring and into the summer. Eventually, however, dissolved Fe concentrations decrease to such low levels as to limit the growth rate of colonial *Phaeocystis*, even under relatively high mean irradiance, and at this point the blooms are terminated. To test this conceptual model requires an understanding of the effect of irradiance on the iron requirements of colonial *P. antarctica*. Towards this goal, we are currently undertaking further dose–response iron-addition experiments with our *P. antarctica* culture at higher levels of irradiance.

With regard to quantitative simulations of the Ross Sea ecosystem, we note that recent numerical models (Arrigo et al. 2003; Tagliabue and Arrigo 2005) have parameterized the growth of colonial *Phaeocystis* using a half-saturation constant of 0.01 nM for iron, which is more than an order of magnitude less than the K_μ value estimated from our laboratory experiment. Even allowing for a significant decrease in K_μ at the higher irradiance levels during late spring and into summer, it seems likely that the growth requirements of colonial *Phaeocystis* are closer to 0.1 nM dissolved Fe, similar to values estimated for Southern Ocean diatoms by Timmermans et al. (2001, 2004). The cellular C/Fe uptake ratio is another critical parameter for such numerical models, with a value of 450,000 (estimated from model diagnostics) used in the recent modelling study of Tagliabue and Arrigo (2005). The measurement of C/Fe uptake ratios in colonial *Phaeocystis* represents a formidable methodological challenge, because dissolved Fe is likely to be partitioned into both the *Phaeocystis* cells and surrounding mucus, and the extent to which mucus-bound Fe is available to the cells is unknown (Schoemann et al. 2005). Techniques that have been used to remove extracellular Fe for cellular Fe uptake measurements (e.g., Hudson and Morel 1989; Tovar-Sanchez et al. 2003; Tang and Morel 2006) might not be applicable to colonial *Phaeocystis*, and sophisticated analytical tools such as the synchrotron X-ray fluorescence microprobe (Twining et al. 2003) may be required to provide accurate estimates of cellular C/Fe uptake ratios.

Acknowledgements The authors gratefully acknowledge the officers and crew of the RV *Nathaniel B. Palmer*, and Raytheon Polar Services Company personnel for their outstanding support in the field. We thank two anonymous reviewers for their helpful criticisms and suggestions. This work has benefited from discussions with Veronique Schoemann, Walker Smith, and Alessandro Tagliabue. Funding for this research was provided by US National Science Foundation grants OPP-0230559 (PNS), OPP-0230513 (GRD) and DGE-0139313 (NSG through M. Vansickle and G. Tempel).

References

Anderson MA, Morel FMM (1982) The influence of aqueous iron chemistry on the uptake of iron by the coastal diatom *Thalassiosira weissflogii*. Limnol Oceanogr 27:789–813

Arrigo KR, van Dijken GL (2004) Annual changes in sea ice, chlorophyll a, and primary production in the Ross Sea, Antarctica. Deep-Sea Res Part II 51:117–138

Arrigo KR, Tagliabue A (2005) Iron in the Ross Sea, Part II: impact of discrete iron addition strategies. J Geophys Res 110 (C03010). DOI: 10.1029/2004JC002568

Arrigo KR, Weiss AM, Smith WO (1998) Physical forcing of phytoplankton dynamics in the southwestern Ross Sea. J Geophys Res 103(C1):1007–1022

Arrigo KR, Robinson DH, Worthen DL, Dunbar RB, DiTullio GR, VanWoert M, Lizotte MP (1999) Phytoplankton community structure and the drawdown of nutrients and CO_2 in the Southern Ocean. Science 283:365–367

Arrigo KR, Worthen DL, Robinson DH (2003) A coupled ocean-ecosystem model of the Ross Sea. Part 2: iron regulation of phytoplankton taxonomic variability and primary production. J Geophys Res 108(C7):3231. DOI: 10.1029/2001JC000856

Blain S, Sedwick PN, Griffiths FB, Queguiner B, Bucciarelli E, Fiala M, Pondaven P, Treguer P (2002) Quantification of algal iron requirements in the Subantarctic Southern Ocean (Indian sector). Deep-Sea Res Part II 49:3255–3273

Boyd PW (2002) Environmental factors controlling phytoplankton processes in the Southern Ocean. J Phycol 38:844–861

Boye M, Nishioka J, Croot PL, Laan P, Timmermans KR, de Baar HJW (2005) Major deviations of iron complexation during 22 days of a mesoscale iron enrichment in the open Southern Ocean. Mar Chem 96:257–271

Brand LE, Sunda WG, Guillard RRL (1983) Limitation of marine phytoplankton reproductive rates by zinc, manganese and iron. Limnol Oceanogr 28:1182–1198

Brown MT, Landing WM, Measures CI (2005) Dissolved and particulate Fe in the western and central North Pacific: results from the 2002 IOC cruise. Geochem Geophys Geosys 6. DOI: 10.1029/2004GC000893

Coale KH, Fitzwater SE, Gordon RM, Johnson KS, Barber RT (1996) Control of community growth and export production by upwelled iron in the equatorial Pacific Ocean. Nature 379:621–624

Coale KH, Wang X, Tanner SJ, Johnson KS (2003) Phytoplankton growth and biological response to iron and zinc addition in the Ross Sea and Antarctic Circumpolar Current along 170°W. Deep-Sea Res Part II 50:635–653

Coale KH, Gordon RM, Wang X (2005) The distribution and behavior of dissolved and particulate iron and zinc in the Ross Sea and Antarctic circumpolar current along 170°W. Deep-Sea Res Part I 52:295–318

De Baar HJW, Buma AGJ, Nolting RF, Cadée GC, Jacques G, Tréguer PJ (1990) On iron limitationof the Southern Ocean: experimental observations in the Weddell and Scotia Seas. Mar Ecol Progr Ser 65:105–122

DiTullio GR, Geesey ME (2002) Photosynthetic pigments in marine algae and bacteria. In: Bitton G (ed) The encyclopedia of environmental microbiology. John Wiley & Sons, pp 2453–2470

DiTullio GR, Smith WO (1995) Relationship between dimethylsulfide and phytoplankton pigment concentrations in the Ross Sea, Antarctica. Deep-Sea Res Part I 42:873–892

DiTullio GR, Smith WO Jr (1996) Spatial patterns in phytoplankton biomass and pigment distributions in the Ross Sea. J Geophys Res 101:18467–18477

DiTullio GR, Hutchins DA, Bruland KW (1993) Interaction of iron and major nutrients controls phytoplankton growth and species composition in the tropical north Pacific Ocean. Limnol Oceanogr 38:495–508

DiTullio GR, Grebmeier JM, Arrigo KR, Lizotte MP, Robinson DH, Leventer A, Barry JP, VanWoert ML, Dunbar RB (2000) Rapid and early export of *Phaeocystis* antarctica blooms in the Ross Sea, Antarctica. Nature 404:595–598

DiTullio GR, Garcia NS, Riseman SF, Sedwick PN (2007) Effects of iron concentration on pigment composition in *Phaeocystis antarctica*. Biogeochemistry (this issue)

Eldridge ML, Trick CG, Alm MB, DiTullio GR, Rue EL, Bruland KW, Hutchins DA, Wilhelm SW (2004) Phytoplankton community response to a manipulation of bioavailable iron in HNLC waters of the subtropical Pacific Ocean. Aquat Microb Ecol 35:79–91

Fitzwater SE, Coale KH, Gordon RM, Johnson KS, Ondrusek ME (1996) Iron deficiency and phytoplankton growth in the Equatorial Pacific. Deep-Sea Res Part II 43:995–1015

Fitzwater SE, Johnson KS, Gordon RM, Coale KH, Smith WO (2000) Trace metal concentrations in the Ross Sea and their relationship with nutrients and growth. Deep-Sea Res Part II 47:3159–3179

Gerringa LJA, de Baar HJW, Timmermans KR (2000) A comparison of iron limitation of phytoplankton in natural oceanic waters and laboratory media conditioned with EDTA. Mar Chem 68:335–346

Gibson JAE, Garrick RC, Burton HR, McTaggart AR (1990) Dimethylsulfide and the alga *Phaeocystis pouchetii* in Antarctic coastal waters. Mar Biol 104:339–346

Gledhill M, van den Berg CMG (1994) Determination of complexation of iron(III) with natural organic complexing ligands in seawater using cathodic stripping voltammetry. Mar Chem 47:41–54

Gordon LI, Codispoti LA, Jennings JC Jr, Millero FJ, Morrison JM, Sweeney C (2000) Seasonal evolution of hydrographic properties in the Ross Sea, Antarctica, 1996–1997. Deep-Sea Res Part II 47:3095–3117

Gregg WW, Carder KL (1990) A simple spectral solar irradiance model for cloudless maritime atmospheres. Limnol Oceanogr 35:1657–1675

Guillard RRL, Hargraves PE (1993) *Stichochrysis immobilis* is a diatom, not a chrysophyte. Phycologia 32:234–236

Hiscock MR (2004) The regulation of primary productivity in the Southern Ocean. Ph.D. Thesis, Duke University, 174 pp

Hong Y, Smith WO, White A-M (1997) Studies on transparent exopolymer particles (TEP) produced by *Phaeocystis* sp. (*Prymnesiophyceae*) in the Ross Sea, Antarctica. J Phycol 33:368–376

Hudson RJM, Morel FMM (1989) Distinguishing between extracellular and intracellular iron in marine-phytoplankton. Limnol Oceanogr 34:1113–1120

Hutchins DA, Franck VM, Brzezinski MA, Bruland KW (1999) Inducing phytoplankton iron limitation in iron-replete coastal waters with a strong chelating ligand. Limnol Oceanogr 44(4):1009–1018

Hutchins DA, Sedwick PN, DiTullio GR, Boyd PW, Quéguiner B, Griffiths FB, Crossley C (2001) Control of phytoplankton growth by iron and silicic acid availability in the subantarctic Southern Ocean Experimental results from the SAZ project. J Geophys Res 106(C12):31559–31572

Knap A, Michaels A, Close A, Ducklow H, Dickson A (1996) Protocols for the Joint Global Ocean Flux Study (JGOFS) core measurements. JGOFS Report 19, 170 pp

Kuma K, Nishioka J, Katsuhiko M (1996) Controls on iron(III) hydroxide solubility in seawater: the influence of pH and natural organic chelators. Limnol Oceanogr 41:396–407

Kuma K, Katsumoto A, Kawakami H, Takatori F, Matsunaga K (1998) Spatial variability of Fe(III) hydroxide solubility in the water column of the northern North Pacific Ocean. Deep-Sea Res Part I 45:91–113

Liu X, Millero FJ (2002) The solubility of iron in seawater. Mar Chem 77:43–54

Maldonado MT, Price NM (1999) Utilization of iron bound to strong organic ligands by plankton communities in the subarctic Pacific Ocean. Deep-Sea Res Part II 46:2447–2473

Maldonado MT, Price NM (2001) Reduction and transport of organically bound iron by *Thalassiosira oceanica* (*bacillariophyceae*). J Phycol 37:298–310

Maldonado MT, Strzepek RF, Sander S, Boyd PW (2005) Acquisition of iron bound to strong organic complexes, with different Fe binding groups and photochemical reactivities, by plankton communities in Fe-limited subantarctic waters. Global Biogeochem Cycles 19. DOI: 10.1029/2005GB002481

Martin JH, Fitzwater SE, Gordon RM (1990) Iron deficiency limits plankton growth in Antarctic waters. Global Biogeochem Cycles 4:5–12

Matrai PA, Vernet M, Hood R, Jennings A, Brody E, Saemundsdottir S (1995) Light dependence of carbon and sulfur production by polar clones of the genus *Phaeocystis*. Mar Biol 124:157–167

Mathot S, Smith WO Jr, Carlson CA, Garrison DL (2001) Estimate of *Phaeocystis* sp. carbon biomass: methodological problems related to the mucilaginous nature of the colonial matrix. J Phycol 36:1049–1056

Measures CI, Yuan J, Resing JA (1995) Determination of iron in seawater by flow injection analysis using in-line preconcentration and spectrophotometric detection. Mar Chem 50:3–12

Moisan TA, Mitchell BG (1999) Photophysiological acclimation of *Phaeocystis antarctica* Karsten under light limitation. Limnol Oceanogr 44:247–258

Nakabayashi S, Kuma K, Sasaoka K, Saitoh S, Mochizuki M, Shiga N, Kusakabe M (2002) Variation in iron(III) solubility and iron concentration in the northwestern North Pacific Ocean. Limnol Oceanogr 47:885–892

Olson RJ, Sosik HM, Chekalyuk AM, Shalapyonok A (2000) Effects of iron enrichment on phytoplankton in the Southern Ocean during late summer: active fluorescence and flow cytometric analyses. Deep-Sea Res Part II 47:3179–3200

Palmisano AC, SooHoo JB, SooHoo SL, Kottmeier ST, Craft LL, Sullivan CW (1986) Photoadaptation in *Phaeocystis pouchetii* advected beneath annual sea ice in McMurdo Sound, Antarctica. J Plankton Res 8:891–906

Raven JA (1990) Predictions of Mn and Fe use efficiencies of phototrophic growth as a function of light availability for growth and C assimilation pathway. New Phytol 116:1–17

Robinson DH, Arrigo KR, DiTullio GR, Lizotte MP (2003) Evaluating photosynthetic carbon fixation during *Phaeocystis antarctica* blooms. In: DiTullio GR, Dunbar RB (eds) Biogeochemistry of the Ross Sea. Antarct Res Ser 78:77–91

Rue EL, Bruland KW (1995) Complexation of iron(III) by natural organic ligands in the Central North Pacific as determined by a new competitive ligand equilibration/adsorptive cathodic stripping voltammetric method. Mar Chem 50:117–138

Rue EL, Bruland KW (1997) The role of organic complexation on ambient iron chemistry in the equatorial Pacific Ocean and the response of a mesoscale iron addition experiment. Limnol Oceanogr 42:901–910

Schoemann V, Becquevort S, Stefels J, Rousseau V, Lancelot C (2005) *Phaeocystis* blooms in the global ocean and their controlling mechanisms: a review. J Sea Res 53:43–66

Sedwick PN, DiTullio GR (1997) Regulation of algal blooms in Antarctic shelf waters by the release of iron from melting sea ice. Geophys Res Lett 24:2515–2518

Sedwick PN, Edwards PR, Mackey DJ, Griffiths FB, Parslow JS (1997) Iron and manganese in surface waters of the Australian subantarctic region. Deep-Sea Res Part I 44:1239–1253

Sedwick PN, DiTullio GR, Mackey DJ (2000) Iron and manganese in the Ross Sea, Antarctica: seasonal iron limitation in Antarctic shelf waters. J Geophys Res 105(C5):11321–11336

Sedwick PN, Blain S, Queguiner B, Griffiths FB, Fiala M, Bucciarelli E, Denis M (2002) Resource limitation of phytoplankton growth in the Crozet Basin, Subantarctic Southern Ocean. Deep-Sea Res Part II 49:3327–3349

Sedwick PN, Church TM, Bowie AR, Marsay CM, Ussher SJ, Achilles KM, Lethaby PJ, Johnson RJ, Sarin MM, McGillicuddy DJ (2005) Iron in the Sargasso Sea (Bermuda Atlantic Time-series Study region) during summer: Eolian imprint, spatiotemporal variability, and ecological implications. Global Biogeochem Cycles 19. DOI: 10.1029/2004GB002445

Smith WO Jr, Gordon LI (1997) Hyperproductivity of the Ross Sea (Antarctica) polynya during austral spring. Geophys Res Lett 24:233–236

Smith WO, van Hilst CM (2003) Effects of assemblage composition on the temporal dynamics of carbon and nitrogen uptake in the Ross Sea. In: DiTullio GR, Dunbar RB (eds) Biogeochemistry of the Ross Sea. Antarct Res Ser 78:197–208

Smith WO Jr, Codispoti LA, Nelson DM, Manley T, Buskey EJ, Niebauer HJ, Cota GF (1991) Importance of *Phaeocystis* blooms in the high latitude ocean carbon cycle. Nature 352:514–516

Smith WO, Marra J, Hiscock MR, Barber RT (2000) The seasonal cycle of phytoplankton biomass and primary productivity in the Ross Sea, Antarctica. Deep-Sea Res Part II 47:3119–3140

Smith WO Jr, Dennett MR, Mathot S, Caron DA (2003) The temporal dynamics of the flagellated and colonial stages of *Phaeocystis antarctica* in the Ross Sea. Deep-Sea Res Part II 50:605–617

Soria-Dengg S, Horstmann U (1995) Ferrioxamines B and E as iron sources for the marine diatom *Phaeodactylum tricornutum*. Mar Ecol Progr Ser 127:269–277

Stefels J, van Leeuwe MA (1998) Effects of iron and light stress on the biochemical composition of antarctic *Phaeocystis* sp. (prymnesiophyceae). I. Intracellular dmsp concentrations. J Phycol 34:486–495

Sunda WG, Swift DG, Huntsman SA (1991) Low iron requirement for growth in oceanic phytoplankton. Nature 351:55–57

Sunda WG, Huntsman SA (1995) Iron uptake and growth limitation in oceanic and coastal phytoplankton. Mar Chem 50:189–206

Sunda WG, Huntsman SA (1997) Interrelated influence of iron, light and cell size on marine phytoplankton growth. Nature 390:389–392

Sweeney C (2003) The annual cycle of surface CO_2 and O_2 in the Ross Sea: a model for gas exchange on the continental shelves of Antarctica. In: DiTullio GR, Dunbar RB (eds) Biogeochemistry of the Ross Sea. Antarct Res Ser 78:295–312

Sweeney C, Smith WO, Hales B, Hansell DA, Carlson CA, Codispoti LA, Gordon LI, Millero FJ, Takahashi T (2000) Nutrient and TCO_2 uptake and export ratios in the Ross Sea. Deep-Sea Res Part II 47:3395–3422

Tagliabue A, Arrigo KR (2005) Iron in the Ross Sea, Part I: impact on CO_2 fluxes via variation in phytoplankton functional group and non-Redfield stoichiometry. J Geophys Res 110(C03009). DOI: 10.1029/2004JC002531

Tang D, Morel FMM (2006) Distinguishing between cellular and Fe-oxide-associated trace elements in phytoplankton. Mar Chem 98:18–30

Timmermans KR, Gerringa LJA, De Baar HJW, van der Wagt B, Veldhuis MJW, De Jong JTM, Croot PL (2001) Growth rates of large and small Southern Ocean diatoms in relation to availability of iron in natural seawater. Limnol Oceanogr 46:260–266

Timmermans KR, van der Wagt B, de Baar HJW (2004) Growth rates, half-saturation constants, and silicate, nitrate, and phosphate depletion in relation to iron availability of four large, open-ocean diatoms from the Southern Ocean. Limnol Oceanogr 49:2141–2151

Tovar-Sanchez A, Sanudo-Wilhelmy SA, Garcia-Vargas M, Weaver RS, Popels LC, Hutchins DA (2003), A trace metal clean reagent to remove surface-bound iron from marine phytoplankton. Mar Chem 82:91–99

Twining BS, Baines SB, Fisher NS, Maser J, Vogt S, Jacobsen C, Tovar-Sanchez A, Sanudo-Wilhelmy SA (2003) Quantifying trace elements in individual aquatic protist cells with a synchrotron X-ray fluorescence microprobe. Anal Chem 75:3806–3816

Van den Berg CMG (1995) Evidence for organic complexation of iron in seawater. Mar Chem 50:139–157

Van Hilst CM, Smith WO (2002) Photosynthesis/irradiance relationships in the Ross Sea, Antarctica and their control by phytoplankton assemblage composition and environmental factors. Mar Ecol Progr Ser 226:1–12

Van Leeuwe MA, Stefels J (1998) Effects of iron and light stress on the biochemical composition of Antarctic *Phaeocystis* sp. (Prymnesiophyceae). II. Pigment Composition. J Phycol 34:496–503

Wells ML (1999), Manipulating iron availability in nearshore waters. Limnol Oceanogr 44:1002–1008

Wells ML, Price NM, Bruland KW (1994) Iron limitation and the cyanobacterium *Synechococcus* in equatorial Pacific waters. Limnol Oceanogr 39:1481–1486

Worthen DL, Arrigo KR (2003) A coupled ocean-ecosystem model of the Ross Sea. Part 1: Interannual variability of primary production and phytoplankton community structure. In: DiTullio GR, Dunbar RB (eds) Biogeochemistry of the Ross Sea. Antarct Res Ser 78:93–105

Wu J, Luther GW III (1995) Complexation of Fe(III) by natural organic ligands in the Northwest Atlantic Ocean by a competitive ligand equilibration method and a kinetic approach. Mar Chem 50:159–177

Wu J, Boyle E, Sunda W, Wen L-S (2001) Soluble and colloidal iron in the oligotrophic North Atlantic and North Pacific. Science 293:847–849

Zivin JA, Waud DR (1982) How to analyze binding, enzyme and uptake data: the simplest case, a single phase. Life Sci 30:1407–1422

Biogeochemistry (2007) 83:99–118
DOI 10.1007/s10533-007-9078-2

REVIEW

The carbohydrates of *Phaeocystis* and their degradation in the microbial food web

Anne-Carlijn Alderkamp · Anita G. J. Buma · Marion van Rijssel

Received: 18 November 2005 / Accepted: 23 March 2006 / Published online: 19 April 2007

Abstract The ubiquity and high productivity associated with blooms of colonial *Phaeocystis* makes it an important contributor to the global carbon cycle. During blooms organic matter that is rich in carbohydrates is produced. We distinguish five different pools of carbohydrates produced by *Phaeocystis*. Like all plants and algal cells, both solitary and colonial cells produce (1) structural carbohydrates, (hetero) polysaccharides that are mainly part of the cell wall, (2) mono- and oligosaccharides, which are present as intermediates in the synthesis and catabolism of cell components, and (3) intracellular storage glucan. Colonial cells of *Phaeocystis* excrete (4) mucopolysaccharides, heteropolysaccharides that are the main constituent of the mucous colony matrix and (5) dissolved organic matter (DOM) rich in carbohydrates, which is mainly excreted by colonial cells. In this review the characteristics of these pools are discussed and quantitative data are summarized. During the exponential growth phase, the ratio of carbohydrate-carbon (C) to particulate organic carbon (POC) is approximately 0.1. When nutrients are limited, *Phaeocystis* blooms reach a stationary growth phase, during which excess energy is stored as carbohydrates. This so-called overflow metabolism increases the ratio of carbohydrate-C to POC to 0.4–0.6 during the stationary phase, leading to an increase in the C/N and C/P ratios of *Phaeocystis* organic matter. Overflow metabolism can be channeled towards both glucan and mucopolysaccharides. Summarizing the available data reveals that during the stationary phase of a bloom glucan contributes 0–51% to POC, whereas mucopolysaccharides contribute 5–60%. At the end of a bloom, lysis of *Phaeocystis* cells and deterioration of colonies leads to a massive release of DOM rich in glucan and mucopolysaccharides. Laboratory studies have revealed that this organic matter is potentially readily degradable by heterotrophic bacteria. However, observations in the field of accumulation of DOM and foam indicate that microbial degradation is hampered. The high C/N and C/P ratios of *Phaeocystis* organic matter may lead to nutrient limitation of microbial degradation, thereby prolonging degradation times. Over time polysaccharides tend to self-assemble into hydrogels. This may have a profound effect on carbon cycling, since hydrogels provide a vehicle to move DOM up the size spectrum to sizes subject to sedimentation. In

A.-C. Alderkamp · A. G. J. Buma · M. van Rijssel
Department of Marine Biology, Centre for Ecological and Evolutionary Studies, University of Groningen, P.O. Box 14, 9750 AA Haren, The Netherlands

Present Address:
A.-C. Alderkamp (✉)
Department of Geophysics, Stanford University, Stanford, CA 94305-2215, USA
e-mail: alderkam@stanford.edu

addition, it changes the physical nature and microscale structure of the organic matter encountered by bacteria which may affect the degradation potential of the *Phaeocystis* organic matter.

Keywords Carbohydrates · Glucan · Hydrogels · Mucilage · *Phaeocystis* · TEP

Introduction

Phaeocystis (Prymnesiophyceae) is a marine phytoplankton genus with a worldwide distribution. It is present from tropical to polar regions and can form blooms in areas with high nutrient concentrations, such as coastal or upwelling regions (Baumann et al. 1994; Lancelot et al. 1998; Schoemann et al. 2005). One of the characteristics of *Phaeocystis* species is the complex polymorphic life cycle. It consists of different stages of free-living cells 3–9 μm in diameter, alternating with colonial stages usually reaching several mm in diameter (Kornmann 1955; Rousseau et al. 1994; Peperzak et al. 2000), sometimes up to 3 cm (Chen et al. 2002). Although free-living *Phaeocystis* cells are ubiquitous in the world's oceans, it is the colonial form that has attracted most attention due to its massive blooms (Lancelot et al. 1987; Schoemann et al. 2005). At present six *Phaeocystis* species have been described (Zingone et al. 1999) of which only three are reported to form blooms; *P. globosa* in temperate regions (English Channel, coastal North Sea, China), *P. pouchetii* in temperate to polar waters of the Northern hemisphere, and *P. antarctica* in (sub)polar waters of the southern hemisphere (Schoemann et al. 2005). During blooms *Phaeocystis* may contribute over 90% to the total phytoplankton abundance and can be responsible for up to 65% of the local annual primary production (Joiris et al. 1982; Veldhuis et al. 1986b; Lancelot and Mathot 1987; Arrigo et al. 1999). The high productivity associated with blooms and its ubiquity make *Phaeocystis* an important contributor to the global carbon cycle (Smith et al. 1991; Arrigo et al. 1999; DiTullio et al. 2000; Schoemann et al. 2005). A significant part of the *Phaeocystis* carbon consists of carbohydrates (Lancelot and Mathot 1987; Rousseau et al. 1990; Fernández et al. 1992; Mathot et al. 2000). This vast carbohydrate injection following *Phaeocystis* blooms has a significant impact on the carbon cycling in the ecosystem and the microbial food web. Carbohydrates are omnipresent in the marine ecosystem, as they are an important constituent of marine DOM (Benner et al. 1992) and particles and therefore play a key role in the marine carbon cycle. Carbohydrates display a remarkable structural diversity as a consequence of the wide variety in aldoses and the different glycosidic bonds between them. These differences in structure and size influence the behavior of polysaccharides in aqueous solution, i.e., solubility and aggregate formation, as well as microbial degradation potential. To understand the impact of the carbohydrate injection after a *Phaeocystis* bloom on the carbon cycling in the ecosystem, it is important to consider the different types of carbohydrates that are produced. In the following sections we will first discuss the current state of knowledge with respect to the characteristics and quantities of different pools of carbohydrates produced during the various life cycle stages. Secondly, mechanisms of DOM release during and following a bloom will be discussed, and finally, the fate of carbohydrates in the microbial food web.

Phaeocystis carbohydrates and their characteristics

Carbohydrates in algae and plants are often classified based on methodological discrimination. The structural carbohydrates are not water-soluble, whereas the other types of carbohydrates are water-soluble and typically extracted by hot water. In *Phaeocystis* five different pools of carbohydrates can be distinguished. Like all algal and plant cells, both solitary and colonial cells produce (1) structural carbohydrates, polysaccharides that are mainly part of the cell wall, (2) mono- and oligosaccharides, which are present as intermediates in the synthesis and catabolism of cell components, and (3) intracellular storage glucan. Colonial cells of *Phaeocystis* excrete (4) mucopolysaccharides, heteropolysaccharides that are the main constituent of the mucous colony matrix and (5) dissolved organic matter (DOM) rich in carbohydrates, which is excreted mainly by colonial cells. A schematic on the procedure to extract the different types of particulate carbohydrates is depicted in Fig. 1.

Structural polysaccharides

In most algae and plants, structural carbohydrates are water-insoluble polysaccharides that are mainly present as cell wall constituents. Although *Phaeocystis* cells do not have a prominent cell wall, flagellate *Phaeocystis* cells are covered with body scales consisting of flat plates with a pattern of radiating ridges (Parke et al. 1971; Moestrup 1979; Hallegraeff 1983; Baumann et al. 1994). In two other haptophytes, *Pleurochrysis scherfelii* and *Chrysochromulina chiton*, the scales contain acidic, sulfated pectin-like polysaccharides, cellulose and glycopeptides (reviewed by Leadbeater 1994). Likely, *Phaeocystis* scales contain similar polysaccharides.

In addition to cellular structural carbohydrates, flagellate cells of *Phaeocystis* excrete filaments composed of α-chitin, an *N*-acetyl-D-glucosamine polymer, forming star-like structures. In electron microscopic studies, five-rayed filamentous stars were observed around flagellate cells of *P. pouchetii*, *P. antarctica*, *P. globosa*, and *P. cordata*, whereas nine-rayed stars were observed around *P. scrobiculata* (Parke et al. 1971; Moestrup 1979; Baumann et al. 1994; Chretiennot-Dinet et al. 1997; Zingone et al. 1999). The biological function of the chitin filaments remains a mystery up to now (Chretiennot-Dinet et al. 1997).

The only quantitative data available on the contribution of structural carbohydrates to *Phaeocystis* biomass are from a culture of *Phaeocystis* sp. in the stationary phase, where structural

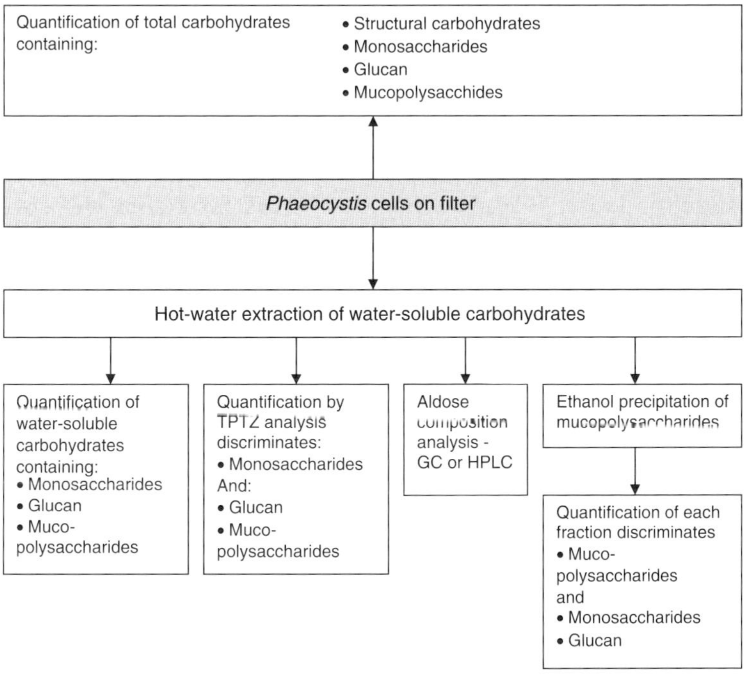

Fig. 1 Schematic representation of procedures to obtain different types of particulate carbohydrates from *Phaeocystis* cells concentrated on a filter. Total carbohydrates can directly be quantified colorimetrically. Water-soluble carbohydrates are typically extracted in hot water. They can be quantified colorimetrically. The 2,4,6-tripyridyl-s-triazine (TPTZ)-analysis (Myklestad et al. 1997) discriminates monosaccharides from larger carbohydrates. The aldose composition of water-soluble carbohydrates can be analyzed by gas chromatography (GC) or anion-exchange high-pressure liquid chromatography (AE-HPLC). Mucopolysaccharides can be separated from glucan and monosaccharides by ethanol precipitation (Alderkamp et al. 2006a), followed by colorimetric quantification of each fraction

carbohydrates made up for 13% of the total organic carbon (TOC) (Biersmith and Benner 1998).

Mono- and oligosaccharides

Monosaccharides are the first products of photosynthesis and form the precursors for biosynthesis of all organic molecules, whereas oligosaccharides are intermediates in polysaccharide synthesis. During catabolism of storage glucan, monosaccharides are formed as intermediates (Granum and Myklestad 1999). In addition to intermediates in metabolic processes, monosaccharides serve a role in osmoregulation (Craigie 1974; Dickson and Kirst 1987a, b). The monosaccharide concentration in phytoplankton cells is quite stable; for example in diatoms monosaccharide concentration was hardly influenced by light (Vårum et al. 1986; Van Oijen et al. 2003). During the growth phase of a bloom of *P. globosa* in the Marsdiep, The Netherlands, the monosaccharides made up for 28–60% of water-extractable particulate carbohydrates (Alderkamp, unpublished data). This ratio was similar to an Antarctic community dominated by diatoms (Van Oijen et al. 2003).

Storage glucan

The principle food reserve in plants and algae is an intracellular food reserve in the form of a glucan, whereas some energy can be stored in lipids. Storage glucans may be α-1,4-glucans like starch or glycogen, or β-1,3-glucans like laminarin. The actual type of glucan is an important taxonomic characteristic. The class of Prymnesiophyceae contains β-1,3-glucan (Van den Hoek et al. 1993). β-1,3-glucans have an essentially linear β-D-1,3-glucan backbone with some branching at position 6 or 2 (Percival 1970; Craigie 1974; Myklestad 1978; Paulsen and Myklestad 1978). They can be classified as laminarin (also spelled laminaran), paramylon, or chrysolaminaran, distinguished mainly on the basis of chain length and presence of mannitol end groups. Laminarin has a degree of polymerization (DP-number of aldoses in the molecule) of approximately 20, contains mannitol-end groups, and is water-soluble. It is found in brown algae. Paramylon has a DP of more then 50, and does not contain mannitol end-groups or side chains. It is water-insoluble, and is found in euglenophytes (Manners et al. 1966). Chrysolaminaran (also called leucosin) has a DP of 16–38, contains no mannitol end-groups and is water-soluble. It is found in chrysophytes, haptophytes and diatoms (Myklestad and Haug 1972; Craigie 1974; Paulsen et al. 1978; Myklestad 1978, 1988). A different glucan structure has been found in the prymnesiophycean *Emiliania huxleyi*, that consists of a β-D-(1,6)-glucan backbone with branching at position 3 (Vårum et al. 1986). *Phaeocystis* cells contain chrysolaminaran as a storage glucan (Janse et al. 1996b). Chrysolaminaran is located in vacuoles, as was observed microscopically in *Phaeocystis* (Kornmann 1955; Parke et al. 1971; Chang 1984) and using immunolocalization of β-1,3-D-glucan in cell sections of the diatoms *Phaeodactylum tricornutum*, *Cilindrotheca fusiformis* and *Thalassiosira pseudonana* (Chiovitti et al. 2004). Due to its presence in ubiquitous bloom-forming phytoplankton chrysolaminaran it is considered to be one of the most abundant types of carbohydrate in the marine environment (Painter 1983).

Chrysolaminaran was characterized as principal storage glucan in a non-colony-forming strain of *P. globosa* that contained mainly non-flagellated cells (Janse et al. 1996b). In this strain, 95% of the water-extractable carbohydrates consisted of glucose and all glucose polymers were digested by the enzyme laminarinase (E.C. 3.2.1.6 from *Trichoderma* sp.). This enzyme preparation contained endo-β-1–3 glucanase activity and some chitinase, cellulase and α-amylase activity. Laminarin from the brown seaweed *Laminaria digitata* was digested to monosaccharides by this enzyme preparation. In colony-forming strains of *P. globosa*, however, no more than 40–70% of glucose polymers were digested (Van Rijssel et al. 2000; Alderkamp et al. 2006a). The reason for this partial digestion may be differences in the type of chrysolaminaran produced by single cells and colonial cells, e.g., with respect to polymer size or number of branches. Therefore, chrysolaminaran from colonial cells may not be fully degraded by the enzyme laminarinase. Alternatively, there may be a different type of glucan produced by colonial *Phaeocystis* cells.

In all plants photosynthetic rates exceed metabolic demands when irradiance is optimal. The excess energy thus harvested is stored mainly as glucan. This glucan can be used as a respiratory substrate to maintain growth when irradiance is suboptimal, for example during the night. Chrysolaminaran is produced in the light and its production is dependent on light intensity. Nocturnal respiration of chrysolaminaran provides energy to maintain cell metabolism and its carbon skeletons can be used for protein synthesis (Cuhel et al. 1984; Lancelot and Mathot 1985; Janse et al. 1996b; Granum et al. 1999; Granum and Myklestad 2001; Van Oijen et al. 2004b; Alderkamp et al. 2006a). Diel rhythms in storage carbohydrate concentrations have been well documented in laboratory cultures of diatoms (Hitchcock 1980; Vårum et al. 1986; Van Oijen et al. 2004b), field populations of pelagic marine phytoplankton dominated by diatoms (Barlow 1982; Hama and Handa 1992, Van Oijen et al. 2004b) and mesocosms dominated by *P. pouchetii* (Alderkamp et al. 2006a). Notably the diel variations in carbohydrate concentrations are a direct result of the variations in light intensity, but diel carbohydrate dynamics may also be the result of an intrinsic circadian clock (Post et al. 1985; Falkowski and Raven 1997). This was shown by Loogman (1982), who found that the diel dynamics in carbohydrate concentrations introduced by a diel light cycle, continued for several days after the light was switched to a continuous regime.

Under normal bloom conditions, *Phaeocystis* will experience varying light conditions during the day. On top of the diel light cycle, vertical mixing in the water column may mix colonies down to suboptimal light conditions, or even below the euphotic zone (Joint and Pomroy 1993; Arrigo et al. 1999; Wassmann et al. 2000). A capability to utilize stored energy is therefore likely to be important during the day as well as during the night. This may be relevant for *Phaeocystis*, since *P. antarctica* blooms occur in regions with deep wind-mixing (Arrigo et al. 1999), and *P. globosa* blooms often occur in turbid waters with high mixing rates, such as the coastal North Sea and estuaries. During sudden transitions from high to low light levels, carbohydrates can provide energy until cells are acclimatized to the new situation, as was shown in the diatom *Thalassiosira weisflogii* (Post et al. 1985). This was, however, not confirmed during a mesocosm study of *P. pouchetii*. Carbohydrate concentrations were unaffected by placing cells at low irradiances or in the dark during the day (Alderkamp et al. 2006a).

Upon nutrient limitation excess energy and carbon tend to accumulate as storage glucan, because the Calvin cycle and the tricarboxylic acid (TCA) cycle are uncoupled. Carbohydrate accumulation has been observed during the stationary growth phase of a single cell culture (Janse et al. 1996b), colonial cultures (Van Rijssel et al. 2000), and a bloom of *P. globosa* (Veldhuis et al. 1986a), and at the end of *P. pouchetii* blooms in mesocosms (Alderkamp et al. 2006a). This so-called overflow metabolism under nutrient limitation is common in phytoplankton (Myklestad 1974, 1988). It leads to an increased contribution of carbohydrates to the organic matter produced during bloom and an increased C/N ratio. Whereas the C/N ratio of exponentially growing free-living *Phaeocystis* cells and colonies is 4–7 (reviewed by Schoemann et al. 2005), C/N ratios as high as 20.5 were reported in nutrient limited mesocosms of *P. pouchetii* (Verity et al. 1988), 17.5 during blooms of *P. pouchetii*, and 30 during blooms of *P. globosa* at low ambient inorganic nitrogen concentrations (Baumann et al. 1994). Energy and carbon produced during overflow metabolism in *Phaeocystis* can be channeled towards both glucan and mucopolysaccharides. During the stationary phase of a *P. pouchetii* bloom in mesocosms the ratio glucan/mucopolysaccharides increased (Alderkamp et al. 2006a), whereas it decreased during the stationary phase of cultures of colonial *P. globosa* (Van Rijssel et al. 2000).

Blooms of *P. antarctica* occur in regions where trace metals, notably iron, are limiting phytoplankton growth (Martin et al. 1990; Coale et al. 2003, Schoemann et al. 2005). The effect of iron limitation on the overflow production by *P. antarctica* is largely unknown. Since iron limitation directly affects the photosynthetic process (Geider and La Roche 1994) the production of carbohydrates is likely to be hampered, thereby decreasing the overflow metabolism. Indeed, the diurnal production of water-extractable carbohydrates was reduced under iron limitation in the Antarctic diatom *Chaetoceros brevis* (Van Oijen et al. 2004a) and in an iron-limited phytoplankton community in the Southern Ocean, dominated by diatoms (Van Oijen et al. 2005). In accordance with these findings, only a minor increase in C/N ratio of *P. antarctica* was reported upon nutrient limitation in enclosures in the Ross Sea, where iron limitation is likely to occur (Smith et al. 1998). Therefore, it is unlikely that the overflow production under iron limitation will be as dramatic as under nitrogen or phosphorus limitation.

From the above it is clear that the concentration of storage glucan in phytoplankton is highly variable; it is dependent on the time of the day, light intensity and nutrient status of the cell. To evaluate the contribution of glucan to *Phaeocystis* carbon, quantitative data were compiled on the contribution of glucose to the aldose composition of particulate water-extractable carbohydrates (Fig. 1; Table 1a). Assuming that mucopolysaccharides consist of 10% glucose (Janse et al. 1999), the remaining glucose can be attributed to glucan. This leads to a highly variable contribution of glucan to the particulate water-extractable carbohydrate fraction of *Phaeocystis* colonies. During the exponential growth phase glucan contributes 14–85% and during the stationary phase 0–85%, both in culture and field samples. Alternatively, glucan and mucopolysaccharides were separated based on polymer size differences by ethanol precipitation (Fig. 1). During the growth phase of a *P. pouchetii* bloom in a mesocosm, glucan contributed 46–78% to particulate water-extractable carbohydrates, depending on the time of the day. This increased to 82% during the stationary phase of the bloom (Alderkamp et al. 2006a). To compute the contribution of glucan to colonial *Phaeocystis* POC, the carbohydrate-C/POC ratio of Alderkamp et al. (2006a) was used (Table 2). This leads to a contribution of glucan-C to POC of 2–11% during exponential growth, dependent on the time of the day. During the stationary phase the contribution ranges between 0% and 51%. The results are summarized in Table 3. In a non-colony-forming, non-flagellate strain of *P. globosa*, chrysolaminaran contributed 40% to TOC during the stationary phase (Janse et al. 1996b).

Mucopolysaccharides

The mucus matrix of *Phaeocystis* colonies consists predominantly of hetero-polysaccharides (Guillard and Hellebust 1971; Painter 1983; Lancelot and Mathot 1985; Eberlein et al. 1985; Van Boekel 1992). The mucopolysaccharides consist of at least nine different aldoses, are water-soluble, and polymers are larger than 1,000 kDa (Janse et al. 1999). Qualitative staining of the colony matrix indicated the absence of lipophilic compounds and chitin, but the presence of amino groups (Hamm et al. 1999). Matrix-assisted laser desorption/ionization (MALDI) analysis showed that N was not proteineous, but probably present in aminosugars (Solomon et al. 2003). The presence of the aminosugar sialic acid was confirmed by aminolabeling (Orellana et al. 2003). The presence of aminosugars may affect the C/N ratio of mucopolysaccharides, leading to a small difference between C/N ratios of single cells and colonial cells (Verity et al. 1988). Verity et al. (1991) concluded that the C/N ratio of the mucus constituents is lower than 9, whereas Biersmith and Benner (1998) measured a C/N ratio of 15.6 in the fraction containing mucopolysaccharides.

The aldose composition of *P. globosa* and *P. pouchetii* carbohydrates has been studied in bloom samples and in strains from different origins (Table 1a). The main non-glucose aldoses are arabinose, galactose, mannose and xylose. In addition, a minor contribution of rhamnose was detected in all samples, whereas fucose, ribose, *O*-methylated hexoses, *O*-methylated pentoses and glucuronic acid were detected in some samples only. The glucosamine detected by two groups may have been derived from copepod chitin contamination in the samples. The aldoses detected in *Phaeocystis* material are the same as those detected in other phytoplankton extracellular carbohydrates (Hoagland et al. 1993; Biersmith and Benner 1998), however, the ratio of constituents seems specific for *Phaeocystis*. There were no major differences in the presence or ratio of non-glucose aldoses between a bloom sample

Table 1a Sugar composition of *Phaeocystis* carbohydrates

Species	% total	% non-glucose sugars												
	glc	ara	fuc	gal	glcA	lyx	man	rib	rha	xyl	*O*mh	*O*mp	heA	glcN	Reference
P. pouchetii dominated mesocosm	26–40	1–8	9–30	13–30			10–31		6–10	15–38				2–6	Metaxatos et al. (2003)
P. globosa colonial cultures, four strains from the North Sea	7–85	25–32		12–25	0–3		12–24		7–12	11–12	0–2	0–3		NP	Van Rijssel et al. (2000)
P. pouchetii bloom, Balsfjord	45	35		10	5		15	2	8	17	2	6			Janse et al. (1999)
Phaeocystis sp. culture	27	13	6	35		2	12	2.5	11	20					Biersmith and Benner (1998)
P. globosa bloom, North Sea	20–75	32–43		17–25	2–8		12–20		3–8	11–20	2–4	0.5–2		NP	Janse et al. (1996a)
Phaeocystis sp. bloom, English Channel		P		P	P				P	P				P	Painter (1983)

ara: arabinose, fuc, fucose, gal: galactose, glc: glucose, lyx, lyxose, man: mannose, rha: rhamnose, rib: ribose, xyl: xylose, gla: glucuronic acid, glcN: glucosamine, *O*mp: *O*-methylated pentoses, *O*mh: *O*-methylated hexoses, Hea: hexuronic acid, glN: glucosamine, P: present, NP: not present and included in analysis

Table 1b Sugar composition of *Phaeocystis* extracolonial DOM[a]

Species	% total carbohydrates													
	glc	ara	fuc	gal	glcA	lyx	man	rib	rha	xyl	*O*mh	*O*mp	heA	glcN	Reference
Phaeocystis sp. culture	4	17	2	16			30	5	21	6					Aluwihare and Repeta (1999)
Phaeocystis sp. culture	17	13	4	28		3	15	1	6	13					Biersmith and Benner (1998)
P. pouchetii cultures, two strains	4.5–30	0–5.2		2–5.6			27–64	3–8	16–20	1.5–3			4–5.5		Guillard and Hellebust (1971)

Note: Abbreviations as in Table 1a

[a] Although classified as DOM, these studies include carbohydrates derived from mucopolysaccharides as DOM fractions were obtained by filtration with the application of pressure

Table 2 Contribution of carbohydrates to POC of *Phaeocystis*

Species	Growth stage	Carbohydrate to POC	Reference
P. pouchetii mesocosm	Exponential	13%	Alderkamp et al. (2006a)
P. pouchetii mesocosm	Stationary	60%	Alderkamp et al. (2006a)
P. globosa culture	Exponential	10%	Van Rijssel et al. (2000)
P. globosa bloom	No info	19–35%	Van Rijssel et al. (1997)
Phaeocystis sp. culture	Stationary	38%	Biersmith and Benner (1998)
Phaeocystis sp. culture	All	20–30%	Biddanda and Benner (1997)
P. globosa single cell culture	Stationary	40%	Janse et al. (1996b)

Table 3 Contribution of glucan-C to POC of *Phaeocystis* colonies

Method	*Phaeocystis* species	Growth phase	Glucan-C to POC	Reference
Aldose composition	*P. pouchetii* and *P. globosa*	Exponential	2–11%	Compiled in this study
		Stationary	0–51%	
Ethanol precipitation	*P. pouchetii*	Exponential	6–10%	Alderkamp et al. (2006a)
		Stationary	49–60%	

from Balsfjord, Norway, dominated by *P. pouchetii* and a sample from the North Sea dominated by *P. globosa* (Janse et al. 1999). No data are available for other species of *Phaeocystis*, including *P. antarctica*. In five different strains of *P. globosa* differences in the ratio of non-glucose constituents were small (Van Rijssel et al. 2000). Nutrient status of the cells does not seem to influence the ratio, as was shown in a batch culture of *P. globosa* (Van Rijssel et al. 2000) or in bloom samples taken at different stages of a bloom (Janse et al. 1996a). The change in colony shape and appearance at the end of *Phaeocystis* blooms (e.g., Verity et al. 1988; Rousseau et al. 1994) is therefore not reflected in the aldose composition of the mucopolysaccharides. This is different from the situation reported in diatoms, where upon nutrient limitation, the benthic diatom *Cylindrotheca closterium* excretes a separate extracellular polysaccharide with a different monosaccharide composition compared to optimal growth conditions (De Brouwer et al. 2002; Underwood et al. 2004).

Phaeocystis acts as a secretory cell; mucopolysaccharides are excreted in a way similar to animal cells and vascular plants in which the regulated export of cellular products takes place via exocytosis. While stored in secretory vesicles inside the cells, the mucopolysaccharides remain in condensed phase. Upon exocytosis, they undergo a characteristic phase transition accompanied by extensive swelling, resulting in the formation of hydrated gels (Chang 1984; Chin et al. 2004). The negatively charged mucopolysaccharides become entangled and form ionic links, such as calcium bridges (Van Boekel 1992). They then form a thin, yet strong, semipermeable colony matrix, with a pore size between 1 and 4.4 nm diameter, with plastic and to a limited extent also elastic properties (Kornmann 1955; Hamm et al. 1999). Electron microscope studies of a *Phaeocystis* sp. from Tasman Bay, New Zealand, show that the matrix has a multilayered gelatinous structure, with an average thickness of 5 μm (Chang 1984). Van Rijssel et al. (1997) calculated a thickness of 7 μm based on carbohydrate measurements and the assumption of a mucous layer with a fixed thickness for various colony sizes, whereas a thickness of 10 μm was determined based on confocal laser scanning imagery. As colonial cells of *Phaeocystis* have similar (Baumann et al. 1994), or even higher growth rates (Veldhuis et al. 2005) compared to single cells, the allocation of the material needed to build the colonial matrix obviously does not hamper cell metabolism.

In addition to a structural function in the mucous matrix, mucopolysaccharides were also suggested to serve as storage polysaccharides as nocturnal consumption of extracellular polysaccharides was reported in *P. globosa* (Lancelot and Mathot 1985; Veldhuis and Admiraal 1985). In both studies cells were separated from mucous matrix by filtration under pressure. Since this treatment may cause cells to break, thus releasing intracellular substances (see above), nocturnal consumption of intracellular chrysolaminaran may have been misreported as consumption of extracellular substances. Using the same method to separate extracellular material from cells, Matrai et al. (1995) observed no nocturnal utilization of extracellular polymers in an Antarctic strain of *Phaeocystis*. Also, when mucopolysaccharides and glucan were separated based on polymer size differences, no utilization of mucopolysaccharides was found when a culture of *P. globosa* was placed in the dark for prolonged times (Alderkamp et al. 2006a). Given the size and structure of mucopolysaccharides it seems unlikely that they serve as storage carbohydrate. Degradation and uptake of these complex polysaccharides would require multiple extracellular enzymes as well as transport systems to take up the different monosaccharides and oligomers. Moreover, since the gel matrix is such a definite structure stabilized by ionic bridges, it will be hard to disentangle polymers from the gel.

Contribution of mucopolysaccharides-C to POC

The contribution of mucus carbon to *Phaeocystis* organic carbon can be assessed in three different ways. In the first, cells are mechanically separated from the mucus matrix by filtration through GF/C filters. Fractionation by filtration yields contributions of 5–80% of mucus carbon to POC (reviewed by Riegman and Van Boekel 1996). As many have pointed out, however, *Phaeocystis* cells are fragile and may be disrupted by filtration under pressure (Veldhuis and Admiraal 1985; Van Rijssel et al. 1997; Mathot et al. 2000). By releasing their water soluble carbon the fraction of mucus-carbon may be overestimated. On the other hand, mucopolysaccharides may form gels on the filter (Chin et al. 1998) as a result of which their contribution may be underestimated. Therefore, all studies in which colonies were fractionated by filtration or centrifugation have to be interpreted with care.

When assuming that carbohydrates are the only significant source of carbon in the mucus matrix, the second method consists of determination of the contribution of mucopolysaccharides to the water-extractable carbohydrates. Mucopolysaccharides are estimated as the combined sum of all non-glucose aldoses and 10% of the glucose. Based on the aldose composition of water extractable carbohydrates (Fig. 1), the contribution of mucopolysaccharides to the water extractable particulate carbohydrates is 15–86% during the exponential growth phase, and 15–100% during the stationary phase (Table 1a and references therein). Alternatively, glucan and mucopolysaccharides were separated based on polymer size (Fig. 1; Alderkamp et al. 2006a). During the exponential growth phase, mucopolysaccharides contributed 38% to the water extractable carbohydrates, decreasing to 9% during the stationary phase. Based on the carbohydrate-C to carbon ratios of Alderkamp et al. (2006a; Table 2) the contribution of mucopolysaccharides to POC is estimated as 2–11% during exponential growth and 5–60% during the stationary phase.

The third method to estimate the carbon of colony matrices is based on conventional conversion factors for determining the carbon content of the average cell biovolume of *Phaeocystis* cells and an additional carbon content of the mucus matrix. In *P. antarctica*, Mathot et al. (2000) found on average 14% of POC devoted to matrix formation, increasing to a maximum of 33%. In *P. globosa* Rousseau et al. (1990) using a colony-volume-based model, calculated up to 90% of POC devoted to matrix formation in large colonies. Van Rijssel et al. (2000), assuming a hollow colony surrounded by a thin mucus layer and using a colony-surface-based model, calculated a contribution up to 35%. Comparisons of estimates obtained by these authors, however, show similar POC contributions for colony sizes predominantly observed in the field (reviewed by Schoemann et al. 2005). The estimates for the contribution of mucous matrix-C to POC are summarized in Table 4.

Extra-colonial DOM

Extra-colonial DOM rich in carbohydrates was reported in cultures of *Phaeocystis* (Table 5 and references therein). Carbohydrates similar in aldose composition to that of mucopolysaccharides form a major contribution to this DOM

Table 4 Contribution of mucous matrix-C to POC of *Phaeocystis* colonies

Method	*Phaeocystis* species	Growth phase	Matrix-C to POC	Reference
Mechanical filtration	*P. pouchetii* and *P. globosa*	All	5–80%	Riegman and Van Boekel (1996)
Aldose composition	*P. pouchetii* and *P. globosa*	Exponential	2–11%	Compiled in this study
		Stationary	10–60%	
Ethanol precipitation	*P. pouchetii*	Exponential	3–5%	Alderkamp et al. (2006a)
		Stationary	5–13%	
Model	*P. antarctica*	Exponential	14 ± 4%	Mathot et al. (2000)
		Stationary	Up to 33%	
Model	*P. globosa*	Exponential (lab.)	Up to 10%	Van Rijssel et al. (1997)
		All (field)[a]	Up to 35%	
Model	*P. globosa*	All[b]	Up to 90%	Rousseau et al. (1990)

[a] Highest values at low ambient nitrate concentrations

[b] Size dependent

Table 5 Characteristics of *Phaeocystis* DOM

Species	Growth stage	carbohydrates to DOC	C/N ratio	DOC to TOC	Reference
Phaeocystis sp. (culture)	Exponential through stationary	25–50%	19.3 (HMW) 6.3 (LMW)	27%	Biddanda and Benner (1997)
Phaeocystis sp. (culture)	Stationary	56%	19.3 (HMW)	14%	Biersmith and Benner (1998)
Phaeocystis sp. (culture)	Late exponential	>67% (HMW)	12 (HMW)		Aluwihare and Repeta (1999)
P. globosa (culture)	Late exponential		8.3 (axenic) 7.4 (xenic)		Solomon et al. (2003)
P. antarctica (culture)	Late exponential		6.9 (axenic) 6.2 (xenic)		Solomon et al. (2003)

(Table 1b). There seem to be two different pools of DOM; (1) a high-molecular-weight (HMW)-DOM pool with a high C/N ratio of over 12 and a carbohydrate content of over 67% and (2) a low-molecular-weight (LMW)-DOM pool with a low C/N ratio of 6.3. Since all studies used filtration techniques applying pressure, it is most likely that the DOM also contained mucopolysaccharides. The overall C/N ratio of the DOM is low, around 7 (Biddanda and Benner 1997; Solomon et al. 2003). Similar to mucopolysaccharides, extra-colonial polymers self assemble in hydrated polymer gels on a time scale of approximately two days (Solomon et al. 2003).

High DOM concentrations have been reported in natural assemblages dominated by *Phaeocystis* (Eberlein et al. 1985), and high DOM excretion rates in cultures (Guillard and Hellebust 1971). Upon careful filtration, however, true extra-colonial DOM excretion was only 14% of primary production in the late exponential phase of a *P. globosa* culture (Veldhuis and Admiraal 1985; Veldhuis et al. 1986a; Biersmith and Benner 1998), and 2–14% of primary production in natural assemblages dominated by *P. globosa* (Laanbroek et al. 1985; Veldhuis et al. 1986a; Lancelot and Mathot 1987). Similar values were observed for *P. pouchetii* (1–19%; Verity et al. 1991), and during a bloom of *P. antarctica* (11%; Carlson et al. 1998).

Mechanisms of DOM release during a *Phaeocystis* bloom

In general 10–25% of phytoplankton primary production is released as DOC (Sharp 1977; Larsson and Hagström 1979; Lignell 1990; Teira et al. 2003; Marañón et al. 2004). The three main mechanisms of DOM production are: (1) release after cell lysis, e.g., as a result of viral infection, zooplankton sloppy feeding and egestion, (2) bacterial cleavage of exopolysaccharides (Smith et al. 1995), or (3) direct release as a physiological process of intact cells (reviewed by Nagata 2000). Two physiological mechanisms of DOM production have been advocated: a passive diffusion of small metabolites through the cell membrane (Bjørnsen 1988) and an overflow mechanism whereby excess photosynthetic products are actively released when the C fixation rate exceeds the rate of macromolecular production (Fogg 1983).

During the exponential growth phase of a *Phaeocystis* bloom DOM release as a result of non-physiological processes will be low. In a *P. globosa* bloom rates of cell lysis were low in the growth phase (Van Boekel et al. 1992; Brussaard et al. 1995, 1996, 2005a), therefore DOM release as a result of cell lysis will be insignificant. Since *Phaeocystis* colonies largely escape grazing control (Breteler and Koski 2003; Nejstgaard et al. 2007), sloppy feeding will not be a cause of significant DOM release either. In a diatom bloom it was shown that bacteria colonized healthy, actively growing *Chaetoceros* and *Thalassiosira* cells with high ectoenzyme activity. When the hydrolysis rate of phytoplankton surface material by bacteria exceeds the uptake rate, DOM will be released into the environment (Smith et al. 1995). Although the growth phase of *P. globosa* blooms are typically accompanied by an increase in bacterial numbers, bacterial production, and enzyme activity (Billen and Fontigny 1987; Van Boekel et al. 1992; Putt et al. 1994; Becquevort et al. 1998; Arrieta and Herndl 2002; Brussaard et al. 1996, 2005b), young healthy colonies are generally free of bacteria (Thingstad and Billen 1994; Lancelot and Rousseau 1994). Therefore, it is unlikely that hydrolysis of the mucous matrix by bacteria will lead to significant DOM production during the growth phase of the bloom. The only significant mechanisms for DOM production is probably direct release of DOM from healthy cells. During all growth stages DOM excretion from healthy cells was observed in cultures of *Phaeocystis* sp. (Biddanda and Benner 1997). Since DOM excretion rates are low (see above) on the whole DOM release during the growth phase of the bloom will be low.

During the stationary phase of a bloom, however, DOM release increases. During the stationary phase colonies become disrupted, and flagellate cells develop within the colonies (Peperzak et al. 2000). Due to the overflow production of mucopolysaccharides and glucan, the ratio of carbohydrates-C to POC increases. During this phase of a bloom DOM production through non-physiological processes will be high. Cell lysis rates of up to 33% day^{-1} were observed at the peak of *P. globosa* blooms (Van Boekel et al. 1992; Brussaard et al. 1995, 1996, 2005a). In this way the carbohydrate rich cell content is released as DOM. When the flagellate cells that develop inside colonies detach from the colony matrix, they are readily grazed upon by microzooplankton (Weisse and Scheffel-Möser 1990; Tang et al. 2001). This leads to DOM release by egestion of the microzooplankton. During the stationary phase of a bloom, bacterial communities attach to *Phaeocystis* colonies and to ghost colonies. Bacteria with high glucosidase activity are present in both free-living and particle-attached communities (Billen and Fontigny 1987; Becquevort et al. 1998). If hydrolysis rates of mucopolysaccharides from the colony matrix exceeds uptake of the sugars, the sugars and oligomers derived from mucopolysaccharides may be released as DOM.

The excess release of DOM under nutrient limitation has not been observed in *Phaeocystis*. In *Phaeocystis* sp. cultures DOC production was coupled to POC production at different growth stages (Biddanda and Benner 1997). Also in enclosures of *P. antarctica*, ratios of recently produced DOC increased only slightly after nutrient depletion (Smith et al. 1998; Carlson et al. 1998). This indicates that nutrient limitation does not lead to increased excretion of DOM resulting from an overflow mechanism.

Formation of hydrogels by *Phaeocystis* carbohydrates

Polysaccharides in aqueous environments may spontaneously assemble into hydrogels. These polymer gels can then penetrate neighboring gels, and anneal them together, increasing the size spectrum. Recent discoveries have highlighted the importance of polymer gels for biogeochemical cycling of organic carbon (see the review by Verdugo et al. 2004). As was discussed above, mucopolysaccharides spontaneously form polyanionic gels in aqueous solution. In addition to gel formation by mucopolysaccharides, β-1,3-glucans are also known to form non-ionic gels in aqueous solutions (Renn 1997; Kim et al. 2000). In general, gelling properties of β-glucans are dependent on the concentration of the molecules, their size, primary and secondary structure and temperature (Bohm and Kulicke 1999; Lazaridou et al. 2003). Therefore, gel-forming properties of laminarin in the marine environment are hard to predict. Preliminary results from our laboratory show that dissolving laminarin in seawater leads to an increase in viscosity, but to our knowledge no further investigations have been conducted to this end. Seuront et al. (2007) showed that *Phaeocystis* derived polymeric matter could increase seawater viscosity up to two orders of magnitude. Because of the high aggregation potential of both glucan and mucopolysaccharides, the carbohydrates released during senescence of the *Phaeocystis* bloom will likely contribute to the increase in seawater viscosity (Fig. 2), thereby changing the physical nature and microscale structure of the organic matter encountered by bacteria.

Gels provide an abiotic vehicle to move organic molecules up the size spectrum to sizes capable of sedimentation, and may also constitute an intermediate in the formation of the foam often observed after *Phaeocystis* blooms (Fig. 2). Large gel particles, also known as transparent exopolymeric particles (TEP) can contribute significantly to sedimentation processes. This may be by direct sedimentation (Wassmann et al. 1990; Riebesell et al. 1995; Hong et al. 1997), or by acting as glue, adhering together organic matter present during a bloom and thus enhancing their contribution to a vertical flux (Passow and Wassmann 1994). Mari et al. (2005) showed that the properties of TEP produced under N or P limitation are different, as a result of which the export versus retention balance of the organic matter may be affected. It is tempting to speculate that the different N/P ratios may determine whether overflow metabolism is directed towards either glucan or mucopolysaccharide production. This would resemble the situation described for the diatom *Cylindrotheca fusiformis*, where P depletion leads to an increase in the ratio of heteropolysaccharides to glucan, compared to N-limitation (Magaletti et al. 2004). In this way the glucan/mucopolysaccharides ratio may influence properties of the hydrogels and TEP that is

Fig. 2 Conceptual representation of the formation of marine gels from different pools of *Phaeocystis* organic matter (after Verdugo et al. 2004). (abbreviations: LMW-DOM: low-molecular-weight DOM, HMW-DOM: high-molecular-weight DOM, POM: particulate organic matter, TEP: transparent exopolymeric particles)

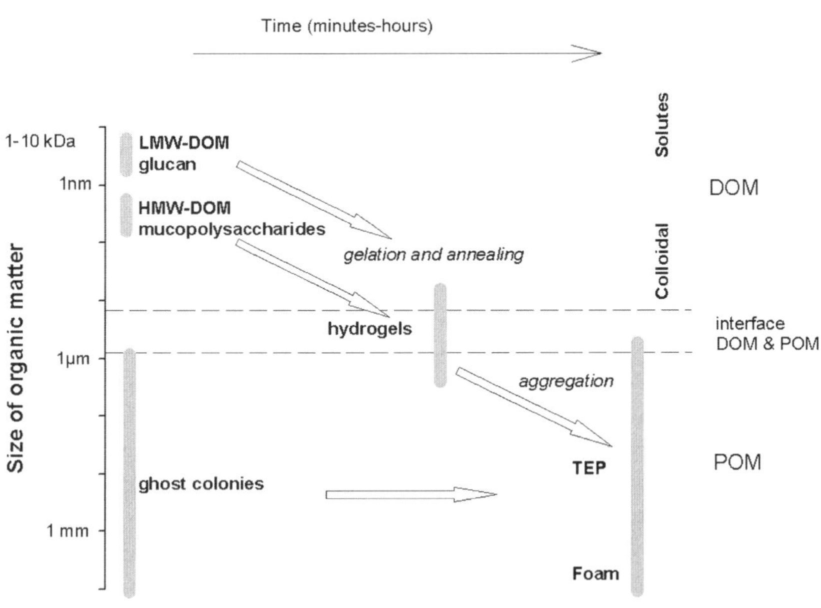

formed, and thereby the degradation potential of *Phaeocystis* organic matter.

Microbial degradation

Generally, most of the biomass produced in *Phaeocystis* blooms is remineralized in the water column (Wassmann 1994, Rousseau et al. 2000) by heterotrophic bacteria. Occasionally, however, accumulation of DOM and foam have been reported (e.g. Eberlein et al. 1985; Lancelot et al. 1987; Seuront et al., 2007). This seems to indicate that bacterial degradation of organic matter produced by in *Phaeocystis* blooms varies between blooms, and sometimes is hampered. There may be various reasons for slow degradation and they may be different in different stages of a bloom.

Microbial degradation of *Phaeocystis* carbohydrates

During the growth phase of a bloom of *P. antarctica* all DOM produced was found to be rapidly degraded by bacteria (Smith et al. 1998). With time, however, the composition and contribution of organic matter that can be utilized by bacteria will change from readily degradable freshly excreted DOM during the growth phase, to the DOM dominated by mucopolysaccharides and glucan during the senescent stage. In laboratory experiments it was shown that carbohydrates derived from *P. globosa* and *P. pouchetii* colonies were readily degraded by bacterial communities under both oxic and anoxic conditions (Osinga et al. 1997; Janse et al. 1999). The degradation rate of glucan was higher than that of mucopolysaccharides (Osinga et al. 1997), but during degradation of the mucopolysaccharides the sugar composition of the mucopolysaccharides remained unchanged. Therefore, there is no indication of refractory parts within the mucopolysaccharide fraction (Janse et al. 1999).

Identification of bacteria involved in mucopolysaccharide degradation in cultures and a mesocosm revealed the presence of bacteria from different phylogenetic groups, such as α- and γ-Proteobacteria, *Cytophaga-Flavobacter* (CF) cluster of the Bacteroidetes and the *Planktomycetes* and *Verrucomicrobiales* clade (Janse et al. 2000; Brussaard et al. 2005b; Alderkamp et al 2006b). During a *P. globosa* bloom in a mesocosm the bacterial community showed distinct changes with time (Brussaard et al. 2005b). The nature of the phytoplankton growth controlling substrate (i.e., N or P) did not influence the microbial community. In addition, since the species composition of the bacterial community remained unchanged during mucopolysaccharide degradation, degradation of mucopolysaccharides in itself does not require a succession of species (Janse et al. 2000). Therefore, the changes in the microbial community were likely induced by the growth stage of the *P. globosa* bloom and enhanced cell lysis and concomitant release of DOM during the senescent stage (Brussaard et al. 2005b). Degradation of polymers requires specialized extracellular enzyme systems to hydrolyze polymers prior to uptake by bacteria. These enzymes can be attached to the bacterial cell (ectoenzymes), or free extracellular enzymes (Chróst 1991). Arrieta and Herndl (2002) revealed a succession of different types of bacterial β-glucosidases during a *P. globosa* bloom, which was related to a change in the bacterial community. Therefore, the copious amounts of glucan and mucopolysaccharides produced during a *Phaeocystis* bloom may shape the composition of bacterial communities specialized in degradation of complex carbohydrates. Bacteria appearing during the senescent stage of the *P. globosa* bloom in the mesocosm mainly belonged to the α-Proteobacteria and CF cluster of the Bacteroidetes (Brussaard et al. 2005b). In the coastal North Sea, bacteria from CF clusters of the Bacteroidetes were dominant during and following a bloom of *P. globosa* (A. C. Alderkamp et al. 2006b). Members of the CF cluster are chemo-organotrophic, and known for their capacity to degrade complex carbohydrates such as pectin, cellulose and chitin (Reichenbach and Dworkin 1991; Cottrell and Kirchman 2000; Kirchman 2002). Therefore, it may be that certain taxa of this cluster have adapted specifically to degradation of complex mucopolysaccharides produced by *Phaeocystis*.

Mineral nutrient limitation of microbial degradation has been put forward as an explanation for accumulation of carbon-rich DOM after a *Phaeocystis* bloom (Thingstad and Billen 1994). The increase in carbohydrate/POC due to overflow metabolism will give rise to a substrate with a C/P and C/N ratio that is unfavorable to bacteria. Since the C/P ratio of bacteria may be considerably lower than that of phytoplankton (Vadstein et al. 1988), especially phosphate limitation may hamper microbial degradation (Thingstad et al. 1997). In addition, since *P. antarctica* was found to remove more CO_2 per mole of phosphate than do diatoms (Arrigo et al. 1999), phosphate limitation during bacterial degradation of *Phaeocystis* material may be more severe than during degradation of diatom material. Indeed, while bacterial growth was carbon limited during the growth phase of a bloom of *P. globosa*, bacterial growth was colimited by carbon and phosphate during the massive release of the glucan and mucopolysaccharides at the end of the bloom (Kuipers and Van Noort, in preperation). This may impede degradation of carbohydrates that are otherwise easily degradable, and prolong the degradation times.

Microbial degradation of hydrogels

Marine gels likely have an effect on microbial degradation of organic matter, yet few studies have examined this effect directly. Based on the current knowledge, gel formation may either enhance or slow down degradation of organic matter. In general HMW-DOC and gels have a higher rate of biodegradation and turnover than LMW-DOC (Amon and Benner 1996). Gels may serve as nutrient and/or attachment surfaces and offer hot-spots of high substrate concentrations for bacteria. They form enclosed microenvironments of substrate for bacteria to colonize, release their extracellular enzymes and efficiently take up the degradation products at high local concentrations (Azam 1998; Azam and Long 2001). Indeed, following a *P. globosa* bloom bacteria attached to particles were more active than the free-living community (Becquevort et al. 1998). On the other hand, assembly of free polymers into gels may complicate degradation. Enzymatic degradation of carbohydrates that form insoluble gels was approximately 50 times slower than degradation of polymers composed of the same monomer linkages, but not forming gels (Alderkamp et al. 2007). If this difference is exemplary of the difference in degradation potential of 'free' polymers and polymers embedded in a gel, turnover times may slow down from days to years.

Microbial degradation dissolves the hydrogels when hydrolysis rates exceed the carbon demand of the bacteria, thereby releasing LMW products from the hydrogels (Smith et al. 1992; Fig. 3). Moreover, the kinetic properties of bacterial enzyme systems may result in the release of oligomers at a higher rate than the release of monomers that can be directly incorporated by the bacteria, resulting in release of LMW and medium molecular-weight (MMW) products (Alderkamp et al. 2007). In addition, polymer photocleavage by ultraviolet radiation can shorten polymer chains making tangled networks unstable (Edwards 1986) and can readily disperse marine polymer gel matrices (Orellana and Verdugo 2003). These three processes produce a random fragmentation, yielding LMW products that may be readily incorporated and metabolized by bacteria (Mopper et al. 1991; Moran and Zepp 1997; Mopper and Kieber 2001), adding to a rapid degradation of the organic matter in hydrogels. On the other hand, polymers may be released that are too large to be transported by bacterial permeases, yet too small to assemble into stable networks that bacteria can colonize and efficiently degrade with their extracellular enzymes. These polymers, indicated as DOM (MMW-DOM) in Fig. 3, may contribute to the formation of refractory DOM.

Bacterial activity may also increase aggregation of gels (Fig. 3). Studies on bacterial exopolymer production suggest that a significant portion of assimilated carbon is incorporated into exopolymer capsular envelopes (Stoderegger and Herndl 1998). These capsular exopolymers increase aggregation probabilities of phytoplankton and other particles and act as stabilizers of already existing aggregates (Allison and Sutherland 1987; Decho 1990; Heissenberger and Herndl 1994). In addition, bacterial exopolymers are

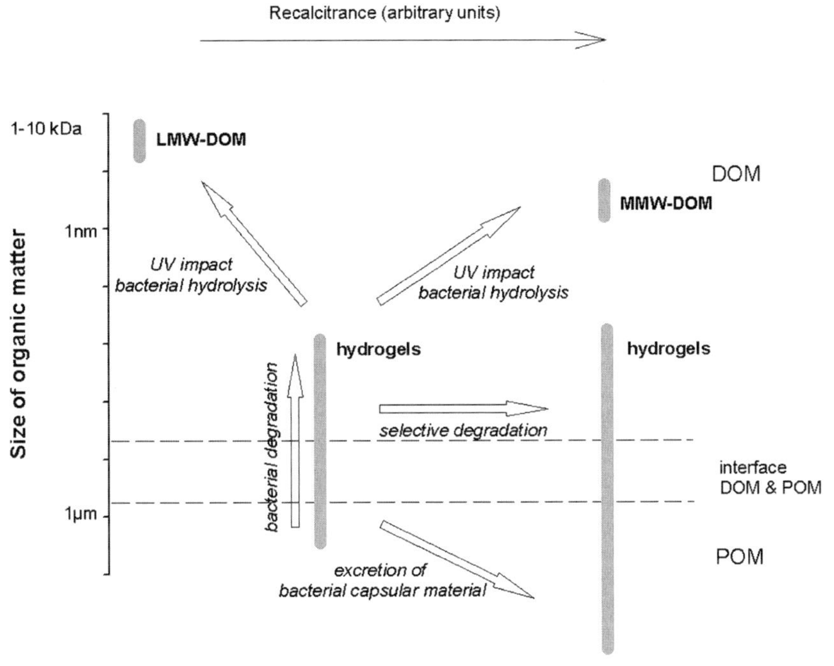

Fig. 3 Conceptual representation of biotic and abiotic processes altering the size and recalcitrance of organic matter. (Abbreviations as in Fig. 1, MMW-DOM: medium-molecular-weight DOM)

resistant against bacterial enzymatic degradation, since a considerable fraction of bacterial extracellular enzymes are embedded in the exopolymer capsule. Therefore, release of bacterial derived exopolymers decreases the degradation potential of hydrogels. Finally, selective degradation of readily degradable compounds may alter the composition of the gels, and decrease the degradation potential of the remaining gel (Sannigrahi et al. 2005). At present it is not known to what extent the mechanisms discussed above influence microbial degradation of hydrogels. Therefore, microbial degradation of *Phaeocystis* derived DOM may either be enhanced or hampered by formation of hydrogels.

Conclusions

Phaeocystis produces at least five different pools of carbohydrates, each with their own characteristics. Since the overflow metabolism during the stationary phase of a bloom can be channeled towards glucan and mucopolysaccharides, these two pools are quantitatively the most important when the impact on the ecosystem is considered. The contribution of the pools is strongly dependent on environmental conditions. High light conditions and nutrient limitation lead to excess carbohydrate production, but thus far the effect of either N or P limitation on the ratio of glucan to mucopolysaccharide production is unclear, as is the effect of Fe limitation on carbohydrate production in *P. antarctica.*

Phaeocystis-derived organic matter appears to be readily degradable, however, the high C/N and C/P ratios may give rise to mineral nutrient limitation of bacteria and therefore hamper degradation. Subsequent assembly into gels could have a dramatic effect on bacterial degradation of *Phaeocystis* derived organic matter, however, the nature of this effect is unclear. Gels may provide a convenient microhabitat to colonize and degrade with extracellular enzymes, thereby enhancing degradation rates of organic matter. Conversely, degradation of polymers embedded in gels may be slowed down by sterical hindrance. Furthermore, bacterial activity may reduce the degradation potential of gels by selective degradation of labile organic matter in the gels, and by production of carbon into bacterial capsular material.

Acknowledgements We thank M.J.E.C. van der Maarel and H. Helmers for the analysis of the viscosity of laminarin. L. Seuront is acknowledged for stimulating comments on an earlier version of the manuscript.

References

Alderkamp AC, Nejstgaard JC, Verity PG, Zirbel MJ, Sazhin AF, Van Rijssel M (2006a) Dynamics in carbohydrate composition of *Phaeocystis pouchetii* colonies during spring blooms in mesocosm. J Sea Res 55:169–181

Alderkamp AC, Sintes E, Herndl GJ (2006b) Abundance and activity of major groups of prokaryotic plankton in the coastal North Sea during the wax and wane of a phytoplankton spring bloom. Aquat Microb Ecol 45:237–246

Alderkamp AC, Van Rijssel M, Bolhuis H (2007) Characterization of marine bacteria and the activity of their enzyme systems involved in degradation of the algal storage glucan laminarin. FEMS Microb Ecol 59:108–117

Allison DG, Sutherland IW (1987) The role of exopolysaccharides in adhesion of fresh-water bacteria. J Gen Microbiol 133:1319–1327

Aluwihare LI, Repeta DJ (1999) A comparison of the chemical characteristics of oceanic DOM and extracellular DOM produced by marine algae. Mar Ecol Prog Ser 186:105–117

Amon RMW, Benner R (1996) Bacterial utilization of different size classes of dissolved organic matter. Limnol Oceanogr 41:41–51

Arrieta JM, Herndl GJ (2002) Changes in bacterial beta-glucosidase diversity during a coastal phytoplankton-bloom. Limnol Oceanogr 47:594–599

Arrigo KR, Robinson DH, Worthen DL, Dunbar RB, DiTullio GR, VanWoert M, Lizotte MP (1999) Phytoplankton community structure and the drawdown of nutrients and CO2 in the Southern Ocean. Science 283:365–367

Azam F (1998) Microbial control of oceanic carbon flux: the plot thickens. Science 280:694–696

Azam F, Long RA (2001) Oceanography – sea snow microcosms. Nature 414:495–498

Barlow RG (1982) Phytoplankton ecology in the Southern Benguela Current .3. Dynamics of a bloom. J Exp Mar Biol Ecol 63:239–248

Baumann MEM, Lancelot C, Brandidni FP, Sakshaugh E, John DM (1994) The taxonomic identity of the cosmopolitan prymnesiophyte *Phaeocystis*: a morphological and ecophysiological approach. J Mar Syst 5:5–22

Becquevort S, Rousseau V, Lancelot C (1998) Major and comparable roles for free-living and attached bacteria in the degradation of *Phaeocystis*-derived organic matter in Belgian coastal waters of the North Sea. Aquat Microb Ecol 14:39–48

Benner R, Pakulski JD, McCarty M, Hedges JI, Hatcher PG (1992) Bulk chemical characteristics of dissolved organic matter in the ocean. Science 255:1561–1564

Biddanda B, Benner R (1997) Carbon, nitrogen, and carbohydrate fluxes during the production of particulate and dissolved organic matter by marine phytoplankton. Limnol Oceanogr 42:506–518

Biersmith A, Benner R (1998) Carbohydrates in phytoplankton and freshly produced dissolved organic matter. Mar Chem 63:131–144

Billen G, Fontigny A (1987) Dynamics of a *Phaeocystis*-dominated spring bloom in Belgian coastal waters. II Bacterioplankton dynamics. Mar Ecol Prog Ser 37:249–257

Bjørnsen PK (1988) Phytoplankton exudation of organic-matter – why do healthy cells do it? Limnol Oceanogr 33:151–154

Bohm N, Kulicke WM (1999) Rheological studies of barley (1->3)(1->4)-beta-glucan in concentrated solution: mechanistic and kinetic investigation of the gel formation. Carbohydr Res 315:302–311

Breteler WCMK, Koski M (2003) Development and grazing of *Temora longicornis* (Copepoda, Calanoida) nauplii during nutrient limited *Phaeocystis globosa* blooms in mesocosms. Hydrobiologia 491:185–192

Brussaard CPD, Riegman R, Noordeloos AAM, Cadée GC, Witte H, Kop AJ, Nieuwland G, Van Duyl FC, Bak RPM (1995) Effects of grazing, sedimentation and phytoplankton cell lysis on the structure of a coastal pelagic food web. Mar Ecol Prog Ser 123:259–271

Brussaard CPD, Gast GJ, Van Duyl FC, Riegman R (1996) Impact of phytoplankton bloom magnitude on a pelagic microbial food web. Mar Ecol Prog Ser 144:211–221

Brussaard CPD, Kuipers B, Veldhuis MJW (2005a) A mesocosm study of *Phaeocystis globosa* population dynamics – 1. Regulatory role of viruses in bloom. Harmful Algae 4:859–874

Brussaard CPD, Mari X, Van Bleijswijk JDL, Veldhuis MJW (2005b) A mesocosm study of *Phaeocystis globosa* (Prymnesiophyceae) population dynamics – II. Significance for the microbial community. Harmful Algae 4:875–893

Carlson CA, Ducklow HW, Hansell DA, Smith WO (1998) Organic carbon partitioning during spring phytoplankton blooms in the Ross Sea polynya and the Sargasso Sea. Limnol Oceanogr 43:375–386

Chang FH (1984) The ultrastructure of *Phaeocystis pouchetii* (Prymnesiophyceae) vegetative colonies with special reference to the production of new mucilaginous envelope. N Z J Mar Fresh Res 18:303–308

Chen YQ, Wang N, Zhang P, Zhou H, Qu LH (2002) Molecular evidence identifies bloom-forming *Phaeocystis* (Prymnesiophyta) from coastal waters of southeast China as *Phaeocystis globosa*. Biochem Syst Ecol 30:15–22

Chin WC, Orellana MV, Verdugo P (1998) Spontaneous assembly of marine dissolved organic matter into polymer gels. Nature 391:568–572

Chin WC, Orellana MV, Quesada I, Verdugo P (2004) Secretion in unicellular marine phytoplankton: demonstration of regulated exocytosis in *Phaeocystis globosa*. Plant Cell Physiol 45:535–542

Chiovitti A, Molino P, Crawford SA, Teng RW, Spurck T, Wetherbee R (2004) The glucans extracted with warm water from diatoms are mainly derived from intracellular chrysolaminaran and not extracellular polysaccharides. Eur J Phycol 39:117–128

Chretiennot-Dinet MJ, Giraud-Guille MM, Vaulot D, Putaux JL, Saito Y, Chanzy H (1997) The chitinous nature of filaments ejected by *Phaeocystis* (Prymnesiophyceae). J Phycol 33:666–672

Chróst RJ (1991) Environmental control of the synthesis and activity of aquatic microbial ectoenzymes. In: Chróst RJ (ed) Microbial enzymes in aquatic environments. Springer Verlag, New York, pp 29–59

Coale KH, Wang XJ, Tanner SJ, Johnson KS (2003) Phytoplankton growth and biological response to iron and zinc addition in the Ross Sea and Antarctic Circumpolar Current along 170 degrees W. Deep-Sea Res II 50:635–653

Cottrell MT, Kirchman DL (2000) Natural assemblages of marine proteobacteria and members of the Cytophaga-Flavobacter cluster consuming low- and high-molecular-weight dissolved organic matter. Appl Environ Microbiol 66:1692–1697

Craigie JS (1974) Storage products. In: Stewart WDP (ed) Algal physiology and biochemistry. Blackwell Scientific Publications, Oxford, pp 206–235

Cuhel RL, Ortner PB, Lean DRS (1984) Night synthesis of protein by algae. Limnol Oceanogr 29:731–744

De Brouwer JFC, Wolfstein K, Stal LJ (2002) Physical characterization and diel dynamics of different fractions of extracellular polysaccharides in an axenic culture of a benthic diatom. Eur J Phycol 37:37–44

Decho AW (1990) Microbial exopolymer secretions in ocean environments – their role(s) in food webs and marine processes. Oceanogr Mar Biol 28:73–153

Dickson DMJ, Kirst GO (1987a) Osmotic adjustment in marine eukaryotic algae – the role of inorganic-ions, quaternary ammonium, tertiary sulfonium and carbohydrate solutes.1. Diatoms and a rhodophyte. New Phytol 106:645–655

Dickson DMJ, Kirst GO (1987b) Osmotic adjustment in marine eukaryotic algae – the role of inorganic-ions, quaternary ammonium, tertiary sulfonium and carbohydrate solutes. 2. Prasinophytes and haptophytes. New Phytol 106:657–666

DiTullio GR, Grebmeier JM, Arrigo KR, Lizotte MP, Robinson DH, Leventer A, Barry JB, VanWoert ML, Dunbar RB (2000) Rapid and early export of *Phaeocystis antarctica* blooms in the Ross Sea, Antarctica. Nature 404:595–598

Eberlein K, Leal MT, Hammer KD, Hickel W (1985) Dissolved organic substances during a *Phaeocystis pouchetii* bloom in the German Bight (North Sea). Mar Biol 89:311–316

Edwards SF (1986) The theory of macromolecular networks. Biorheology 23:589–603

Falkowski PG, Raven JA (1997) Aquatic photosynthesis. Blackwell Science, Malden, MA

Fernández E, Serret P, Demadariaga I, Harbour DS, Davies AG (1992) Photosynthetic carbon metabolism and biochemical composition of spring phytoplankton assemblages enclosed in microcosms: the diatom *Phaeocystis* sp. succession. Mar Ecol Prog Ser 90:89–102

Fogg GE (1983) The ecological significance of extracellular products of phytoplankton photosynthesis. Bot Mar 26:3–14

Geider RJ, La Roche J (1994) The role of iron in phytoplankton photosynthesis and the potential for iron-limitation of primary productivity in the sea. Photosynth Res 39:275–301

Granum E, Myklestad SM (1999) Effects of NH4+ assimilation on dark carbon fixation and beta-1,3-glucan metabolism in the marine diatom *Skeletonema costatum* (Bacillariophyceae). J Phycol 35:1191–1199

Granum E, Myklestad SM (2001) Mobilization of beta-1,3-glucan and biosynthesis of amino acids induced by NH4+ addition to N-limited cells of the marine diatom *Skeletonema costatum* (Bacillariophyceae). J Phycol 37:772–782

Guillard RRL, Hellebust JA (1971) Growth and production of extracellular substances by two strain of *Phaeocystis pouchetii*. J Phycol 7:330–338

Hallegraeff GM (1983) Scale-bearing and loricate nanoplankton from the East Australian Current. Bot Mar 16:515

Hama J, Handa N (1992) Diel variation of water-extractable carbohydrate composition of natural phytoplankton populations in Kinu-ura Bay. J Exp Mar Biol Ecol 162:159–176

Hamm CE, Simson DA, Merkel R, Smetacek V (1999) Colonies of *Phaeocystis globosa* are protected by a thin but tough skin. Mar Ecol Prog Ser 187:101–111

Heissenberger A, Herndl GJ (1994) Formation of high molecular weight material by free-living marine bacteria. Mar Ecol Prog Ser 111:129–135

Hitchcock GL (1980) Diel variation in chlorophyll a, carbohydrate and protein content of the marine diatom *Skeletonema costatum*. Mar Biol 57:271–278

Hoagland KD, Rosowski JR, Gretz MR, Roemer SC (1993) Diatom extracellular polymeric substances: function, fine structure, chemistry, and physiology. J Phycol 29:537–566

Hong Y, Smith WO, White AM (1997) Studies on transparent exopolymer particles (TEP) produced in the Ross Sea (Antarctica) and by *Phaeocystis antarctica* (Prymnesiophyceae). J Phycol 33:368–376

Janse I, Van Rijssel M, Gottschal JC, Lancelot C, Gieskes WWC (1996a) Carbohydrates in the North Sea during spring blooms of *Phaeocystis*: a specific fingerprint. Aquat Microb Ecol 10:97–103

Janse I, Van Rijssel M, Van Hall PJ, Gerwig GJ, Gottschal JC, Prins RA (1996b) The storage glucan of *Phaeocystis globosa* (Prymnesiophyceae) cells. J Phycol 32:382–387

Janse I, Van Rijssel M, Ottema A, Gottschal JC (1999) Microbial breakdown of *Phaeocystis* mucopolysaccharides. Limnol Oceanogr 44:1447–1457

Janse I, Zwart G, Maarel MJEC, Gottschal JC (2000) Composition of the bacterial community degrading *Phaeocystis* mucopolysaccharides in enrichment cultures. Aquat Microb Ecol 22:119–133

Joint I, Pomroy A (1993) Phytoplankton biomass and production in the southern North-Sea. Mar Ecol Prog Ser 99:169–182

Joiris C, Billen G, Lancelot C, Daro MH, Mommaerts JP, Bertels A, Bossicart M, Nijs J, Hecq JH (1982) A budget of carbon cycling in the Belgian coastal zone:

relative roles of zooplankton, bacterioplankton and benthos in the utilization of primary production. Neth J Sea Res 16:260–275

Kim YT, Kim EH, Cheong C, Williams DL, Kim CW, Lim ST (2000) Structural characterization of beta-D-(1 fwdarw 3, 1 fwdarw 6)-linked glucans using NMR spectroscopy. Carbohydr Res 328:331–341

Kirchman DL (2002) The ecology of *Cytophaga-Flavobacteria* inaquatic environments. FEMS Microbiol Ecol 39:91–100

Kornmann P (1955) Beobachtungen an *Phaeocystis*-Kulturen. Helgol Wiss Meeresunters 5:218–233

Laanbroek HL, Verplanke JC, De Visscher PRM, De Vuyst R (1985) Distribution of phyto- and bacterioplankton growth and biomass parameters, dissolved inorganic nutrients and free amino acids during a spring bloom in the Oosterschelde basin, The Netherlands. Mar Ecol Prog Ser 25:1–11

Lancelot C, Mathot S (1985) Biochemical fractionation of primary production by phytoplankton in Belgian coastal waters during short- and long-term incubations with 14C-bicarbonate. II. *Phaeocystis poucheti* colonial population. Mar Biol 86:227–232

Lancelot C, Mathot S (1987) Dynamics of a *Phaeocystis*-dominated spring bloom in Belgian coastal waters. I Phytoplanktonic activities and related parameters. Mar Ecol Prog Ser 37:239–248

Lancelot C, Rousseau V (1994) Ecology of *Phaeocystis*: the key role of colony forms. In: Green JC, Leadbeater BSC (eds) The haptophyte algae. Clarendon Press, Oxford, pp 229–245

Lancelot C, Billen G, Sournia A, Weisse T, Colijn F, Veldhuis MJW, Davies A, Wassmann P (1987) *Phaeocystis* blooms and nutrient enrichment in the continental coastal zones of the North Sea. Ambio 16:38–46

Lancelot C, Keller MD, Rousseau V, Smith WO, Mathot S (1998) Autoecology of the marine haptophyte *Phaeocystis* sp. In: Anderson DM, Cembella AD, Hallegraeff GM (eds) Physiological ecology of harmful algal blooms. Springer-Verlag, Berlin, pp 209–224

Larsson U, Hagström A (1979) Phytoplankton exudate release as an energy-source for the growth of pelagic bacteria. Mar Biol 52:199–206

Lazaridou A, Biliaderis CG, Izydorczyk MS (2003) Molecular size effects on rheological properties of oat beta-glucans in solution and gels. Food Hydrocolloids 17:693–712

Leadbeater BSC (1994) Cell coverings. In: Green JC, Leadbeater BSC (eds) The haptophyte algae. Clarendon Press, Oxford, pp 23–46

Lignell R (1990) Excretion of organic-carbon by phytoplankton – its relation to algal biomass, primary productivity and bacterial secondary productivity in the Baltic Sea. Mar Ecol Prog Ser 68:85–99

Loogman JG (1982) Influence of photoperiodicy on algal growth kinetics. Ph.D. University of Amsterdam, The Netherlands

Magaletti E, Urbani R, Sist P, Ferrari CR, Cicero AM (2004) Abundance and chemical characterization of extracellular carbohydrates released by the marine diatom *Cylindrotheca fusiformis* under N- and P-limitation. Eur J Phycol 39:133–142

Manners DJ, Ryley JF, Stark JR (1966) Studies on the metabolism of the protozoa – the molecular structure of the reserve polysaccharide from *Astasia ocellata*. Biochem J 101:323–327

Marañón E, Cermeño P, Fernández E, Rodríguez J, Zabala L (2004) Significance and mechanisms of photosynthetic production of dissolved organic carbon in a coastal eutrophic ecosystem. Limnol Oceanogr 49:1652–1666

Mari X, Rassoulzadegan F, Brussaard CPD, Wassmann P (2005) Dynamics of transparent exopolymeric particles (TEP) production by *Phaeocystis globosa* under N- or P-limitation: a controlling factor of the retention/export balance. Harmful Algae 4:895–914

Martin JH, Gordon RM, Fitzwater SE (1990) Iron in Antarctic Waters. Nature 345:156–158

Mathot S, Smith WO, Carlson CA, Garrison DL, Gowing MM, Vickers CL (2000) Carbon partitioning within *Phaeocystis antarctica* (Prymnesiophyceae) colonies in the Ross Sea, Antarctica. J Phycol 36:1049–1056

Matrai PA, Vernet M, Hood R, Jennings A, Brody E, Saemundsdottir S (1995) Light-dependence of carbon and sulfur production by polar clones of the genus *Phaeocystis*. Mar Biol 124:157–167

Metaxatos A, Panagiotopoulos C, Ignatiades L (2003) Monosaccharide and aminoacid composition of mucilage material produced from a mixture of four phytoplanktonic taxa. J Exp Mar Biol Ecol 294:203–217

Moestrup O (1979) Identification by electron microscopy of marine nanaoplankton from New Zealand, including the description of four new species. N Z J Bot 17:61–95

Mopper K, Kieber DJ (2001) Marine photochemistry and its impact on carbon cycling. In: De Mora S, Demers S, Vernet M (eds) The effect of UV radiation in the marine environment. Cambridge University Press, Cambridge, UK, pp 102–129

Mopper K, Zhou XL, Kieber RJ, Kieber DJ, Sikorski RJ, Jones RD (1991) Photochemical degradation of dissolved organic-carbon and its impact on the oceanic carbon-cycle. Nature 353:60–62

Moran MA, Zepp RG (1997) Role of photoreactions in the formation of biologically labile compounds from dissolved organic matter. Limnol Oceanogr 42:1307–1316

Myklestad SM (1974) Production of carbohydrates by marine planktonic diatoms. I. Comparison of nine different species in culture. J Exp Mar Biol Ecol 15:261–274

Myklestad SM (1978) Beta-1,3-glucans in diatoms and brown seaweeds. In: Hellebust JA, Craigie JS (eds) Handbook of phycological methods. Cambridge University Press, Cambridge, pp 133–141

Myklestad SM (1988) Production, chemical structure, metabolism, and biological function of the (1,3)-linked, beta-D-glucans in diatoms. Biol Oceanogr 6:313–326

Myklestad SM, Haug A (1972) Production of carbohydrates by the marine diatom *Chaetoceros affinis* var. *Willei* (Gran) Hustedt. II Preliminary investigation of the extracellular polysaccharide. J Exp Mar Biol Ecol 9:137–144

Myklestad SM, Skanoy E, Hestmann S (1997) A sensitive and rapid method for analysis of dissolved mono- and polysaccharides in seawater. Mar Chem 56:279–286

Nagata T (2000) Production mechanisms of dissolved organic matter. In: Kirchman DL (ed) Microbial ecology of the oceans. Wiley-Liss, pp 121–152

Nejstgaard JC, Tang KW, Steinke M, Dutz J, Antajan E, Long JD (2007) Zooplankton grazing on *Phaeocystis*: a critical review and future challenges. Biogeochemistry this issue

Orellana MV, Verdugo P (2003) Ultraviolet radiation blocks the organic carbon exchange between the dissolved phase and the gel phase in the ocean. Limnol Oceanogr 48:1618–1623

Orellana MV, Lessard EJ, Dycus E, Chin WC, Foy MS, Verdugo P (2003) Tracing the source and fate of biopolymers in seawater: application of an immunological technique. Mar Chem 83:89–99

Osinga R, De Vries KA, Lewis WE, Van Raaphorst W, Dijkhuizen L, Van Duyl FC (1997) Aerobic degradation of phytoplankton debris dominated by *Phaeocystis* sp. in different physiological stages of growth. Aquat Microb Ecol 12:11–19

Painter TJ (1983) Algal polysaccharides. In: Aspinall GO (ed) The polysaccharides. Academic Press, New York, pp 195–285

Parke M, Green JC, Manton I (1971) Observations on the fine structure of zoids of the genus *Phaeocystis* (Haptophyceae). J Mar Biol Assoc UK 51:927–941

Passow U, Wassmann P (1994) On the trophic fate of *Phaeocystis pouchetii* (Hariot): IV. The formation of marine snow by *P. pouchetii*. Mar Ecol Prog Ser 104:153–161

Paulsen BS, Myklestad SM (1978) Structural studies of the reserve glucan produced by the marine diatom *Skeletonema costatum*. Carbohydr Res 62:386–388

Peperzak L, Colijn F, Vrieling EG, Gieskes WWC, Peeters JCH (2000) Observations of flagellates in colonies of *Phaeocystis globosa* (Prymnesiophyceae); a hypothesis for their position in the life cycle. J Plankt Res 22:2181 2203

Percival E (1970) Algal polysaccharides. In: Pigman W, Horton D (eds) The carbohydrates, vol IIB. Academic Press, New York, pp 541–544

Post AF, Dubinsky Z, Wyman K, Falkowski PG (1985) Physiological-responses of a marine planktonic diatom to transitions in growth irradiance. Mar Ecol Prog Ser 25:141–149

Putt M, Miceli G, Stoecker DK (1994) Association of bacteria with *Phaeocystis* sp. in McMurdo Sound, Antarctica. Mar Ecol Prog Ser 105:179–189

Reichenbach H, Dworkin M (1991) The order Cytophagales. In: Balows A, Truber HG, Dworkin M, Harder W, Schleifer KH (eds) The prokaryotes. Springer-Verlag, New York, pp 356–379

Renn DW (1997) Purified curdlan and its hydroxyalkyl derivatives: preparation, properties and applications. Carbohydr Pol 33:219–225

Riebesell U, Reigstad M, Wassmann P, Noji T, Passow U (1995) On the trophic fate of *Phaeocystis pouchetii* (Hariot): VI. Significance of *Phaeocystis*-derived mucus for vertical flux. Neth J Sea Res 33:193–203

Riegman R, Van Boekel W (1996) The ecophysiology of *Phaeocystis globosa*: a review. J Sea Res 35:235–242

Rousseau V, Mathot S, Lancelot C (1990) Calculating carbon biomass of *Phaeocystis* sp. from microscopic observations. Mar Biol 107:305–314

Rousseau V, Vaulot D, Casotti R, Cariou V, Lenz J, Gunkel J, Baumann MEM (1994) The life cycle of *Phaeocystis* (Prymnesiophyceae): evidence and hypotheses. J Mar Syst 5:23–40

Rousseau V, Becquevort S, Parent JY, Gasparini S, Daro MH, Tackx M, Lancelot C (2000) Tropic efficiency of the planktonic food web in a coastal ecosystem dominated by *Phaeocystis* colonies. J Sea Res 43:357–372

Sannigrahi P, Ingall ED, Benner R (2005) Cycling of dissolved and particulate organic matter at station Aloha: insights from C-13 NMR spectroscopy coupled with elemental, isotopic and molecular analyses. Deep-Sea Res I 52:1429–1444

Schoemann W, Becquevort S, Stefels J, Rousseau W, Lancelot C (2005) *Phaeocystis* blooms in the global ocean and their controlling mechanisms: a review. J Sea Res 53:43–66

Seuront L, Lacheze C, Doubell MJ, Seymour JR, Van Dongen-Vogels V, Newton K, Alderkamp AC, Mitchell JG (2007) The influence of *Phaeocystis globosa* on microscale spatial patterns of chlorophyll *a* and bulk-phase seawater viscosity. Biogeochemistry this issue

Sharp JH (1977) Excretion of organic matter by marine phytoplankton: do healthy cells do it? Limnol Oceanogr 22:381–399

Smith DC, Simon M, Alldredge AL, Azam F (1992) Intense hydrolytic enzyme activity on marine aggregates and implications for rapid particle dissolution. Nature 359:139–141

Smith DC, Steward GF, Long RA, Azam F (1995) Bacterial mediation of carbon fluxes during a diatom bloom in a mesocosm. Deep Sea Res II 42:75–97

Smith WO, Codispoti LA, Nelson DM, Manley T, Buskey EJ, Niebauer HJ, Cota GF (1991) Importance of *Phaeocystis* blooms in the high-latitude ocean carbon cycle. Nature 352:514–516

Smith WO, Carlson CA, Ducklow HW, Hansell DA (1998) Growth dynamics of *Phaeocystis antarctica*-dominated plankton assemblages from the Ross Sea. Mar Ecol Prog Ser 168:229–244

Solomon CM, Lessard EJ, Keil RG, Foy MS (2003) Characterization of extracellular polymers of *Phaeocystis globosa* and *P. antarctica*. Mar Ecol Prog Ser 250:81–89

Stoderegger K, Herndl GJ (1998) Production and release of bacterial capsular material and its subsequent utilization by marine bacterioplankton. Limnol Oceanogr 43:877–884

Tang KW, Jakobsen HH, Visser AW (2001) *Phaeocystis globosa* (Prymnesiophyceae) and the planktonic food web: feeding, growth, and trophic interactions among grazers. Limnol Oceanogr 46:1860–1870

Teira E, Pazo MJ, Quevedo M, Fuentes MV, Niell FX, Fernández E (2003) Rates of dissolved organic carbon production and bacterial activity in the eastern North Atlantic Subtropical Gyre during summer. Mar Ecol Prog Ser 249:53–67

Thingstad TF, Billen G (1994) Microbial degradation of *Phaeocystis* material in the water column. J Mar Syst 5:55–66

Thingstad TF, Hagström A, Rassoulzadegan F (1997) Accumulation of degradable DOC in surface waters: is it caused by a malfunctioning microbial loop? Limnol Oceanogr 42:398–404

Underwood GJC, Boulcott M, Raines CA, Waldron K (2004) Environmental effects on exopolymer production by marine benthic diatoms: dynamics, changes in composition, and pathways of production. J Phycol 40:293–304

Vadstein O, Jensen A, Olsen Y, Reinertsen H (1988) Growth and phosphorus status of limnetic phytoplankton and bacteria. Limnol Oceanogr 33:489–503

Van Boekel WHM (1992) *Phaeocystis* colony mucus components and the importance of calcium ions for colony stability. Mar Ecol Prog Ser 87:301–305

Van Boekel WHM, Hansen FC, Riegman R, Bak RPM (1992) Lysis-induced decline of a *Phaeocystis* spring bloom and coupling with the microbial foodweb. Mar Ecol Prog Ser 81:269–276

Van den Hoek C, Jahns HM, Mann DG (1993) Algen. Georg Thieme Verlag Stuttgard, New York

Van Oijen T, Van Leeuwe MA, Gieskes WWC (2003) Variation of particulate carbohydrate pools over time and depth in a diatom-dominated plankton community at the Antarctic Polar Front. Polar Biol 26:195–201

Van Oijen T, Van Leeuwe MA, Gieskes WWC, De Baar HJW (2004a) Effects of iron limitation on photosynthesis and carbohydrate metabolism in the Antarctic diatom *Chaetoceros brevis* (Bacillariophyceae). Eur J Phycol 39:161–171

Van Oijen T, Van Leeuwe MA, Granum E, Weissing FJ, Bellerby RGJ, Gieskes WWC, De Baar HJW (2004b) Light rather than iron controls photosynthate production and allocation in Southern Ocean phytoplankton populations during austral autumn. J Plankt Res 26:885–900

Van Oijen T, Veldhuis MJW, Gorbunov MY, Nishioka J, Van Leeuwe MA, De Baar HJW (2005) Enhanced carbohydrate production by Southern Ocean phytoplankton in response to *in-situ* iron fertilisation. Mar Chem 93:33–52

Van Rijssel M, Hamm CE, Gieskes WWC (1997) *Phaeocystis globosa* (Prymnesiophyceae) colonies: hollow structures built with small amounts of polysaccharides. Eur J Phycol 32:185–192

Van Rijssel M, Janse I, Noordkamp DJB, Gieskes WWC (2000) An inventory of factors that affect polysaccharide production by *Phaeocystis globosa*. J Sea Res 43:297–306

Vårum KM, Kvam BJ, Myklestad SM (1986) Structure of a food-reserve beta-D-glucan produced by the haptophyte alga *Emiliania huxleyi* (Lohmann) Hay and Mohler. Carbohydr Res 152:243–248

Veldhuis MJW, Admiraal W (1985) Transfer of photosynthetic products in gelatinous colonies of *Phaeocystis pouchetii* (Haptophyceae) and its effect on the measurement of excretion rate. Mar Ecol Prog Ser 26:301–304

Veldhuis MJW, Admiraal W, Colijn F (1986a) Chemical and physiological-changes of phytoplankton during the spring bloom, dominated by *Phaeocystis pouchetii* (Haptophyceae) – observations in Dutch coastal waters of the North Sea. Neth J Sea Res 20:49–60

Veldhuis MJW, Colijn F, Venekamp LAH (1986b) The spring bloom of *Phaeocystis pouchetii* (Haptophyceae) in Dutch coastal waters. Neth J Sea Res 20:37–48

Veldhuis MJW, Brussaard CPD, Noordeloos AAM (2005) Living in a *Phaeocystis* colony: a way to be a successful algal species. Harmful Algae 4:841–858

Verdugo P, Alldredge AL, Azam F, Kirchman DL, Passow U, Santschi PH (2004) The oceanic gel phase: a bridge in the DOM-POM continuum. Mar Chem 92:67–85

Verity PG, Villareal TA, Smayda TJ (1988) Ecological investigations of blooms of colonial *Phaeocystis pouchetii*. I Abundance, biochemical composition and metabolic rates. J Plankt Res 10:219–248

Verity PG, Smayda TJ, Sakshaug E (1991) Photosynthesis, excretion, and growth rates of *Phaeocystis* colonies and solitary cells. Polar Res 10:117–128

Wassmann P (1994) Significance of sedimentation for the termination of *Phaeocystis* blooms. J Mar Syst 5:81–100

Wassmann P, Vernet M, Mitchell BG, Rey F (1990) Mass sedimentation of *Phaeocystis pouchetii* in the Barents Sea. Mar Ecol Prog Ser 66:183–196

Wassmann P, Reigstad M, Oygarden S, Rey F (2000) Seasonal variation in hydrography, nutrients, and suspended biomass in a subarctic fjord: applying hydrographic features and biological markers to trace water masses and circulation significant for phytoplankton production. Sarsia 85:237–249

Weisse T, Scheffel-Möser U (1990) Growth and grazing loss rates in single-celled *Phaeocystis* sp. (Prymnesiophyceae). Mar Biol 106:153–158

Zingone A, Chretiennot-Dinet MJ, Lange M, Medlin L (1999) Morphological and genetic characterization of *Phaeocystis cordata* and *P. jahnii* (Prymnesiophyceae), two new species from the Mediterranean Sea. J Phycol 35:1322–1337

Biogeochemistry (2007) 83:119–135
DOI 10.1007/s10533-007-9079-1

ORIGINAL PAPER

The role of iron in the bacterial degradation of organic matter derived from *Phaeocystis antarctica*

S. Becquevort · C. Lancelot · V. Schoemann

Received: 5 May 2006 / Accepted: 16 November 2006 / Published online: 24 April 2007

Abstract In high-nutrient low-chlorophyll areas, bacterial degradation of organic matter may be iron-limited. The response of heterotrophic bacteria to Fe addition may be directly controlled by Fe availability and/or indirectly controlled through the effect of enhanced phytoplankton productivity and the subsequent supply of organic matter suitable for bacteria. In the present study, the role of Fe on bacterial carbon degradation was investigated through regrowth experiments by monitoring bacterial response to organic substrates derived from *Phaeocystis antarctica* cultures set up in <1 nM Fe (LFe) and in Fe-amended (HFe) Antarctic seawater. Results showed an impact of Fe addition on the morphotype dominance (colonies vs. single cells) of *P. antarctica* and on the quality of *Phaeocystis*-derived organic matter. Fe addition leaded to a decrease of C/N ratio of *Phaeocystis* material. The bacterial community composition was modified as observed from denaturing gradient gel electrophoresis (DGGE) profiles in LFe as compared to HFe bioassays. The percentage of active bacteria as well as their specific metabolic activities (ectoenzymatic hydrolysis, growth rates and bacterial growth efficiency) were enhanced in HFe bioassays. As a consequence, the lability of *Phaeocystis*-derived organic matter was altered, i.e., after seven days more than 90% was degraded in HFe and only 9% (dissolved) and 55% (total) organic carbon were degraded in LFe bioassays. By inducing increased bacterial degradation and preventing the accumulation of dissolved organic carbon, the positive effect of Fe supply on the carbon biological pump may partly be counteracted.

Keywords Bacterioplankton · Iron · Organic matter · *Phaeocystis antarctica* · Remineralisation

S. Becquevort (✉) · C. Lancelot · V. Schoemann
Ecologie des Systèmes Aquatiques, Université Libre de Bruxelles, Campus de la Plaine, Bd. du Triomphe, CP 221, 1050 Bruxelles, Belgium
e-mail: sbecq@ulb.ac.be

Introduction

Phaeocystis antarctica recurrently produces large blooms in the Southern Ocean such as in Prydz Bay and the southern Ross Sea (e.g., El-Sayed et al. 1983; Davidson and Marchant 1992; Smith et al. 1996). In the Ross Sea, *P. antarctica* blooms early in the season, in waters characterized by relatively high Fe levels (Sedwick et al. 2000). Among the phytoplankton community, *P. antarctica* showed indeed the strongest response to Fe addition, increasing its relative abundance from <5 up to 20% in 2.5 nM Fe-enriched seawater (Coale et al. 2003).

The question of the fate of the carbon produced by *P. antarctica* in the Southern Ocean is still unresolved. Bacterial degradation has been proposed to be a major pathway in polar waters. However, several cases of dissolved organic carbon (DOC) accumulation have been observed in Prydz Bay (e.g., Bölter and Dawson 1982; Davidson and Marchant 1992). This DOC accumulation can either results from the refractory nature of the organic matter and/or from the limitation of its consumption by bacteria (Thingstad et al. 1997).

The biodegradability of organic carbon depends on the physico-chemical properties of the organic matter itself such as molecular size, chemical structure and elemental composition. In high-nutrient low-chlorophyll (HNLC) regions, Fe addition can have a positive effect on primary production and consequently increases the production of organic substrates. In addition, it can have an impact on the quality of the organic matter for example the quality of phytoplankton-derived organic matter is related to nutrient growth conditions with more proteins synthesized under nonlimiting conditions (Lancelot et al. 1986). In the case of *P. antarctica*, it can also control the dominant morphotype, colonies versus flagellated cells (Smith et al. 2003), generally characterized by contrasted elemental ratio (Schoemann et al. 2005).

Bacterial degradation of organic carbon may be limited by temperature (Pomeroy et al. 1991), predation (Thingstad et al. 1997), but also by ambient inorganic nutrients. For example, results reported for northern European *Phaeocystis* blooms suggested that the degradation of *Phaeocystis*-derived organic carbon could be nutrient-limited by phosphorus and nitrogen (Thingstad and Billen 1994). In HNLC waters, inorganic major nutrient limitation of bacterial degradation is unlikely, whereas Fe can be limiting. Dissolved Fe concentration are extremely low in January (<0.2 nM), at the senescent stage of *P. antarctica* bloom in the Ross Sea (Sedwick et al. 2000). Heterotrophic bacteria, with their high surface-to-volume ratio and ability to produce siderophores (e.g., Granger and Price 1999), are probably very efficient competitors for Fe acquisition even in environments experiencing picomolar concentrations of Fe commonly encountered in HNLC areas. However, the large Fe requirement of the respiratory system requiring 60% more Fe per mol of cytochrome-*c* than the photosynthetic respiratory system (Raven 1988), suggest that these organisms can still be Fe-limited *in situ*.

Bacterial processes such as the ectoenzymatic hydrolysis of polymeric organic matter, the uptake of monomeric substrates and respiration can be directly controlled by Fe availability and/or indirectly controlled through the effect of enhanced phytoplankton productivity and the subsequent supply of organic matter suitable for bacteria. Moreover, Fe availability can affect the bacterial community structure and also the proportion of metabolically active cells playing a role in the degradation of organic matter. During Fe enrichment bottle experiments conducted in the coastal Southern Ocean (Gerlache Strait), Pakulski et al. (1996) found that Fe induces increases of both heterotrophic bacterial abundance and cell-specific growth rates. These experiments, conducted in the dark and in the absence of phytoplankton and bacterivores, suggest a direct stimulation of heterotrophic bacterial growth by Fe enrichment. In other controlled experiments, carbon seemed to be the first growth-limiting factor of the bacteria although Fe quickly became limiting when carbon limitation was alleviated (Church et al. 2000; Kirchman et al. 2000). Fe may also produce an increase in bacterial carbon metabolism efficiency. Indeed, enhanced bacterial growth efficiency (BGE) in Fe-enriched conditions has been previously observed (Tortell et al. 1996; Kirchman et al. 2003). Although increasing Fe availability alone seems to have little effect on bacterial diversity, changes in the organic matter source derived from Fe-enhanced algal production may affect bacterial community structure in favour of species having appropriate enzymes (Hutchins et al. 2001; Arrieta et al. 2004).

In this work we conducted laboratory-controlled experiments to investigate the role of Fe in the bacterial degradation of *P. antarctica*-derived organic matter. Our specific objectives were to determine how Fe affects the production and quality of *Phaeocystis* organic matter and to assess the role of Fe in bacterial processes involved in the degradation of this organic matter. The experimental strategy was to mimic post-phytoplankton bloom events by running bacterial regrowth experiments enriched with *Phaeocystis*-derived organic matter obtained under limiting and nonlimiting Fe conditions. Finally, the effects of Fe availability on the organic matter degradability are discussed in terms of significance for the carbon biological pump.

Materials and methods

Experimental procedure

Regrowth experiments were performed by mixing *P. antarctica*-derived organic matter with an inoculum of Antarctic bacteria. *Phaeocystis*-derived organic matter was prepared from pure cultures of *P. antarctica* grown under Fe-limiting (LFe) and enriched conditions (HFe).

All the experimental procedures were carried out under trace-metal clean conditions as described in Schoemann et al. (2001). Except for the organic matter sampling, flasks and filtration devices used for *Phaeocystis* cultures, the regrowth experiments and seawater sampling were in polycarbonate or high-density polyethylene. For at least 48 h, the polycarbonate labware was soaked in 1 N HCl, whereas the more-resistant polyethylene containers were decontaminated with 7 N HNO_3. Nutrient samples were collected in 1 N HCl acid–cleaned polyethylene vials. The laboratory material was thoroughly rinsed with ultra-high-purity water (Milli-Q Element system, Millipore) before use. All procedures prior to the incubations and during sampling were performed in a class-100 laminar flow bench. The material for organic carbon sampling and measurements, were either in glass or Teflon and carefully cleaned by ashing (4 h at 550°C) or by washing with chromic–sulphuric acid (Merck) or concentrated 6 N HCl.

Phaeocystis cultures

Phaeocystis antarctica strain CCMP1871, originating from the Bellingshausen Sea was obtained from the Provasoli-Guillard National Center for Cultures of Marine Phytoplankton (CCMP, USA). Cultures were performed in natural (low-Fe, high-macronutrient) Antarctic seawater, without the addition of any artificial chelator (e.g., ethylediaminetetraacetic acid, EDTA) in order to preserve the equilibrium and kinetics of the various natural chemical forms of Fe (Gerringa et al. 2000). The seawater was filtered on a 0.2 μm porosity Sartobran cartridge and sterilized by γ irradiation. *Phaeocystis antarctica* were grown and acclimated for at least five generations to LFe (<1 nM) and HFe (+2 nM) seawater. Cultures were grown at 2°C with a day:night cycle of 16:8 h at 180 μmole quanta m^{-2} s^{-1}. An inoculum (15 ml) of *P. antarctica* pre-adapted to LFe and HFe conditions was, respectively, added to 20 l LFe and HFe Antarctic seawater. Surface Antarctic seawater was collected in the southwestern Pacific Antarctic Ocean (135°E–150°E) during CLIVAR SR3 cruise in December 2001. For HFe conditions, Antarctic seawater was enriched with 2 nM $FeCl_3$ 72 h before the start of the culture. Another spike of 2 nM $FeCl_3$ was added to the culture after 8 days. In total 4 nM Fe was added in two times to prevent direct precipitation of the added Fe and ensure >2 nM concentrations of Fe in the batch culture. After 15 days, samples were collected during the exponential phase of growth for the regrowth experiments and to measure nutrients, Chl *a*, *P. antarctica* cell number and biomass of both single cells and colonies, particulate organic carbon (POC), particulate nitrogen (PON) and DOC.

Bacteria cultures

The bacteria inoculum was prepared from 0.8 μm prefiltered Antarctic seawater collected four months before the experiments at McMurdo Sound (Antarctica). These bacteria were pre-incubated in LFe (<1 nM) and HFe (+2 nM) Antarctic seawaters for at least five generations. They were grown in the dark at 2°C.

Regrowth experiments

Two pools of organic matter were prepared from exponentially growing LFe and HFe *P. antarctica*: (i) dissolved organic matter (DOM) obtained by gentle filtration (<0.3 atm) on 0.1 μm polycarbonate Nuclepore filters, (ii) total organic matter (TOM), unfiltered, including DOM and POM obtained after phytoplankton lysis by sonication (10 min of sonication at 22 W with pulse of 0.2 s/s) and freeze/thaw (−80°C deep freezer/60°C water bath cycles, repeated three times) methods. The broken cells were observed by epifluorescence microscopy, evidencing the efficiency of this method.

Bacteria were inoculated in a 1:10 volume ratio to each prepared media of organic matter (2 l). Bioassays were then incubated at 2°C in the dark for about 30 days. Sub-samples were collected to monitor the time evolution of bacterial abundance, biomass, genetic diversity, 5-cyano-2,3-ditotyl tetrazolium chloride (CTC)-positive cells, ectoenzymatic activities, production and organic carbon utilisation.

Analytical procedures

Nutrients

Nitrate and phosphate concentrations were determined according to Koroleff (1983a, b). Fe samples were filtered on 0.1 μm Nuclepore polycarbonate membrane and acidified to pH < 2 with ultraclean quartz distilled concentrated nitric acid (Ultrex, JT Baker). Dissolved Fe concentrations were measured according to a chemiluminescence method adapted from Obata et al. (1993) as applied by Sarthou et al. (2003). The blank is in average equal to 0.06 ± 0.02 nM. The detection limit is equal to 0.03 ± 0.01 nM.

Chlorophyll a and Phaeocystis antarctica biomass

Chl *a* was measured fluorimetrically following Yentsch and Menzel (1963) after 90% v/v acetone extraction (12 h) in the dark at 4°C of the particulate material retained on 0.8 μm Nuclepore polycarbonate membrane filters. The relative standard deviation of the method was better than 2%.

Samples to determine *P. antarctica* biomass were preserved with glutaraldehyde (0.5% final concentration) for free-living cells and with gluteraldehyde–Lugol's solution (1% final concentration) for colonies. Free-living cells were enumerated by epifluorescence microscopy (400× magnification) after 4′-6-diamidino-2-phenylindole (DAPI) staining (Porter and Feig 1980). Colonies were enumerated by inverted light microscopy (100× magnification) according to the method of Utermöhl (1958). Colonies and cell biomass were estimated according to Mathot et al. (2000).

Organic carbon

For phytoplankton particulate organic carbon (POC) and particulate nitrogen (PON) determination, some suspended matter was collected on precombusted (450°C) Whatman GF/F filters, dried at 60°C and stored in polystyrene Petri dishes until analysis. POC and PON were analyzed with a Carlo Erba NA 2000 elemental analyzer. The analytical error of the method is ± 0.8 μM for POC and ± 0.04 μM N for PON.

Samples for dissolved (DOC) and total organic carbon (TOC) were stored in glass tubes, poisoned with azide (final concentration 0.5%) and sealed with Teflon-lined screw caps. They were kept in the dark at 4°C until analysis. Organic carbon was measured by high-temperature catalytic oxidation (HTCO; procedure of Sugimura and Suzuki 1988) with a Dohrmann Apollo 9000 analyzer. Carbon concentration was determined using a five-point calibration curve performed with standards prepared by diluting a stock solution of potassium phthalate in Ultra pure water (Milli-Q Element system, Millipore). Each value corresponds to the average of at least five injections. Samples were measured in duplicate and the relative standard deviation never exceeded 2%. The accuracy of our DOC measurements was tested by measuring reference materials provided by Hansell laboratory (University of Miami). We obtained an average concentration of 45.1 ± 0.7 μM C (n = 10) for deep-ocean reference material (Sargasso Sea Deep water, 2,600 m) and 1.4 ± 0.7 μM C (n = 10) for low-carbon reference water. Our values are within the nominal values provided by the reference laboratory (44.0 ± 1.5 and 2.0 ± 1.5 μM C, respectively).

Bacterial biomass and activities

Samples for bacterial analyses were preserved with 40% buffered formaldehyde (final concentration 2%). Bacteria were filtered through 0.2-μm black Nuclepore polycarbonate filters of 25-mm diameter and stained by with 4′-6-diamidino-2-phenylindole (DAPI, final concentration 2.5 μg ml^{-1}) according to Porter and Feig (1980). A minimum of 1,000 cells were counted with a Leica epifluorescence microscope in at least 10 different fields at 1,000× magnification. A relative standard deviation of 15% (n = 20) was estimated on bacterial abundance determination. Bacterial biovolumes were determined by image analysis (Lucia 4.6 software) and calculated by treating rods and cocci as cylinders and spheres, respectively (Watson et al. 1977). They were converted to carbon biomass by using the relation established from data measured by Simon and Azam (1989):

$$\mathrm{C} = 0.12\, V^{0.72},$$

where C is the carbon per cell (pg C μm^{-3}) and V is the biovolume (μm^3).

The number of 5-cyano-2,3-ditotyl tetrazolium chloride (CTC)-reducing bacteria were determined by using a final concentration of 2.5 mM and incubating the sample at 2°C for 2 h (Rodriguez et al. 1992). The reaction was stopped with 40% buffered formaldehyde (final concentration 2%). The samples were further filtered, DAPI stained and observed as described above.

Ectoenzymatic activity was determined fluorimetrically (Kontron SFM 25 fluorimeter) as the maximum hydrolysis rate (H_m) of model substrates for β-glucosidase (4-methylumbelliferyl-β-D-glucoside) and leucine-aminopeptidase (L-leucine-7-amido-4-methyl-coumarin) added at saturating concentration (40 mM solution of L-Leucine-7-amino-4-methyl-coumarin for ectoprotease activity (EPA) and 6 mM solution of 4-methylumbelliferyl-β-D-glucoside). Wavelengths for excitation and emission were 360 and 445 nm for 7-amino-4-methylcoumarine (MCA), and 365 and 455 nm for 4-methylumbelliferone (MUF). Fluorescence in the samples was measured as a function of time over 60 min at 2°C in the dark. Increase of fluorescent units with time was converted into activity from a standard curve prepared with the end product of the reaction, 7-amino-4-methylcoumarin (Sigma) for EPA and 4-methylumbelliferone (Sigma) for EGA. The relative standard deviation was 12% (n = 20) on aminopeptidase and 11% (n = 20) on glucosidase activities.

Bacterial production was estimated by incorporation of ^{3}H-thymidine (Fuhrman and Azam 1982). Duplicate 10 ml subsamples were incubated for 1 h in the dark at 2°C in the presence of 20 nM of ^{3}H-thymidine (Amersham; 40–50 Ci mol^{-1}), filtered on 0.2-µm cellulose acetate membrane filters (Sartorius) and extracted with ice-cold 5% Trichloroacetic acid (TCA) (Becquevort and Smith 2001). Thymidine incorporation was converted into bacterial production using conversion factors of Ducklow et al. (1999) established for the Ross Sea bacterial communities (i.e., 8.6×10^{17} bacteria produced per mole of thymidine incorporated in the cold TCA insoluble material). For bacterial production, the relative standard deviation was 9.3% (n = 20). The specific growth rate was estimated from bacterial production and bacterial biomass.

Bacterial growth efficiency (BGE), i.e., the efficiency at which bacterioplankton convert OC into bacterial biomass was calculated from the rate of decline in substrate (OC) and the rate of increase in bacterial biomass (BB) estimated from bacterial production as follow:

$$\mathrm{BGE} = \Delta \mathrm{BB} / -\Delta \mathrm{OC}.$$

Based on this equation, we derived BGE by property–property linear regression of ΔBB vs. −ΔOC from the model II regression which could be determined significantly when $P < 0.05$.

The genetic diversity was determined at 0, 6 and 30 days using denaturing gradient gel electrophoresis (DGGE) of polymerase chain reaction (PCR)-amplified 16S rRNA genes. The DNA extraction and PCR-DGGE were performed according to Schäfer and Muizer (2001) and using 341F-GC/907R primers. DGGE fingerprints were processed by image analysis (DOC-PRINT, Proxylab). The distance between lanes on the gels was calculated by simply matching distance and the cluster analysis was performed using the SYSTAT Package (V.11, SPSS) [Euclidean distance and unweight pair group method with arithmetic mean (UPGMA)].

Statistical analysis

We used three-way analysis of variations (ANOVA) analysis without replication to test for the separate and interactive effect of Fe enrichment of the *Phaeocystis* culture (LFe and HFe), organic matter fractionation (DOM and TOM), and incubation time on the various parameters measured. Data were initially tested for normal distribution and equality of variances to verify ANOVA assumptions. When these conditions were not met, data were log-transformed. All statistical analyses were conducted using the SYSTAT Package (V.11, SPSS), using a significance level of 0.05. Fitting of least-squares regression was performed using Sigmaplot (V.7, SPSS) to determine whether OC changed with time. A Pearson's correlation was done for ectoenzymatic activities and bacteria growth rates among the various treatments to determine if linear relationships existed among these variables.

Results

Characteristics of LFe and HFe *Phaeocystis* cultures

At the time of subsampling for regrowth experiments, dissolved Fe concentrations were 0.85 nM in the LFe culture and 2.76 nM in the HFe culture. The Chl *a* cell content was lower in the LFe (0.05 pg Chl *a* cell^{-1} in LFe and 0.24 pg Chl *a* cell^{-1} in HFe) reflecting Fe stress of *Phaeocystis* cells in the LFe (Table 1). In the LFe culture, 81% of total phytoplanktonic biomass was composed of *Phaeocystis* single cells while in the HFe culture *Phaeocystis* colonies represented 68% of the total biomass. Mucus carbon of *Phaeocystis* colonies represented only a minor fraction of total phytoplanktonic carbon, representing only 2.3% of the C in HFe bioassay. In the LFe conditions, the mucus contribution in total phytoplanktonic carbon was even lower (0.3%). In the latter, the colonies observed were usually composed of two cells embedded in a mucilaginous matrix.

POC and PON concentrations were higher in the HFe culture, but the increases of POC and PON were not proportional (Table 1). This results in contrasted C:N molar ratios with the highest value (8.3) recorded in the LFe culture as compared to the HFe culture (5.0). DOC was higher in the LFe culture than in the HFe culture. DOC initially present in the Antarctic seawater (before *Phaeocystis* inoculation) was 107 μM and thus freshly produced DOC (DOC_{fresh}) during the phytoplanktonic growth, was 73 μM in the LFe culture and 22 μM in the HFe culture. These DOC_{fresh} values contributed to 71 and 20% of freshly produced TOC (TOC_{fresh} = (DOC_{fresh} + POC). TOC_{fresh} amounted to 103 and 111 μM in the LFe and HFe culture, respectively.

Table 1 Fe concentrations, Chl *a* per cell, *P. antarctica* cell abundance, particulate organic carbon (POC), particulate organic nitrogen (PON) and dissolved organic carbon (DOC) in the *P. antarctica* cultures under low Fe (<1 nM Fe, LFe) and Fe amended (HFe) conditions used for bacterial regrowth experiments

Parameters	LFe	HFe
Dissolved Fe (nM)	0.85	2.76
Chl *a* cell^{-1} (pg cell^{-1})	0.05	0.24
Cell abundance (cells l^{-1})	7.44×10^6	3.69×10^7
POC (μM)	30	89
PON (μM)	3.6	17.9
DOC (μM)	179	128

Bacterial regrowth experiments

Time evolution of organic carbon

The initial concentrations of organic carbon ranged between 125 and 208 μM (Fig. 1). Organic carbon was similar in the DOM and TOM LFe bioassays (DOM_{LFe} and TOM_{LFe}, Fig. 1a) showing little contribution of POC to total organic carbon (7%). On the contrary, in the HFe bioassays, a larger fraction of the organic matter (41%) was in the particulate form (89 μM, Fig. 1b) due to the high contribution of *Phaeocystis* particulate material in the TOM_{HFe} bioassay. The lowest initial values of organic carbon were found in the DOM_{HFe} bioassay (Fig. 1b).

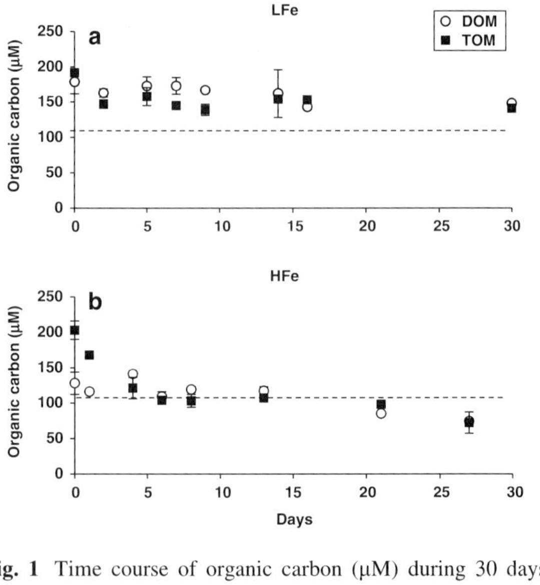

Fig. 1 Time course of organic carbon (μM) during 30 days incubation in low Fe (<1 nM Fe) bioassay (**a**) and Fe-amended bioassay (**b**). Dashed line indicates the initial DOC value in the culture seawater. Error bars represent the standard deviation on two replicates. An exponential decay function ($y = a\,e^{-bx}$) was estimated in each bioassays. DOM_{LFe} bioassay: $a = 174.73$ and $b = 0.0066$, $r^2 = 0.64$, $P < 0.05$; TOM_{LFe} bioassay $a = 163.41$ and $b = 0.0061$, $r^2 = 0.26$, n.s.; DOM_{HFe} bioassay: $a = 131.65$ and $b = 0.0182$, $r^2 = 0.74$, $P < 0.01$; TOM_{HFe} bioassay: $a = 166.77$ and $b = 0.0369$, $r^2 = 0.70$, $P < 0.01$

As reported in Table 2, LFe–HFe conditions as well as time significantly affected ($P < 0.05$) the OC concentrations. In the LFe bioassays, organic carbon decreased very slowly to about 142 μM at day 30 in both DOM_{LFe} and TOM_{LFe} bioassays. After five days of incubation in the HFe bioassays, the organic carbon in the TOM_{HFe} bioassay decreased to similar concentrations as in the DOM_{HFe} bioassay (Fig. 1b), underlining the high lability of fresh *Phaeocystis*-derived particulate organic carbon. Lower concentrations of ~75 μM were reached in both HFe bioassays compared to LFe bioassays. These values were less than the initial DOC (107 μM) measured in the culture seawater, suggesting that the addition of fresh organic matter could stimulate the degradation of the organic matter being otherwise semilabile as previously observed in some other cases (Cherrier et al. 1996).

The percentage of biodegradable organic carbon ($BOC_{30\ days}$), estimated by subtracting the OC at day 30 from the OC at day 0 and expressed as a percentage of the OC at day 0, ranged between 18 (DOM_{LFe}) and 64% (TOM_{HFe}; Table 3). By taking into account only freshly produced OC (DOC_{fresh} and POC), these percentages increased and ranged between 44 and >100%. The >100% data were likely related to the stimulation of semilabile organic matter degradation, as suggested above. $BOC_{30\ days}$ was enhanced by a factor of 2 in the HFe bioassays, where 100% of the freshly organic matter was degraded after 30 days. After seven days, 100% of TOC_{fresh} and 90% of DOC_{fresh} were already degraded. In the

Table 2 *P*-values of a three-way ANOVA without replication performed on organic carbon, bacterial abundance, CTC-positive cell ratio, biovolume, ectoprotease activity (EPA), specific EPA, ectoglucosidase activity (EGA), specific EGA, EPA/EGA ratio, production, specific growth rate and specific (CTC+) growth rate measured in the four bioassays

Effects parameters	Time	HFe–LFe	DOM–TOM	HFe–LFe × DOM–TOM
Organic carbon	<0.0001	<0.0001	0.986	<0.0001
Bacterial abundance	0.914	<0.0001	0.052	0.104
CTC-positive cell ratio	<0.0001	<0.0001	<0.0001	<0.0001
Biovolume	0.015	0.158	0.001	0.001
EPA	0.001	0.008	<0.0001	<0.0001
Specific EPA	0.044	0.011	<0.0001	<0.0001
EGA	<0.0001	<0.0001	<0.0001	<0.0001
Specific EGA	0.020	0.045	0.008	0.044
EPA/EGA ratio	0.006	0.043	0.008	0.044
Production	0.000	0.000	0.238	0.040
Specific growth rate	0.001	0.045	0.003	0.035
Specific (CTC+) growth rate	0.001	0.083	0.008	0.392

Grouping factors were time (from 0 to 30 days), HFe–LFe and DOM–TOM conditions

Table 3 Biodegradable organic carbon (BOC) after 30 days ($BOC_{30\ days}$, absolute value and percentage of initial organic carbon), biodegradable freshly produced organic carbon after 30 days ($B_{fr}OC_{30\ days}$, percentage of initial organic carbon) and after seven days ($B_{fr}OC_{7\ days}$, percentage of initial organic carbon) in dissolved (DOM) and total organic matter (TOM) under low Fe (<1 nM Fe, LFe) and Fe amended (HFe) conditions used for bacterial regrowth experiments

	LFe		HFe	
	DOM_{LFe}	TOM_{HFe}	DOM_{HFe}	TOM_{HFe}
$BOC_{30\ days}$ (μM)	32	51	53	131
$BOC_{30\ days}$ (%)	18	26	42	64
$B_{fr}OC_{30\ days}$ (%)	44	60	102	105
$B_{fr}OC_{7\ days}$ (%)	9	55	90	100

The standard deviation for BDOC concentrations was estimated to amount less than 5%

LFe bioassay, TOC (TOM bioassay) was also more biodegradable than DOC (DOM bioassay; Table 3). Some 44 and 60% of freshly produced OC were degraded after 30 days in LFe DOM and TOM bioassays, respectively. The TOC_{fresh} was quickly degraded but only 9% of DOC_{fresh} was degraded after seven days in LFe bioassays.

Bacterial communities

Bacterial abundances ranged from 0.3 to 1.5×10^9 cells l^{-1} in the LFe bioassays and from 0.1 to 10.1×10^9 cells l^{-1} in the HFe bioassays (Fig. 2a, b). Indeed, a significant ($P < 0.05$) effect of LFe–HFe conditions on the bacterial abundance was found (Table 2). Bacterial abundance in HFe bioassays increased rapidly during the first 6–8 days of the experiment and then decreased to value close to initial conditions (Fig. 2b). During the first 6 days, the evolution of bacterial abundance was quite similar in the DOM_{HFe} and TOM_{HFe} bioassays, increasing by a factor 10 compared to day 0 (Fig. 2b). The highest abundance in the TOM_{HFe} bioassay was reached at day 8, whereas the abundance decreases rapidly after day 6 in the DOM_{HFe} bioassay (Fig. 2b). No comparable increase in bacterial cells was observed in the LFe bioassays (Fig. 2a). The percentage of CTC positive cells was always below 23% in the LFe bioassays, while the percentage increased to 41% (day 6) and 61% (day 8) in the DOM_{HFe} and TOM_{HFe} bioassays, respectively (Fig. 2c, d). As reported in Table 2, LFe–HFe as well as DOM–TOM conditions affected significantly ($P < 0.05$) the percentage of CTC positive cells. In the LFe bioassays (Fig. 2e), biovolumes decreased from 0.4 to 0.07 μm^3. In contrast, the biovolumes in the TOM_{HFe} bioassay (Fig. 2f) reached 0.71 μm^3 at day 6 and decreased to 0.12 μm^3. The statistical analysis reported a significant ($P < 0.05$) effect of time and fractionation of organic matter (OM) on the biovolumes (Table 2). LFe–HFe condition did not affect significantly bacterial biovolume (Table 2).

DGGE of PCR-amplified 16S rDNA revealed between 7 and 20 (typically 11) bands in each sample, of which two were detected in all community DNA samples. The maximum number of bands was observed in the HFe bioassays. DGGE fingerprints corresponding to day 0 were similar for DOM and TOM bioassays in the both Fe conditions, whereas five bands were similar in bacterial communities present in HFe compared to those in LFe conditions. As illustrated by the dendrogram in Fig. 3, the assemblage did not change until day 6 in DOM bioassays but was immediately modified in the TOM

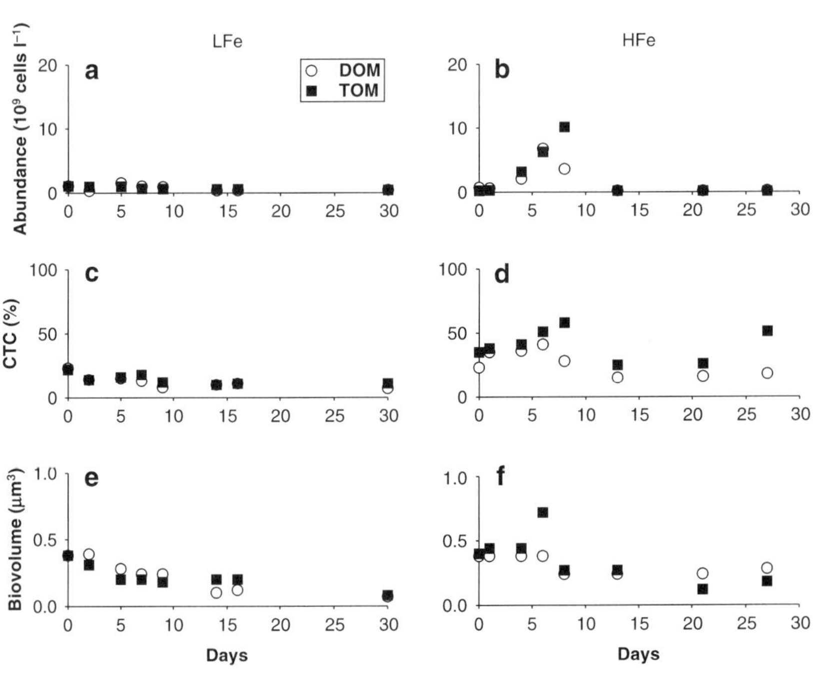

Fig. 2 Time course of bacterial abundance (10^9 cells l^{-1}) (**a**, **b**), CTC-positive cell ratio (%) (**c**, **d**) and bacterial biovolume (μm^3) (**e**, **f**) during 30 days incubation in low Fe (<1 nM Fe) bioassay and Fe amended bioassay

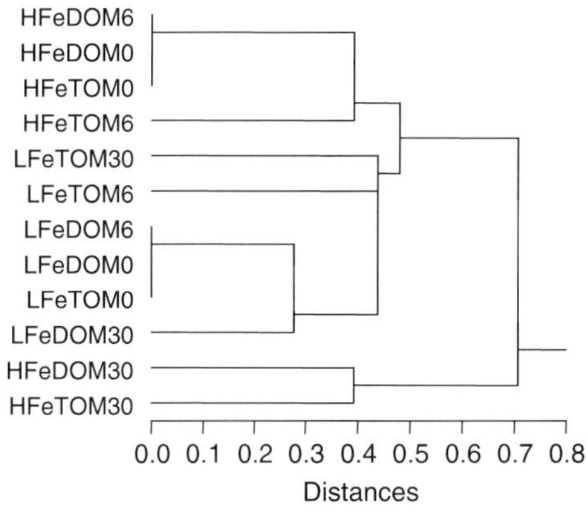

Fig. 3 Dendrograme showing the similarity of DGGE patterns from different sampling times (0, 6 and 30 days) in low Fe (<1 nM Fe) bioassay and Fe-amended bioassay

bioassays. Final detected assemblages, particularly in HFe bioassays, were always significantly different from the initial ones in all bioassays.

Bacterial activities

The ectoenzymatic activities were significantly ($P < 0.05$) affected by LFe–HFe and DOM–TOM conditions (Table 2). The highest ectoenzymatic activities, leucine-peptidase and β-glucosidase, were reported in the HFe bioassays (Fig. 4d). At day 5, EPA and EGA drastically increased, concomitantly to the bacterial abundance (Fig. 2b). The specific activities (per cell) were not always higher in the HFe bioassay than in the LFe conditions (Table 4). As reported in Table 2, the specific EGA was significantly affected by LFe–HFe conditions. In fact, in the presence of particulate organic matter such as in the TOM bioassay, the specific activities in the LFe condition drastically increased. In this experiment, the EPA as well as the EGA showed the highest values. The EPA/EGA ratios ranged between 104 and 3,449 (Table 4). This ratio was significantly regulated by the HFe–LFe and DOM–TOM conditions (Table 2).

LFe–HFe conditions were a highly significant ($P < 0.0001$) regulating factor for bacterial production (Table 2). Indeed, bacterial production in the HFe bioassays increased after day 4 (Fig. 4f) and rapidly decreased to background levels observed in all bioassays. The growth rates ranged from 0.0004 to 0.0066 h^{-1} in the LFe bioassays and from 0.0001 to 0.0326 h^{-1} in the HFe bioassays (Table 4). LFe–HFe as well as DOM–TOM conditions affected significantly ($P < 0.05$) the specific growth rates (Table 2).

Fig. 4 Time course of ectoprotease activities (EPA, µmoles l^{-1} h^{-1}) (**a**, **b**), ecto-β-glucosidase activities (EGA, nmoles l^{-1} h^{-1}) (**c**, **d**) and bacterial production (10^6 cells l^{-1} h^{-1}) (**e**, **f**) during 30 days incubation in low Fe (<1 nM Fe) bioassay and Fe-amended bioassay

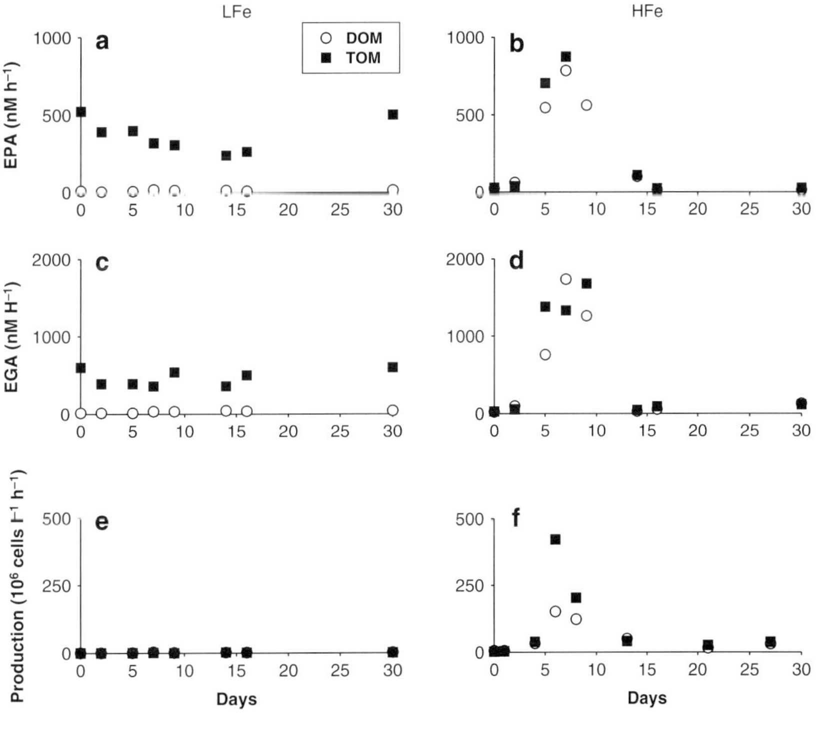

Table 4 Bacterial activities in presence of dissolved (DOM) and total organic matter (TOM) in low-Fe (<1 nM Fe, LFe) and amended-Fe (HFe) conditions; specific ectoenzymatic activities [ectoprotease (EPA) and β-glucosidase (EGA) activities] and their ratio, specific growth rates for total bacteria and for CTC-positive (CTC+) cells and bacterial growth efficiencies (BGE)

Parameters	LFe	HFe
DOM EPA (10^{-15} moles cell^{-1} h^{-1})	4.4–37.7	31.5–453.7
TOM EPA (10^{-15} moles cell^{-1} h^{-1})	397.8–1194.1	118.5–535.4
DOM EGA (10^{-18} moles cell^{-1} h^{-1})	8.4–125.2	29.5–609.4
TOM EGA (10^{-18} moles cell^{-1} h^{-1})	388.3–4137.1	123.4–984.0
DOM EPA/EGA ratio	220–624	104–3,449
TOM EPA/EGA ratio	289–1,025	224–2,423
DOM growth rate (h^{-1})	0.0004–0.0066	0.0011–0.0326
TOM growth rate (h^{-1})	0.0002–0.0048	0.0001–0.0134
DOM growth rate (CTC+) (h^{-1})	0.0017–0.0771	0.0030–0.0812
TOM growth rate (CTC+) (h^{-1})	0.0007–0.0476	0.0001–0.0892
DOM BGE	0.05 (0.03)	0.29 (0.03)*
TOM BGE	0.15 (0.04)*	0.30 (0.04)*

Bacterial growth efficiency is determined by the slope of the model II regression and values in parentheses represent standard error of the slope

* The values reported were significant at the 0.05 significance level

By taking into account only CTC-positive cells in the estimation of cell specific growth rates, higher values were obtained for all bioassays as expected, and no difference between values in HFe and LFe conditions was observed (Table 2). BGE ranged from 0.05 (n.s. $P > 0.05$) to 0.30 (Table 4). LFe–HFe conditions also affected bacterial growth efficiency, particularly when organic matter was only provided in the dissolved form (Table 4).

Discussion

Fe control on *P. antarctica*-derived organic matter concentration and quality

There are numerous potential sources of organic matter for bacteria which can result from a variety of processes (Nagata 2000). Two major sources are photosynthetic extracellular release (ER) by the algae and products from phytoplankton lysis (Nagata 2000). These two organic pools were represented in our DOM and TOM bioassays, respectively. The relative significance of ER and phytoplankton lysis varies in function of different environmental factors and notably due to inorganic nutrients (Nagata 2000). The biochemical composition of the organic matter derived from ER and/or phytoplankton lysis is different, consequently they have their specific lability.

In our study, POC concentrations, derived from *P. antarctica* cultures, increased by a factor 3 in the HFe compared to the LFe culture, although TOC_{fresh} (DOC_{fresh} and POC) was in the same range for both cultures. POC corresponded to 80% of TOC_{fresh} in the HFe conditions and only 29% of TOC_{fresh} in LFe conditions. The value in the HFe culture is comparable to those measured in the Ross Sea where 70–99% of the accumulated organic matter was partitioned in POM during a *P. antarctica* bloom, with modest net accumulation of DOM stocks (Carlson et al. 1998). In bottle experiments using natural seawater assemblages (Southern Ross Sea), *P. antarctica* did not produce large amounts of DOC during nutrient-replete growth (Smith 1998). In mesoscale Fe fertilization experiments, reports on DOC are very scarce and no consistent trend has been evidenced so far (de Baar et al. 2005). Increase in DOC excretion by phytoplankton was previously observed under phosphate and nitrate limitation and saturating light conditions (Lancelot 1983; Nagata 2000). Phytoplankton excretion of organic matter is interpreted as an overflow mechanism of recently photosynthetized compounds that occurs during nutrient limitation (Fogg 1983). However, as Fe limits photosynthesis directly, an increase in the excretion under Fe limitation is less probable. Oliver et al. (2004) showed a strong correlation between primary production (PP) and bacterial production (BP) in the Southern Ocean Fe experiment (SOFeX), suggesting a strong dependence of bacterial activity on phytoplankton production. However, the ratio BP:PP stayed low and did not change, indicating that ER did not increase as a result of Fe addition. To our knowledge, there is no available data on excretion under Fe limitation in the literature so far. On the other hand, Fe limitation could enhance lysis of phytoplankton cells and subsequent release of organic compounds (Gobler et al. 1997). Poorvin et al. (2004) showed that lysate-associated Fe can be more available than inorganic Fe. Consequently, lysis could be a mechanism of survival for the whole phytoplankton community allowing Fe recycling and its availability for the surviving cells.

Fe limitation also induces significant changes in phytoplankton growth characteristics, cell metabolism and therefore in the biochemical composition of its organic matter (e.g., Sunda and Huntsman 1995; van Leeuwe and Stefels 1998; Kudo et al. 2000; Lewandowska and Kosakowska 2004). Therefore Fe has an effect on the concentration and quality of organic matter available for bacteria. In our study, Fe controlled the morphotype dominance of *P. antarctica*, i.e., mucilaginous colonies dominated in the HFe culture whereas free-living cells represented most of the biomass in the LFe culture. As a general rule, amendment of Fe was shown to induce a shift in phytoplankton community structure from nanoplankton to microplankton taxa dominating biomass (de Baar et al. 2005). The particularity of *P. antarctica* is that the size range shift is observed within the same species. This change in the morphological form could have resulted in higher C/N ratio for colonies, the mucus exopolymers having higher C/N ratio than the cells (Solomon et al. 2003). However, the mucus carbon contribution to total carbon biomass was only minor, 0.3 and 2.3% in LFe and HFe conditions, respectively. In the Ross Sea, Mathot et al. (2000) also reported relatively low mean mucus carbon contribution of C_{col} of 14% (3–33% of C_{col}). This suggests that the observed difference in C/N could better result from the direct effect of Fe on phytoplankton metabolism rather than from the effect of Fe on the morphology of *P. antarctica*. Some studies report deviations in the Redfield ratios of marine algae as a function of Fe status. The nature and strength of the deviations vary greatly among species (Tremblay and Price 2002). The most commonly observed response is a decrease of the N/P and C/P ratios under Fe-limited growth, whereas C/N ratio can remain relatively constant over a wide range of Fe concentrations (Geider and La Roche 2002) or increase as a result of Fe limitation under nitrate-replete conditions (Muggli et al. 1996). In our study C/N was below the Redfield value of 6.6 (Redfield et al. 1963) for the organic matter derived from *P. antarctica* in the HFe culture, in agreement with the range of 4–6.4 reported for *P. antarctica* grown under nonlimiting conditions (Schoemann et al. 2005). Accordingly, Smith et al. (2000) estimated C/N values between 5.9 and 6.8 during the exponential phase of *P. antarctica* blooms in the Ross Sea. The C/N increased above Redfield value in the LFe culture. During *P. antarctica* blooms decay in the Ross Sea, increased C/N values of 7.3–8.7 were observed by Smith et al. (2000), likely attributable to Fe limitation (Vaillancourt et al. 2003).

Fe control on the bacterial community composition and activities

The few studies that exist on Fe control of bacterial degradation in HNLC waters agreed that bacterioplankton is first C-limited as a consequence of low primary production due to Fe limitation and not directly limited by Fe (Hutchins et al. 1998; Church et al. 2000; Kirchman et al. 2000). As reported above, the amount of TOC, potential OC available for bacteria, was similar in our HFe and LFe experiments and therefore bacteria could be colimited by two major factors: the availability of Fe and the quality of the organic matter. These factors cannot be discriminated in our experiments. HFe bioassays contained, in addition to increased Fe, a better quality of OM than the LFe bioassays. Moreover, as the distribution in dissolved and particulate organic carbon was affected by Fe availability in the phytoplanktonic cultures, the response of bacteria to different size pool of organic matter (DOM–TOM) was also investigated.

Regrowth experiments show typically a lag phase, an exponential bacterial growth followed by a stationary phase (Ammerman et al. 1984). In our study, the evolution of bacteria abundance stayed very constant along the time in LFe bioassays. In the presence of OM derived from HFe *Phaeocystis* culture, the bacteria grew and reached a maximal abundance at about 6–8 days, just after the initial decline in DOC and TOC. The exponential phase is nevertheless followed by a sharp decline of bacterial abundance and associated activities such as ectoenzymatic activities and production. Sempéré et al. (1998) attributed the observed bacteria decline in bioassays to bacterial mortality processes.

The degradation of the organic matter depends initially on the presence of bacterial species in the environment able to use it. Indeed, bacterial variability can be comparable to that of phytoplankton, reflecting the ultimate dependence of bacteria on substrates generated by primary production (Giovannoni and Sting 2005). In our experiments, bacterial community structure was modified in HFe bioassays as compared to LFe bioassays. Only six species seemed well adapted to both HFe and LFe conditions. These results seem in contradiction with the first published studies in the field showing that Fe addition had little or no effect on DGGE profiles (Hutchins et al. 2001; Arrieta et al. 2004), suggesting that field bacterial communities are well adapted to episodic Fe inputs. However, in our experiments, the modification of the bacterial community structure could result from the longer time period of bacterial pre-adaptation (>30 days) in the different bioassays. This long pre-adaptation could favour the development of best adapted species for specific Fe/organic ligand complexes (Weaver et al. 2003).

Some <10 to >75% of the bacteria are not active in marine environments (Ducklow 2000). The degradation of organic matter depends on active bacteria. The proportion of active cells in the bacterial community, determined from CTC-positive cells, was also controlled by LFe–HFe condition. The percentages of CTC-positive bacteria in our experiments was in the higher range of the data observed in marine ecosystems (Schumann et al. 2003), particularly in Fe-enriched bioassays where up to 61% of bacterial cells were CTC-positive after one week of incubation. The reduction of CTC to red fluorescing formazan crystals via electron transport activity is used to detect respiring bacterial cells (Rodriguez et al. 1992) and the abundance of the CTC-positive cells is correlated with in situ respiration rates (Smith 1998). As most of the cellular Fe in heterotrophic bacteria is being dedicated to the respiratory electron chain (Tortell et al. 1999), the percentage of CTC-positive cells could therefore be regulated by Fe availability. However, the highest percentage was measured in the presence of TOM, suggesting that the quality of the organic matter could also have had an impact on the relative abundance of respiring cells.

Little attention has been paid so far to the role of Fe on bacterial organic matter hydrolysis activities. The hydrolysis of organic matter is a necessary step prior to bacterial uptake and remineralization of organic matter. Basically, bacterial growth is directly dependent on the concentration of monomers, the only organic molecules able to be directly transferred inside bacteria. These monomers are supplied either directly by phytoplankton ER or indirectly after bacterial ectoenzymatic hydrolysis of high molecular weight organic matter. In LFe bioassay, a very high activity (protease and β-glucosidasic activities) was measured in the presence of TOM. Under LFe conditions, it could be advantageous for bacteria to increase its enzymatic hydrolytic activity to release and allow the availability of Fe associated to high-molecular-weight organic matter. In the LFe bioassays, significant correlations between specific ectoenzymatic activities and specific growth rates were obtained (Table 5), suggesting that all substrates produced by hydrolytic activities are simultaneously used for growth.

Moreover, our study shows that specific ectoenzymatic activities (per cell) are stimulated in presence of TOM as compared to DOM especially in LFe bioassays. During the in situ Fe fertilization experiment EISENEX, the ectoenzymatic activities were highest inside the Fe patch (Arrieta et al. 2004). The highest specific ectoenzymatic activities in the HFe bioassay could result (i) from a significant input of high-molecular-weight organic matter due to increase in phytoplankton production, and/or (ii) from difference in the percentage of bacteria actively expressing a distinct enzyme. Moreover, a direct role of Fe on the ectoenzymatic hydrolysis cannot be excluded. For example, some aminopeptidases require a metal as cofactor (Lowther and Matthews 2000), suggesting that low metal concentrations in open-ocean seawater could potentially limit protein hydrolysis.

Table 5 Correlation (*r*) between specific ectoenzymatic activities [ectoprotease (EPA) and ectoglucosidase (EGA) activities] and growth rates in low-Fe (<1 nM, LFe) and Fe-amended (HFe), dissolved (DOM) and total organic matter (TOM) bioassays

	LFe		HFe	
	EPA	EGA	EPA	EGA
DOM	0.81*	0.85*	n.s.	0.66**
TOM	0.91*	0.82*	n.s.	0.61**

*$P < 0.05$, **$P < 0.1$, n.s., $P > 0.1$

As reported for Antarctic seawaters (e.g., Christian and Karl 1995; Becquevort and Smith 2001), protease activity was 2–3 orders of magnitude higher than that of β-glucosidasic activity in all bioassays. Interestingly, the ratio of protease to β-glucosidasic activities increased in HFe bioassays, mostly due to increase of protease activity. Proteases hydrolyze peptides and proteins, which comprise a part of the utilizable fraction of the marine organic N pool (Kirchman 2000). Consequently, the decrease of elemental organic composition (C/N ratio) observed in HFe conditions could induce protease production.

Our results showed that bacterial abundance and specific growth rate respond to Fe enrichment. In all Fe enrichment experiments, shipboard incubation as well as in situ fertilization, specific growth rates of bacteria increased significantly in response to Fe enrichment (Price et al. 1994; Pakulski et al. 1996; Hutchins et al. 1998; Church et al. 2000; Kirchman et al. 2000; Cochlan 2001; Hall and Safi 2001; Oliver et al. 2004; Arrieta et al. 2004; Suzuki et al. 2005). However, a significant increase in bacterial abundance is not always observed (Cochlan 2001; Hall and Safi 2001) due to a close coupling between bacteria and bacterivorous protozoa. Interestingly, in the present study, the specific growth rates estimated from the CTC-positive cells were on the same range in both Fe conditions, suggesting an actively growing subpopulation of bacteria specifically adapted to low-Fe condition.

Fe enrichment impacts directly and/or indirectly on the bacterial growth efficiency (BGE), i.e., the ratio between growth and carbon uptake. Nevertheless a different limitation was observed in presence of DOM or TOM. Fe conditions affected bacterial growth efficiency, especially when organic matter was provided in the only dissolved form (Table 4). In LFe conditions, a large fraction of the consumed organic matter was catabolized and the energy produced was primarily used for maintenance rather than for growth, with resulting low growth efficiency (Tortell et al. 1996; Church et al. 2000; Kirchman et al. 2003). As reported above, in LFe bioassays, respiratory electron transport activity seems to be reduced, as suggested by the low relative abundance of CTC-positive cells. The respiratory electron transport system is essential for adenosine triphosphate (ATP) synthesis in aerobic heterotrophic bacteria, and low Fe can result in a significant decrease of BGE with a subsequent effective co-limitation by Fe and C.

The estimation of bacterial growth efficiency from the variation of bacterial biomass and organic carbon consumption could be underestimated due to continual recycling of organic matter by successive bacterial lysis and consumption of this lysed organic matter in our bioassays. In both Fe conditions, the increase of bacterial abundance is controlled by mortality processes. No protozoa were observed by microscopy. Mortality by protozoan grazing can thus be excluded.

On the contrary, lysis and particularly viral lysis could be a major process in these bioassays. In cultures of natural marine bacteria inoculated into 0.2 μm filtered sea water, Wilcox and Furhman (1994) reported that virus abundance increased after few days of bacterial growth. The high abundance and production of bacteria in the HFe bioassay could enhance viral activity and consequently increase specific mortality rates. In agreement with this, during the mesoscale Fe fertilization EISENEX, a higher viral infection of bacterioplankton was estimated in the Fe-enriched patch (Weinbauer et al. 2003). Lysis could be of significance in Fe-limited ecosystems, as Fe released via lysis can be highly bioavailable (Poorvin et al. 2004).

Fe control on organic matter remineralization in a *Phaeocystis*-dominated ecosystem

In nutrient replete bioassays, the organic matter derived from *Phaeocystis* is generally quickly degraded (see review Schoemann et al. 2005). In natural conditions, large accumulation of DOC have been occasionally reported after *Phaeocystis* blooms (Eberlein et al. 1985; Lancelot et al. 1987; Davidson and Marchant 1992), suggesting a temporal uncoupling between DOM production and bacterial assimilation. Mechanistic explanations are that inorganic nutrients can limit bacterial uptake capacity of otherwise biodegradable DOM and/or alternatively that a part of the produced DOM escapes fast remineralization due to low biochemical biodegradability (Thingstad and Lignell 1997). As reported above and in other studies (Verity et al. 1988; Baumann et al. 1994), the OM derived from nutrient-limited *Phaeocystis* is characterized by high C/N and C/P, giving rise to a substrate for bacteria with unfavourable high carbon content. In temperate ecosystems, bacteria seem first limited by P availability (Thingstad et al. 1997). In HNLC environment, Fe could control bacterial activities and/or organic matter lability and therefore the accumulation of OM in the upper water column.

In HFe condition, *P. antarctica*-derived organic carbon is quickly degraded. As estimated from our experiments, more than 90% of freshly produced organic carbon, for dissolved as well as for total organic carbon, was degraded by bacteria after seven days. These results are the highest percentages of *Phaeocystis*-derived biodegradable organic carbon reported in the literature. For *P. globosa*-derived organic carbon between 57 and 93% could be degraded after seven days, under conditions of nonlimiting nutrients (see Schoemann et al. 2005 and reference herein). Only 7–24% of *P. antarctica*-derived organic carbon was degraded in *in situ* conditions after 7 days (Carlson et al. 1999). In LFe conditions, the degradation is slower; nevertheless up to 54% of TOC could be degraded after 7 days but only 9% of DOC was degraded after 7 days. The high lability of TOM is in agreement with the high microbial hydrolysis of OM measured in LFe TOM bioassay. Nevertheless, a significant part of *P. antarctica*-derived OM was degraded before 30 days in low and high Fe conditions.

Conclusion

Fe has been shown to play a key role in limiting phytoplankton growth and carbon uptake, and in determining the phytoplankton community structure in the Southern Ocean. Therefore, the carbon transfer efficiency to deep water by the biological carbon pump depends on Fe availability. Heterotrophic bacteria may also influence the biological carbon pump efficiency through the remineralization of POM originating from phytoplankton but also of DOM in the surface and mesopelagic waters. As shown in this study, the degradation of OM is directly and/or indirectly influenced by Fe availability. The quality of dissolved as well as of particulate matter derived from *P. antarctica* improved and was rapidly degraded by bacteria in HFe conditions. A high fraction of organic carbon consumed by bacteria was incorporated in biomass, as suggested by the high bacterial growth efficiency. Very little accumulation of DOM was observed and therefore the potential C export will mainly be due to *Phaeocystis* colonies and aggregates sedimentation. Contrastingly, in LFe conditions, the relative proportion of DOM in the *P. antarctica*-derived organic matter was high. This organic matter was slowly degraded, but even so, after one month, 44% was degraded by bacteria. The particulate organic carbon was as in the HFe conditions, quickly degraded. In this case, the persistence of DOC could participate to carbon exportation by the advection of surface water masses to greater depth. A large fraction of carbon consumed by bacteria was remineralized and respired as suggested by the low bacterial growth efficiency, suggesting a lower relative potential export of C in this case.

Acknowledgments We are especially grateful to Geraldine Sarthou (LEMAR, Brest, France) for measuring Fe concentrations. We thank Natacha Brion and Nicolas Savoye for POC and PON measurements, Jeroen de Jong and Delphine Lannuzel for their help during the experiments and Elsa Breton for her help in the statistical analysis. We also thank three anonymous reviewers for their comments and suggestions, which greatly improved the manuscript. This research was supported by the Belgian Science policy (contract no. EV/11/7B—BELCANTO II) and the Belgian French Community (ARC-contract 02/07-287-SIBCLIM). This is also a contribution to the SOLAS international research initiative, the European Network of Excellence EUR-OCEANS (contract no. 511106-2).

References

Ammerman JW, Fuhrman JA, Hagström Å, Azam F (1984) Bacterioplankton growth in seawater. I. Growth kinetics and cellular characteristics in seawater culture. Mar Ecol Prog Ser 18:31–39

Arrieta JM, Weinbauer MG, Lute C, Herndl GJ (2004) Response of bacterioplankton to iron fertilization in the Southern Ocean. Limnol Oceanogr 49(3):799–808

Baumann MEM, Lancelot C, Brandini FP, Sakshaug E, John DM (1994) The taxonomic identity of the cosmopolitan prymnesiophyte *Phaeocystis*: a morphological and ecophysiological approach. J Mar Syst 5:5–22

Becquevort S, Smith WO Jr (2001) Aggregation, sedimentation and biodegradability of phytoplankton-derived material during spring in the Ross Sea, Antarctica. Deep-Sea Res II 48:4155–4178

Bölter M, Dawson R (1982) Heterotrophic utilization of biochemical compounds in Antarctic waters. Neth J Sea Res 16:315–332

Carlson CA, Ducklow HW, Smith WO, Hansel DA (1998) Carbon dynamics during spring blooms in the Ross Sea polynya and the Sargasso Sea: contrasts in dissolved and particulate organic carbon portioning. Limnol Oceanogr 43:375–386

Carlson CA, Bates NR, Ducklow HW, Hansell DA (1999) Estimation of bacterial respiration and growth efficiency in the Ross Sea, Antarctica. Aquat Microb Ecol 19:229–244

Cherrier J, Bauer JE, Druffel ERM (1996) Utilization and turnover of labile dissolved organic matter by bacterial heterotrophs in eastern North Pacific surface waters. Mar Ecol Prog Ser 139:267–279

Christian JR, Karl DM (1995) The bacterial ectoenzyme activities in marine waters: activity ratios and temperature responses in three oceanographic provinces. Limnol Oceanogr 40:1042–1049

Church MJ, Hutchins DA, Ducklow HW (2000) Limitation of bacterial growth by dissolved organic matter and iron in the Southern Ocean. Appl Environ Microbiol 66:455–466

Coale KH, Wang X, Tanner SJ, Johnson KS (2003) Phytoplankton growth and biological response to iron and zinc addition in the Ross Sea and Antarctic Circumpolar Current along 170°W. Deep Sea Res II 50:635–653

Cochlan WP (2001) The heterotrophic bacterial response during a mesoscale iron enrichment experiment (IronEx II) in the eastern Equatorial Pacific Ocean. Limnol Oceanogr 46:428–435

Davidson AT, Marchant HJ (1992) Protists abundance and carbon concentration during a *Phaeocystis*-dominated bloom at an Antarctic coastal site. Polar Biol 12:387–395

de Baar HJW, Boyd PW, Coale KH, Landry MR, Tsuda A, Assmy P, Bakker DCE, Bozec Y, Barber RT, Brzezinski MA, Buesseler KO, Boye M, Croot PL, Gervais F, Gorbunov MY, Harrison PJ, Hiscock WT, Laan P, Lancelot C, Law CS, Levasseur M, Marchetti A, Millero FJ, Nishioka J, Nojiri Y, van Oijen T, Riebesell U, Rijkenberg MJA, Saito H, Takeda S, Timmermans KR, Veldhuis MJW, Waite AM, Wong C (2005) Synthesis of iron fertilization experiments: from the iron age in the age of enlightenment. J Geophys Res 110, C09S16. DOI: 10.1029/2004JC002601

Ducklow H (2000) Bacterial production and biomass in the oceans. In: Kirchman DL (ed) Microbial ecology of the oceans. Wiley, New-York, pp 85–120

Ducklow HW, Carlson CA, Smith WO (1999) Bacterial growth in experimental plankton assemblages and seawater cultures from the *P. antarctica* bloom in the Ross Sea, Antarctica. Aquat Microb Ecol 19:215–227

Eberlein K, Leal MT, Hammer KD, Hickel W (1985) Dissolved organic substances during a *Phaeocystis pouchetii* bloom in the German Bight (North Sea). Mar Biol 89:311–316

El-Sayed SZ, Biggs DC, Holm-Hansen O (1983) Phytoplankton standing crop, primary productivity, and near surface nitrogenous nutrient fields in the Ross Sea, Antarctica. Deep-Sea Res 30:871–886

Fogg GE (1983) The ecological significance of extracellular products of phytoplankton photosynthesis. Bot Mar 256:3–14

Furhman JA, Azam F (1982) Thymidine incorporation as a measure of heterotrophic bacterioplankton production in marine surface waters: evaluation and field results. Mar Biol 166:109–120

Geider RJ, LaRoche J (2002) Redfield revisited: variability of C:N:P in marine microalgae and its biochemical basis. Eur J Phycol 37:1–17

Gerringa LJA, de Baar HJW, Timmermans KR (2000) A comparison of iron limitation of phytoplankton in natural oceanic waters and laboratory media conditioned with EDTA. Mar Chem 68:335–346

Giovannoni SJ, Sting U (2005) Molecular diversity and ecology of microbial plankton. Nature 437:343–347

Gobler CJ, Hutchins DA, Fisher NS, Cosper EM, Sanùdo-Wilhelmy SA (1997) Release and bioavailability of C, N, P, Fe, and Se following viral lysis of a marine chrysophyte. Limnol Oceanogr 42:1492–1504

Granger J, Price NM (1999) The importance of siderophores in iron nutrition of heterotrophic marine bacteria. Limnol Oceanogr 44:541–555

Hall JA, Safi K (2001) The impact of *in situ* Fe fertilisation on the microbial food web in the Southern Ocean. Deep-Sea Res II 48:2591–2613

Hutchins DA, DiTullio GR, Zhang Y, Bruland W (1998) An iron limitation mosaic in the California upwelling regime. Limnol Oceanogr 43:1037–1054

Hutchins DA, Campbell BJ, Cottrell MT, Takeda S (2001) Response of marine bacterial community composition to iron additions in three iron-limited regimes. Limnol Oceanogr 46(6):1535–1545

Kirchman DL (2000) Uptake and regeneration of inorganic nutrients by marine heterotrophic bacteria. In: Kirchman DL (ed) Microbial ecology of the oceans. Wiley, New-York, pp 261–288

Kirchman DL, Meon B, Cottrell MT, Hutchins DA (2000) Carbon versus iron limitation of bacterial growth in the California upwelling regime. Limnol Oceanogr 45(8):1681–1688

Kirchman DL, Hoffman KA, Weaver R, Hutchins DA (2003) Regulation of growth and energetics of a marine bacterium by nitrogen source and iron availability. Mar Ecol Progr Ser 250:291–296

Koroleff F (1983a) Determination of ammonia. In: Grasshoff K, Ehrhardt M, Kremling K (eds) Methods of seawater analysis. Verlag-Chemie, Basel, pp 150–157

Koroleff F (1983b) Determination of phosphorus. In: Grasshoff K, Ehrhardt M, Kremling K (eds) Methods of seawater analysis. Verlag-Chemie, Basel, pp 125–139

Kudo I, Miyamoto M, Noiri Y, Maita Y (2000) Combined effects of temperature and iron on the growth and physiology of the marine diatom *Phaeodactylum tricornutum* (Bacillariophyceae). J Phycol 36(6):1096–1102

Lancelot C (1983) Factors affecting phytoplankton extracellular release in the Southern Bight of the North Sea. Mar Ecol Prog Ser 12:115–121

Lancelot C, Mathot S, Owens NJP (1986) Modelling protein synthesis, a step to an accurate estimate of net primary production: the case of *Phaeocystis pouchetii* colonies in Belgian coastal waters. Mar Ecol Prog Ser 32:193–202

Lancelot C, Billen G, Sournia A, Weisse T, Colijin F, Veldhuis MJW, Davies A, Wassman P (1987) *Phaeocystis* blooms and nutrient enrichment in the continental coastal zones of the North Sea. AMBIO 16(1):38–46

Lewandowska J, Kosakowska A (2004) Effect of iron limitation on cells of the diatom *Cyclotella meneghiniana Kützing*. Oceanologia 46(2):269–287

Lowther WT, Matthews BW (2000) Structure and function of the methionine aminopeptidases. Biochim Biophys Acta 1477:157–167

Mathot S, Smith WO Jr, Carlson CA, Garrison DL (2000) Estimate of *Phaeocystis* sp. carbon biomass: methodological problems related to the mucilaginous nature of the colonial matrix. J Phycol 36:1049–1056

Muggli DL, Lecourt M, Harrison PJ (1996) Effects of iron and nitrogen source on the sinking rate, physiology and metal composition of an oceanic diatom from the subarctic Pacific. Mar Ecol Prog Ser 132:215–227

Nagata T (2000) Production mechanisms of dissolved organic matter. In: Kirchman L (ed) Microbial ecology of the oceans. Wiley, New York, pp 121–152

Obata H, Karatani H, Nakayama E (1993) Automated determination of iron in seawater by chelating resin concentration and chemiluminescence detection. Anal Chem 5:1524–1528

Oliver JL, Barber RT, Smith WO, Ducklow HW (2004) The heterotrophic bacterial response during the southern ocean iron experiment (SOFeX). Limnol Oceanogr 49:2129–2140

Pakulski JD, Coffin RB, Kelley CA, Holder SL, Downer R, Aas DP, Lyons MM, Jeffrey WH (1996) Iron stimulation of Antarctic bacteria. Nature 383:133–143

Pomeroy LR, Wiebe WJ, Deibel D, Thompson RJ, Rowe GT, Pakulski JD (1991) Bacterial responses to temperature and substrate concentration during the Newfoundland spring bloom. Mar Ecol Prog Ser 75:143–159

Poorvin L, Rinta-Kanto JM, Hutchins SW, Wilhem SW (2004) Viral release of iron and its bioavailability to marine plankton. Limnol Oceanogr 49:1734–1741

Porter KG, Feig YS (1980) The use of DAPI for identifying and counting aquatic microflora. Limnol Oceanogr 25:943–948

Price NM, Ahner BA, Morel FMM (1994) The equatorial Pacific ocean: grazer controlled phytoplankton populations in an iron-limited ecosystem. Limnol Oceanogr 39:520–534

Raven J (1988) The iron and molybdenum use efficiencies of plant growth with different energy, carbon and nitrogen sources. New Phytol 109:279–287

Redfield AC, Ketchum BH, Richards FA (1963) The influence of organisms on the composition of seawater. In: Hill MN (ed) The sea. Wiley, New York, pp 26–77

Rodriguez GG, Phipps D, Ishiguro K, Ridgeway HF (1992) Use of fluorescent redox probe for direct visualization of actively respiring bacteria. Appl Environ Microbiol 58:1801–1808

Sarthou G, Baker AR, Blain S, Achterberg EP, Boye M, Bowie AR, Croot P, Laan P, de Baar HJW, Jickells TD, Worsfold PJ (2003) Atmospheric iron deposition and sea-surface dissolved iron concentrations in the eastern Atlantic Ocean. Deep-Sea Res 50:1339–1352

Schäfer H, Muyzer G (2001) Denaturing gradient gel electrophoresis in marine microbial ecology. In: Paul J (ed) Methods in microbiology, vol 30. Academic Press, London, pp 425–468

Schoemann V, Wollast R, Chou L, Lancelot C (2001) Effects of photosynthesis on the accumulation of Mn and Fe by *Phaeocystis* colonies. Limnol Oceanogr 46:1065–1076

Schoemann V, Becquevort S, Stefels J, Rousseau V, Lancelot C (2005) *Phaeocystis* blooms in the global ocean and their controlling mechanisms: a review. J Sea Res 53:43–66

Schumann R, Rieling T, Görs S, Hammer A, Selig U, Schiever U (2003) Viability of bacteria from different aquatic habitats. I. Environmental conditions and productivity. Aquat Microb Ecol 32:121–135

Sedwick PN, Ditullio GR, Mackey DJ (2000) Iron and manganese in the Ross Sea: seasonal iron limitation in Antarctic shelf waters. J Geophys Res 105:11321–11336

Sempéré R, Van Wambeke F, Azourmanian H, Chambaut AL, Ferrière L, Bianchi M (1998) On the use of batch systems to determine DOC bacterial lability and bacterial growth efficiency in seawater samples. In: Baeyens J, Dehairs F, Goyens L (eds) Integrated marine system analysis. European Network for integrated Marine System Analysis, pp 233–238

Simon M, Azam F (1989) Protein content and protein synthesis rate of planktonic marine bacteria. Mar Ecol Progr Ser 51:201–213

Smith EM (1998) Coherence of microbial respiration rate and cell-specific bacterial activity in a coastal planktonic community. Aquat Microb Ecol 16:27–35

Smith WO Jr, Nelson DM, DiTullio GR, Leventer AR (1996) Temporal and spatial patterns in the Ross Sea: phytoplankton biomass, elemental composition, productivity and growth rates. J Geophys Res 101:18455–18466

Smith WO Jr, Marra J, Hiscock MR, Barber RT (2000) The seasonal cycle of phytoplankton biomass and primary productivity in the Ross sea, Antarctica. Deep-Sea Res II 47:3119–3140

Smith WO Jr, Dennett MR, Mathot S, Caron DA (2003) The temporal dynamics of the flagellated and colonial stages of *Phaeocystis antarctica* in the Ross Sea. Deep-Sea Res II 50:605–617

Solomon CM, Lessard EJ, Keil RG, Foy MS (2003) Characterization of extracellular polymers of *Phaeocystis globosa* and *P. antarctica*. Mar Ecol Prog Ser 250:81–89

Sugimura Y, Suzuki Y (1988) A high temperature catalytic oxidation method for the determination of non-volatile dissolved organic carbon in seawater by direct injection of a liquid sample. Mar Chem 24:105–131

Sunda WG, Huntsman SA (1995) Iron uptake and growth limitation in oceanic and coastal phytoplankton. Mar Chem 50:189–206

Suzuki K, Hinuma A, Saito H, Kiyosawa H, Liu H, Saino T et al (2005) Response of phytoplankton and heterotrophic bacteria in the northwest subarctic Pacific to *in situ* iron fertilization as estimated by HPLC pigment analysis and flow cytometry. Prog Oceanogr. DOI: 10.1016/j.pocean.2005.02.007

Thingstad F, Billen G (1994) Microbial degradation of *Phaeocystis* material in the water column. J Mar Syst 5(1):55–65

Thingstad TF, Lignell R (1997) Theoretical models for the control of bacterial growth rate, abundance, diversity and carbon demand. Aquat Microb Ecol 13:19–27

Thingstad TF, Hagström A, Rassoulzadegan F (1997) Accumulation of degradable DOC in surface waters: is it caused by a malfunctioning microbial loop? Limnol Oceanogr 42(2):398–404

Tortell PD, Maldonado MT, Price NM (1996) The role of heterotrophic bacteria in the iron-limited ocean ecosystems. Nature 383:330–332

Tortell PD, Maldonado MT, Granger J, Price NM (1999) Marine bacteria and biogeochemical cycling of iron in the oceans. FEMS Microb Ecol 29:1–11

Tremblay J-E, Price NM (2002) The effect of iron in the C/N/P/Si composition of phytoplankton: does Fe-deficiency affect structural or labile, soluble pools? Ocean Sciences Meeting, Hawaii, Feb. 11–15

Utermölh H (1958) Zur Vervelkommnung der quantitativen Phytoplankton-Methodik. Mitt Int Verein Theor Angew Limnol 9:1–38

Vaillancourt RD, Marra J, Barber RT, Smith WO Jr (2003) Primary productivity and *in situ* quantum yields in the Ross Sea and Pacific Sector of the Antarctic Circumpolar Current. Deep-Sea Res II 50:559–578

van Leeuwe MA, Stefels J (1998) Effects of iron and light stress on the biochemical composition of Antarctic *Phaeocystis* sp. (Prymnesiophycea). II. Pigment composition. J Phycol 34:496–503

Verity PG, Villareal TA, Smayda TJ (1988) Ecological investigations of blooms of colonial *Phaeocystis pouchetti* – 1. Abundance, biochemical composition, and metabolic rates. J Plankton Res 10(2):219–248

Watson SW, Novitsky TJ, Quinby HL, Valois FW (1977) Determination of bacterial number and biomass in the marine environment. Appl Environ Microbiol 33:940–946

Weaver RS, Kirchman DL, Hutchins DA (2003) Utilization of iron/organic ligand complexes by marine bacterioplankton. Aquat Microb Ecol 31:227–239

Weinbauer MG, Arrieta JM, Herndl GJ (2003) Stimulation of viral infection of bacterioplankton during a mesoscale iron fertilization experiment in the Southern ocean. Geophys Res abstracts 5:12280

Wilcox RM, Fuhrman JA (1994) Bacterial viruses in coastal seawater: lytic rather than lysogenic production. Mar Ecol Progr Ser 114:35–45

Yentsch CS, Menzel DW (1963) A method for the determination of phytoplankton chlorophyll and phaeophytin by fluorescence. Deep-Sea Res 10:221–231

Biogeochemistry (2007) 83:137–145
DOI 10.1007/s10533-007-9086-2

BRIEF COMMUNICATION

The colonization of two *Phaeocystis* species (*Prymnesiophyceae*) by pennate diatoms and other protists: a significant contribution to colony biomass

Andrey F. Sazhin · L. Felipe Artigas · Jens C. Nejstgaard · Marc E. Frischer

Received: 2 October 2005 / Accepted: 12 June 2006 / Published online: 14 March 2007

Abstract The association of *Phaeocystis* spp. with small pennate diatoms during three *Phaeocystis*-dominated spring blooms were investigated in the Eastern English Channel (2003 and 2004) and in coastal waters of Western Norway during a mesocosm experiment (2005). In each of these studies, colonization of the surface of large *Phaeocystis* spp. colonies by small needle-shaped diatoms (*Pseudo-nitzschia* spp.) were observed. In the English Channel the diatom *Pseudo-nitzschia delicatissima* colonized the surface of large (>100 μm) *Phaeocystis globosa* colonies. The abundance of *Pseudo-nitzschia delicatissima* reached 130 cells per colony and formed up to 70% of the total carbon associated with *Phaeocystis* cells during late bloom stages. In Norwegian waters, the surface of large (>250 μm) *Phaeocystis pouchetii* colonies were colonized by *Pseudo-nitzschia* cf. *granii* var. *curvata* and to a lesser degree by other phytoplankton and protist species, although the abundance of these diatoms was never greater than 40 cells per colony. Based on these observations we suggest that diatoms utilize *Phaeocystis* colonies not only as habitat, but that they are able to utilize the colonial matrix as a growth substrate. Furthermore, these observations indicate that a considerable fraction of biomass (chlorophyll) associated with *Phaeocystis* colonies, especially large colonies concerned with intense and prolonged blooms, are due to co-occurring plankton species and not exclusively *Phaeocystis* cells.

Keywords Biomass estimate, colonies · Colonization · *Phaeocystis* bloom · *Pseudo-nitzschia* species

A. F. Sazhin (✉)
P.P. Shirshov Institute of Oceanology RAS, Nakhimovsky Prospect, Moscow, 117851, Russia
e-mail: asazhin@com2com.ru

L. F. Artigas
UMR 8013 ELICO (FORTEC), MREN-Universite du Littoral Cote d'Opale, Wimereux, France

J. C. Nejstgaard
Department of Biology, University of Bergen, Bergen, Norway

M. E. Frischer
Skidaway Institute of Oceanography, Savannah, GA, 31411, USA

Introduction

The coexistence of *Phaeocystis* species with pennate diatoms and other protists, although not universally observed, is a well-known and common phenomenon (Hasle 1964; Rousseau et al. 1994; Hasle and Syvertsen 1997; Peperzak et al. 1998; Wassmann et al. 1999; Throndsen et al. 2003; Hamm and Rousseau 2003). In general, the colonization of the microphytobenthos is well known in shelf waters and in the near-shore regions of seas and oceans, pennate diatoms are also known to dominate in the community of nano- and microalgae on the surface of macrophytes and zoobenthos (e.g. Proshkina-Lavrenko 1963; Sapozhnikov 2003 and many others).

Since the genus *Phaeocystis* was first described over 100 years ago (Pouchet 1892), a large number of observations and studies have reported the conspicuous bloom-forming *Phaeocystis* spp. (see, e.g., review of Schoemann et al. 2005). However, for a long time the presence of the small needle-shaped *Nitzschia* species on and/or in *Phaeocystis* colonies was reported only by Hasle and co-workers (Hasle 1964; Hasle and Syvertsen 1997). These investigators described two diatoms species in association with the surface of *Phaeocystis pouchetii* colonies: *Pseudo-nitzschia delicatissima* and *Pseudo-nitzschia granii* var. *curvata* (Throndsen et al. 2003). Wassmann et al. (1999) reported abundant populations of *Pseudo-nitzschia* cf. *pseudodelicatissima* and the cryptophyte flagellate *Plagioselmis* sp., associated with colonies of *P. pouchetii* in the Barents Sea. In other studies microscopic examination of senescent *Phaeocystis* colonies and foam revealed the presence of large numbers of the pennate diatoms (*Nitzschia* species) on the surface of *Phaeocystis globosa* (Peperzak et al. 1998; Hamm and Rousseau 2003).

During our studies, we observed an abundance of the small needle-shaped *Pseudo-nitzschia* species on *Phaeocystis* colonies provoking interest in both qualitative and quantitative analysis of this phenomenon. If *Pseudo-nitzschia* species comprise a significant fraction of total *Phaeocystis* colony biomass, it is essential to take this fact into consideration in the studies of food webs, vertical fluxes, biogeochemical element fluxes, etc. since *Phaeocystis* is a widely distributed phytoplankter and it often develops massive blooms (Schoemann et al. 2005).

Material and methods

During a bloom of *P. globosa* in March–May 2003 and in February–April 2004 water samples were collected at several stations in the Eastern English Channel in the coastal waters off Boulogne—Wimereux, France (Fig. 1). Water samples were collected from three depth ranges including surface waters (0.5–2 m), mid-depth waters (10–12 m), and water from just above the bottom (20–22 m) using Niskin bottles during several cruises of the RVs "Sepia II" and "Côtes de la Manche". Coastal-offshore transects and 24 h drifting experiments were carried out at two sites, one located off the Wimereux-Slack estuaries and another located southward in the Bay of Somme.

Samples of *P. pouchetii* were also collected from blooms in a mesocosm experiment conducted at the marine biological field station, University of Bergen, Western Norway (60°16′ N, 05°14′ E), on 01–27 April 2005 (Fig. 2). The experiment was conducted essentially as described by Nejstgaard et al. (2006) in floating 11 m^3 polyethylene enclosures (4.5 m deep, 2 m diameter). The mesocosms were transparent with 90% penetration of photosynthetically active radiation (PAR). Mesocosms were filled on 31 March by pumping fjord water from a depth of 5 m. The water column was well mixed with an airlift-system, pumping 40 l water min^{-1}. In order to allow the introduction of new species, to avoid substantial pH changes due to primary production, and to replace the water sampled during the mesocosm experiment, 10% of the mesocosm water was renewed daily starting from April 1 by pumping fjord water from outside the mesocosm from a depth of 2.5 m. An intense bloom of *P. pouchetii* was stimulated after fertilization with NO_3 (16 μm) and PO_4 (1 μm).

Whole colonies and cells within colonies (non-motile stage) of *Phaeocystis* were identified and enumerated by light microscopy. The samples were either live or preserved with 1% glutaraldehyde-lugol solution (Rousseau et al. 1990). In addition, we used epifluorescence microscopy to enumerate and identify flagellate forms (motile stage) of *Phaeocystis* and microplankton (Sherr et al. 2000). In our modified procedure, the samples were stained with primulin, fixed with 3.6% glutaraldehyde solution with 10% glycerol added for better preservation, and gently filtered onto black-stained Nucleopore filters (0.4 μm). Identical microscopy procedures were applied for samples

Fig. 1 Map of the sampling area in the Eastern English Channel

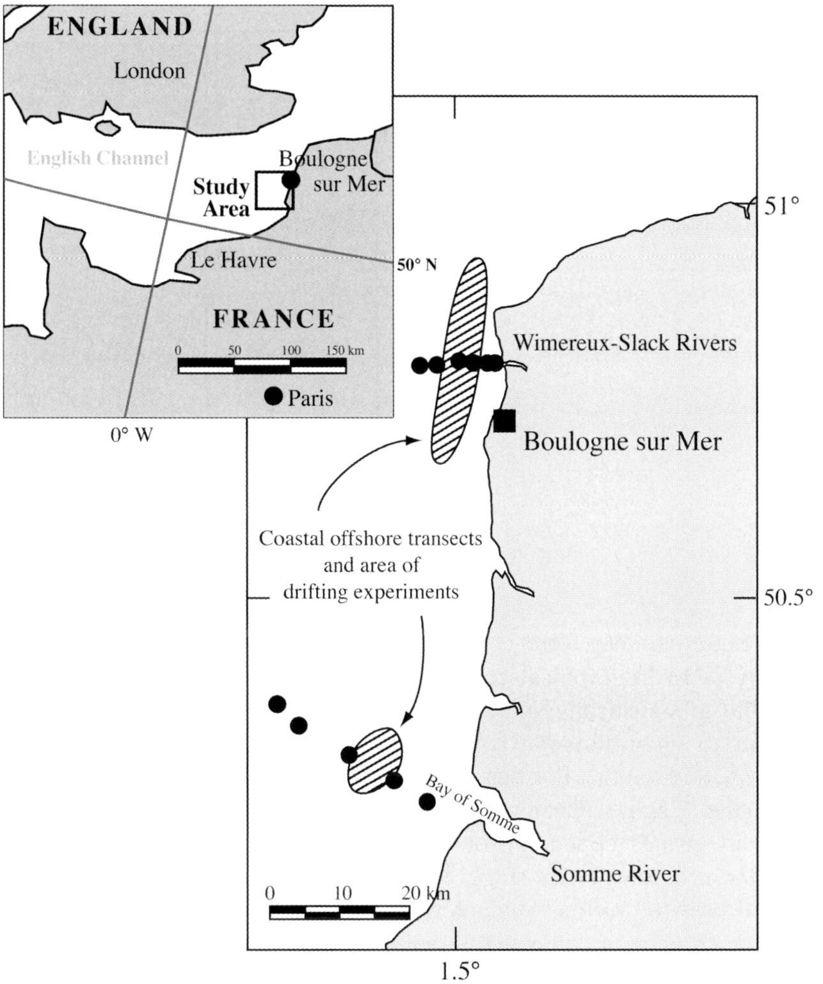

from the Eastern English Channel and Western Norway. Our microscopic estimation of the number of nonmotile cells of *Phaeocystis* inside colonies of different volumes corresponded to the results obtained by Rousseau et al. (1994), i.e., according to the following regression equation: $\log N_{cell} = 0.51 \log V + 3.67$ where N_{cell} is the colony cell number and V is the colonial volume expressed in mm^3. For the identification of diatom species, we used a taxonomic key based on light microscope observations (Hasle et al. 1996; Hasle and Syvertsen 1997; Throndsen et al. 2003); and electron microscope observations (Hasle et al. 1996; Priisholm et al. 2002). For *Pseudo-nitzschia delicatissima* from the English Channel, scanning electron microscopy (SEM) electron micrographs were obtained using a LEO 438 VP scanning electron microscope. Cell volume was calculated by approximation to the closest sample 3D shapes and converted into C according to Menden-Deuer and Lessard (2000). The volume of diatoms was generally less than 3000 μm^3, so we applied the following volume-to-carbon conversion formula for protist plankton: pg C $cell^{-1} = 0.216 \times volume^{0.939}$.

From the English Channel, 84 samples were counted. Most of the samples were preserved and colonies of *Phaeocystis* were enumerated in the total sample volume collected (100–250 ml). The samples were settled for minimum 24 h, then gently concentrated by removing surface water with a plastic syringes (ca. 10 mm diameter), passed through a 5 μm nylon mesh. All volume of concentrated subsamples (5–10 ml) was observed in 1 ml-Naumann counting chamber (Naumann 1922).

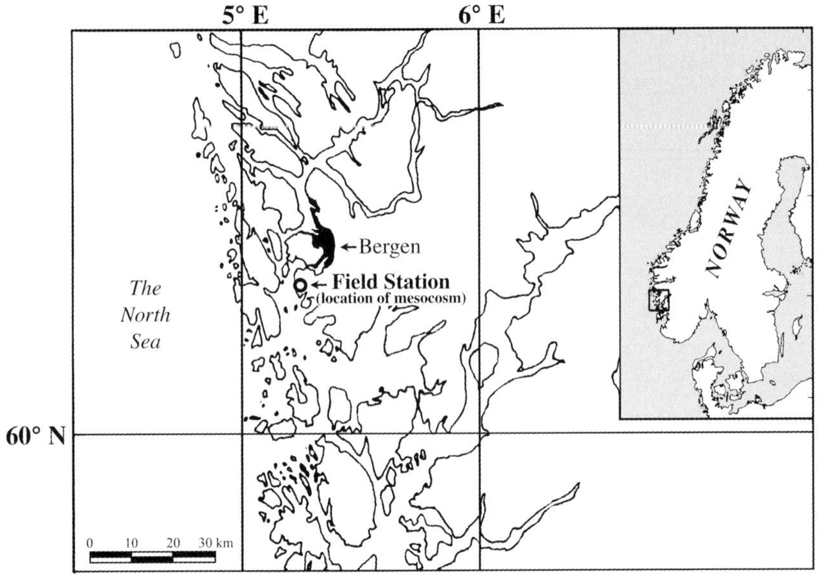

Fig. 2 Map of the location of the Norwegian mesocosm experiment

Because samples were collected from a large variety of hydrographical conditions, it was not possible to statistically analyze the samples with respect to environmental conditions (i.e., time, location, and depth). For this reason, the samples from the English Channel were divided into different groups depending on the stage of the *P. globosa* bloom. These stages were determined on the basis of various characteristics: the presence or absence of colonies, their size and shape and the number of cells within the colonies. An additional indication of one or another stage of the *P. globosa* bloom was provided by other algae, mainly diatoms which were also counted in the samples. Succession changes in phytoplankton community allowed us to make more precise grouping of the data. Three groups were distinguished. The first group included the samples collected in March and several samples collected at the end of February (start of the *P. globosa* bloom), the second group included the samples collected in April (middle of the bloom), and the third included the samples collected in May (end of the bloom). Within each group, independently of collection date, location and depth, the data were counted as one set.

In the Western Norway mesocosm experiment only live samples were analyzed. The development of the *P. pouchetii* bloom in the fertilized mesocosm was followed daily from 1 April–27 April, 2005. During this period about 30 samples were analyzed. In each sample at least 500 colonies were counted and at least 20 colonies were measured.

Results

Pseudo-nitzschia delicatissima[1] (usually as single or formed pairs) colonized the surface of the *P. globosa* colonies at all depths from the Eastern English Channel (Fig. 3a–c). Diatom growth usually started when colony size was over 100 μm (typical biomass values for *P. globosa* cells colony^{-1} are about 380 pg C) and the bloom was rather intense. At this stage, diatoms were observed on 5–14% of the *P. globosa* colonies, with about 5–7 *Pseudo-nitzschia delicatissima* cells per colony (88–124 pg C colony^{-1}) (Fig. 4). Over the course of bloom development, the frequency of colonization increased to about 30%. After one month, when colony size reached 300–600 μm (typical biomass values for *P. globosa* cells colony^{-1} about 1140–1795 pg C), the number of *Pseudo-nitzschia delicatissima* increased

[1] In a recent article (a combination of the morphological and molecular findings) *Pseudo-nitzschia delicatissima* was shown to be a complex of three different species (*Pseudo-nitzschia delicatissima*, *P. decipiens* sp. nov. and *P. dolorosa* sp. nov. (Lundholm et al. 2006).

to 25–50 cells per colony (442–883 pg C colony^{-1}). Two months after the beginning of the bloom, nearly 100% of *P. globosa* colonies were colonized by diatoms, the average colony size was above 1000 μm (biomass values for *P. globosa* cells colony^{-1} were above 3560 pg C), and the number of *Pseudo-nitzschia delicatissima* varied from a few to 120–130 cells per colony (2119–2296 pg C colony^{-1}). Similar observations were made both in 2003 and 2004. In these studies, over the duration of the bloom, the average contribution of *Pseudo-nitzschia delicatissima* to the total biomass associated with *P. globosa* colonies was 46% of total carbon (nonmotile *P. globosa* cells and diatoms). However, during late bloom stages, the biomass of *Pseudo-nitzschia delicatissima* accounted for up to 70% of the total carbon of the *Phaeocystis* cells (nonmotile stage) within the colonies. In these estimates carbon from the colony matrix was not included since noncellular material associated with the colonies contains very little carbon (Rijssel et al. 1997).

During the bloom of *P. pouchetii* in the mesocosm experiment (Norway) the first diatoms appeared when the colonies reached a size of 250 × 180 μm containing about 50 nonmotile cells (544 pg C colony^{-1}). At this stage in colony development the colonies of *P. pouchetii* were transitioning from a spherical to ellipsoid shape and we began to observe their colonization by the diatom *Pseudo-nitzschia* cf. *granii* var. *curvata* (Fig. 5a, b). During the following two weeks, the mean size of the *P. pouchetii* colonies continued to increase to about 370 × 350 μm containing about 200 cells colony^{-1} (2176 pg C of cells/colony) and the number of *Pseudo-nitzschia* cf. *granii* var. *curvata* varied from 4 to 59 cells per colony (136–2011 pg C of cells colony^{-1}), with a mean of 18 cells per colony (613 pg C of cells colony^{-1}). However, fewer than 1% (0.22–0.92%) of *Phaeocystis* colonies was colonized by diatoms (Fig. 6). In addition to *Pseudo-nitzschia* cf. *granii* var. *curvata*, *P. pouchetii* colonies were occasionally colonized by other diatom species. For example, we observed a *P. pouchetii* colony with 12 cells of *Nitzschia frigida* on the surface and other colonies with 1–3 cells of *Cylindrotheca closterium* (Fig. 7). In two samples we observed *P. pouchetii* colonies with attached suctorian ciliates (*Acineta tuberosa*) (Fig. 8). Ciliates of the genus *Acineta* have been reported to be epibionts of decapods, cladocerans, copepods, ostracods isopods and amphipods (Fernandez-Leborans and Tato-Porto 2000; Arndt et al. 2005).

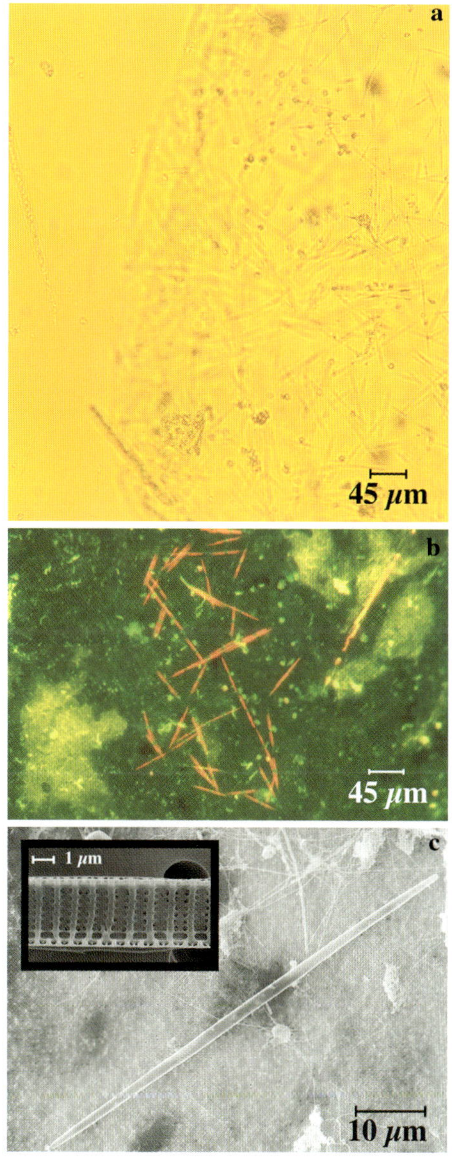

Fig. 3 (**a–c**) Diatom *Pseudo-nitzschia delicatissima* inhabiting the surface of the *Phaeocystis globosa* colonies from the Eastern English Channel (**a** light microscopy, **b** epifluorescence microscopy, **c** scanning electron microscopy)

Discussion

In this report we present observations of *Phaeocystis* colonies being colonized by different small

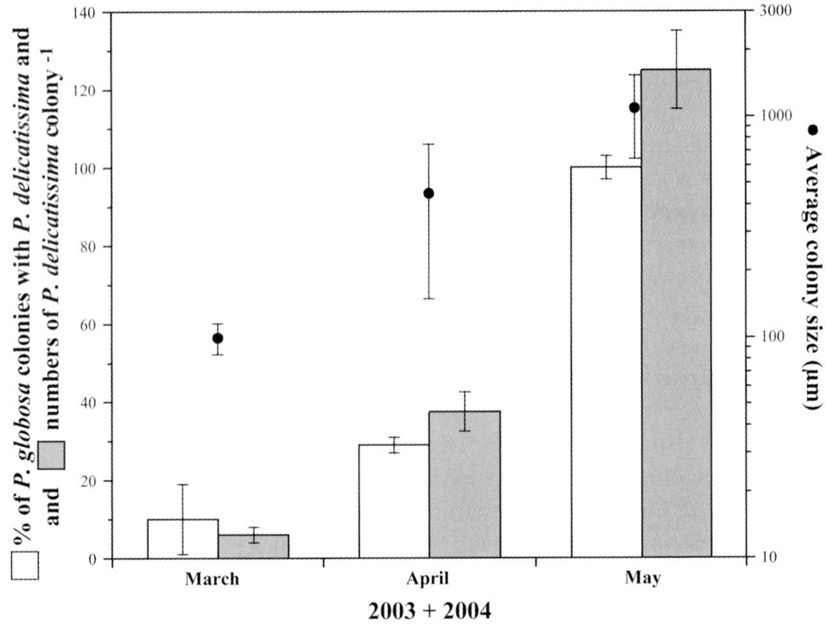

Fig. 4 Bloom and evolution of *Phaeocystis globosa* colonies and diatom *Pseudo-nitzschia delicatissima* from the Eastern English Channel in 2003 and 2004

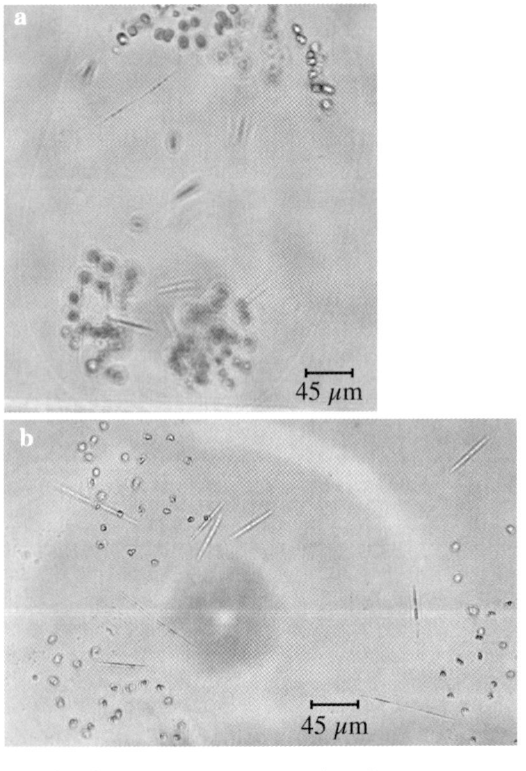

Fig. 5 (**a**, **b**) Diatom *Pseudo-nitzschia* cf. *granii* var. *curvata* inhabiting the surface of *Phaeocystis pouchetii* colonies from Western Norway (light microscopy)

needle-shaped *Pseudo-nitzschia* species to such an extent that the estimated diatom carbon at times was similar to or even exceeded the contribution of the nonmotile *Phaeocystis* cells within colonies. The colonization of *P. globosa* by *Pseudo-nitzchia delicatissima* during two successive years in the English Channel, of *P. pouchetii* by other *Pseudo-nitzschia* species in the Norwegian mesocosm experiment, and earlier reports by other authors, confirm that colonization of *Phaeocystis* colonies by diatoms and other protists appears to be widespread and common phenomenon. For example, the presence of *Calliacantha natans* (Choanoflagellida) in high abundance on senescent colonies of *P. pouchetii* was reported from the Eastern Bering Sea shelf (Sukhanova and Flint 2001), ovoid heterotrophic gymnodinoid dinoflagellates were found on senescent *Phaeocystis* colonies (Peperzak et al. 1998), and preliminary observations of *Phaeocystis antarctica* colonies collected in the Ross Sea show that small (less than 10 µm) heterotrophic protists were associated with the colonial matrix (Shields and Smith 2005). Since the colonies of *Phaeocystis* species may be intensely colonized by pigment-containing algae such as diatoms, care should be taken when estimating *Phaeocystis* colony biomass using pigment-based methods. For example, quantification by Chl *a*, or even high-performance liquid chromatography (HPLC) and chemical taxonomy (CHEMTAX) without direct inspection of the algal material (cf. Irigoien et al. 2004) may lead to significant overestimation of the *Phaeocystis* biomass. As seen in

Fig. 6 Bloom and evolution of *Phaeocystis pouchetii* colonies (**a**) and diatom *Pseudo-nitzschia* cf. *granii* var. *curvata* (**b**) from Western Norway. The squares refer to the number of colonies and the circles to the average size of the colonies

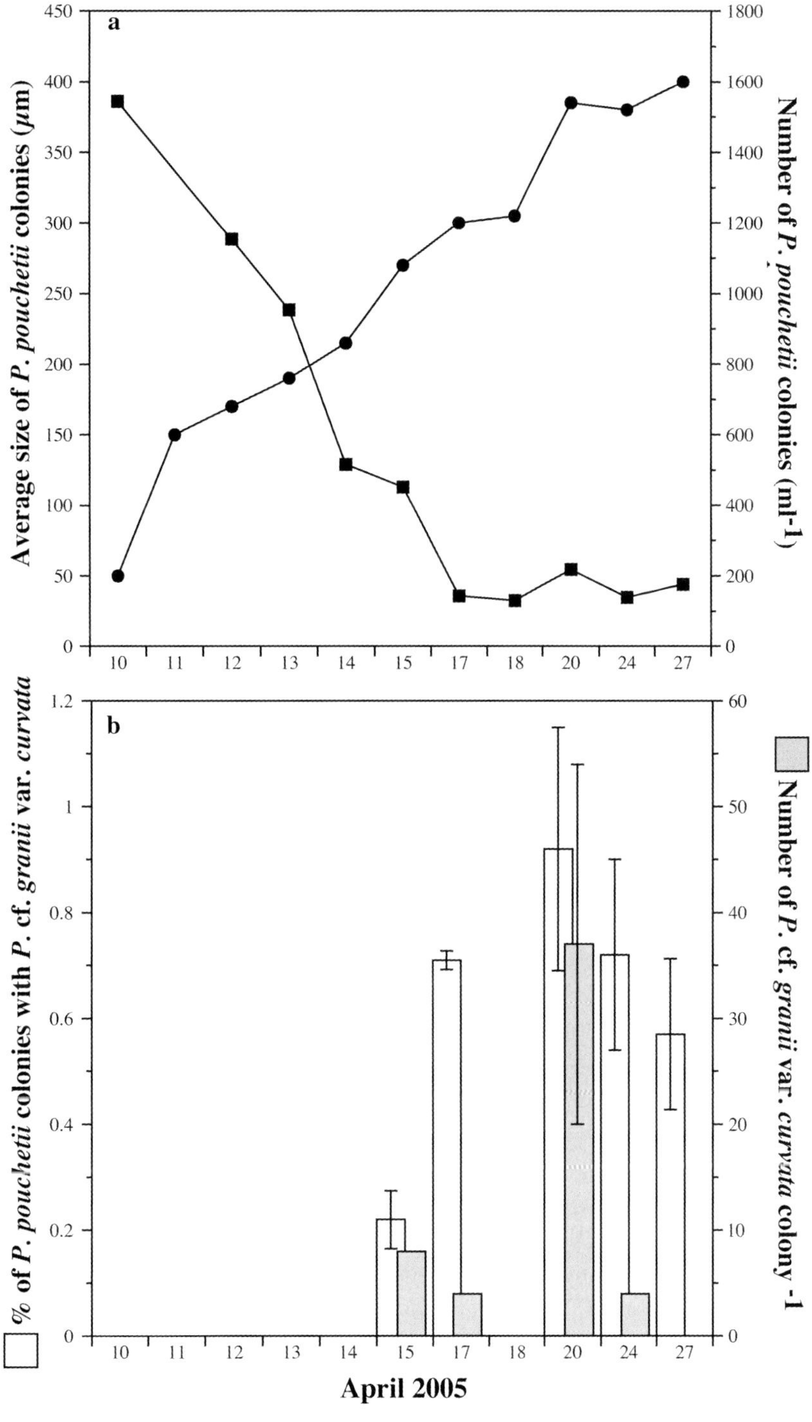

Fig. 3b, the majority of total chlorophyll associated with a *Phaeocystis* colony can be derived from diatoms on the colony rather than *Phaeocystis* cells within the colony. Therefore to accurately quantify *Phaeocystis* biomass, particularly during intense and prolonged blooms, it is necessary to consider the accompanying algae on the colonies and unambiguous methods such as microscopy should be routinely utilized.

The underlying processes influencing the association between *Phaeocystis* colonies and diatoms remains incompletely understood. The relationship

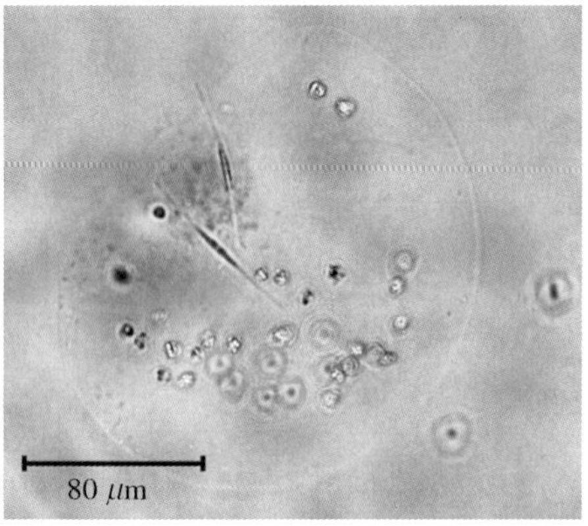

Fig. 7 Diatom *Cylindrotheca closterium* inhabiting the surface of a *Phaeocystis pouchetii* colony from Western Norway (light microscopy)

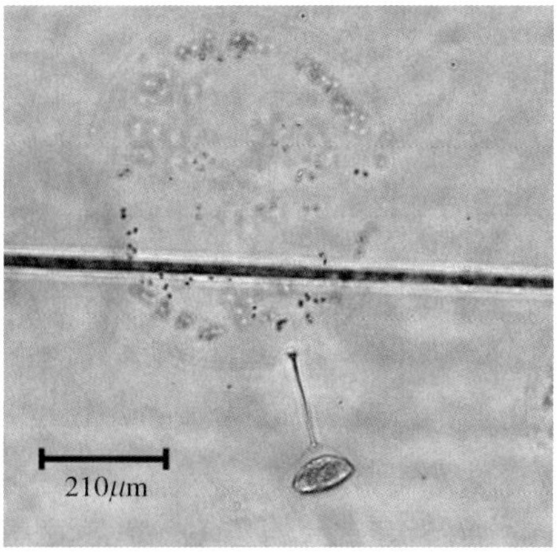

Fig. 8 *Acineta tuberosa* (Suctoria) attached to the surface of a *Phaeocystis pouchetii* colony from Western Norway (light microscopy)

between colonized *Phaeocystis* and diatoms may represent either an active or passive association. Recent studies in the Southern Ocean have demonstrated that *P. antarctica* and diatoms have similar photosynthetic capabilities and that growth of both can co-occur under similar light conditions (Hilst and Smith 2002). If this is also the case for *P. globosa*, *P. pouchetii*, and associated *Pseudo-nitzschia* species, the colonization of *Phaeocystis* colonies by diatoms might be expected to occur without advantage or disadvantage to either species under similar favorable growth conditions. Alternatively, it is well known that diatoms are able to use various organic substances, especially when photosynthesis is limited (Lewin and Hellebust 1976; Molloy and Syrett 1988; Petterson and Sahlsten 1990; Graham and Wilcox 2000). It is therefore possible that diatoms might use *Phaeocystis* spp. colonies not only as habitat, but may be able to utilize organic and inorganic substances derived from or associated with the colonial matrix. Such a commensalism could be increased as the *Phaeocystis* colony becomes older and less defended due to deterioration of the colony membrane (Hamm et al., 1999; Hamm, 2000) or becomes less allelopathic (Long et al. submitted). Still another possibility is that *Phaeocystis* might benefit from an association with *Pseudo-nitzschia* in which case their association would represent a true symbiosis. It has been reported that *Pseudo-nitzschia* species can produce specific aldehydes that reduce copepod fecundity (Miralto et al. 1999) and thus, the presence of *Pseudo-nitzschia* spp. on *Phaeocystis* colonies might provide a level of grazing protection to aging colonies. However, in our opinion the interactions of *Pseudo-nitzschia* species and *Phaeocystis* colonies are most likely to be of the facultative commensalism type, although additional studies are needed to investigate this hypothesis. Regardless, conceptual models of the life cycle of *Phaeocystis* spp. (Whipple et al. 2005) especially ecological significance of the colonial life form should now take into account *Pseudo-nitzschia*/*Phaeocystis* association.

Acknowledgments We are indebted to Dr. Lucie Courcot from Universite du Littoral Cote d'Opale, Wimereux, France for SEM photographs that were useful for identification of the *Pseudo-nitzschia delicatissima*. We thank the crews of R.V. Sepia II and Côtes de la Manche for their help during cruises in the Eastern English Channel, as well as staff of Université of Littoral and University of Bergen for their technical support and help. Anna Boyette (Skidaway Institute of Oceanography) prepared the figures. This work was part of CPER "Bloom of *Phaeocystis*" and "PNEC Manche Orientale" French regional and national programs for Eastern Channel Monitoring. Jens C. Nejstgaard was supported by the Norwegian Research Council (152714/120). This study was also funded by the US National Science Foundation Office of Polar Programs grant (OPP–00-83381) and the US Department of Energy Biotechnology Investigations–Ocean Margins Program (FG02-98EF 62531).

References

Arndt CE, Fernandez-Leborans G, Seuthe L, Berge J, Gulliksen B (2005) Ciliated epibionts on the Arctic

sympagic amphipod *Gammarus wilkitzkii* as indicators for sympago-benthic coupling. Mar Biol 147:643–652
Fernandez-Leborans G, Tato-Porto ML (2000) A review of the species of protozoan epibionts on crustaceans II. Suctorian ciliates. Crustaceana 73:1205–1237
Graham LE, Wilcox LW (2000) Algae. Prentice-Hall, NY, 700 p
Hamm CE (2000) Architecture, ecology and biogeochemistry of *Phaeocystis* colonies. J Sea Res 43:307–315
Hamm CE, Rousseau V (2003) Composition, assimilation and degradation of *Phaeocystis globosa*-derived fatty acids in the North Sea. J Sea Res 50:271–283
Hamm CE, Simson D, Merkel L, Smetacek V (1999) Colonies of *Phaeocystis globosa* are protected by a thin but tough skin. Mar Ecol Prog Ser 187:101–111
Hasle GR (1964) *Nitzschia* and *Fragilariopsis* species studied in the light and electron microscopes. 1. Some marine species of the groups *Nitzschiella* and *Lanceolatae*. Skrifter utgitt av Det Norske Videnskaps-Akademi i Oslo. I. Matematisk-Naturvidenskapelig Klasse. Ny Serie 16:1–48
Hasle GR, Syvertsen EE (1997) Marine Diatoms. In: Tomas CR (ed) Identifying marine phytoplankton. Academic Press, San Diego, pp 5–385
Hasle GR, Lange CB, Syvertsen EE (1996) A review of *Pseudo-nitzschia*, with special reference to the Skagerrak, North Atlantic, and adjacent waters. Helgol Meeres 50:131–175
Hilst CM, Smith WO Jr (2002) Photosynthesis/irradiance relationships in the Ross Sea, Antarctica, and their control by phytoplankton assemblage composition and environmental factors. Mar Ecol Prog Ser 226:1–12
Irigoien X, Meyer B, Harris R, Harbour D (2004) Using HPLC pigment analysis to investigate phytoplankton taxonomy: the importance of knowing your species. Helgol Mar Res 58:77–82
Lewin J, Hellebust JA (1976) Heterotrophic nutrition of the marine pennate diatom *Nitzschia angularis* var. *affinis*. Mar Biol 36(4):313–320
Lundholm N, Moestrup Ø, Kotaki Y, Hoef-Emden K, Scholin C, Miller P (2006) Inter- and intraspecific variation of the *Pseudo-nitzschia delicatissima* complex (*Bacillariophyceae*) illustrated by RRNA probes, morphological data and phylogenetic analyses. J Phycol 42:464–481
Menden-Deuer S, Lessard EJ (2000) Carbon to volume relationship for dinoflagellates, diatoms and other protists plankton. Limnol Oceanogr 45(3):569–579
Miralto A, Barrone G, Romano G, Poulet SA, Ianora A, Russo GL, Buttino I, Mazzarella G, Laabir M, Cabrini M, Giacobbe MG (1999) The insidious effect of diatoms on copepod reproduction. Nature 42:174–176
Molloy CJ, Syrett PJ (1988) Interrelationships between uptake of urea and uptake of ammonium by microalgae. J Exp Mar Biol Ecol 118:85–95
Naumann E (1922) Zwei neue typen von planktonkammern. Z Wiss Mikrosk 39:163–164
Nejstgaard JC, Frischer ME, Verity PG, Anderson JT, Jacobsen A, Zirbel MJ, Larsen A, Martinez-Martinez J, Sazhin AF, Walters T, Bronk DA, Whipple SJ, Borrett SR, Patten BC, Long JD (2006) Plankton development and trophic transfer in sea water enclosures added nutrients and *P. pouchetii.* Mar Ecol Prog Ser (in press, forthcoming publications, No M 6307)
Peperzak L, Colijn F, Gieskes WWC, Peeters JCH (1998) Development of the diatom-*Phaeocystis* spring bloom in the Dutch coastal zone of the North Sea: the silicon depletion versus the daily irradiance threshold hypothesis. J Plankton Res 20:517–537
Petterson K, Sahlsten E (1990) Diet patterns of combined nitrogen uptake and intracellular storage of nitrate by phytoplankton in the open Skagerrak. J Exp Mar Biol Ecol 138:167–182
Pouchet MG (1892) Sur une algue pélagique nouvelle. CR Séances Soc Biol 44:34–36
Priisholm K, Moestrup Ø, Lundholm N (2002) Taxonomic notes on the marine diatom genus *Pseudo-nitzschia* in the Andaman Sea near the island of Phuket, Thailand, with a description of *Pseudo-nitzschia micropora* sp. nov. Diatom Res 17(1):153–175
Proshkina-Lavrenko AI (1963) Benthic Diatoms of the Black Sea. Academy of Sciences of the USSR, Moscow-Leningrad, 237 p (in Russian)
Rijssel M, Hamm CE, Gieskes WWC (1997) *Phaeocystis globosa* (Prymnesiophyceae) colonies: hollow structures built with small amounts of polysaccharides. Eur J Phycol 32:185–192
Rousseau V, Mathot S, Lancelot C (1990) Calculating carbon biovolume of *Phaeocystis* sp. from microscopic observations. Mar Biol 107(2):305–314
Rousseau V, Vaulot D, Casotti D, Cariou V, Lenz J, Gunkel J, Baumann M (1994) The life cycle of *Phaeocystis* (Prymnesiophyceae): evidence and hypotheses. J Mar Syst 5:23–39
Sapozhnikov FV (2003) Colonial Diatoms – satellites of Mussels-filtrators. Ecol Sea 64:13–17 (in Russian)
Schoemann V, Becquevort S, Stefels J, Rousseau V, Lancelot C (2005) *Phaeocystis* blooms in the global ocean and their controlling mechanisms: a review. J Sea Res 53(1–2):43–66
Sherr EB, Caron DA, Sherr BF (2000) Staining of heterotrophic Protists for visualization via Epifluorescence Microscopy. In: Kemp PF et al (eds) Handbook of methods in Aquatic microbial ecology. Levis publishers, pp 213–227
Shields AR, Smith WO Jr (2005) A novel technique to examine the role of colonial *Phaeocystis antarctica* in the microbial loop of the Ross Sea, Antarctica. In: Thesis SCOR working group (120. *Phaeocystis*, major link in the biogeochemical cycling of elements, 30 August–3 September 2005. University of Groningen, Haren, The Netherlands, pp 54–55
Sukhanova IN, Flint MV (2001) *Phaeocystis pouchetii* at the Eastern Bering Sea shelf. Oceanology 41(1):75–85
Throndsen J, Hasle GR, Tangen K (2003) Norsk Kystplankton Flora. Almater Forlag AS, Oslo, 341 p
Wassmann P, Rat'kova T, Andreassen I, Vernet M, Pedersen G, Rey F (1999) Spring bloom development in the Marginal ice zone and the Central Barents Sea. Mar Ecol 20(3–4): 321–346
Whipple SJ, Patten BC, Verity PG (2005) Life cycle of the marine alga *Phaeocystis*: a conceptual model to summarize literature and guide research. J Mar Syst 57:83–110

Biogeochemistry (2007) 83:147–172
DOI 10.1007/s10533-007-9098-y

REVIEW

Zooplankton grazing on *Phaeocystis*: a quantitative review and future challenges

Jens C. Nejstgaard · Kam W. Tang · Michael Steinke · Jörg Dutz · Marja Koski · Elvire Antajan · Jeremy D. Long

Received: 23 May 2006 / Accepted: 16 November 2006 / Published online: 20 April 2007

Abstract The worldwide colony-forming haptophyte phytoplankton *Phaeocystis* spp. are key organisms in trophic and biogeochemical processes in the ocean. Many organisms from protists to fish ingest cells and/or colonies of *Phaeocystis*. Reports on specific mortality of *Phaeocystis* in natural plankton or mixed prey due to grazing by zooplankton, especially protozooplankton, are still limited. Reported feeding rates vary widely for both crustaceans and protists feeding on even the same *Phaeocystis* types and sizes. Quantitative analysis of available data showed that: (1) laboratory-derived crustacean grazing rates on monocultures of *Phaeocystis* may have been overestimated compared to feeding in natural plankton communities, and should be treated with caution; (2) formation of colonies by *P. globosa* appeared to reduce predation by small copepods (e.g., *Acartia*, *Pseudocalanus*, *Temora* and *Centropages*), whereas large copepods (e.g., *Calanus* spp.) were able to feed on colonies of *Phaeocystis pouchetii*; (3) physiological differences between different growth states, species, strains, cell types, and laboratory culture versus natural assemblages may explain most of the variations in reported feeding rates; (4) chemical signaling between predator and prey may be a major factor controlling grazing on *Phaeocystis*; (5) it is unclear to what extent different zooplankton, especially protozooplankton, feed on the different life forms of *Phaeocystis* in situ. To better understand the mechanisms controlling zooplankton grazing in situ, future studies

J. C. Nejstgaard (✉)
UNIFOB AS, Department of Biology, University of Bergen, Bergen High Technology Centre, 5020 Bergen, Norway
e-mail: jens.nejstgaard@bio.uib.no

K. W. Tang
Virginia Institute of Marine Science, 1208 Greate Road, Gloucester Point, VA, 23062, USA

M. Steinke
Department of Biological Sciences, University of Essex, Wivenhoe Park, Colchester, CO4 3SQ, UK

J. Dutz
Institute for Baltic Sea Research, Seestraße 15, 18119 Rostock Warnemunde, Germany

M. Koski
Danish Institute for Fisheries Research, Kavalergården 6, 2920 Charlottenlund, Denmark

E. Antajan
Laboratoire d'Océanographie de Villefranche, Université Pierre et Marie Curie-Paris6, 06230 Villefranche-sur-Mer, France

E. Antajan
Laboratoire d'Océanographie de Villefranche, CNRS, 06230 Villefranche-sur-Mer, France

J. D. Long
Northeastern University Marine Science Center, 430 Nahant Road, Nahant, MA, 01908, USA

should aim at quantifying specific feeding rates on different *Phaeocystis* species, strains, cell types, prey sizes and growth states, and account for chemical signaling between the predator and prey. Recently developed molecular tools are promising approaches to achieve this goal in the future.

Keywords Colony formation · DMS · Gut pigment · Molecular methods · Microzooplankton · *Phaeocystis* · *antarctica* · Predator defense

Introduction

The haptophyte *Phaeocystis* is a dominant phytoplankton genus in tropical to polar seas (Baumann et al. 1994). They are key species in marine food webs and biogeochemical cycles, e.g., as major producers of carbon and climatically important sulfide compounds (Liss et al. 1994; Alderkamp et al. this volume; Stefels et al. this volume). Three major species: *P. globosa* Scherffel, *P. pouchetii* (Hariot) Lagerheim and *P. antarctica* Karsten (Medlin and Zingone this volume) exist in two main morphotypes: small single cells and mucilaginous colonies (see Rousseau et al. this volume for details on the different cell and morphotypes).

The success of *Phaeocystis* has been ascribed to a number of factors, including escaping from grazing by its dramatic ability to shift morphotype between solitary flagellates of a few micrometers to large mucous colonies up to several cm in diameter (Weisse et al. 1994; Chen et al. 2002; Schoemann et al. 2005; Veldhuis and Wassmann 2005; Rousseau et al. this volume). For a phytoplankton bloom to form, the sum of growth and accumulation must be larger than the sum of loss due to horizontal and vertical advection, sinking, lysis and predation (e.g., Smayda 1997; Banse 1994). Advection is beyond the control of all phytoplankton, and thus not likely to be a strong selective force for *Phaeocystis* (but see Seuront et al. this volume). Neither does the growth rate of *Phaeocystis* appear to be exceptionally high relative to other bloom-forming phytoplankton, such as diatoms (Hegarty and Villareal 1988). Indeed, in many field and mesocosm studies *Phaeocystis* blooms co-occurred with, or followed the demise of, diatom blooms (Peperzak et al. 1998; Goffart et al. 2000; Rousseau et al. 2002; Tungaraza et al. 2003; Larsen et al. 2004). Sinking loss of *Phaeocystis* also seems to be small for the large colonies (Reigstad and Wassmann this volume). This could be partly due to the balloon-like characteristics of the colonies: colonial cells are embedded in a thin mucous skin whereas the interior of the colonies is hollow (van Rijssel et al. 1997). Although single cells are susceptible to viral lysis, an intact mucous skin may protect colonial *Phaeocystis* cells from viral lysis and other infections (reviewed by Brussaard et al. this volume). A remaining possible explanation for the success of *Phaeocystis* despite its moderate growth rate is its ability to reduce grazing mortality, which is the focus of this review article.

Grazing on *Phaeocystis* (mainly *P. globosa* and *P. pouchetii*) has been studied since the beginning of the last century (e.g., Lebour 1920, 1922), and a wide range of organisms have been reported to be able to ingest *Phaeocystis* (Table 1). Unfortunately, most of the early studies were based on light microscopy of gut contents of the grazers, and neither feeding rate nor detailed description of the prey species was available (see Notes in Table 1, and further discussion below). It was not until the 1980s that direct quantitative studies on zooplankton feeding on *Phaeocystis* were reported (reviewed in Peperzak 2002; Rousseau et al. 2000; Schoemann et al. 2005; Weisse et al. 1994). Published grazing rates span wide ranges even for the same predators feeding on the same *Phaeocystis* species, morphotype and food concentration. This indicates that the widely accepted view of grazing vulnerability as a function of predator-to-prey size ratio and prey abundance (Frost 1972; Hansen et al. 1994a) for a given predator type (Hansen et al. 1997) may be compounded by other factors controlling grazing on *Phaeocystis*.

Although copepods and other organisms may ingest *Phaeocystis* spp. (Table 1), at least when they are in palatable condition (Estep et al. 1990; Long and Hay 2006), there are also reports of reduced zooplankton feeding and abundance during blooms of *Phaeocystis* (Table 2) and low reproductive output in copepods, even when feeding rates are relatively high (Verity and Smayda 1989; Turner et al. 2002; Klein Breteler and Koski 2003; Long and Hay 2006). Several authors have previously reviewed various negative effects of *Phaeocystis* on different organisms (e.g., Weisse et al. 1994; Turner et al. 2002; Schoemann et al. 2005); and we provide an updated summary in Table 2. However,

Table 1 Organisms recorded to ingest *Phaeocystis* spp.

Grazer	Developmental stage	*Phaeocystis* species	type/stage	Obervation method	References
Dinoflagellates					
Amphidinium sp.	–	sp.	"remains of spores"	in situ gut microscopy	Lebour (1922)
Gymnodinium rhomboides Schütt	–	sp.	–	in situ gut microscopy	Lebour (1922)
Gymnodinium sp.	–	*globosa*	single cells	incubation experiments	Tang et al. (2001)
Gymnodinium triangularis Lebour	–	sp.	–	in situ gut microscopy	Lebour (1922)
Gyrodinium dominans Hulburt	–	*globosa*	single cells	incubation experiments	Tang et al. (2001)
Gyrodinium cf. *spirale*	–	*globosa*	small colony	photomicrograph	Stelfox-Widdicombe et al. (2004)
Noctiluca scintillans (Macartney) Kofoid & Swezy	–	cf. *globosa*	single cells and colonies	incubation experiments and microscopy	Hansen (1992, 1995), Jakobsen and Tang (2002), Weisse et al. (1994)
Oxyrrhis marina Dujardin	–	cf. *globosa*	single cells	incubation experiments	Hansen et al. (1993)
Ciliates					
Euplotes sp.	–	*antarctica*	single cells and colonies	fluorescently labeled algae	Shields and Smith (2005)
Euplotes sp.	–	*globosa*	single cells and colonies	incubation experiments	Long (2004)
Helicostomella subulata (Ehrenberg)	–	*globosa*	single cells	live microscopy	Admiraal and Venekamp (1986)
Lohmanniella oviformis Leegaard	–	*globosa*	single cells	incubation experiments	Tang et al. (2001)
Rimostrombidium conicum (Kahl) Petz & Foissner	–	*globosa*	single cells	incubation experiments	Tang et al. (2001)
Strombidinopsis acuminatum Fauré-Fremiet	–	cf. *globosa*	single cells and colonies	incubation experiments	Hansen (1995, 1993)
Strombidium sp.	–	*globosa*	single cells	incubation experiments	Verity (2000)
Strombidium vestitum Leegaard	–	*globosa*	single cells	incubation experiments	Tang et al. (2001)
Tintinnopsis beroidea Stein	–	*globosa*	single cells	live microscopy	Admiraal and Venekamp (1986)
Rotifers					
Syncaheta vorax Rousselet	–	*globosa*	eating at colonies	live microscopy	Hollowday (1949)
Polychaets					
Polydora pulchra Carazzi	nectochaeta larvae	cf. *globosa*	single cells and colonies	incubation experiments	Hansen (1992, 1995)
Tomopteris sp.	Larvae	cf. *globosa*	single cells and colonies	incubation experiments	Hansen (1992, 1995)
Molluscs					
Macoma balthica (L)	Mussels??	sp.	ingest single cells	laboratory incubation	Kamermans (1994)
Mytilus edulis L.	Mussels 33–58 mm	cf. *globosa*	cultured, ingested single cells	flow chamber, cell counts	Smaal and Twisk (1997)

Table 1 continued

Grazer	Developmental stage	*Phaeocystis* species	type/stage	Obervation method	References
Mytilus edulis L.	Mussels > 3 cm	*globosa*	able to filter colonies	?	Petri and Vareschi (1997)
Nassarius reticulatus (L)	veliger larvae	*pouchetii*	cultured, probably single cells	incubation, visual inspection	Fretter and Montgomery (1968)
Rissoa inconspicua Alder	veliger larvae	*pouchetii*	cultured, probably single cells	incubation, visual inspection	Fretter and Montgomery (1968)
Cladocerans					
Evadne nordmanni Lovén	–	sp.	cells and "spores"	in situ gut microscopy	Lebour (1922)
Copepods, calanoids					
Acartia clausi (Gunn.)	copepodites ?	sp.	"green remains, probably *Phaeocystis* in many"	in situ gut microscopy	Lebour (1922)
Acartia hudsonica Pinhey	CVI female	*pouchetii*	single cells and colonies	incubation experiments	Verity and Smayda (1989)
Acartia longiremis (Lilljeborg)	CVI female	*pouchetii*	ingested only colonies < 100 μm	incubation experiments	Hansen et al. (1994b)
Acartia tonsa Dana	CVI female	*globosa, pouchetii*	single cells and colonies	incubation experiments	Tang et al. (2001), Tang and Simó (2003), Verity (2000), Verity and Smayda (1989)
Anomalocera pattersoni Tempelton	copepodites ?	sp.	"*Phaeocystis*"	in situ gut microscopy	Lebour (1922)
Calanoides sp.	Copepodites	*Antarctica*	fluorescently labeled single cells	gut microscopy	K. Tang unpublished
Calanus finmarchicus (Gunn.)	copepodites ?	sp.	cells and "remains"	in situ gut microscopy	Lebour (1922)
Calanus finmarchicus (Gunn.)	CI-V	*pouchetii*	single cells and colonies	incubation experiments	Hansen et al. (1990), Tande and Båmstedt (1987)
Calanus glacialis Jaschnov	CIII-CV	*pouchetii*	colonies	incubation experiments	Estep et al. (1990)
Calanus helgolandicus (Claus)	copepodites > 280 μm	cf. *globosa*	single cells and colonies	incubation experiments	Hansen (1992, 1995), Turner et al. (2002)
Calanus hyperboreus (Krøyer)	CIV-CVI female	*pouchetii*	single cells and colonies	incubation experiments	Estep et al. (1990), Huntley et al. (1987), Tande and Båmstedt (1987)
Centropages hamatus Lilljeborg	copepodites > 280 μm	cf. *globosa*	single cells and colonies	incubation experiments	Hansen (1995)
Centropages hamatus Lilljeborg	copepodites ?	cf. *globosa*	colonies? "*Phaeocystis* bladders"	in situ gut microscopy	Jones and Haq (1963)
Centropages typicus Krøyer	copepodites ?	sp.	"spores"	in situ gut microscopy	Lebour (1922)
Eucalanus pileatus Giesbrecht	CVI female	*globosa*	single cells and colonies	incubation experiments	Long and Hay (2006)
Metridia gerlachi Giesbrecht	copepodites ?	*pouchetii* (in Antarctica)	colonies 0.05–1.5 mm	incubation experiments	Schnack (1983)

Table 1 continued

Grazer	Developmental stage	*Phaeocystis* species	type/stage	Obervation method	References
Microcalanus sp.	Copepodites	*antarctica*	fluorescently labeled single cells	gut microscopy	K. Tang unpublished
Pareuchaeta antarctica (Giesbrecht)	copepodites ?	*pouchetii* (in Antarctica)	colonies 0.05–1.5 mm	incubation experiments	Schnack (1983)
Pseudocalanus elongatus (Boeck)	copepodites > 280 µm	cf. *globosa*	single cells and colonies	incubation experiments	Hansen (1992, 1995)
Pseudodiaptomus pelagicus Herrick	CVI female	*globosa*	single cells and colonies	incubation experiments	Long and Hay (2006)
Temora longicornis Müller	nauplii to adults	*globosa*	single cells and colonies	incubation experiments	Dutz and Koski (2006), Koski et al. (2005), Hansen (1992, 1995)
Temora longicornis Müller	copepodites ?	cf. *globosa*	colonies? "*Phaeocystis* bladders"	in situ gut microscopy	Jones and Haq (1963)
Temora longicornis Müller	copepodites ?	sp.	"green cells, probably *Phaeocystis*"	in situ gut microscopy	Lebour (1922)
Temora stylifera Dana	adult female	appeared similar to *globosa*	cultured single cells and colonies	incubation experiments	Turner et al. (2002)
Copepods, cyclopoids					
Corycaeus anglicus Lubbock	copepodites ?	sp.	"*Phaeocystis* in several"	in situ gut microscopy	Lebour (1922)
Oithona nana Giesbrecht	copepodites ?	cf. *globosa*	colonies? "*Phaeocystis* bladders"	in situ gut microscopy	Jones and Haq (1963)
Oithona sp.	copepodites	*antarctica*	fluorescently labeled single cells	gut microscopy	K. Tang unpublished
Oncaea curvata (Gunn.)	adult female	cf. *pouchetii* (in Antarctica)	colonies	incubation experiments	Metz (1998)
Copepods, harpacticoids					
Euterpina acutifrons (Dana)	copepodites ?	sp.	"flagellates...especially *Phaeocystis*"	in situ gut microscopy	Lebour (1922)
Cirripedes					
Balanus crenatus Bruguière	nauplii	cf. *globosa*	single cells and colonies	incubation experiments	Hansen (1992, 1995)
"Cirripedes"	nauplii	sp.	"probably *Phaeocystis*"	in situ gut microscopy	Lebour (1922)
Longipeda minor T. & A. Scott	adult female	sp.	colonies	incubation, visual inspection	Nicholls (1935)
Longipeda scotti G.O.Sars	adult female	sp.	colonies	incubation, visual inspection	Nicholls (1935)

Table 1 continued

Grazer	Developmental stage	*Phaeocystis* species	type/stage	Obervation method	References
Amphipods					
Pontogeneia antarctica Chevreux	hatchilings (2 mm)-adults	*antarctica*	"green plastidic debris... probably indicate ingestion of *Phaeocystis antarctica*"	in situ gut microscopy	Richardson and Whitaker (1979)
Euphausiids					
Euphausia superba Dana		*sp.*		in situ gut microscopy	Sieburth (1960)
Euphausia superba Dana	35–45 mm	*antarctica*	single cells and colonies < 500 μm	incubation experiments	Haberman et al. (2003)
Thysanoessa inermis (Krøyer)	carapace ca 5 mm	*pouchetii*	ingested only colonies > 20 μm	incubation experiments	Hansen et al. (1994b)
Thysanoessa raschii (M. Sars)	carapace ca 5 mm	*pouchetii*	ingested only colonies > 20 μm	incubation experiments	Hansen et al. (1994b)
Decapods					
Crab (unidentified)	zoëa	sp.	"*Phaeocystis*"	in situ gut microscopy	Lebour (1922)
Galathea intermedia Lilljeborg	megalopa	cf. *globosa*	single cells and colonies	incubation experiments	Hansen (1992, 1995)
Porcellana sp.	"larva"	sp.	"green spores, probably *Phaeocystis*"	in situ gut microscopy	Lebour (1922)
Portunus depurator (Linnaeus)	zoëa III-V	cf. *globosa*	single cells and colonies	incubation experiments	Hansen (1992, 1995)
Portunus holsatus (Fabricius)	megalopa	cf. *globosa*	single cells and colonies	incubation experiments	Hansen (1992, 1995)
Appendicularians					
Megalocercus huxleyi (Ritter)	-	sp.	small colony	in situ faecal pellet microscopy	Gorsky et al. (1999)
Fish					
Gadus morhua L.	larvae < 0.2 mg/ 4 mm	*pouchetii*	colonies	in situ gut microscopy	Krogstad (1989), Løken (1990)
Pleuronectes flesus L.	larvae < 11 mm	sp.	colonies	incubation, visual inspection and in situ gut microscopy	Lebour (1920)
Scomber scombrus L.	large from fish catches	*globosa*	"*phytoplankton chiefly P. globosa*" associated with jelly probably colonies	in situ gut microscopy	Bullen (1908)

(–) denotes no information/not applicable. (?) denotes likely but unclear. Note that species composition within the genus *Phaeocystis* has been debated (Baumann et al. 1994; Sournia 1988). Because the solitary cells of *P. globosa*, *P. pouchetii* and others may not be distinguishable by light microscopy, reports before 1980 could be misleading [e.g., the species in Antarctica reported by Schnack (1983) and Metz (1998) may likely be *P. antarctica*, rather than *P. pouchetii*]. Where multiple references are available (mostly for calanoid copepods) only selected representative references are shown here. See text for more references.

some of these data were derived from ecologically unrealistic organisms or assays and should be interpreted with caution (see discussion below). Further, positive correlations between *Phaeocystis* abundance and mesozooplankton in the field have also been reported, and some of the negative effects have been debated and ascribed to other organisms (see Notes in Table 2). Furthermore, results from fatty acid analysis indicate that dominating crustacean zooplankton may derive a major part of their diet from *P. pouchetii* in northern latitudes (Sargent et al. 1985) and from *P. globosa* in lower latitudes (Hamm and Rousseau 2003), suggesting that *Phaeocystis* do enter the food web.

A number of factors have been shown to control predation on this enigmatic genus: predator-to-prey size ratio, predator species and stage, prey species, morphotype, growth state and abundance (reviewed in Weisse et al. 1994; Rousseau et al. 2000; Peperzak 2002; Schoemann et al. 2005). However, to our knowledge, a quantitative analysis of the interactive effect of these factors has not been attempted. Using available literature, we attempt to quantify the relative importance and interactions among some of these factors in controlling zooplankton grazing on *Phaeocystis* spp. We also review recent findings on various unique aspects of the life history and physiology of *Phaeocystis* spp., and how they affect zooplankton grazing. Last, we discuss the need for new methodological approaches to address some of the remaining questions concerning zooplankton grazing on *Phaeocystis*.

Grazing on *Phaeocystis*: quantitative patterns in published data

Reports on copepods and other crustaceans dominate available zooplankton grazing data on *Phaeocystis* spp. These data show a wide range of rates, even for similar predator-prey combinations. We therefore attempt to summarize available quantitative data in relation to some of the mechanisms proposed to control the feeding on *Phaeocystis*. The shortage of quantitative feeding studies on microzooplankton and *Phaeocystis* was pointed out already a decade ago (Weisse et al. 1994), and the number of such studies is still limited, especially for protozoan microzooplankton, whether using laboratory cultures (Table 3) or natural plankton assemblages (Table 4) as food. Due to the shortage of protozooplankton data a general statistical analysis of the factors controlling their grazing on *Phaeocystis* cannot yet be performed, but the available data will be discussed further below.

Crustacean zooplankton grazing on *Phaeocystis*: is there a general pattern?

It is generally assumed that a shift from single cell to colony may be part of *Phaeocystis* defense against grazing by microzooplankton (Weisse et al. 1994; Rousseau et al. 2000; Peperzak 2002; Schoemann et al. 2005). Because of the ability of *Phaeocystis* to vary its functional prey size by forming colonies, it is often considered as dual species (sensu Turner et al. 2002): colonies can be referred to as macroplankton and are assumed to be suitable food for zooplankton and nekton, while the solitary nanoplankton are assumed to be vigorously grazed by microzooplankton (Lebour 1922; Hollowday 1949; Admiraal and Venekamp 1986; Weisse and Scheffel-Möser 1990). On the other hand an increase in size may result in a more suitable prey size range for larger predators such as crustacean zooplankton. For palatable prey, ingestion rates generally increase from small prey to the optimum prey size before the prey becomes too large to handle or ingest (Frost 1972; Hansen et al. 1994a). Thus, size may be a major factor controlling predation on *Phaeocystis*.

Many studies showed that the size of *Phaeocystis* strongly influences the feeding rates of differently sized metazoan predators. Hansen et al. (1990) reported that *Calanus finmarchicus* copepodites (CI-V) showed lower feeding rate on *P. pouchetii* colonies of >100 μm than similarly sized diatoms, while in the 30–100 μm size range feeding rates on diatoms and colonies were similar. Tande and Båmstedt (1987) showed that copepodite stage V of *C. finmarchicus* fed equally well on *P. pouchetii* single cells and the diatom *Chaetoceros furcellatus*, whereas Huntley et al. (1987), using the larger copepod *C. hyperboreus* CIV-V, reported much higher grazing rates on colonies 25–200 μm and >200 μm than on single cells of *P. pouchetii* or the diatom *Chaetoceros socialis*. In a study with grazers of different sizes feeding on *P. pouchetii*, Hansen et al. (1994b) observed an optimal predator-to-prey size ratio for feeding and an upper size ratio above which no ingestion

Table 2 Negative effects on various organisms ascribed to *Phaeocystis* spp., with notes on contradictory reports

Effect	Note	References
Protists and other microbes		
Defence against viral attacks by intact colony membranes		Reviewed by Brussaard et al. (this volume)
Antibiotic to bacteria in the acid environment of penguin guts.	A	Sieburth (1960, 1961)
Haemolytic activity in laboratory, or correlated with abundance of *P. pouchetii* during blooms of natural planktonin mesocosms		Stabell et al. (1999), van Rijssel et al. (this volume)
Suggested allelopathy, e.g., possible negative intercations by *P. pouchetii* on *T. nordenskioeldii* and *Skeletonema costatum* in vitro. Rapid decrease in abundance of *Thalassiosira nordenskioeldii* and *Ebria* sp. during a bloom of *P. pouchetii*.		Barnard et al. (1984), Smayda (1973)
Allelopathy, filtrates from mesocosm blooms of *P. pouchetii* and laboratory cultures of *P. globosa* resulting in reduced growth and lysis of the cryptophyte phytoplankton *Rhodomonas baltica*.		Long (Unpublished data)
Lack of feeding of the ciliates *Mesodinium pulex* and *Strombidium elegans* in laboratory experiments despite high concentrations of *P. globosa* and no alternative food.		Hansen et al. (1993), Tang et al. (2001)
Low biomass of bacteria and protozooplankton during blooms of colonial *Phaeocystis*	B	van Boekel et al. (1992)
Metazooplankton		
Many copepods show low abundances, development, gut content and/or much lower feeding activity compared to alternative food such as diatoms, during blooms of *Phaeocystis*	C	Bautista et al. (1992), Breton et al. (1999), Daro (1985), Davies et al. (1992), Frangoulis et al. (2001), Gasparini et al. (2000), Hansen and van Boekel (1991), Turner (1994), Weisse et al. (1994)
Low feeding by *Calanus* spp. and *Metridia longa* on actively growing colonies of *P. pouchetii*	D	Estep et al. (1990)
Low feeding by nauplii on some cell types of *P. globosa*		Dutz and Koski (2006)
Low reproductive output from *Acartia* spp., *Temora longicornis, Pseudodiaptomus pelagicus* and *Eucalanus pileatus* feeding on *Phaeocystis*, despite sometimes high ingestion rates.	E	Klein Breteler and Koski (2003), Long and Hay (2006), Tang et al. (2001), Turner et al. (2002), Verity and Smayda (1989)
Inhibiting copepod feeding by transparent exopolymer particles (TEP) derived from a *Phaeocystis globosa*.		Dutz et al. (2005)
Suggested reduction of zooplankton feeding due to mucus secretion from colonies of *P. globosa*.		Seuront et al. (this volume)
Not a good food organism for the barnacle *Balanus balanoides*		Cook and Gabbott (1972)
Other organisms		
Oxygen deficiency due to sedimenting *Phaeocystis* killing less mobile fauna in the sediments.		Rogers and Lockwood (1990)
Colonies clogging gills and thus reducing feeding in several mussels, especielly smaller, e.g., *P. globosa* colonies reduced feeding in *Mytilus edulis* <3 cm, while larger where able to filter colonies.		Blauw (this volume), Kopp (1978), Meixner (1981), Petri et al. (1999), Pieters et al. (1980), Smaal and Twisk (1997)
Inhibited growth in the bryozoan *Electra pilosa* possibly due to *Phaeocystis*.		Jebram (1980)
Toxic to sea urchin larvae		Hansen et al. (2003, 2004)
Toxic to blowflies (*Calliphora omitoria*)		Stabell et al. (1999)

Table 2 continued

Effect	Note	References
Migrating herring hindered by a "barrier" of *Phaeocystis* blooms	F	Savage (1930, 1932)
Toxic to cod larvae		Aanesen et al. (1998), Eilertsen and Raa (1995), Hansen et al. (2004)
Skin and eye irritation of humans		Dunne et al. (1984)

Notes:

A: It has also been reported that the high concentrations of acrylate in *Phaeocystis* colonies do not inhibit surrounding bacteria and may be absorbed to the mucus matrix (Noordkamp et al. 2000)

B: But when solitary cells are abundant during blooms (Admiraal and Venekamp 1986; Weisse and Scheffel-Möser 1990), or during the breakdown (van Boekel et al. 1992), microzooplankton such as ciliates may be numerous

C: There are also some reports on apparently increased abundances of mesozooplankton during periods of blooms of *Phaeocystis* (Fransz and Gieskes 1984; Fransz et al. 1992; Weisse et al. 1986)

D: But in a later bloom stage the copeods switch to selectively feed upon the senescent *P. pouchetii*, rather than the diatoms (Estep et al. 1990)

Table 3 Quantitative laboratory studies on protozooplankton grazing on *Phaeocystis*

Grazer	Alternative prey	Feeding on *P. globosa*	References
Oxyrrhis marina	No	Yes	Hansen et al. (1993)
Strombidinopsis acuminatum	No	Yes	Hansen et al. (1993)
Strombidium elegans	No	No	Hansen et al. (1993)
Strombidium sp.	No	Yes	Verity (2000)
Noctiluca scintillans	No	Yes	Jakobsen and Tang (2002)
Gyrodinium dominans	No	Yes	Tang and Simó (2003)
Lohmanniella oviformis	*Isochrysis galbana*	Lower	Tang et al. (2001)
Rimostrombidium conicum	*Isochrysis galbana*	Higher	Tang et al. (2001)
Strombidium vestitum	*Isochrysis galbana*	Higher	Tang et al. (2001)
Gyrodinium dominans	*Isochrysis galbana*	Higher	Tang et al. (2001)
Gymnodinium sp.	*Isochrysis galbana*	Lower	Tang et al. (2001)
Mesodinium pulex	*Gymnodinium* sp.	No	Tang et al. (2001)

All experiments are based on particle disappearance and cell counts of cultured *P. globosa*, or cf. *globosa* in Hansen et al. (1993). Notes are given on whether the grazer feed (yes) or not at all (no) on *Phaeocystis*, or showed higher or lower feeding rates on Phaeocystis, compared to alternative prey offered in parallel single prey species incubations

occurs. These results were also supported by Levinsen et al. (2000), who reported ingestion of single *P. pouchetii* by the smaller females of *C. finmarchus* but not by the larger *C. glacialis* and *C. hyperboreus*. Also, the krill *Euphausia superba* (35–40 mm) showed similar feeding on 50–100 μm *P. antarctica* colonies compared to the diatom *Thalassiosira antarctica*, whereas both single cells and larger colonies of *P. antarctica* were grazed at lower rates compared to the diatom (Haberman et al. 2003). Thus, data such as these support the notion that substantial shifts in size between *Phaeocystis* life forms affect grazing by differently sized predators. However, there is still a considerable variation in feeding rates on various forms of *Phaeocystis*, ranging from zero to expected maximum rates (sensu Hansen et al. 1997) for a number of predators (reviewed in Weisse et al. 1994; Schoemann et al. 2005).

In order to evaluate the relative importance of factors other than size and how they may help explain the observed variations in grazing rates, we compiled the available quantitative data, converted them into carbon units, and performed a series of correspondence and cluster analyses described below.

Table 4 Literature inferring quantitative evidence of microzooplankton community grazing (g) and grazing impact (percentage of average standing stock (SS) removed d^{-1}) on *Phaeocystis* solitary cells in natural plankton

Phaeocystis spp.	Area	Time	Analysis	μ (d^{-1})	g (d^{-1})	Impact (% SS d^{-1})	Note	References
MESOCOSM (natural plankton indoor tank)								
globosa	South North Sea	Sep	Cell Counts	? – 0.3 – ?	0.3–0.7	26–48		Brussard et al. (2005)
FIELD								
cf. globosa?	South North Sea	Apr–May	Cell Counts	0.8–2.3	0.9–4.2	59–98	A	Weisse and Scheffel-Möser (1990)
cf. globosa?	South North Sea	Feb–Jul	Total Chl *a*	−0.3–0.4	0.0–0.6	0–46	B	Brussard et al. (1995)
globosa	South North Sea	April	Total Chl *a*	0.1–0.7	0.3–1.1	24–68		Stelfox-Widdicombe et al. (2004)
pouchetii	North Norway fjord	April	Total Chl *a*	0.3–0.5	0.21–0.36	19–30	C	Archer et al. (2000)
pouchetii	Labrador Sea	May–Jun	Total Chl *a*	0.2–0.9	0.0–0.3	0–27	D	Wolfe et al. (2000)
antarctica	Ross Sea	Oct–Apr	Chl *a* HPLC FC	?	0–0.23	0–21	E	Caron et al. (2000)
antarctica	Antarctica	Nov–Mar	Chl *a* HPLC FC	0.1–0.6	0.1–0.3	14–26	E	Landry et al (2002)
antarctica	Antarctica	Dec	Chl *a* HPLC FC	0.2–0.9	0.1–1.0	10–63	E	Selph et al. (2001)

Specific phytoplankton growth rate (μ) is also shown. All rates are quantified by the dilution technique (Landry and Hassett 1982). Prey concentrations were analyzed by cell counts, total chlorophyll *a* fractions (Total Chl *a*), high-performance liquid chromatography (HPLC), or flow cytometry (FC)

A: Incubation time only 5–14 h during the day

B: Instantaneous grazing (g) max $\leq 0.1\ d^{-1}$ when the phytoplankton was dominated by *Phaeocystis*

C: Additional experiment with FLA (*Chlorella*). Only 5 of 13 tested ciliates ingested the labeled phytoplankton

D: Also measured dimethyl sulphide (DMS) dynamics. During *P. pouchetii* bloom g was 0.0 and 0.3

E: No notes on specific *Phaeocystis* rates

Treatment of literature data

Many reports of crustacean grazing on *Phaeocystis* spp. are based on non-prey-specific gut fluorescence during blooms of *Phaeocystis*. An unknown fraction of such fluorescence may stem from prey other than *Phaeocystis*. Indeed, the often co-occurring diatoms may even be a preferred prey (Gasparini et al. 2000). In addition, more than half the total biomass (and chlorophyll) of the *Phaeocystis* colony may be due to diatoms and other organisms living in and on the colonies during late bloom stages (Sazhin et al. this volume). Therefore, we exclude data based on non-specific gut pigments of grazers feeding on mixed or natural food suspensions. We converted all grazing rate measurements into common currency of carbon units, using either (in the preferred order): (1) directly reported data, (2) data reported in accompanying papers, (3) personal communications with the authors, or (4) common conversion factors (such as those used in Hansen et al. 1997). For quantitative and statistical analyses we included only grazing data from studies that reported experiment location (field or laboratory study), grazer species, stage or size, and *Phaeocystis* average size (or size range). To be included in the statistical analysis the sources also must have reported *Phaeocystis* species, abundance, morphotype and growth state (exponential or stationary—either given in the text or assumed from the state of the bloom). See Table 5 for further details on the variables. The sources used are listed in Table 6.

Statistical methods

To investigate relationships between crustacean grazing rates on *Phaeocystis* and experimental conditions, a multiple correspondence analysis (MCA) followed by a hierarchical cluster analysis (HCA) was performed using SPAD 3.5 software (Lebart et al. 1988). The combination of MCA and cluster analysis is a common way to explore relationships among a large number of variables and to facilitate interpretation of the correspondence analysis results (Lebart et al. 2000). MCA uses a contingency table as data, which provides a simultaneous representation of the observations (rows) and variables (column) in a factorial space. This form of multivariate analysis describes the total inertia (or variability) of a multidimensional

Table 5 Active structural variables used in the multiple correspondence analysis (MCA). *n* denotes the number of observations from each source

Description of variables: modalities	Label	*n*
Experiment location		
Field	Field	91
Laboratory	Laboratory	205
Crustacean predators		
Acartia clausi	Acl	13
Acartia hudsonica	Ahu	5
Acartia tonsa	Ato	12
Pseudocalanus elongatus	Pel	13
Temora longicornis	Tlo	35
Temora stylifera	Tst	15
Metridia longa	Mlo	1
Centropages hamatus	Cha	14
Calanus finmarchicus	Cfi	105
Calanus glacialis	Cgl	8
Calanus hyperboreus	Chy	75
Phaeocystis *species*		
globosa	*P. globosa*	107
pouchetii	*P. pouchetii*	189
Phaeocystis growth		
Exponential	Exponential	277
Stationary	Stationary	19
Phaeocystis *abundance* (Ab, μg C l^{-1})		
0.1–125	Ab1	81
125–250	Ab2	91
250–500	Ab3	37
500–1000	Ab4	71
1000–1810	Ab5	16
C-specific ingestion (CI, d^{-1})		
0–0.01	CI1	83
0.01–0.1	CI2	151
0.1–1.5	CI3	62
Phaeocystis *selection*		
Positive selection	For	14
Negative selection	Against	76
Null selection	Null	5
Single prey	Single	201
Predator-to-prey size ratio (P:p)		
0.8–4	P:p1	57
4–16	P:p2	90
16–64	P:p3	49
64–256	P:p4	25
256–1120	P:p5	75

Table 6 Sources of data used for the multiple correspondence analysis (MCA) and hierarchical cluster analysis (HCA) presented in Fig. 1, and quantitative summary in Fig. 2, respectively

References	Location	*Phaeocystis*	Fig. 1 *n*	Fig. 2 *n*
Antajan unpublished	Field	*globosa*	6	6
Dutz and Koski (2006)	Lab	*globosa*	7	7
Estep et al. (1990)	Field	*pouchetii*	25	25
Gasparini et al. (2000)	Field	*globosa*	55	55
Haberman et al. (2003)	Lab	*antartica*		3
Hansen et al. (1990)	Lab	*pouchetii*	62	62
Hansen et al. (1993)	Lab	*globosa*	5	5
Hansen et al. (1994b)	Lab	*pouchetii*		20
Huntley et al (1987)	Lab	*pouchetii*	4	4
Klein Breteler and Koski (2003)	Field	*globosa*		4
Koski et al. (2005)	Field	*globosa*	2	6
Koski unpublished	Field	*pouchetii*	3	3
Tande and Båmstedt (1987)	Lab	*pouchetii*	95	95
Tang et al (2001)	Lab	*globosa*		4
Turner et al. (2002)	Lab	*globosa*	15	15
Verity and Smayda (1989)	Lab	*globosa*	9	9
Verity (2000)	Lab	*globosa*	8	12
Weisse (1983)	Lab	*globosa*		18

n denotes the number of observations from each source

set of data, in a sample of fewer dimensions (or axes) that is the best summary of the information contained in the data (Greenacre 1984; Everitt and Dunn 2001). Our data set contains both categorical and continuous variables that were used in the MCA as active (Table 5) and illustrative variables (PL, ESD, CI and Ab defined below), respectively. The factorial axes of the MCA were computed using active (categorical) variables whereas the continuous illustrative (or supplementary) variables were simply projected into this factorial plane without participating in its computation. The location of these continuous illustrative variables, is shown as arrows along the factorial axes in Fig. 1A, and expresses their linkage to the pattern of the categorical active variables displayed by the factorial axes (Lebart et al. 2000). Observations with missing data for carbon-specific ingestion and experimental characteristics were excluded from the MCA. Only four observations with *Phaeocystis antartica* were available, and were therefore also excluded. A data matrix of the remaining 296 observations with eight active and four illustrative variables were used for the analyses (Table 5). The size of predators (prosome length; PL) and prey (equivalent spherical diameter; ESD) were treated as illustrative variables. These continuous variables were used to calculate a predator-to-prey size ratio (P:p) with five categories (or modalities) used in the statistical treatment. Similarly, C-specific ingestion (CI) and *Phaeocystis* abundance (Ab) were transformed into categorical variables of three and five categories, respectively, to be tested both as illustrative and active variables in the MCA. The metric used in the MCA is based on the χ^2. This was also the metric used in the following cluster analysis. The type of HCA used here is an agglomerative clustering, i.e., a procedure that successively groups the closest objects into clusters, which then are grouped into larger clusters of higher rank (Legendre and Legendre 1998). The programme identifies: (i) the cluster (group of observations) which has the smallest within-group variance and the greater variance between groups, and (ii) the descriptors (variables) that are highly representative of each cluster. The descriptors of the observations used in the clustering were their factorial coordinates on the first six axes obtained in the MCA. These first six axes explained about 58% of the total variability (inertia) in the data, and the additional amount of variability explained by the axes decreased markedly after the first six axes.

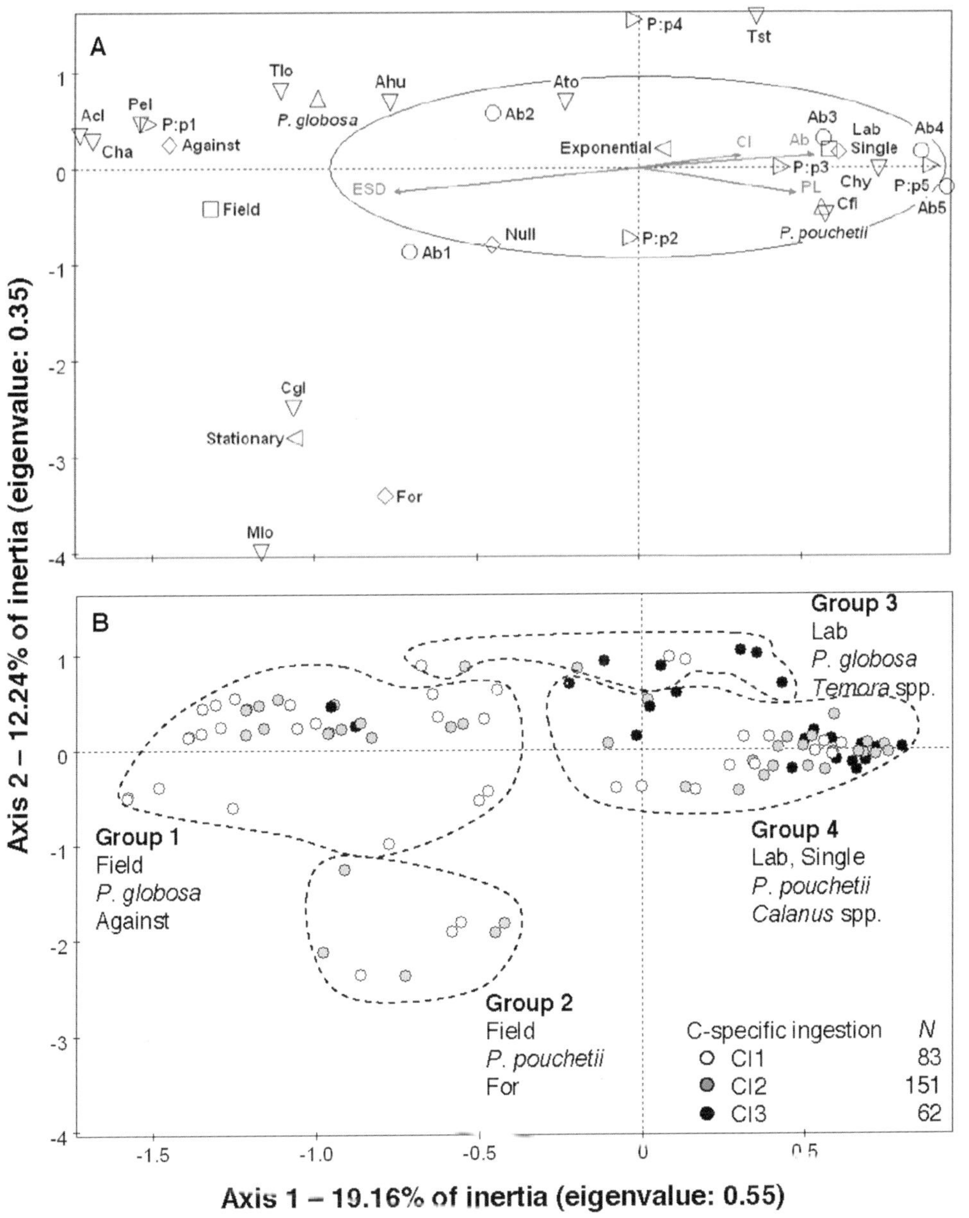

Fig. 1 First factorial plane of MCA of data on crustacean grazing experiment on *Phaeocystis*. (**A**) Projections of continuous illustrative variables in the correlation circle (radius 1) and ordination of active variables: *Phaeocystis* species (Δ), growth (◁) and abundance (○), crustacean species (∇), predator-to-prey size ratio (▷), experiment location (□), selection (◇). (**B**) Ordination of data (○) labelled according to their C-specific ingestion characteristic and delimitation of the four groups designed by the hierarchical clustering (χ^2 distance). See Table 5 for label identification

Results from statistical analysis

The result of the MCA indicated that the first and second axes accumulate about 19% and 12% of the total variability, respectively (Fig. 1A). Two variables basically contribute to the first axis: the selection ("Against" and "Single", cumulated 18% of the contribution to axis 1) and the experiment location ("Lab" and "Field", cumulated 17.5% of the contribution). Thus, this axis differentiates laboratory experiments where *Phaeocystis* was the only prey ("Single"), as opposed to field experiments where

Table 7 Characterization of the most typical properties of the four groups designed by the hierachical cluster analysis defined in Fig. 1 panel B

Group *n*	1			2			3			4		
	82			15			27			172		
	Modality	gr/mod	mod/gr	Modality	gr/mod	mod/gr	Modality	gr/mod	mod/gr	Modality	gr/mod	mod/gr
Group characteristics	Against	95	88	For	100	93	P:p4	96	89	Single	84	98
	Field	80	89	Stationary	79	100	Tst	100	56	*P. pouchetii*	87	95
	P:p1	96	67	Ab1	19	100	*P. globosa*	25	100	Lab	82	98
	P. globosa	66	87	Field	16	100	Tlo	34	44	P:p5	100	44
	CI1	55	56	P:p2	14	87	Lab	13	100	Chy	96	42
	Cha	100	17	Cgl	38	20	CI3	19	44	Cfi	88	53
	Acl	100	16	*P. pouchetii*	7	93				Ab4	86	35
	Pel	100	16							Exponential	62	100
	Ab1	48	48							P:p3	88	25
	Tlo	63	27							Ab5	100	9
	Ab2	44	49							CI3	77	28
										CI2	66	58

Gr/mod corresponds to the percentage of the modality belonging to the group, and mod/gr corresponds to the percentage of the group belonging to the modality. Modality label as in Table 5. *n* denotes the number of observations

alternative prey were available and *Phaeocystis* was selected against ("Against"). The predator-to-prey size ratio modality "P:p1" (size ratio < 4) also has an important contribution to the first axis (10%), showing that small copepods (such as *Centropages hamatus*, *Acartia clausi* and *Pseudocalanus elongatus*) tend to reject *Phaeocystis* in situ. On axis 2, the two modalities "Stationary" and "For" have the greatest contribution (18% and 20%, respectively). This axis separates grazing experiments where *Phaeocystis* in a stationary growth phase was selected positively by predators in situ. In the first factorial plane we also observed that *Phaeocystis* concentration and predator-to-prey size ratio in lab experiments were higher than what was generally observed in situ (Fig. 1A) and that higher carbon-specific ingestions were mostly related to laboratory experiments (Fig. 1B).

The combined analysis of the MCA and HCA clearly distinguished four groups of observations (Fig. 1B and Table 7). Groups 1 and 2 bring together 97% of the grazing experiments performed in the field and are separated from groups 3 and 4, which comprise 96% of the lab experiments with *Phaeocystis* provided as single prey. Group 1 consists of field experiments with small-size copepods selecting against *P. globosa*. Group 2 includes all field experiments with positive selection for *P. pouchetii* (93% of the group), and is characterized by a slightly higher predator-to-prey size ratio (87% of them had a P:p-ratio of 4–16, "P:p2"). In these experiments *Phaeocystis* was in the stationary growth phase and in relatively low abundance ("Ab1" = abundance between 0.1 and 125 μg C l^{-1}). The group 3 is characterized by lab grazing experiments of *Temora* spp. on *P. globosa*. The predator-to-prey size ratio in these laboratory experiments was higher (89% of them had a P:p ratio of 64–256, "P:p4") than what was commonly observed in grazing experiments on natural plankton. Similarly we observed higher carbon-specific ingestion than what was observed in situ for this copepod genus (group 1). Group 4 brings together the lab grazing experiments on *P. pouchetii* by larger copepods (*Calanus* spp.). This group includes the highest *Phaeocystis* abundance and predator-to-prey size ratio tested in lab experiments and grouped the highest carbon-specific ingestion estimates.

In conclusion, the MCA and the HCA analyses suggested that: (1) the lowest grazing rates from the field (often zero) were recorded for small copepods (*Acartia*, *Pseudocalanus*, *Temora* and *Centropages*) in blooms of *Phaeocystis globosa* colonies, (2) large copepods (e.g., *Calanus*) feeding on *Phaeocystis pouchetii* had higher grazing rates, especially in lab studies, when no alternative food was present and (3)

P. pouchetii in a stationary growth phase was positively selected by large copepods.

It is important to note that grazing data for large copepods (*Calanus* spp. and *Metridia longa*) were only available for *P. pouchetii*, whereas grazing studies with small copepods (*Acartia*, *Centropages*, *Pseudocalanus* and *Temora*) were limited to *P. globosa*. Thus, the presently available data do not allow us to determine whether the different results for *P. pouchetii* and *P. globosa* are due to differences between the two species, or differences between the predators. In order to test for species-specific differences, we need to compare grazing on both species simultaneously using similar methodologies in future studies.

Quantitative results from data on crustacean grazing

Because the statistical analysis suggested a large difference between feeding rates in laboratory and field investigations, we compared the average specific ingestion rates in all available field and laboratory experiments (Table 6), and for five groups of predator-to-prey size ratios separately (Fig. 2). These comparisons revealed that: (1) overall feeding (carbon specific ingestion) on *Phaeocystis* spp. was significantly lower [$p < 10^{-8}$, analysis of variation (ANOVA)] in all field (average 2.5% d^{-1}) studies with natural plankton, compared to laboratory studies (average 11% d^{-1}), (2) the highest average carbon-specific ingestion rate (23% d^{-1}) on *Phaeocystis* spp. was found when the predator-to-prey size ratio (P:p) was 4–16 in lab studies, whereas there was no clear size ratio trend in field studies, and (3) ingestion was low ($\leq$2% d^{-1}) when P:p was < 4 in both lab and field, indicating that the upper effective size limit for prey equivalent spherical diameter was ¼ of the predator prosome length.

Thus, crustacean grazers show a much lower grazing on *Phaeocystis* in the field than in the laboratory. One reason for this discrepancy could be that fewer than 5% of the laboratory studies offered the copepods alternative prey to *Phaeocystis* whereas alternative prey were available in the field. Alternatively, laboratory cultures might not display the same chemical grazing cues (and possible grazing deterrents) as *Phaeocystis* growing in natural plankton. Haptophytes may lose their inhibitory effects in vitro, and toxicity may be species-, strain- or growth-condition-

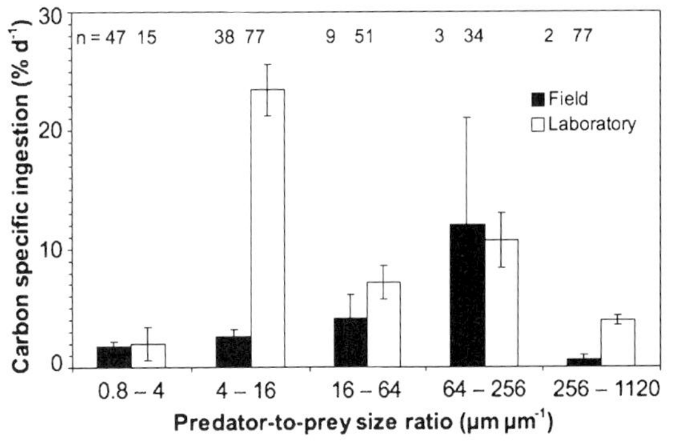

Fig. 2 Crustaceans (all stages of copepods and krill) grazing on *Phaeocystis pouchetii*, *P. globosa* or *P. antarctica*, in laboratory and field experiments. Average daily carbon-specific ingestion rates (percent µg C µg C^{-1} d^{-1}) are presented for five different predator-to-prey size categories [derived from predator prosome or carapace length and average equivalent spherical diameter (ESD) of the *Phaeocystis* prey, solitary cells or colonies]. Error bars denote the standard error (SE) of the average; the number of observations (*n*) for each category are given above each column. Note the high daily ration for P:p size ratio 64–256 in the field is due to a single (30%) value (with that value excluded, the rate is 3.0 ± 1.0 %)

specific (Edvardsen and Paasche 1998). Support for this hypothesis comes from laboratory studies that failed to recreate toxicity in the lab that was observed in the field, even with haptophytes isolated from highly toxic blooms, such as the *Chrysochromulina polylepis* bloom in 1988 (Nielsen et al. 1990). Likewise, the antipredation effect revealed by Estep et al. (1990) in actively growing field-collected *Phaeocystis pouchetii* apparently disappeared during the first 12 h in experimental containers. It may be that laboratory or other in vitro studies underestimate the negative effects of potentially toxic haptophytes in situ, but this requires explicit evaluation.

Ideally, future investigations on the ecology of zooplankton feeding should preferably be conducted in situ, although laboratory experiments are needed to investigate the effects of potentially important signaling substances between *Phaeocystis* and different predators in presence of realistic alternative prey.

Grazing by protozooplankton and other microzooplankton

General trends in grazing by protozoan microzooplankton on *Phaeocystis* are difficult to assess since the number of quantitative investigations is limited (Tables 3 and 4). Very few studies have investigated

Table 8 Microzooplankton grazing on *Phaeocystis globosa* solitary cells during blooms in the English Channel off northen France May 2003 and April 2004 (JC Nejstgaard, AF Sazhin and LF Artigas unpubl.)

Date	Prey type (μm)	r^2	μ (d^{-1})	G (d^{-1})	Grazing impact (% SS d^{-1})
Apr 2003	Chl *a* (>0.45)	0.58	0.24 ± 0.07**	0.61 ± 0.14***	46
	Diatoms (5-99)	0.45	1.34 ± 0.18***	1.24 ± 0.40**	71
	Phaeocystis (2-8)	0.04	0.08 ± 0.16 ns	0.21 ± 0.30 ns	19
May 2004	Chl *a* (>0.45)	0.13	−1.17 ± 0.12***	0.18 ± 0.18 ns	17
	Diatoms (5-99)	0.18	0.16 ± 0.09 ns	0.19 ± 0.13 ns	18
	Phaeocystis (2-8)	0.04	−1.14 ± 0.20***	−0.23 ± 0.34 ns	0

Dilution experiments were performed and analyzed as decribed in Nejstgaard et al. (2001). Specific phytoplankton growth rate (μ), microzooplankton grazing coefficient (g), grazing impact (percentage of average standing stock removed d^{-1}). ±SE for the mean. * $p < 0.05$, ** $p < 0.01$, and *** $p < 0.001$ for μ or $g = 0$. Prey types are given with equivalent spherical diameter size ranges (μm)

microzooplankton as potential grazers on colonial *Phaeocystis* spp. The colonies are protected by a tough skin (Hamm et al. 1999) and microzooplankters are generally assumed to be too small to prey actively on colonial forms of *Phaeocystis* (Hansen et al. 1994a; Weisse et al. 1994; Tang 2003). In accordance, Irigoien et al. (2003) reported that nauplii of *Calanus finmarchicus* did not ingest colonies of *Phaeocystis* in the Irminger sea. However, the protists *Noctiluca scintillans* (Weisse et al. 1994; Jakobsen and Tang 2002) and apparently *Gyrodinium cf. spirale* (Stelfox-Widdicombe et al. 2004) are able to ingest small colonies, while some tintinnids and rotifers "attack" and "hover around colonies", possibly ingesting released cells (Hollowday 1949; Admiraal and Venekamp 1986). The abundance of microzooplankton has also been reported to decline or remain low during natural blooms of colonial *Phaeocystis*, but increases rapidly during the breakdown of the colonies at the end of the bloom (e.g., Peperzak et al. 1998; van Boekel et al. 1992). This suggests that actively growing colonies are inferior prey for microzooplankton, while single cells released from decaying colonies may be a suitable food source. However, an alternative explanation for the low microzooplankton abundance during *Phaeocystis* blooms is the strong selective predation pressure from larger zooplankton such as copepods (Hansen et al. 1993; Gasparini et al. 2000).

It is clear from the literature that at least some microzooplankton may readily ingest single-celled *Phaeocystis* (Tables 3 and 4), and chemical cues from microzooplankton grazing have been shown to induce colony formation (Long and Hay 2006) and enlargement in *P. globosa* (Tang 2003), supporting the assumption that microzooplankton graze on the single cells but not the colonies. However, some of the reported grazing rates in laboratory studies were low, with or without alternative prey present (Table 3). Also, in several incubation experiments with natural plankton the microzooplankton community grazing rates were very low ($\leq$0.1 d^{-1}) during dominance of *Phaeocystis* (Table 4). It is well known that some microzooplankton feed selectively (e.g., Verity 1988; González et al. 1990; Strom and Loukos 1998; Archer et al. 2000), and it was recently shown that copepod nauplii ingest some clones and cell morphotypes of *P. globosa* at low rates (Dutz and Koski 2006). Results showing selection against single cells by microzooplankton in the sea was obtained in recent dilution experiments in the English Channel (Table 8). These experiments yielded zero or close to zero microzooplankton grazing rates on single-celled *P. globosa*, but high microzooplankton grazing rates on diatoms based on cell counts, while bulk measurements of chlorophyll *a* (as used in most of the experiments in Table 4) indicated intermediate grazing values. This suggests that microzooplankton feeding on *Phaeocystis* may be highly variable in the sea, and reliable grazing rate measurements will therefore require taxon-specific quantitative methods, rather than bulk measurements (see further discussion under Future challenges).

Cell-type and life-stage-specific interactions with grazers

In the species *P. globosa*, three solitary, flagellated cell types (micro-, meso, and macroflagellates) differing in size and morphology have been described (Kornmann 1955; Peperzak et al. 2000). The ecological roles of these different cell types are largely unknown, partly due to difficulties in the identification and quantification of the flagellated cell type in field samples (Rousseau et al. 1994; Peperzak et al. 2000; Schoemann et al. 2005; Rousseau et al. this volume). A recent study showed that the grazing mortality of single cells may depend on the cell type, with moderate to high ingestion by *Temora longicornis* nauplii on non-flagellated cell types, but rejection of a flagellate type (Dutz and Koski 2006). This rejection was ascribed to the production of chitinous threads and the cohesion of threads into pentagonal star-like structures which are a typical feature of mesoflagellates (Chretiennot-Dinet et al. 1997; Peperzak et al. 2000).

The variable cell-type-specific grazing mortality of *P. globosa* offers important insight into the natural history of this phytoplankton species. Although flagellated and non-flagellated cell types co-occur throughout the year, blooms of *Phaeocystis* in temperate waters are generally dominated by the colonial cell type and by colonies (Peperzak et al. 2000; Rousseau et al. this volume). At the end of such blooms, single cells are released from the colonies presumably due to nutrient deficiency or irradiance limitation. The liberated cells appear to suffer a high mortality due to microzooplankton grazing, cell lysis and/or perhaps sedimentation (van Boekel et al. 1992; Brussaard et al. 1995; Riebesell et al. 1995; Brussaard et al. this volume; Reigstad & Wassmann this volume). Concurrent to the liberation and disappearance of single cells, an increasing abundance of micro/mesoflagellates has been observed in mesocosms and in the field (Veldhuis et al. 1986; Peperzak et al. 2000; Escaravage et al. 1995; Nejstgaard et al. 2006). The formation of intracolonial flagellated cells may bridge the disappearance of colonial cells and the appearance of flagellated cells (Peperzak et al. 2000). These haploid micro/mesoflagellates probably function as gametes or spores to survive unfavourable conditions during the warm summer months (Veldhuis et al. 1986; Peperzak et al. 2000). Reduced vulnerability of these cells to microzooplankton grazing could explain the accumulation of single flagellated cells despite a high grazing pressure at the end of *Phaeocystis* blooms, and be part of these phytoplankters' strategy to increase survival and foster bloom formation when favorable conditions return. Future studies should investigate the relevance of cell type- and life-stage-specific selection on wider range of zooplankton organisms.

Colony formation and its potential role in morphological defense

The presence of the two distinctly different morphotypes in *Phaeocystis* has long intrigued scientists, and much remains unknown about their respective biological roles and regulation of transition between morphotypes (Rousseau et al. 1994; Lancelot et al. 1998; Rousseau et al. this volume). The prevalence of both morphotypes in natural *Phaeocystis* blooms in contrasting water types prompts the idea that *Phaeocystis* colony development is regulated by common factors (Lancelot and Rousseau 1994) and contributes to the success of the genus in marine systems (Rousseau et al. 1994; Lancelot et al. 1998; Rousseau et al. this volume). The production of the mucilaginous structure represents a substantial energy investment by the cells; the mucilaginous matrix may account for >50% of the total organic carbon of a *Phaeocystis* population (Rousseau et al. 1990; Mathot et al. 2000). The large colony size may also pose a problem of diffusion limitation to nutrient and oxygen uptake, although a modeling study shows that the colonies are unlikely to be limited by diffusion (Ploug et al. 1999). Thus, compared to solitary *Phaeocystis* cells, it is likely that the formation of colonies involves some additional costs, but that the drastic increase in size may be an effective defense mechanism against (relatively) small grazers, not unlike the highly successful freshwater colonial chlorophyte *Scenedesmus* (Lürling 2003; e.g., Lürling and Van Donk 1997), cyanobacteria *Microcystis aeruginosa* and *Sphaerocystis* sp. (e.g., De Bernardi and Giussani 1990), and for different copepods feeding on the large colony-forming diatom *Thalassiosira partheneia* (Schnack 1983).

Some studies suggest that microzooplankton prefer and perform better on single cells than on colonies. When incubated with a mixture of *Phaeocystis globosa* single cells and colonies, microzooplankton reduced the single cells to very low abundance, and the grazer population subsequently declined even when colonies were abundant, presumably due to starvation (Jakobsen and Tang 2002). When restricted to a single diet of either single cells or colonies, the ciliate *Euplotes* grew three times faster on single cells (Long 2004). In contrast, the copepod *Acartia tonsa* and mixed mesozooplankton had higher feeding rates on colonies compared to single cells (Long and Hay 2006). The differential susceptibility of morphologies to grazers could allow *Phaeocystis* to use transformations as an adaptive, inducible response towards grazers. Two recent studies tested the role of chemical cues for inducing transformations using either incubators separated by a permeable membrane (Tang 2003) or filtrate experiments (Long 2004). In one study chemical cues from grazers triggered colony enlargement (Tang 2003). In the other study microzooplankton cues enhanced colony formation while cues from macrozooplankton suppressed colony formation (Long 2004). These observations point to the involvement of a grazing-related chemical signal in morphological defense, but the chemical characteristics of this signal remain unknown (Long and Hay 2006; Tang 2003). Colony formation by *Phaeocystis* could affect grazing even after the demise of a bloom. For example, the degrading colony matrix, making the surrounding water gelatinous, may continue to prevent grazing losses of single cells by microzooplankton (Seuront et al. this volume), and similarly, transparent exopolymer particles formed by coagulation of colony-derived carbohydrates could also inhibit copepod grazing (Dutz et al. 2005).

Chemical defense

Macroalgae commonly use chemical defense against grazers (Hay and Fenical 1988; Pohnert 2004), and similar examples may exist in phytoplankton (see review by Pohnert 2004; and further examples in Yoshida et al. 2004; Pohnert 2005). A few observations suggest that chemical defense could be important for survival of *Phaeocystis* during exponential growth (Weisse et al. 1994; Estep et al. 1990; Long and Hay 2006). Recently, a number of toxins have been isolated from *Phaeocystis* (reviewed by van Rijssel et al. this volume), and *Phaeocystis* is reportedly toxic to some aquatic organisms, including phytoplankton, sea urchin and cod larvae (Table 2). However, specific chemical zooplankton feeding deterrents from *Phaeocystis* have not yet been isolated and characterized so their existence is still hypothetical. Several of the reported toxic effects were obtained using organisms (or even extracts) that do not interact in nature, and the ecological relevance of such results is unknown.

Does DMS affect grazing on *Phaeocystis*?

Phaeocystis is a prominent producer of the secondary metabolite dimethylsulphoniopropionate (DMSP) that can be enzymatically cleaved to acrylate and the volatile trace gas dimethyl sulphide (DMS; Keller et al. 1989; Schoemann et al. 2005; Stefels et al. this volume). DMSP is a multifunctional compound that probably assists several physiological processes related to salinity-, temperature- and light-stress in algal cells (Stefels 2000). This compound and its cleavage products have also been suggested to function as grazing deterrents (Strom et al. 2003; Wolfe et al. 1997) and it is possible that they provide a competitive advantage to *Phaeocystis* when non-DMSP-producing alternative prey are available to potential predators. Microzooplankton grazing greatly increases the production of DMS in *Emiliania huxleyi* (Wolfe and Steinke 1996) and microzooplankton are suspected to have caused the conversion of *Phaeocystis*-DMSP to DMS in the southern North Sea (Archer et al. 2003). The grazing-induced production of DMS may be analogous to the production of volatile infochemicals during herbivore grazing on terrestrial plants that utilize such trace gases to attract carnivores (Dicke and Sabelis 1988; Steinke et al. 2002; Hay et al. 2004). Such indirect defense mechanisms involve interactions on three trophic levels that include plants, herbivores and carnivores. Seabirds are attracted by gradients of DMS in the atmosphere (Nevitt et al. 1995; Nevitt and Bonadonna 2005), and copepods are chemosensitive to DMS and react with search behavior when encountering plumes of this compound (Steinke et al. 2006). If DMS produced during grazing makes the microzooplankton grazers more susceptible to predation by copepods, then investment into such an indirect defense mechanism could be beneficial for the phytoplankton.

However, the ecological consequences of the production of DMS and associated compounds are controversial and explicit tests of possible defensive functions for *Phaeocystis* are required. For example, Tang and Simó (2003) did not observe negative effects when the heterotrophic dinoflagellate *Gyrodinium dominans* grazed on *P. globosa* single cells, and the grazers retained >40% of the ingested DMSP. Furthermore, the grazing-induced production of DMS by unicellular phytoplankton is a result of grazing on individual cells and, as a consequence, these ingested cells will not directly benefit from the potential signaling to carnivorous enemies of the microzooplankton. It is possible that such indirect defense mechanisms are beneficial for *Phaeocystis* populations but they would also be of benefit to other members of the phytoplankton community that may not invest in chemical defenses (the cheater problem: Lewis 1986). One may argue that the cheater problem does not exist within a bloom of genetically closely related cells. However, molecular studies have shown that the genetic variation in phytoplankton populations is high, even within a monospecific diatom bloom (Rynearson and Armbrust 2005). It is impossible to predict functional diversity from such molecular data and there is very little information on the effect of genetic variation on the ecophysiological fitness in *Phaeocystis*.

Evidence is accumulating that chemical communication influences the composition and dynamics of pelagic communities (Pohnert 2004). The production of DMSP in *Phaeocystis* is probably only one example where a secondary metabolite benefits the physiology of individual cells and its catabolic products could affect the structure and function of *Phaeocystis*-dominated food webs. This is a complex research area that needs to be addressed with more detailed studies in the future.

Does nutritional value affect grazing on *Phaeocystis*?

Copepods may feed selectively based on nutritional quality of the prey (e.g., Houde and Roman 1987). However, it is not clear how food quality of the different forms of *Phaeocystis* affects zooplankton grazing. It has been suggested that *Phaeocystis* spp. are of low nutritional value due to their low content of polyunsaturated fatty acids (PUFA) (Al-Hasan et al. 1990; Claustre et al. 1990; Rogers and Lockwood 1990; Nichols et al. 1991; Cotonnec et al. 2001; Tang et al. 2001; Turner et al. 2002). This contrasts with the high survival and development rates observed in other studies (Verity and Smayda 1989), at least for some strains of *P. globosa* (Dutz and Koski 2006). High amounts of C_{18}-PUFA were detected in a *P. globosa* bloom in the North Sea (Hamm and Rousseau 2003) and in a *P. pouchetii* bloom in Balsfjorden in northern Norway (Hamm et al. 2001). The mucus produced by colonies is generally considered to be of low nutritional value, refractory (Thingstad and Billen 1994), and of low carbohydrate content (van Rijssel et al. 1997). However, the composition of the mucus material may vary during a bloom cycle (Alderkamp et al. 2006), and colonization of this material by pennate diatoms and other organisms on and inside older colonies (Hamm and Rousseau 2003; Sazhin et al. this volume) may increase its nutritional value. Besides PUFA and carbohydrates, the effects of other important nutritional components of *Phaeocystis* such as amino acids, sterols and vitamins, on copepod feeding and reproduction are still poorly known.

Survival of gut passage?

Some phytoplankton can survive gut passage, a trait that is often overlooked in zooplankton studies. For instance, large fractions of the chlorophytes *Chlamydomonas reinhardtii* and *Selenastrum capricornutum* survive passing the guts of *Daphnia*. This is considered a defense mechanism by the algae to reduce grazing pressure when their growth rates are low (van Donk et al. 1997). Further, the colonial chlorophyte *Sphaerocystis schroeteri* survives gut passage in *Daphnia* and benefits from grazer-released nutrients. This interaction is hypothesized to affect bloom formation in this species (Porter 1976). If cells of *Phaeocystis* may survive gut passage of zooplankton, this would have significant implications for the understanding of the grazing interactions. Because, net removal rates of the prey from the population will decrease, and if cells may produce defense chemicals upon predation and surviving gut passage it would affect the chemical antipredatory mechanisms suggested above. But, to our knowledge there are no such studies for *Phaeocystis* or any other haptophyte phytoplankton. However, small intact colonies of *P. pouchetii* have been observed fluorescing inside copepod pellets in the Barents Sea (P. Verity and P. Wassmann, personal communication) suggesting that surviving gut passage might be possible for *Phaeocystis*.

Conclusions and future challenges

Although over 100 publications have examined zooplankton grazing on *Phaeocystis* spp., more than 90% report on copepods or other crustacean grazers, so quantitative data for other grazers are limited. Grazing on *Phaeocystis* was often determined by feeding on only single types of cultured phytoplankton in the laboratory. Field reports often do not provide enough information to evaluate even some of the most basic factors known to affect zooplankton grazing simultaneously, such as predator-to-prey size ratio, food concentration and feeding rates on alternative prey. Further, most reports on crustacean grazing in the field are based on non-prey-specific grazing estimations such as bulk gut pigments, that cannot be used to determine specific feeding on *Phaeocystis* in the mixed natural plankton assemblages.

Nevertheless, our analysis of the present literature suggests that: (1) laboratory-derived crustacean grazing rates on monocultures of *Phaeocystis* may have been overestimated compared to feeding in natural plankton communities, and should be treated with caution; (2) formation of colonies by *P. globosa* appeared to reduce predation by small copepods (e.g., *Acartia*, *Pseudocalanus*, *Temora* and *Centropages*), whereas large copepods (e.g., *Calanus* spp.) were able to feed on colonies of *Phaeocystis pouchetii*; (3) physiological differences between different growth states, species, strains, cell types, and laboratory culture versus natural assemblages may explain most of the variations in reported feeding rates; (4) chemical signaling between predator and prey may be a major factor controlling grazing on *Phaeocystis*; (5) it is unclear to what extent different zooplankton, especially protozooplankton, feed on the different life forms of *Phaeocystis* in situ.

In the present literature there is a dichotomy between data showing high feeding rates on both single cells and colonies of *Phaeocystis pouchetii* by larger copepods, especially in the laboratory studies, and avoidance of colonies of *Phaeocystis globosa* by other smaller copepods in the field. This imbalance in basic quantitative data could initially be remedied by applying classical bottle incubation methods with a range of predators, preferably using natural plankton in situ, analyzed with methods that resolve the feeding rates on the different life forms (and prey sizes) of different species of *Phaeocystis* from alternative prey types. Such analyses could be done by microscopy (Verity and Paffenhöfer 1996; Båmstedt et al. 2000) corrected for food web cascades (Nejstgaard et al. 2001), and/or possibly analyzed by a combination of more recent methods such as flow cytometry (Collier and Campbell 1999; Jonker et al. 2000), computer/video-aided plankton counting devices (See et al. 2005), and specific molecular probing (Caron 2005) to increase the taxonomic precision. However, the ability of chemical cues to rapidly alter colony formation indicates that grazing estimates based on cell counts could grossly misrepresent actual grazing rates.

For example, zooplankton grazing may induce or suppress colony formation and enlargement (Tang 2003; Long 2004; Long and Hay 2006), and accumulating data suggests that chemical signaling between different levels of predators of prey can play an important role influencing the dynamics in pelagic communities (e.g., reviewed by Pohnert 2004; Yoshida et al. 2004). This suggests that chemical signaling in artificially increased grazer densities over long incubations could confound data from bottle experiments by changing the relative abundance of different life forms in the grazing bottle due to factors other than direct ingestion of the cells. Further, recent results suggesting that feeding on *Phaeocystis* may even be cell type- or life-stage-specific (Dutz and Koski 2006), and that new methods accounting for such differences are necessary.

Thus, the ideal way to assess feeding on *Phaeocystis* would be a direct quantification of feeding rates on specific prey by assessing individual predators that have been feeding undisturbed in situ. Molecular quantitative analysis of the prey in the guts and faecal pellets of the predators could potentially achieve this. Such approaches are already common in studies of e.g., terrestrial arthropods (Sheppard and Harwood 2005; Symondson 2002), and it has been shown that prey DNA from haptophyte phytoplankton can be detected in copepod guts and fecal pellets by standard polymerase chain reaction (PCR) targeting 18S ribosomal DNA (Nejstgaard et al. 2003). More recently, it was also shown that carnivorous diet may be inferred from the content of copepod fecal pellets using standard PCR (Vestheim et al. 2005), and that different phytoplankton DNA, including *Phaeocystis*, may be quantified in guts and fecal pellets from both appendicularians and copepods using quantitative real-time PCR (Nejstgaard et al. 2005; Troedsson et al. 2007).

Thus the rapidly developing molecular approaches may help us obtain more-realistic ingestion rates of *Phaeocystis* and other live prey in situ. However, to further reveal and quantify mechanisms controlling the complex trophic interactions between *Phaeocystis* and its predators, new approaches need to be developed. In concert with the recent review by Pohnert (2004) we conclude that such development will probably be as complex as the interactions it tries to reveal. Developing successful innovative approaches and better quantitative analytical tools will depend largely upon multidisciplinary efforts between chemists, molecular biologists and plankton ecologists.

Acknowledgements We thank S. Gasparini for his valuable suggestions concerning MCA and HCA analyses. This work was supported by the Norwegian Research Council project 152714/120 30 to JCN, the US-NSF OPP award ANT-0440478 to KWT, the UK Natural Environment Research Council grants NER/I/S/2000/00897 and NE/B500282/1 to MS, the German Federal Ministry for Education and Research grants within the GLOBEC Germany project 03F0418C to JD, the Carlsberg Foundation to MK, the Belgian Research Program for sustainable management of the North Sea within the AMORE project MN/DD/20/21/22 to EA, and the EPA Star Fellowship U-91599501-0 and a National Parks Ecological Research Fellowship to JDL.

References

Aanesen RT, Eilertsen HC, Stabell OB (1998) Light-induced toxic properties of the marine alga *Phaeocystis pouchetii* towards cod larvae. Aquat Toxicol 40:109–121

Admiraal W, Venekamp LAH (1986) Significance of tintinnid grazing during blooms of *Phaeocystis pouchetii* (haptophyceae) in Dutch coastal waters. Neth J Sea Res 20:61–66

Al-Hasan RH, Ali AM, Radwan SS (1990) Lipids, and their constituent fatty-acids, of *Phaeocystis* sp. from the Arabian Gulf. Mar Biol 105:9–14

Alderkamp A-C, Buma AGJ and van Rijssel M (this volume) The carbohydrates of *Phaeocystis* and their degradation in the microbial food web. Biogeochemistry. doi:10.1007/s10533-007-9078-2

Alderkamp A-C, Nejstgaard JC, Verity PG, Zirbel MJ, Sazhin AF, van Rijssel M (2006) Dynamics in carbohydrate composition of *Phaeocystis pouchetii* colonies during spring blooms in mesocosms. J Sea Res 55:169–181

Archer SD, Stelfox-Widdicombe CE, Malin G, Burkill PH (2003) Is dimethyl sulphide production related to microzooplankton herbivory in the southern North Sea? J. Plankton Res 25:235–242

Archer SD, Verity PG, Stefels J (2000) Impact of microzooplankton on the progression and fate of the spring bloom in fjords of northern Norway. Aquat Microb Ecol 22: 27–41

Båmstedt U, Gifford DJ, Irigoien X, Atkinson A, Roman M (2000) Feeding. In: Harris R, Wiebe P, Lenz J, Skjoldal HR, Huntley M (eds) ICES Zooplankton Methodology Manual. Academic Press, London, pp 297–399

Banse K (1994) Grazing and zooplankton production as key controls of phytoplankton production in the open ocean. Oceanography 7:13–20

Barnard WR, Andreae MO, Iverson RL (1984) Dimethylsulfide and *Phaeocystis poucheti* in the Southeastern Bering Sea. Cont Shelf Res 3:103–113

Baumann MEM, Lancelot C, Brandini FP, Sakshaug E, John DM (1994) The taxonomic identity of the cosmopolitan prymnesiophyte *Phaeocystis*: a morphological and ecophysiological approach. J Mar Syst 5:5

Bautista B, Harris RP, Tranter PRG, Harbour D (1992) In situ copepod feeding and grazing rates during a spring bloom dominated by *Phaeocystis* sp. in the English Channel. J Plankton Res 14:691–703

Blauw AN (this volume) *Phaeocystis* blooms forecasting in Dutch coastal waters, the impact of suspended matter. Biogeochemistry

Breton E, Sautour B, Brylinski J-M (1999) No feeding on *Phaeocystis* sp. as solitary cells (post bloom period) by the copepod Temora longicornis in the coastal waters of the English Channel. Hydrobiologia 414:13–23

Brussaard CPD, Bratbak G, Baudoux A-C and Ruardij P (this volume) *Phaeocystis* and its interaction with viruses. Biogeochemistry. doi:10.1007/s10533-007-9096-0

Brussaard CPD, R. R, Noordeloos AAM, Cadée GC, Witte H, Kop AJ, Nieuwland G, van Duyl FC and Bak RPM (1995) Effects of grazing, sedimentation and phytoplankton cell lysis on the structure of a coastal pelagic food web. Mar Ecol Prog Ser 123

Bullen GE (1908) Plankton studies in relation to the western mackerel fishery. J Mar Biol Ass UK 8:269–302

Caron DA (2005) Introductory remarks: advances in the molecular ecology of protists. J Eukaryot Microbiol 52:81–82

Caron DA, Dennett MR, Lonsdale DJ, Moran DM, Shalapyonok L (2000) Microzooplankton herbivory in the Ross Sea, Antarctica. Deep Sea Res Part II Top Stud Oceanogr 47:3249–3272

Chen Y-Q, Wang N, Zhang P, Zhou H, Qu L-H (2002) Molecular evidence identifies bloom-forming *Phaeocystis* (Prymnesiophyta) from coastal waters of southeast China as *Phaeocystis globosa*. Biochem Syst Ecol 30:15–22

Chretiennot-Dinet MJ, Giraud-Guille MM, Vaulot D, Putaux JL, Saito Y, Chanzy H (1997) The chitinous nature of filaments ejected by Phaeocystis (*Prymnesiophyceae*). J Phycol 33:666–672

Claustre H, Poulet SA, Williams R, Marty JC, Coombs S, Benmlih F, Hapette AM, Martinjezequel V (1990) A biochemical investigation of a *Phaeocystis* sp. bloom in the Irish Sea. J Mar Biol Ass U K 70:197–207

Collier JL, Campbell L (1999) Flow cytometry in molecular aquatic ecology. Hydrobiologia 401:33–53

Cook PA, Gabbott PA (1972) Seasonal changes in the biochemical composition of adult barnacle, *Balanus balanoides*, and the possible relationships between biochemical composition and cold-tolerance. J Mar Biol Ass U K 52:805–825

Cotonnec G, Brunet C, Sautour B, Thoumelin G (2001) Nutritive value and selection of food particles by copepods during a spring bloom of *Phaeocystis* sp. in the English Channel, as determined by pigment and fatty acid analyses. J Plankton Res 23:693–703

Daro M-H (1985) Field study of the diel, selective and efficiency feeding of the marine copepod *Temora longicornis* in the Southern Bight of the North Sea. In: Van Grieken R, Wollast R (eds) Proc progress in Belgian oceanographic research, Brussels, pp 250–263

Davies AG, Demadariaga I, Bautista B, Fernandez F, Harbour DS, Serret P, Tranter PRG (1992) The ecology of a coastal *Phaeocystis* bloom in the North-Western English-Channel in 1990. J Mar Biol Ass UK 72:691–708

De Bernardi R, Giussani G (1990) Are blue-green algae suitable food for zooplankton? A review. Hydrobiologia 200/201:29–41

Dicke M, Sabelis MW (1988) How plants obtain predatory mites as bodyguards. Neth J Zool 38:148–165

Dunne T, Doyle J, Parker M (1984) A review of the incidence and effects of unusual marine blooms in Irish coastal waters. ICES Spec Meet ICES CM 1984/D 9:4

Dutz J, Klein Breteler WCM, Kramer G (2005) Inhibition of copepod feeding by exudates and transparent exopolymer particles (TEP) derived from a *Phaeocystis globosa* dominated phytoplankton community. Harmful Algae 4:929

Dutz J, Koski M (2006) Trophic significance of solitary cells of the prymnesiophyte *Phaeocystis globosa* depends on cell type. Limnol Oceanogr 51:1230–1238

Edvardsen B, Paasche E (1998) Bloom dynamics and physiology of *Prymnesium* and *Chrysochromulina*. In: Anderson DM, Cembella AD, Hallegraef GM (eds) The physiological ecology of harmful algal blooms. Springer Verlag, Heidelberg, pp 193–208

Eilertsen HC, Raa J (1995) Phytoplankton toxins in sea-water. J Mar Biotechnol 3:115–119

Escaravage V, Peperzak L, Prins TC, Peeters JCH, Joordens JCA (1995) The development of a *Phaeocystis* bloom in a mesocosm experiment in relation to nutrients, irradiance and coexisting algae. Ophelia 42:55–74

Estep KW, Nejstgaard JC, Skjoldal HR, Rey F (1990) Predation by copepods upon natural populations of *Phaeocystis pouchetii* as a function of the physiological state of the prey. Mar Ecol Prog Ser 67:235–249

Everitt BS, Dunn G (2001) Applied multivariate data analysis, 2nd edn. Oxford University Press, New York

Frangoulis C, Belkhiria S, Goffart A, Hecq J-H (2001) Dynamics of copepod faecal pellets in relation to a *Phaeocystis* dominated phytoplankton bloom: characteristics, production and flux. J Plankton Res 23:75–88

Fransz HG, Gieskes WWC (1984) The unbalance of phytoplankton and copepods in the North Sea. J Cons Int Explor Mer 183:218–225

Fransz HG, Gonzalez SR, Cadée GC, Hansen FC (1992) Long-term change of *Temora longicornis* (copepoda, Calanoida) abundance in a Dutch tidal inlet (Marsdiep) in relation to eutrophication. Neth J Sea Res 30:23–32

Fretter V, Montgomery MC (1968) The treatment of food by prosobranch veligers. J Mar Biol Ass UK 48:275–544

Frost BW (1972) Effect of size and concentration of food particles on the feeding behaviour of the marine planktonic copepod *Calanus pacificus*. Limnol Oceanogr 17:805–815

Gasparini S, Daro MH, Antajan E, Tackx M, Rousseau V, Parent J-Y, Lancelot C (2000) Mesozooplankton grazing during the *Phaeocystis globosa* bloom in the southern bight of the North Sea. J Sea Res 43:345–356

Goffart A, Catalano G, Hecq JH (2000) Factors controlling the distribution of diatoms and *Phaeocystis* in the Ross Sea. J Mar Syst 27:161

González JM, Sherr EB, Sherr BF (1990) Size-selective grazing on bacteria by natural assemblages of estuarine flagellates and ciliates. Appl Environ Microbiol 56:583–589

Gorsky G, Chretiennot-Dinet M-J, Blanchot J, Palazzoli I (1999) Picoplankton and nanoplankton aggregation by appendicularians: fecal pellet contents of *Megalocercus huxleyi* in the equatorial Pacific. J Geophys Res C 104:3381–3390

Greenacre MJ (1984) Theory and applications of correspondence analysis. Academic Press, London

Haberman KL, Quetin LB, Ross RM (2003) Diet of the Antarctic krill (*Euphausia superba* Dana): I. Comparisons of grazing on *Phaeocystis antarctica* (Karsten) and *Thalassiosira antarctica* (Comber). J Exp Mar Biol Ecol 283:79–95

Hamm CE, Reigstad M, Wexels Riser C, Mühlebac A, Wassmann P (2001) On the trophic fate of *Phaeocystis pouchetii*. VII. Sterols and fatty acids reveal sedimentation of *P. puchetii*-derived organic matter via krill fecal strings. Mar Ecol Prog Ser 209:55–69

Hamm CE, Rousseau V (2003) Composition, assimilation and degradation of *Phaeocystis globosa*-derived fatty acids in the North Sea. J Sea Res 50:271

Hamm CE, Simson DA, Merkel R, Smetacek V (1999) Colonies of *Phaeocystis globosa* are protected by a thin but tough skin. Mar Ecol Prog Ser 187:101–111

Hansen B, Bjørnsen PK, Hansen PJ (1994a) The size ratio between planktonic predators and their prey. Limnol Oceanogr 39:395–403

Hansen B, Tande KS, Berggreen UC (1990) On the trophic fate of *Phaeocystis pouchetii* (Hariot). III. Functional responses in grazing demonstrated on juvenile stages of *Calanus finmarchicus* (Copepoda) fed diatoms and *Phaeocystis*. J Plankton Res 12:1173–1187

Hansen B, Verity P, Falkenhaug T, Tande KS, Norrbin F (1994b) On the trophic fate of *Phaeocystis pouchetti* (Hariot). V. Trophic relationships between *Phaeocystis* and zooplankton: an assessment of methods and size dependence. J Plankton Res. 16:487–511

Hansen E, Eilertsen HC, Ernstsen A, Geneviere A-M (2003) Anti-mitotic activity towards sea urchin embryos in extracts from the marine haptophycean *Phaeocystis pouchetii* (Hariot) Lagerheim collected along the coast of northern Norway. Toxicon 41:803–812

Hansen E, Ernstsen A, Eilertsen HC (2004) Isolation and characterisation of a cytotoxic polyunsaturated aldehyde from the marine phytoplankter *Phaeocystis pouchetii* (Hariot) Lagerheim. Toxicology 199:207–217

Hansen FC (1992) Zooplankton-Grazing an *Phaeocystis* mit besonderer Berücksichtigung der Calanoiden Copepoden.

Institut für Meereskunde Kiel, Christian-Albrechts-Universität, Kiel, Kiel, p 137

Hansen FC (1995) Trophic interactions between zooplankton and *Phaeocystis* cf *globosa*. Helgol Meeresunters 49: 283–293

Hansen FC, Reckermann M, Klein Breteler WCM, Riegman R (1993) *Phaeocystis* blooming enhanced by copepod predation on protozoa: evidence from incubation experiments. Mar Ecol Prog Ser 102:51–57

Hansen FC, van Boekel WHM (1991) Grazing pressure of the calanoid copepod *Temora longicornis* on a *Phaeocystis* dominated spring bloom in a Dutch tidal inlet. Mar Ecol Prog Ser 78:123–129

Hansen PJ, Bjørnsen PK, Hansen BW (1997) Zooplankton grazing and growth: scaling within the 2–2,000-μm body size range. Limnol Oceanogr 42:687–704

Hay ME, Fenical W (1988) Marine plant-herbivore interactions: the ecology of chemical defense. Annu Rev Ecol Syst 19:111–145

Hay ME, Parker JD, Burkepile DE, Caudill CC, Wilson AE, Hallinan ZP, Chequer AD (2004) Mutualisms and aquatic community structure: the enemy of my enemy is my friend. Annu Rev Ecol Evol Syst 35:175–197

Hegarty SG, Villareal TA (1988) Effects of light level and N:P supply ratio on the competition between *Phaeocystis* cf. *pouchetii* (Hariot) Lagerheim (Prymnesiophyceae) and five diatom species. J Exp Mar Biol Ecol 226:241–258

Hollowday ED (1949) A preliminary report on the Plymouth marine and brackish-water rotifera. J Mar Biol Ass UK 28:239–253

Houde SEL, Roman MR (1987) Effects of food quality on the functional ingestion response of the copepod *Acartia tonsa*. Mar Ecol Prog Ser 40:69–77

Huntley M, Tande K, Eilertsen HC (1987) On the trophic fate of *Phaeocystis pouchetii* (Hariot). II. Grazing rates of *Calanus hyperboreus* (Krøyer) on diatoms and different size categories of *Phaeocystis pouchetii*. J Exp Mar Biol Ecol 110:197–212

Irigoien X, Titelman J, Harris RP, Harbour D, Castellani C (2003) Feeding of *Calanus finmarchicus* nauplii in the Irminger Sea. Mar Ecol Prog Ser 262:193–200

Jakobsen HH, Tang KW (2002) Effects of protozoan grazing on colony formation in *Phaeocystis globosa* (Prymnesiophyceae) and the potential costs and benefits. Aquat Microb Ecol 27:261–227

Jebram D (1980) Prospection for a sufficient nutrition for the cosmopolitic marine bryozoan *Electra pilosa* (Linnaeus). Zool Jahrb Syst 107:368–390

Jones PGV, Haq SM (1963) The distribution of *Phaeocystis* in the Eastern Irish Sea. J Cons Int Explor Mer 28:8–20

Jonker R, Groben R, Tarran G, Medlin L, Wilkins M, Garcia L, Zabala L, Boddy L (2000) Automated identification and characterisation of microbial populations using flow cytometry: the AIMS project. Sci Mar 64:225–234

Kamermans P (1994) Nutritional value of solitary cells and colonies of *Phaeocystis* sp. for the bivalve *Macoma balthica* (L). Ophelia 39:35–44

Keller MD, Bellows WK, Guillard RRL (1989) Dimethyl sulfide production in marine phytoplankton. Acs Symp Ser 393:167–182

Klein Breteler WCM, Koski M (2003) Development and grazing of *Temora longicornis* (Copepoda, Calanoida) nauplii during nutrient limited *Phaeocystis globosa* blooms in mesocosms. Hydrobiologia 491:185–192

Kopp J (1978) Les phénomènes d'eaux colorées ayant affecté le littoral Bas-Normand au cours du mois de juin 1978. Inst Sci Tech Pêches Mar, Centre de Quistreham: 30

Kornmann VP (1955) Beobachtungen an *Phaeocystis*-Kulturen. Helgoländer wiss Meeresunters 5:218–233

Koski M, Dutz J, Klein Breteler WCM (2005) Selective grazing of *Temora longicornis* in different stages of a *Phaeocystis globosa* bloom: a mesocosm study. Harmful Algae 4:915

Krogstad PK (1989) Produksjon av planteplankton til bruk som fór i intensivt oppdrett av torskelarver (*Gadhus morhua* L.) med referanse i ekstensivt oppdrettsystem (poll). Institute of Fisheries, University of Tromsø, Tromsø, p 147

Lancelot C, Keller MD, Rousseau V, Smith WOJ, Mathot S (1998) Autecology of the marine haptophyte *Phaeocystis* sp. In: Anderson DM, Cembella AD, Hallegraef GM (eds) The physiological ecology of harmful algal blooms. Springer Verlag, Berlin, pp 209–224

Lancelot C, Rousseau V (1994) Ecology of *Phaeocystis*: the key role of colony forms. In: Green JC, Leadbeater BSC (eds) The haptophyte algae, Clarendon Press, Oxford, pp 229–245

Landry MR, Hassett RP (1982) Estimating the grazing impact of marine micro-zooplankton. Mar Biol 67:283–288

Landry MR, Selph KE, Brown SL, et al (2002) Seasonal dynamics of phytoplankton in the Antarctic Polar Front region at 170? W Deep Sea Res Part II Top Stud Oceanogr 49:1843–1865

Larsen A, Fonnes Flaten GA, Sandaa R-A, Castberg T, Thyrhaug R, Erga SR, Jacquet S, Bratbak G (2004) Spring phytoplankton bloom dynamics in Norwegian coastal waters: Microbial community succession and diversity. Limnol Oceanogr 49:180–190

Lebart L, Morineau A, Lambert T (1988) SPAD.N Système portable pour l'analyse des données. Manuel de référence. (SPAD.N a software for Correspondence Analysis User's. guide). CISIA, Saint-Mandé, France

Lebart L, Morineau A, Piron M (2000) Statistique exploratoire multidimentionnelle. Dunod, Paris

Lebour MV (1920) The food of young fish, No. III (1919). J Mar Biol Ass UK 12:261–324

Lebour MV (1922) The food of planktonic organisms. J Mar Biol Ass UK 12:644–677

Legendre P, Legendre L (1998) Numerical Ecology. Elsevier, Amsterdam

Levinsen H, Turner JT, Nielsen TG, Hansen BW (2000) On the trophic coupling between protists and copepods in arctic marine ecosystems. Mar Ecol Prog Ser 204:65–77

Lewis WMJ (1986) Evolutionary interpretations of allelochemical interactions in phytoplankton algae. Am Nat 127:184–194

Liss PS, Malin G, Turner SM, Holligan PM (1994) Dimethyl sulphide and *Phaeocystis*: a review. J Mar Syst 5:41

Long JD (2004) Plasticity of consumer-prey interactions in the sea: chemical signaling, learned aversions, and ecological consequences. Biology. Georgia Institute of Technology, Atlanta, p 111

Long JD, Hay ME (2006) When intraspecific exceeds interspecific variance: effects of phytoplankton morphology and growth phase on copepod feeding and fitness. Limnol Oceanogr 51:988–996

Løken S (1990) Næringsøkologi og vekst hos torskelarver (*Gadus morhua* L.) i poll og intensive system. Institute of Fisheries. University of Tromsø, Tromsø, p 86

Lucas CE, Henderson GTD (1936) On the association of jellyfish and other organisms with catches of herring. J Mar Biol Ass U K 21:293–304

Lürling M (2003) Phenotypic plasticity in the green algae *Desmodesmus* and *Scenedesmus* with special reference to the induction of defensive morphology. Ann Limnol 39:85–101

Lürling M, Van Donk E (1997) Morphological changes in *Scenedesmus* induced by infochemicals released in situ from zooplankton grazers. Limnol Oceanogr 42:783–788

Mathot S, Smith WO, Carlson CA, Garrison DL, Gowing MM, Vickers CL (2000) Carbon partitioning within *Phaeocystis antarctica* (Prymnesiophyceae) colonies in the Ross Sea, Antarctica. J Phycol 36:1049–1056

Medlin L and Zingone A (this volume) A review: the genus *Phaeocystis* and its species. Biogeochemistry

Meixner R (1981) Herzmuschel-Suche mit FA-Niobe im ostfriesischen Wattenmeer. Inf Fischwirtsch 28:133–134

Metz C (1998) Feeding of *Oncaea curvata* (Poecilostomatoida, Copepoda). Mar Ecol Prog Ser 169:229–235

Nejstgaard JC, Frischer ME, Raule CL, Gruebel R, Kohlberg KE, Verity PG (2003) Molecular detection of algal prey in copepod guts and faecal pellets. Limnol Oceanogr Methods 1:29–38

Nejstgaard JC, Frischer ME, Simonelli P, Brakel M, Sazhin A, Artigas LF (2005) Comparison of molecular and traditional methods to quantify selective copepod feeding in situ ASLO, Summer Meeting. Santiago de Compostela, Spain

Nejstgaard JC, Frischer ME, Verity PG, Anderson JT, Jacobsen A, Zirbel MJ, Larsen A, Martínez-Martínez J, Sazhin AF, Walters T, Bronk DA, Whipple SJ, Borett SR, Patten BC, Long JD (2006) Temporal patterns in planktonic food web development in mesocosms with added nutrients and *Phaeocystis*. Mar Ecol Prog Ser 321:99–121

Nejstgaard JC, Naustvoll L-J, Sazhin A (2001) Correcting for underestimation of microzooplankton grazing in bottle incubation experiments with mesozooplankton. Mar Ecol Prog Ser 221:59–75

Nevitt GA, Bonadonna F (2005) Sensitivity to dimethyl sulphide suggests a mechanism for olfactory navigation by seabirds. Biol Let 1:303–305

Nevitt GA, Veit RR, Kareiva P (1995) Dimethyl sulfide as a foraging cue for Antarctic Procellariiform seabirds. Nature 376:680–682

Nicholls AG (1935) The larval stages of *Longipedia coronata* Claus, *L. scotti* G. O. Sars, and *L. minor* T. and A. Scott, with a description of the male of *L. scotti*. J Mar Biol Ass UK 20:29–45

Nichols PD, Skerratt JH, Davidson A, Burton H, McMeekin TA (1991) Lipids of cultured *Phaeocystis pouchetii*: signatures for food-web, biogeochemical and environmental studies in Antarctica and the Southern-Ocean. Phytochemistry 30:3209–3214

Nielsen TG, Kiørboe T, Bjørnsen PK (1990) Effects of a *Chrysochromulina polylepis* subsurface bloom on the planktonic community. Mar Ecol Prog Ser 62:1–2

Noordkamp DJB, Gieskes WWC, Gottschal JC, Forney LJ, van Rijssel M (2000) Acrylate in *Phaeocystis* colonies does not affect the surrounding bacteria. J Sea Res 43:287–296

Peperzak L (2002) The wax and wane of *Phaeocystis globosa* blooms. Wiskunde en Natuurwetenschappen. Rijksuniversitet Groningen, Groningen, p 254

Peperzak L, Colijn F, Vrieling EG, Gieskes WWC, Peeters JCH (2000) Observations of flagellates in colonies of *Phaeocystis globosa* (Prymnesiophyceae); a hypothesis for their position in the life cycle. J Plankton Res 22:2181–2203

Peperzak LM, Colijn F, Gieskes WWC, Peeters JCH (1998) Development of the diatom-*Phaeocystis* spring bloom in the Dutch coastal zone of the North Sea: the silicon depletion versus the daily irradiance threshold hypothesis. J Plankton Res 20:517

Petri G, Donat FH, Vareschi E (1999) Investigations on uptake and utilization of *Phaeocystis globosa* colonies by the blue mussel *Mytilus edulis*. Senckenb Marit 29:117–118

Petri G, Vareschi E (1997) Utilization of *Phaeocystis globosa* colonies by young *Mytilus edulis*. Arch Fish Mar Res 45:77–91

Pieters H, Kluytmans JH, Zandee DI, Cadée GC (1980) Tissue composition and reproduction of *Mytilus edulis* in relation to food availability. Neth J Sea Res 14:349–361

Ploug H, Stolte W, Jørgensen BB (1999) Diffusive boundary layers of the colony-forming plankton alga *Phaeocystis* sp—implications for nutrient uptake and cellular growth. Limnol Oceanogr 44:1959–1967

Pohnert G (2004) Chemical defense strategies of marine organisms. In: Schulz SE (ed) Topics in current chemistry. Springer-Verlag GmBH, Berlin, pp 179–219

Pohnert G (2005) Diatom/copepod interactions in plankton: the indirect chemical defense of unicellular algae. ChemBioChem 6:946–959

Porter KG (1976) Enhancement of algal growth and productivity by grazing zooplankton. Science 192:1332–1334

Reigstad M and Wassmann P (this volume) Does *Phaeocystis* spp. contribute significantly to vertical export of biogenic matter? Biogeochemistry

Richardson MG, Whitaker TM (1979) An Antarctic fast-ice food chain: observations on the interaction of the amphipod *Pontogeneia antarctica* Chevreux with ice-associated micro-algae. Br Antarct Surv Bull 47:107–115

Riebesell U, Reigstad M, Wassmann P, Noji T, Passow U (1995) On the trophic fate of *Phaeocystis pouchetii* (Hariot): 6. Significance of Phaeocystis-derived mucus for vertical flux. Neth. J Sea Res 33:193–203

Rogers SI, Lockwood SJ (1990) Observations on coastal fish fauna during a spring bloom of *Phaeocystis pouchetii* in the eastern Irish Sea. J Mar Biol Ass UK 70:249–253

Rousseau V, Becquevort S, Parent J-Y, Gasparini S, Daro M-H, Tackx M, Lancelot C (2000) Trophic efficiency of the planktonic food web in a coastal ecosystem dominated by *Phaeocystis* colonies. J Sea Res 43:357–372

Rousseau V, Chrétiennot-Dinet M-J, Jacobsen A, Verity PG, Whipple SJ (this volume) The life cycle of *Phaeocystis*: state of knowledge and presumptive role in ecology. Biogeochemistry. doi:10.1007/s10533-007-9085-3

Rousseau V, Leynaert A, Daoud N, Lancelot C (2002) Diatom succession, silicification and silicic acid availability in Belgian coastal waters (Southern North Sea). Mar Ecol Prog Ser 236:61–73

Rousseau V, Mathot S, Lancelot C (1990) Calculating carbon biomass of *Phaeocystis* sp. from microscopic observations. Mar Biol 107:305–314

Rousseau V, Vaulot D, Casotti R, Cariou V, Lenz J, Gunkel J, Baumann M (1994) The life cycle of *Phaeocystis* (Prymnesiophycaea): evidence and hypotheses. J Mar Syst 5:23

Rynearson TA, Armbrust EV (2005) Maintenance of clonal diversity during a spring bloom of the centric diatom *Ditylum brightwellii*. Mol Ecol 14:1631–1640

Sargent JR, Eilertsen HC, Falk-Petersen S, Taasen JP (1985) Carbon assimilation and lipid production in phytoplankton in northern Norwegian fjords. Mar Biol 85:109–116

Savage RE (1930) The influence of *Phaeocystis* on the migrations of the herring. Fish Invest II 12:1–14

Savage RE (1932) *Phaeocystis* and herring shoals. J Ecol 20:326–340

Sazhin AF, Artigas LF, Nejstgaard JC, Frischer ME (this volume) The colonization of two *Phaeocystis* species (Prymnesiophyceae) by pennate diatoms and other protists: a significant contribution to colony biomass. Biogeochemistry

Schnack SB (1983) On the feeding of copepods on *Thalassiosira partheneia* from the Northwest African upwelling area. Mar Ecol Prog Ser 11:49–53

Schoemann V, Becquevort S, Stefels J, Rousseau V, Lancelot C (2005) *Phaeocystis* blooms in the global ocean and their controlling mechanisms: a review. J Sea Res 53:43

See JH, Campbell L, Richardson TL, Pinckney JL, Shen R, Guinasso NL (2005) Combining new technologies for determination of phytoplankton community structure in the northern Gulf of Mexico. J Phycol 41:305–310

Selph KE, Landry MR, Allen CB, et al (2001) Microbial community composition and growth dynamics in the Antarctic Polar Front and seasonal ice zone during late spring 1997. Deep Sea Res Part II Top Stud Oceanogr 48:4059–4080

Seuront L, Lacheze C, Doubell MJ, Seymour JR, Mitchell JG (this volume) The influence of *Phaeocystis globosa* bloom dynamics on microscale spatial patterns of phytoplankton biomass and bulk-phase seawater viscosity. Biogeochemistry

Sheppard SK, Harwood JD (2005) Advances in molecular predator–prey ecology. Funct Ecol 19:751–762

Shields AR, Smith WOJ (2005) A novel technique to examine the role of colonial *Phaeocystis antarctica* in the microbial loop of the Ross Sea, Antarctica. SCOR working group #120. Final meeting: *Phaeocystis*, major link in the biogeochemical cycling of climate-relevant elements. University of Groningen, Haren, the Netherlands, p 14

Sieburth JM (1960) Acrylic acid, an "antibiotic" principle in *Phaeocystis* blooms in antarctic waters. Science 132:676–677

Sieburth JM (1961) Antibiotic properties of acrylic acid, a factor in the gastrointestinal antibiosis of polar marine animals. J Bacteriol 82:72–79

Smaal AC, Twisk F (1997) Filtration and absorption of *Phaeocystis* cf. *globosa* by the mussel *Mytilus edulis* L. J Exp Mar Biol Ecol 209:33–46

Smayda TJ (1973) The growth of *Skeletonema costatum* during a winter-spring bloom in Narragansett Bay, Rhode Island. Norw J Bot 20:219–247

Smayda TJ (1997) What is a bloom? A commentary. Limnol Oceanogr 42:1132–1136

Sournia A (1988) *Phaeocystis* (Prymnesiophyceae): how many species? Nova Hedwigia 47:211–217

Stabell OB, Aanesen RT, Eilertsen HC (1999) Toxic peculiarities of the marine alga *Phaeocystis pouchetii* detected by in vivo and in vitro bioassay methods. Aquat Toxicol 44:279–288

Stefels J (2000) Physiological aspects of the production and conversion of DMSP in marine algae and higher plants. J Sea Res 43:183–197

Stefels J, Steinke M, Turner S, Gill M and Belviso S (this volume) Environmental constraints on the production of the climatically active gas dimethylsulphide (DMS) and implications for ecosystem modelling. Biogeochemistry. doi:10.1007/s10533-007-9091-5

Steinke M, Malin G, Liss PS (2002) Trophic interactions in the sea: an ecological role for climate relevant volatiles? J Phycol 38:630–638

Steinke M, Stefels J, Stamhuis E (2006) Dimethyl sulfide triggers search behavior in copepods. Limnol Oceanogr 51:1925–1930

Stelfox-Widdicombe CE, Archer SD, Burkill PH, Stefels J (2004) Microzooplankton grazing in *Phaeocystis* and diatom-dominated waters in the southern North Sea in spring. J Sea Res 51:37

Strom S, Wolfe G, Slajer A, Lambert S, Clough J (2003) Chemical defense in the microplankton II: Inhibition of protist feeding by ß-dimethylsulfoniopropionate (DMSP). Limnol Oceanogr 48:230–237

Strom SL, Loukos H (1998) Selective feeding by protozoa: model and experimental behaviors and their consequences for population stability. J Plankton Res 20:831–846

Symondson WOC (2002) Molecular identification of prey in predator diets. Mol Ecol 11:627–641

Tande KS, Båmstedt U (1987) On the trophic fate of Phaeocystis pouchetii 1. Copepod feeding rates on solitary cells and colonies of *P. pouchetii*. Sarsia 72:3–4

Tang KW (2003) Grazing and colony size development in *Phaeocystis globosa* (Prymnesiophyceae): the role of a chemical signal. J Plankton Res 25:831–842

Tang KW, Jakobsen HH, Visser AW (2001) *Phaeocystis globosa* (Prymnesiophyceae) and the planktonic food web: feeding, growth, and trophic interactions among grazers. Limnol Oceanogr 46:1860–1870

Tang KW, Simó R (2003) Trophic uptake and transfer of DMSP in simple planktonic food chains. Aquat Microb Ecol 31:193–202

Thingstad F, Billen G (1994) Microbial degradation of *Phaeocystis* material in the water column. J Mar Syst 5:55

Troedsson CG, Frischer ME, Nejstgaard JC, Thompson EM (2007) Molecular quantification of differential ingestion and particle trapping rates by the appendicularian *Oikopleura dioica* as a function of prey size and shape. Limnnol Oceanogr 52:416–427

Tungaraza C, Rousseau V, Brion N, Lancelot C, Gichuki J, Baeyens W, Goeyens L (2003) Contrasting nitrogen uptake by diatom and *Phaeocystis*-dominated phytoplankton assemblages in the North Sea. J Exp Mar Biol Ecol 292:19–41

Turner JT (1994) Planktonic Copepods of Boston Harbor, Massachusetts Bay and Cape Cod Bay, 1992. Hydrobiologia 293:405–413

Turner JT, Ianora A, Esposito F, Carotenuto Y, Miralto A (2002) Zooplankton feeding ecology: does a diet of *Phaeocystis* support good copepod grazing, survival, egg production and egg hatching success? J Plankton Res 24:1185–1195

van Boekel WHM, Hansen FC, Riegman R, Bak RPM (1992) Lysis-induced decline of a *Phaeocystis* spring bloom and coupling with the microbial foodweb. Mar Ecol Prog Ser 81:269–276

van Donk E, Luerling M, Hessen DO, Lokhorst GM (1997) Altered cell wall morphology in nutrient-deficient phytoplankton and its impact on grazers. Limnol Oceanogr 42:357–364

van Rijssel M, Alderkamp A-C, Nejstgaard JC, Sazhin AF, Verity PG (this volume) Haemolytic activity of living *Phaeocystis pouchetii* during mesocosm blooms. Biogeochemistry. doi:10.1007/s10533-007-9095-1

van Rijssel M, Hamm CE, Gieskes WWC (1997) *Phaeocystis globosa* (Prymnesiophyceae) colonies: hollow structures built with small amounts of polysaccharides. Eur J Phycol 32:185–192

Veldhuis MJW, Colijn F, Venekamp LAH (1986) The spring bloom of *Phaeocystis pouchetii* (Haptophyceae) in Dutch coastal waters. Neth J Sea Res 20:37–48

Veldhuis MJW, Wassmann P (2005) Bloom dynamics and biological control of a high biomass HAB species in European coastal waters: a *Phaeocystis* case study. Harm Algae 4:805

Verity PG (1988) Chemosensory behavior in marine planktonic ciliates. Bull Marine Sci 43:772–782

Verity PG (2000) Grazing experiments and model simulations of the role of zooplankton in *Phaeocystis* food webs. J Sea Res 43:317–343

Verity PG, Paffenhöfer G-A (1996) On assessment of prey ingestion by copepods. J Plankton Res 18:1767–1779

Verity PG, Smayda TJ (1989) Nutritional value of Phaeocystis pouchetii (Prymnesiophyceae) and other phytoplankton for *Acartia* spp. (Copepoda): Ingestion, egg production, and growth of nauplii. Mar Biol 100:161–171

Vestheim H, Edvardsen B, Kaartvedt S (2005) Assessing feeding of a carnivorous copepod using species specific PCR. Mar Biol 147:381–385

Weisse T, Grimm N, Hickel W, Martens P (1986) Dynamics of *Phaeocystis pouchetii* blooms in the Wadden Sea of Sylt (German Bight, North Sea). Est Coast Shelf Sci 23:171

Weisse T, Scheffel-Möser U (1990) Growth and grazing loss rates in single-celled *Phaeocystis* sp. (Prymnesiophyceae). Mar Biol 106:153–158

Weisse T, Tande K, Verity P, Hansen F, Gieskes W (1994) The trophic significance of *Phaeocystis* blooms. J Mar Syst 5:67–79

Wolfe GV, Levasseur M, Cantin G, Michaud S (2000) DMSP and DMS dynamics and microzooplankton grazing in the Labrador Sea: application of the dilution technique. Deep-Sea Res Part I Oceanogr Res Pap 47:2243–2264

Wolfe GV, Steinke M (1996) Grazing-activated production of dimethyl sulfide (DMS) by two clones of *Emiliania huxleyi*. Limnol Oceanogr 41:1151–1160

Wolfe GV, Steinke M, Kirst GO (1997) Grazing-activated chemical defence in a unicellular marine alga. Nature 387:894–897

Wulff A (1934) Über Hydrographie und Oberflächenplankton nebst Verbreitung von *Phaeocystis* in der Deutschen Bucht im Mai 1933. Ber Deutsch Wiss Komm Meeresforsch NF 7:343–350

Yoshida T, Hairston NG, Ellner SP (2004) Evolutionary trade-off between defence against grazing and competitive ability in a simple unicellular alga, *Chlorelia vulgaris*. Proc Roy Soc Lond Ser B Biol Sci 271:1947–1953

Biogeochemistry (2007) 83:173–188
DOI 10.1007/s10533-007-9097-z

ORIGINAL PAPER

The influence of *Phaeocystis globosa* on microscale spatial patterns of chlorophyll *a* and bulk-phase seawater viscosity

L. Seuront · C. Lacheze · M. J. Doubell · J. R. Seymour · V. Van Dongen-Vogels · K. Newton · A. C. Alderkamp · J. G. Mitchell

Received: 17 October 2005 / Accepted: 8 May 2006 / Published online: 27 April 2007

Abstract A two-dimensional microscale (5 cm resolution) sampler was used over the course of a phytoplankton spring bloom dominated by *Phaeocystis globosa* to investigate the structural properties of chlorophyll *a* and seawater excess viscosity distributions. The microscale distribution patterns of chlorophyll *a* and excess viscosity were never uniform nor random. Instead they exhibited different types and levels of aggregated spatial patterns that were related to the dynamics of the bloom. The chlorophyll *a* and seawater viscosity correlation patterns were also controlled by the dynamics of the bloom with positive and negative correlations before and after the formation of foam in the turbulent surf zone. The ecological relevance and implications of the observed patchiness and biologically induced increase in seawater viscosity are discussed and the combination of the enlarged colonial form and mucus secretion is suggested as a competitive advantage of *P. globosa* in highly turbulent environments where this species flourishes.

Keywords Eastern English Channel · Patchiness · *Phaeocystic globosa* · Plankton rheology · Turbulence

L. Seuront (✉) · C. Lacheze
CNRS FRE 2816 ELICO, Station Marine de Wimereux, University of Sciences and Technologies of Lille-Lille 1, 28 avenue Foch, 62930 Wimereux, France
e-mail: Laurent.Seuront@univ-lille1.fr

L. Seuront · M. J. Doubell · K. Newton · J. G. Mitchell
School of Biological Sciences, Flinders University, GPO Box 2100, Adelaide, SA 5001, Australia

J. R. Seymour
Department of Civil and Environmental Engineering, Massachusetts Institute of Technology, 77 Massachussets Avenue, Cambridge, MA 02139-4307, USA

V. Van Dongen-Vogels
Marine Biology Laboratory, Catholic University of Louvain, Batiment Kellner, 3 Place Croix du Sud, Louvain-La-Neuve 1348, Belgium

A. C. Alderkamp
Department of Marine Biology, University of Groningen, PO Box 14, Haren 9750AA, The Netherlands

Introduction

Plankton patchiness is widely acknowledged as a ubiquitous and key feature of marine ecosystems (Martin 2003). Many organisms have been shown to exploit patches of food (e.g., Tiselius 1992), and patch formation may be important in the foraging success of many marine invertebrates (Seuront et al. 2001) and vertebrates (Cartamil and Lowe 2004), as well as for the sexual encounters among individuals of relatively rare species (Buskey 1998). While the quantification of the spatial and temporal structure of phytoplankton distributions has for the most part focussed on empirical observations at scales greater than 10 m to several kilometres (Martin 2003), there is a growing evidence for the existence of both vertical and horizontal phytoplankton patchiness at scales smaller than one metre (e.g., Dekshenieks et al. 2001). Because the analysis of large-scale patterns must integrate processes occurring at much smaller scales (Levin 1992), investigating the physical and biological dynamics controlling microscale plankton patchiness may be an absolute prerequisite to improve our understanding of the scales, intensities, durations and ecological relevance of these patterns.

The cosmopolitan genus *Phaeocystis* is a key organism in driving global geochemical cycles, climate regulation and fisheries yield (Schoemann et al. 2005), and has recently been suggested as a potential source of microscale phytoplankton patchiness (Seuront 2005). Of particular importance is the formation of large, gel-like colonies (from several mm to several cm; Verity and Medlin 2003) which form thick brown jelly layers (Al-Hasan et al. 1990) that modify the rheological properties of seawater, produce foam that resembles whipped egg white (Dreyfuss 1962), and clog plankton and fish nets (Peperzak 2002). Large quantities of foam are generated in the turbulent surf zone of beaches along the North Sea and the Eastern English Channel (Lancelot et al. 1987; Seuront et al. 2006). Quantitative descriptions of the bulk-phase properties of seawater during phytoplankton blooms have estimated the changes in the viscous properties of seawater that are induced by phytoplankton mucus secretion (Jenkinson 1986, 1993; Jenkinson and Biddanda 1995; Seuront et al. 2006), and found a positive correlation between seawater viscosity and chlorophyll *a* concentration during *Phaeocystis* sp. blooms in the German Bight and the North Sea (Jenkinson 1993; Jenkinson and Biddanda 1995). More recent work (Seuront et al. 2006) showed a significant increase in seawater viscosity with the development of a *Phaeocystis globosa* bloom in the Eastern English Channel, and a positive correlation between chlorophyll concentration and seawater viscosity before foam formation as well as a negative correlation after foam formation. These results suggest that seawater viscosity is influenced by extracellular materials associated with colony formation and colony maintenance rather than by cell composition and standing stock.

In this context, the aim of this study is to use a microscale (5 cm resolution) two-dimensional sampler to investigate: (i) the heterogeneity of the microscale distributions of phytoplankton biomass and bulk-phase seawater viscosity, (ii) the potential changes in these distributions over the course of a *P. globosa* spring bloom and (iii) the nature of the correlation between seawater viscosity and phytoplankton before, during and after the formation of foam in the turbulent surf zone.

Materials and methods

Study site

Samples were taken from the end of a coastal pier situated in the intertidal waters of the French coast of the Eastern English Channel, approximately 20 m from the shoreline (50°45′896″ N, 1°36′364″ E), near Boulogne-sur-Mer (France). The Eastern English Channel is characterised by 3 to 9 m tides. These tides are characterised by a residual translation parallel to the coast, with nearshore coastal waters drifting from the English Channel into the North Sea. Coastal waters are influenced by freshwater run-off from the Seine estuary to the Straits of Dover. This coastal flow (Brylinski et al. 1991) is separated from offshore waters by a tidally maintained frontal area. The coastal flow water mass is characterised by its low salinity, high turbidity, high phytoplankton richness and high productivity compared to the oceanic offshore waters (Seuront 2005). This sampling site was chosen because the physical and hydrological properties are representative of the inshore water masses of the Eastern English Channel (Seuront 2005). Sampling was conducted at high tide, before, during and after the *Phaeocystis globosa* bloom in the Eastern English Channel on 26 September 2003, 5 March 2004, 18 March 2004, 1 April 2004, 20 April 2004, 10 May 2004, 25 May 2004 and 20 June 2004.

Microscale sampling device

A purpose-designed two-dimensional (2D) sampling device was developed to investigate the microscale distributions of chlorophyll *a* and seawater viscosity.

The design of the device was based on sampling systems previously employed to measure the microscale spatial distributions of bacteria and phytoplankton (Seymour et al. 2000; Waters et al. 2003) and consisted of a 10 × 10 array of 50 ml syringes, separated by 5 cm and with a 3 mm wide opening. The pneumatically operated device was attached to a series of electric pumps allowing for the simultaneous collection of 100 samples across an area of 0.2 m^2. On retrieval, a polyvinyl chloride (PVC) rack holding 60 ml vials was inserted below the syringe array on stainless-steel rails, and the samples were transferred into the vials, which were immediately brought back to the laboratory in a cooler, stored in the dark in a room at the in situ temperature; all subsequent analysis were conducted within 2 h to avoid potential biases related to exudation during temperature changes and in high light. Samples were taken from a depth of 1 m, and the temperature and salinity of each water sample was measured using a Hydrolab probe at each of the sampling dates. The presence and the size of *P. globosa* colonies were investigated from bulk seawater samples taken simultaneously to the microscale samples.

Chlorophyll *a* analysis

Chlorophyll concentrations were estimated from 40 ml subsamples following Suzuki and Ishimaru (1990). Samples were filtered on Whatman GF/F glass-fibre filters (pore size 0.45 μm). Chlorophyllous pigments were extracted by direct immersion of the filters in 5 ml of N,N-dimethylformamide, and actual extractions were made in the dark at −20°C. Concentrations of chlorophyll *a* in the extracts were determined following Strickland and Parsons (1972) using a Turner 450 fluorometer previously calibrated with chlorophyll *a* extracted from *Anacystis nidulans* (Sigma Chemicals, St Louis).

Bulk-phase seawater viscosity measurements

Viscosity measurements were conducted using a portable ViscoLab400 viscometer (Cambridge Applied Systems Inc., Boston) from 10 ml subsamples following Seuront et al. (2006). Viscosity was measured in triplicate from 3 ml water subsamples poured into a small chamber, where a low-mass stainless-steel piston is magnetically forced back and forth, with a 230 μm piston-cylinder gap size. The force driving the piston is constant, and the time required for the piston to move back and forth into the measurement chamber is proportional to the viscosity of the fluid. The more viscous the fluid the longer it takes the piston to move through the chamber, and vice versa. As viscosity is influenced by temperature and salinity (Miyake and Koizumi 1948), the measured viscosity η_m (cP) can be thought of as the sum of a physically controlled viscosity component $\eta_{T,S}$ (cP) and a biologically controlled viscosity component η_{Bio} (cP) (Jenkinson 1986):

$$\eta_m = \eta_{T,S} + \eta_{\mathrm{Bio}} \tag{1}$$

The physically controlled component $\eta_{T,S}$ was estimated from viscosity measurements conducted on subsamples after passing through 0.20 μm pore size filters. The biologically induced excess viscosity η_{Bio} (cP) was subsequently defined from each water sample as:

$$\eta_{\mathrm{Bio}} = \eta_m - \eta_{T,S} \tag{2}$$

The relative excess viscosity η (%) is finally given as (Seuront et al. 2006):

$$\eta = \left(\eta_m - \eta_{T,S}\right)/\eta_{T,S} \tag{3}$$

Before each viscosity measurement, temperature and salinity of the water sample were simultaneously measured using a Hydrolab probe, and did not exhibit any significant difference with the theoretical viscosity estimated from temperature and salinity (Wilcoxon–Mann–Whitney U-test, $P > 0.05$). This ensures the relevance of our $\eta_{T,S}$ estimates. Between each viscosity measurement, the viscometer chamber was carefully rinsed first with deionised water and then with bulk-phase seawater filtered through 0.2 μm pore size filters to avoid any potential dilution of the next sample.

Potential biases and limitations

Due to the difficulty in handling *Phaeocystis* sp. colonies without disrupting them, our viscosity measurements could have been biased through the disruption of colonies: (i) during the sampling process and/or (ii) the viscosity measurement in the chamber of the viscometer. This would result in creating additional mucus particles and releasing additional macromolecules into solution, and thus a biased increase in seawater viscosity. These two potential issues have been carefully addressed to ensure the relevance of our viscosity measurements.

Any *P. globosa* colony larger than the opening of the syringes (i.e. 3 mm) would be destroyed during the sampling. This is unlikely to have occurred during our sampling for two reasons. First, the size of *P. globosa* colonies in bulk samples taken simultaneously to our 2D sampling, observed by inverted microscopy and measured using an ocular micrometer, ranged in size from 0.1 to 1.3 mm at concentrations ranging from 500 to 5,000 per litre (Table 1). Control measurements were carried out at each of our sampling dates, during which we compared the viscosity measured from water samples taken by the syringes of the 2D sampler and from bulk seawater samples. As no significant differences were found between the two estimates (Wilcoxon–Mann–Whitney U-test, $P > 0.05$, $N = 100$), the syringe sampler has not been considered as a potential source of bias in subsequent viscosity measurements.

Fragile colonies, and more specifically the colonies larger than 230 μm (as the piston-cylinder gap is only 230 μm; see above) would still be destroyed in the measurement chamber, and would result in a biased increase in seawater viscosity. Therefore all visible *P. globosa* colonies were systematically removed from the 3 ml samples with a Pasteur pipette under a dissecting microscope (magnification × 100), under temperature controlled conditions before viscosity measurements were conducted. Considering the relatively large colonies at relatively low concentrations (500 to 5000 per litre; Table 1) we can assume that most though maybe not all colonies were removed. It could also be argued here that this could have been achieved through a careful screening of the water samples through a 200 μm mesh before doing the measurements. However, a 200 μm screening would still not remove the colonies smaller than 200 μm that are still likely to be destroyed by the high shear occurring in the measurement chamber and to create additional mucus particles and release additional macromolecules into solution. The two methods are then both potentially biased, but in similar ways, as preliminary testing conducted

Table 1 Mean abundance (colonies l^{-1}) and size (μm) of *P. globosa* colonies observed at each sampling date

Date	*Phaeocystis globosa* colonies			
	Size (μm)	N_1	Abundance (col l^{-1})	N_2
26/09/03	–	–	0	–
05/03/04	–	–	0	–
10/03/04	–	–	0	–
01/04/04	180 (100–220)	100	1400(1200–1600)	5
20/04/04	270(230–660)	100	1200(990–1350)	4
10/05/04	500 (300–1300)	100	5000(4820–5250)	4
25/05/04	390 (300–645)	100	500(290–560)	4
20/06/04	–	–	0	–

The values given in parenthesis are the minimum and maximal values. N_1 is the number of colonies considered for size measurements and N_2 is the number of bulk samples used to estimate colony abundance

during the 2002 and 2003 spring blooms did not exhibit any significant difference (Fig. 1). Finally, we have already shown a distinct increase in excess seawater viscosity following the mechanical disruption of *P. globosa* colonies (Seuront et al. 2006). This would not have been possible if just the act of making a measurement increased viscosity.

Data analyses

Quantifying microscale spatial variability. A simple quantification of the degree of microscale variability observed for a given 2D microscale distribution was expressed as the ratio between maximum and minimum values of the distributions, considered as an estimate of the maximum variability (Seuront and Spilmont 2002), and the coefficient of variation CV ($CV = SD/\bar{x}$, where $\bar{x}$ and SD are the mean and the standard deviation, respectively) estimated for that range.

Identifying spatial structure

Spatial autocorrelation analysis was used to quantify patterns in two-dimensional microscale distribution of chlorophyll *a* concentrations (μgChl-a l^{-1}) and relative seawater excess viscosity η (%). The objective is to link these parameters to a common patch size to quantify their spatial linkage.

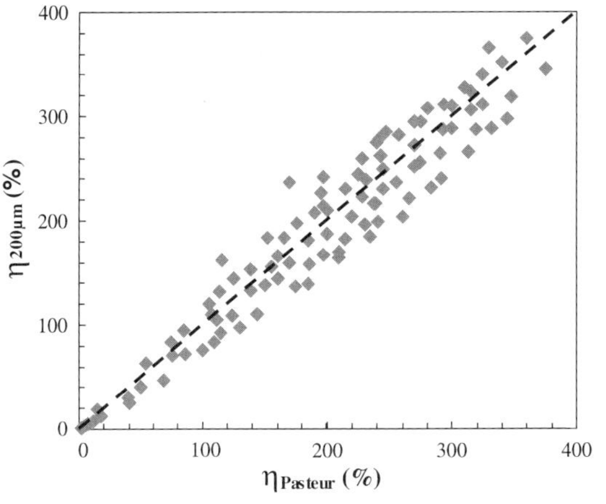

Fig. 1 Comparisons of the excess viscosity measurements η (%) obtained from bulk phase seawater samples after removing all the visible colonies from the water samples with a Pasteur pipette under a dissecting microscope (magnification × 100), $\eta_{Pasteur}$, and after carefully screening the water samples through a 200 µm mesh ($\eta_{200\mu m}$) before doing the measurements. A modified *t*-test (Zar 1996) did not exhibit a significant difference between our measurements and the theoretical expectation $\eta_{Pasteur} = \eta_{200\mu m}$ ($P > 0.05$, $N = 100$)

Generally speaking, positive spatial autocorrelation indicates that abundances in adjacent localities are similar (aggregation) while negative-autocorrelation signifies that adjacent localities alternate between high and low values, i.e. peaks and valleys, and hot spots and cold spots. The theory and use of spatial autocorrelation procedures to analyse spatial associations has been extensively discussed (Sokal and Oden 1978a, b), and the reader is referred to these papers for more-detailed descriptions.

Moran's *I* and Geary's *c* spatial autocorrelation statistics (Moran 1950; Geary 1954) are defined as:

$$I(d) = \frac{\frac{1}{W}\sum_{i=1}^{n}\sum_{j=1}^{n} w_{ij}(y_i - \bar{y})(y_j - \bar{y})}{\frac{1}{n}\sum_{i=1}^{n}(y_j - \bar{y})^2} \text{ for } i \neq j \tag{4}$$

and

$$c(d) = \frac{\frac{1}{2W}\sum_{i=1}^{n}\sum_{j=1}^{n} w_{ij}(y_i - y_j)^2}{\frac{1}{(n-1)}\sum_{j=1}^{n}(y_j - \bar{y})^2} \text{ for } i \neq j \tag{5}$$

where y_i and y_j are the values of the observed variable at sites *i* and *j*, $\bar{y}$ is their mean and *W* is defined as $W = \sum_i^n \sum_j^n w_{ij}$ where $w_{ij} = d_{ij}^{-2}$ when sites *i* and *j* are at a distance *d* and $w_{ij} = 0$ otherwise. Only the pairs of site (*i*, *j*) within the stated distance class (*d*) are taken into account.

Moran's *I* Eq. 4 and Geary's *c* Eq. 5 are two related, but conceptually different, autocorrelation coefficients. Moran's *I*, which ranges from +1 to −1, is sensitive to extreme abundances in nearby localities while Geary's *c*, bounded between 0 and an unspecified value larger than 1, indicates whether nearby localities exhibit similar abundances. A positive autocorrelation (i.e. positive values of *I*) indicates dependence between samples that are spatially close. In contrast, a negative autocorrelation (i.e. negative values of *I*) indicates dependence between distant samples. The no-correlation values is thus $I = 0$. A Geary's *c* correlogram varies as the reverse of a Moran's *I* correlogram: strong autocorrelation produces high values of *I* and low values of *c*. Positive autocorrelation translates into values of *c* between 0 and 1 whereas negative autocorrelation produces values larger than 1. Hence, the reference no-correlation value is $c = 1$.

Under the hypothesis of a random spatial distribution, the expected values of the Moran's *I* and Geary's *c* are $I = -(n-1)^{-1}$ and $c = 1$, respectively. When combined with the Fisher's dispersion index ($S^2/\bar{x}$ or variance-to-mean ratio), Moran's *I* and Geary's *c* were used to characterise and classify in an objective way the spatial pattern of a particular descriptor (see Table 2). In addition, patch sizes were determined from spatial correlograms, i.e. Moran's *I* plotted against the separation distance *d*. As a minimum number of 30 pairs are necessary to estimate autocorrelation statistics with confidence (Legendre and Legendre 1998), nine distance classes were used, with distances ranging from 5 to 45 cm. The significance of spatial correlograms was tested using the Bonferroni-corrected significance procedure (Legendre and Legendre 1998). For each 2D data set, the distance at which *I* changes sign was used as an estimate of the average patch radius (Sokal and Oden 1978a).

Statistical analyses

For each date, the chlorophyll *a* distribution and seawater viscosity data were significantly non-normally distributed (Kolmogorov–Smirnov test,

Table 2 Types of distribution patterns based on values of Moran's I, Geary's c and variance-to-mean ratio. S indicates that the value is significantly different from the expected value, while NS denotes no significant difference

Type	I	c	$S^2/\bar{x}$	Distribution
A	S	S	–	Transition from high to low abundance occurs within the sampling area, i.e. gradient in abundance
B	S–	S–	S–	Uniform
C	S	NS	–	Aggregative, based on a few extreme peaks in abundance
D	NS	S	–	Aggregative, based on several similar peaks in abundance
E	NS	NS	S+	Aggregative, due to high abundance in a single, isolated location
F	NS	NS	NS	Random

The + and – signs indicate whether the value is higher or lower than the expected value, respectively. Conclusions are based on discussions contained in Sokal and Oden (1978a), and Decho and Fleeger (1988)

$P < 0.01$), so nonparametric statistics were used throughout this work. Simple comparisons between two parameters were conducted using the Wilcoxon–Mann–Whitney (WMW) test. Correlation between variables was investigated using Kendall's coefficient of rank correlation, τ. Kendall's coefficient of correlation was used in preference to Spearman's coefficient of correlation ρ because Spearman's ρ gives greater weight to pairs of ranks that are further apart, while Kendall's τ weights each disagreement in rank equally (Sokal and Rohlf 1995).

Results

Environmental conditions

Salinity did not exhibit any characteristic pattern and fluctuated between 35.1 and 35.3 PSU (35.20 ± 0.02 PSU; $\bar{x} \pm$ SD). Temperature exhibited a clear seasonal cycle, ranging from 6.1°C on March 5 to 17.2°C on September 26.

Microscale spatial variability

The temporal patterns of chlorophyll *a* concentration and seawater excess viscosity are summarised in Table 3. Chlorophyll concentration and seawater excess viscosity were below 3.0 μg Chl-*a* l^{-1} on September 26 and March 5 (Table 3). Chlorophyll concentration then exhibited a sharp increase, corresponding to the initiation and the development of the spring bloom, to reach a maximum value of 57.4 ± 6.5 μg Chl-*a* l^{-1} ($\bar{x} \pm$ SD) on May 10, followed by a 3.5-fold decrease between May 10 and May 25 that coincided with the formation of foam in the turbulent surf zone. In contrast, seawater excess viscosity reached its peak (28 ± 25%;$\bar{x} \pm$ SD) on May 25 after the beginning of foam formation, and finally sharply decreased down to 5 ± 3% on June 20.

The values of the coefficient of variation (CV) and the ratios between maximum and minimum values (max./min.) demonstrate the high variability of microscale 2D distributions of chlorophyll *a* biomass and seawater excess viscosity (Table 3). The variability of chlorophyll and excess viscosity 2D distributions were minimum during the formation of foam in the turbulent surf zone. As further illustration of this variability, the 5 cm resolution, two-dimensional distributions of chlorophyll *a* concentration and seawater excess viscosity are shown for the pre-bloom condition (Fig. 2a), during the spring bloom, before (Fig. 2b) and after (Fig. 2c) the formation of foam in the turbulent surf zone, and after the bloom (Fig. 2d). Chlorophyll concentration and excess viscosity show alternating high- and low-density areas separated by sharp gradients and characterised by localised hot spots and cold spots (Fig. 2). The differential effect of foam formation on the two-dimensional distributions of chlorophyll *a* and excess viscosity is also clear from the transition observed between the pre- and post-foam distributions (Fig. 2b, c). Before the formation of foam, chlorophyll *a* and excess viscosity distributions are dominated by a high-density background and a few cold spots (Fig. 2b), while after the formation of foam, the chlorophyll *a* distributions were dominated by a

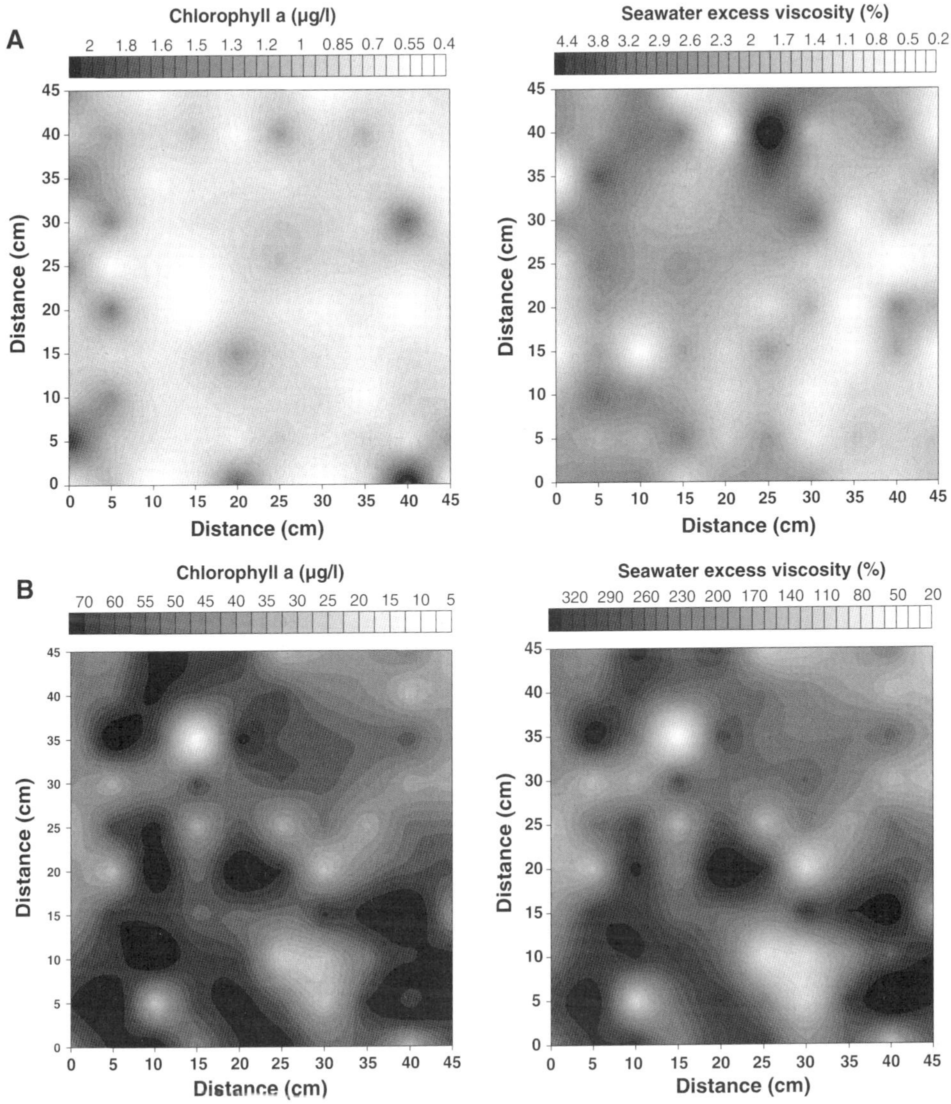

Fig. 2 Two-dimensional distributions of chlorophyll *a* concentrations (µg l^{-1}) and seawater excess viscosity (%) before the spring phytoplankton bloom (**A**) and during the spring bloom before foam formation in the turbulent surf zone (**B**). Two-dimensional distributions of chlorophyll *a* concentrations (µg l^{-1}) and seawater excess viscosity (%) during the spring phytoplankton bloom after foam formation in the turbulent surf zone (**C**), and after the bloom (**D**)

low-density background and a few hot spots (Fig. 2c). Finally, the chlorophyll biomass and excess viscosity do not show any clear correlation pattern in pre- and post-bloom conditions (Fig. 2a, d). In contrast, they exhibit positive and negative correlation before and after the formation of foam, respectively (Fig. 2b, c).

Microscale spatial correlation

To identify any possible correlation between chlorophyll concentration and excess viscosity, excess viscosity was plotted against chlorophyll *a* concentration (Fig. 3). In pre-bloom conditions (September 26), no relationship is visible between chlorophyll concentration and excess viscosity. The development of the spring bloom is subsequently clearly visible, with simultaneous increases in chlorophyll *a* concentration and excess viscosity from March 5 to April 20. After the formation of foam, a progressive decoupling between chlorophyll concentration and excess viscosity occurred. It was initiated on May 10, when positive and negative correlations were observed

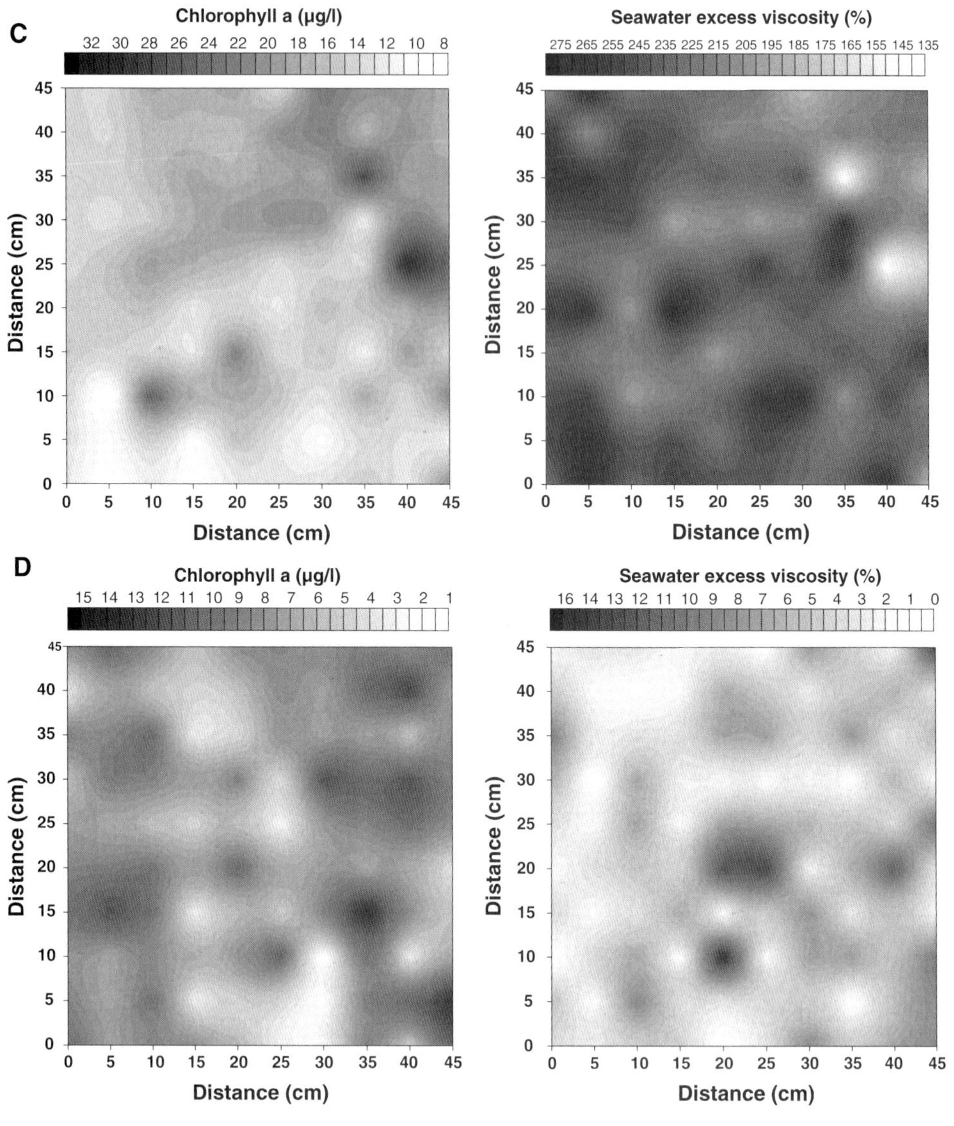

Fig. 2 continued

between chlorophyll concentration and excess viscosity for chlorophyll concentrations higher and lower than 55 µg Chl-*a* l^{-1}, respectively. On May 25, accompanying a sharp decrease in Chl-*a* concentration, excess viscosity was negatively related to chlorophyll concentration. Finally, in post-bloom conditions (June 20), chlorophyll concentration and excess viscosity appeared independently distributed. These observations are specified by the values of Kendall rank correlation coefficients (Fig. 4) estimated at each date between the two distributions. No significant correlation was observed in pre- and post-bloom conditions, while significant positive and negative correlations were observed during the spring bloom before and after the formation of foam, respectively (Fig. 4).

Microscale spatial structure

The results of spatial autocorrelation analysis showed that none of the investigated patterns were uniform nor random (Table 4), indicating the existence of structural complexity in 2D microscale patterns of chlorophyll *a* concentration and seawater viscosity. Except on June 20, consistent spatial patterns were found for chlorophyll and excess

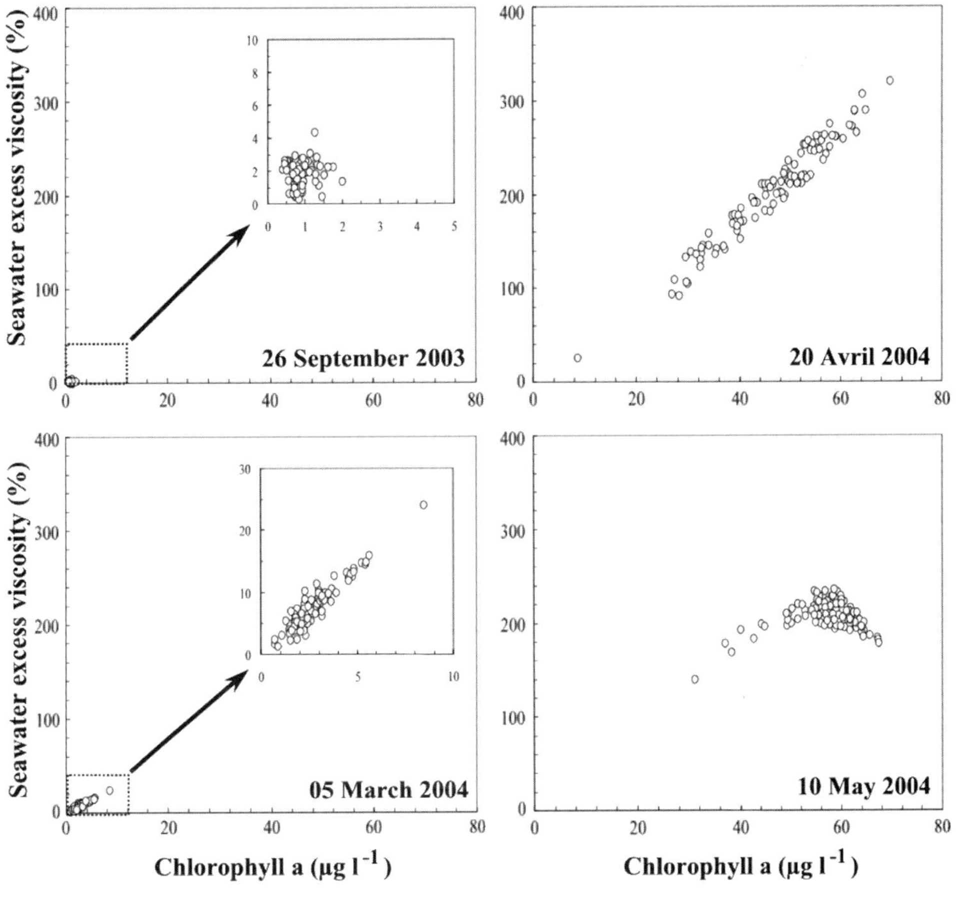

Fig. 3 Scatterplots of seawater excess viscosity η (%) as a function of chlorophyll *a* concentration (µg l^{-1}), showing the two different regimes observed over the course of our survey. The sampling conducted on 26 September has been used as a separate reference sample, characteristic of non-bloom conditions. Scatterplots of seawater excess viscosity η (%) as a function of chlorophyll *a* concentration (µg l^{-1}), showing the two different regimes observed over the course of our survey

viscosity. More specifically, pre- and post-bloom conditions were mainly characterised by three types of aggregative distributions (Table 4). In contrast, the spatial patterns observed during the bloom were either aggregative (i.e. dominated by the presence of a few extreme peaks) or characterised by a variety of transitions from high to low values, the latter being only observed during the formation of foam. Patch sizes were finally estimated from correlograms constructed for each parameter at each date to determine spatial autocorrelation as a function of increasing distance between samples. Because the distance intervals were based on the distance intervals between the centre points of samples, patch sizes were expressed in terms of the distance interval in which the autocorrelation value changed sign (positive to negative autocorrelation or vice versa). Patch sizes for chlorophyll concentration and excess viscosity were between 5 and 10 cm in pre- and post-bloom conditions (Table 4). Larger patch sizes were found for chlorophyll concentration and excess seawater viscosity in bloom conditions. During the formation of foam viscous patches were larger than chlorophyll patches (Table 4).

Discussion

Microscale spatial patterns and Phaeocystis globosa bloom dynamics

Chlorophyll, viscosity and foam formation. The shift from significant positive correlations to significant

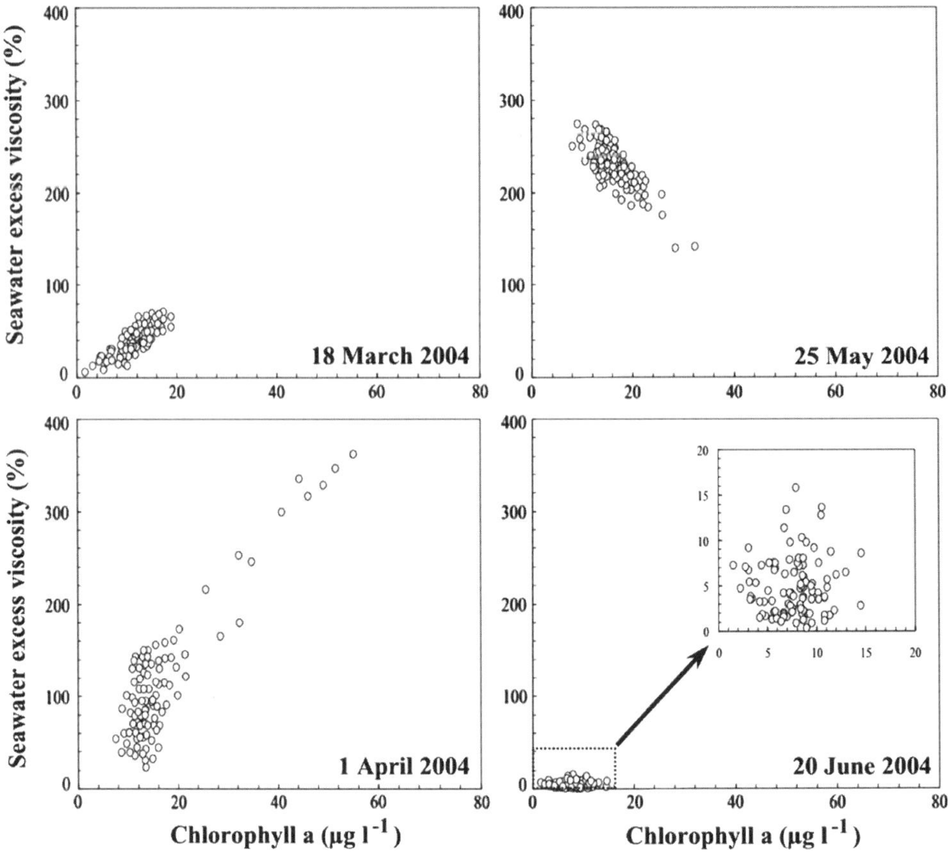

Fig. 3 continued

negative correlations before and after the foam formation is congruent with a recent mechanistic explanation (Seuront et al. 2006), suggesting that the disruption of the mucilaginous colonial matrix by turbulent mixing in the surf zone leads: (i) to the formation of foam and to the transformation of colonial cells into flagellated ones (Peperzak 2002), (ii) to a decrease in chlorophyll *a* concentration (Fig. 5) as a significant proportion of cells are entrained within the foam during the emulsion process (Seuront et al. 2006), and finally (iii) to the decoupling between the viscous (i.e. colonial polymeric materials) and nonviscous (flagellated cells) contribution of *P. globosa* to bulk-phase seawater properties in intertidal (Fig. 5a, b) and inshore (Fig. 5c, d) water masses, located respectively 20 m and 2 nautical miles from the shoreline. However, the relationship observed between chlorophyll concentration and excess viscosity on May 10 through our microscale sampling strategy (Fig. 3) indicates that the aforementioned decoupling process could be more complex than initially thought. As the negative correlation occurs for the chlorophyll *a* concentrations higher than 55 μg l^{-1}, the high-density chlorophyll patches, likely to be the largest and/or the oldest ones, may be more fragile and thus the first ones to be destroyed by turbulent mixing. This hypothesis is consistent with the strong decrease in variability and the changes in the distribution patterns observed in both chlorophyll and excess viscosity spatial patterns between April 20 and May 10 (see Tables 3 and 4, Fig. 5a, b).

The mechanistic explanation proposed above for the dynamics of chlorophyll *a* concentration and seawater excess viscosity is also consistent with recent work conducted on the dynamics of transparent exopolymeric particles (TEP) produced by *Phaeocystis globosa* (Mari et al. 2005). Two phases for the dynamics of TEP were then identified: (i) a production phase during the growth phase of *P. globosa* where TEP and chlorophyll *a* concentration were

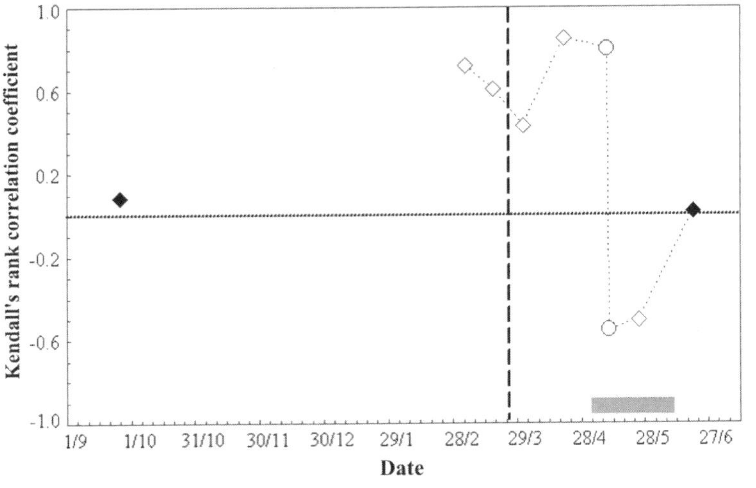

Fig. 4 The time course of the Kendall's correlation coefficient τ, estimated between the 100 simultaneous measurements of chlorophyll concentrations and seawater viscosity at each sampling date. The open and black diamonds represent significant ($P < 0.05$) and nonsignificant Kendall's τ, and the open dots the two correlation patterns observed on May 10 (see Fig. 2). The grey bar indicates the period of foam formation (May 3 to June 10). The sampling conducted on 26 September has been used as a separate reference sample, characteristic of non-bloom conditions

positively correlated, and (ii) a release of large TEP from the mucilaginous matrix of *P. globosa* colonies subsequent to colony disruption (caused by nutrient depletion) where TEP and chlorophyll *a* concentration were negatively correlated. The causes of colony disruption observed by Mari et al. (2005) and the area described in the present work might diverge. The similarity between the dynamics of chlorophyll *a* concentration-excess viscosity shown in the present work and chlorophyll *a* concentration-TEP concen-

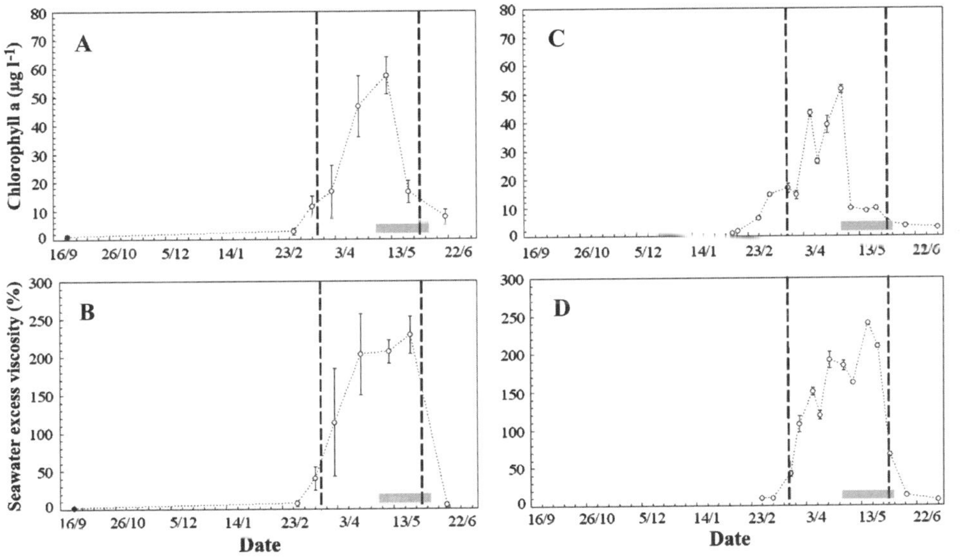

Fig. 5 Time course of chlorophyll *a* concentration (µg l^{-1}; **A**, **C**) and seawater excess viscosity η (%; **B**, **D**), in the intertidal (**A**, **B**; present work) and coastal (**C**, **D**; modified from Seuront et al. 2006) waters of the Eastern English Channel. The grey bar indicates the period of foam formation (May 3 to June 10), and the dashed vertical lines the appearance and disappearance of *P. globosa* in the phytoplankton assemblage. The sampling conducted on 26 September (black dot; **A**, **B**) has been used as a separate reference sample, characteristic of non-bloom conditions

Table 3 Descriptive statistics of the two-dimensional microscale distributions of chlorophyll *a* concentration (μg l^{-1}) and seawater excess viscosity (%) for each sampling date

Date	Mean	SD	CV	Max./min.
Chlorophyll concentration				
26/09/2003	0.9	0.3	31.4	5.0
05/03/2004	2.7	1.2	44.8	11.9
18/03/2004	11.5	3.6	31.2	11.4
01/04/2004	16.7	9.3	55.8	7.4
20/04/2004	46.6	10.6	22.8	8.1
10/05/2004	57.4	6.5	11.3	2.2
25/05/2004	16.6	4.0	24.0	4.0
20/06/2004	7.7	2.6	33.9	9.7
Excess viscosity				
26/09/2003	1.8	0.7	39.1	16.1
05/03/2004	7.5	3.6	47.6	18.5
18/03/2004	40.1	15.5	38.6	12.1
01/04/2004	113.7	70.9	62.4	15.6
20/04/2004	203.5	53.4	26.3	12.4
10/05/2004	206.8	15.4	7.5	1.7
25/05/2004	228.4	24.9	10.9	2.0
20/06/2004	4.9	3.1	63.6	47.5

SD—standard deviation; CV—coefficient of variation; and max./min.—ratio between the maximum and minimum values of a distribution

Table 4 Results of spatial autocorrelation analyses of chlorophyll *a* and seawater excess viscosity two-dimensional microscale distributions

Date	Moran's *I*	Geary's *c*	S^2/ Mean	Distribution	Patch size (cm)
Chlorophyll concentration					
26/09/2003	NS	S+	NS	D	5–10
05/03/2004	NS	NS	S+	E	5–10
18/03/2004	S+	NS	S+	C	5–10
01/04/2004	S+	NS	S+	C	5–10
20/04/2004	S+	S−	S+	A	10–15
10/05/2004	S+	S−	S+	A	10–15
25/05/2004	S+	S−	S+	A	10–15
20/06/2004	NS	S+	S+	D	5–10
Escess viscosity					
26/09/2003	NS	S+	NS	D	5–10
05/03/2004	NS	NS	S+	E	5–10
18/03/2004	S+	NS	S+	C	5–10
01/04/2004	S+	NS	S+	C	5–10
20/04/2004	S+	S−	S+	A	15–20
10/05/2004	S+	S−	S+	A	15–20
25/05/2004	S+	S−	S+	A	15–20
20/06/2004	NS	S+	S+	D	5–10

tration recently investigated, strongly suggests, however, that TEP could play a critical role in the increase in seawater viscosity. The variety of dissolved particulate carbohydrates that are differentially produced during a *Phaeocystis globosa* bloom (see Alderkamp et al. 2005), contribute (for some of them) to the formation of microscopic polymer gels in seawater (Chin et al. 1998) and large sedimenting particles (e.g. TEP; Logan et al. 1995). They may also form an intermediate in the formation of the foam observed after *P. globosa* blooms. The issue of relating quantitative seawater viscosity measurements to the quality and the quantity of dissolved and particulate carbohydrates and transparent exopolymeric particles should then be carefully addressed in the future.

Type of spatial patterns and patch sizes

Spatial autocorrelation and correlograms were used to describe the microscale spatial patterns of chlorophyll *a* concentration and seawater excess viscosity. Identical distribution patterns may or may not produce identical correlograms, while different spatial patterns result in different correlograms (Sokal & Oden 1978b). Both distributions exhibited positive spatial autocorrelation at all the sampling dates. This type of spatial pattern is characterised by an association between extreme and similar per sample values and a transition from high to low values within the sampling area (0.2 m^2). Before and after the spring *P. globosa* bloom the observed spatial patterns (Fig. 2a, d) exclusively exhibited aggregative distributions characterised by the alternation of hot spots and cold spots. The similar increases in patch sizes suggest that the aggregative character of the spatial patterns observed before and after the bloom were mainly due to localised hot spots that may be related to local phytoplankton aggregates (e.g. Koike et al. 1990). In contrast, during the *P. globosa* bloom, and mainly after the formation of foam, these distributions were characterised by a variety of smooth transitions from high to low values. The smoother, more continuous and structured distributions observed during the bloom and characterised by larger patch sizes may be thought of as: (i) the result of the strong biomodification of the fluid properties mainly by *P. globosa*-related polymeric materials and (ii) a two-dimensional description of the organic matter continuum (Chin et al. 1998; Azam 1998). This is also consistent with the similar patch sizes observed for chlorophyll concentration and excess viscosity before and after the bloom, and the larger excess-viscosity patch sizes observed after the foam formation, when phytoplankton biomass and seawater viscosity appear to be uncoupled.

More generally, microscale hot spots of phytoplankton abundance and/or organic matter concentration may well represent the basis of microhabitats for the development of microbial foodwebs (Thingstad and Billen 1994). Microscale patchiness of organisms attached to aggregates and in the surrounding water column will also fundamentally alter predator-prey and virus-host dynamics, and may subsequently modify the flow of matter through the microbial loop (Azam et al. 1983). The intensity and frequency of these hotspots will also influence the sedimentation processes characterising most *Phaeocystis* sp. blooms (e.g. Andreassen and Wassman 1998) and ultimately determine the proportion of carbon that is sequestered to the sediments or retained in the pelagic food web.

Small-scale versus microscale variability

The temporal patterns observed in the present work for chlorophyll *a* concentration and seawater excess viscosity (Fig. 5a, b) are very similar to those obtained in the coastal waters of the Eastern English Channel, roughly 2 nautical miles offshore, from a standard sampling strategy based on 5 replicate samples taken at three different depth (subsurface, intermediate and bottom waters) using Niskin bottles (Fig. 5c, d; Seuront et al. 2006). This ensures the relevance of the previously proposed mechanisms and generalises them from the scale of the water column down to the microscale. The main difference is the significantly higher chlorophyll concentrations and excess viscosity observed at each sampling date from our intertidal station than from the coastal station (WMW test, $P < 0.05$), which is consistent with the previously observed differences in hydrological properties between intertidal and coastal waters of the Eastern English Channel (Seuront and Spilmont 2002).

The variability observed in chlorophyll *a* concentration and seawater excess viscosity from our microscale (5 cm resolution) two-dimensional sampling is significantly higher than the variability previously estimated at the scale of the water column with a resolution of a few meters (Seuront et al. 2006; see Fig. 5). The coefficient of variation and the ratios between maximum and minimum values are, thus significantly higher at the microscale than at the small-scale (WMW test, $P < 0.01$). More specifically, the ratios between the coefficients of variation (CV) obtained at the microscale and the small-scale range from 2.6 to 42.3 for chlorophyll concentration and from 4.5 to 243.2 for seawater excess viscosity. The ratios between maximum and minimum values are 2.1 to 7.4-fold larger at the microscale than at the small-scale for chlorophyll concentration and 3.1- to 14.3-fold larger at the microscale than at the small scale for excess viscosity. These results are in accordance with previous work showing increased variability and structural complexity with decreasing scale (Seuront and Spilmont 2002; Waters et al. 2003), and indicate that extent of microscale variability amongst marine phytoplankton communities can be greater than the variability observed at larger scales.

These findings are particularly pertinent within the Eastern English Channel as the intense tidal mixing, characterised by turbulent kinetic energy dissipation rates ranging from 10^{-16} to 10^{-4} m^2 s^{-3} (Seuront 2005), is expected to homogenise phytoplankton distributions for scales typically smaller than the depth of the vertically well-mixed water column. Both the extent and the non-random spatial structure of the microscale variability observed here in terms of chlorophyll biomass and seawater excess viscosity imply that phytoplankton organisms, and more specifically *Phaeocystis globosa*, can be independent of the surrounding turbulent flows and induce high levels of spatial heterogeneity, even below the 5 cm scale resolution achieved here.

On the potential role of biologically increased seawater viscosity in *P. globosa* ecology

The impact of turbulence on microorganisms is dependent on its relative size to the smallest

Kolmogorov turbulent eddies defined as $l_k = (\nu^3/\varepsilon)^{0.25}$, where ν is the kinematic viscosity ($m^2\ s^{-1}$) related to the measured viscosity η_m (see Eq. 1) by $\nu = \eta_m/\rho$ (where η_m and ρ are the fluid viscosity and density, respectively) and ε is the turbulent energy dissipation rate ($m^2\ s^{-3}$). Above this scale the flow is turbulent, while below it, viscosity dominates resulting in a laminar shear. The coastal waters of the Eastern English Channel and the Southern North Sea where *P. globosa* typically flourishes are characterised by elevated turbulent dissipation rates (typically bounded between 10^{-7} and $10^{-4}\ m^2\ s^{-3}$; Seuront 2005). These high turbulence intensities and the related sub-millimetre Kolmogorov eddies, ranging from 3.2×10^{-1} to 1.8 mm, are not compatible with the size nor the breakability of *P. globosa* colonies (Peperzak 2002). Mucous secretion, and any subsequent increase in seawater viscosity and Kolmogorov eddy size, may be an environmental engineering strategy that *P. globosa* uses to dampen turbulence, create a favourable, turbulent-free physical habitat to protect colony integrity.

At then end of the bloom, the release of nanoplanktonic solitary cells under nutrient limitation would lead to dramatic losses to small protozoan grazers, viruses and zooplankton that would easily offset population growth (Turner et al. 2002). However, excess viscosity remains high, or even increases (Seuront et al. 2006; present work), after colony destruction. This is consistent with the excretion of mucus and dissolved organic carbon observed under nutrient limitation observed in a single cell culture of *P. globosa* (Janse et al. 1999), in colonial cultures of *P. globosa* (van Rijssel et al. 2000) and at the end of *P. pouchetii* blooms in mesocosms (Alderkamp et al. 2005) as well as in many phytoplankton species (Myklestad 1974, 1988). The released flagellated cells may still avoid grazing and viral infection via an alteration of motility and diffusion processes, and a decrease in encounter probabilities and/or in grazing rates attributed, although never demonstrated, to a mechanical hindrance due to increased viscosity (Schoemann et al. 2005). The high-shear environment related to suspensions of aggregates may also be used by *P. globosa* flagellates released from colonies to minimise predation as high shear will decrease their conspicuousness to predators, and not all the perceived preys will be accessible (Seuront et al. 2006). In addition to colony formation, the exudates released by *P. globosa* and the subsequent increase in viscosity might then also be considered as a potential antipredator adaptive strategy that ultimately ensures the completion of its life cycle in highly turbulent environments.

In the sea, the viscous properties of most polymers and suspensions of aggregates are generally dependent on deformation rate. The biologically induced excess viscosity η_{Bio} (see Eq. 3) is then related to the shear γ (s^{-1}) as (Jenkinson 1986) $\eta_{Bio} = k\gamma^{-P}$, where k is a constant, $\gamma = (\varepsilon/\nu)^{0.5}$ with $\nu = 10^{-6}\ m\ s^{-1}$, and P has so far been found to lie between 0 and 1.6 (e.g. Jenkinson and Biddanda 1995), or even as low as −0.2 (Jenkinson et al. 1998). A single value of P ($P = 1.11$) has been derived from 15 samples taken in the German Bight when *Phaeocystis* sp. was blooming (Jenkinson 1993). Considering the lack of information related to the value of P for *Phaeocystis globosa*, the intrinsic plurispecific and dynamic nature of the phytoplankton assemblages, the range of P values proposed in the literature and the range of turbulence intensities found in *P. globosa* natural environment (i.e. $\varepsilon = 10^{-7}$to $10^{-4}\ m^2\ s^{-3}$, Seuront 2005), any attempt to quantify the nontrivial effect of excess viscosity on the Kolmogorov scale and related microscale processes is still unreasonable at this time. Future investigations should, however, focus on the potential differential contribution of the different carbohydrates that are differentially produced during a *Phaeocystis globosa* bloom (see Alderkamp et al., 2007) to the observed increase in seawater viscosity.

Conclusions

The present work suggests that *Phaeocystis globosa* has developed specific adaptive strategies to favourably modify its immediate microenvironment. The consequences of this strategy are far-reaching as they have been shown to modify the microscale variability of chlorophyll concentration and seawater excess viscosity for scales ranging from 5 to 45 cm. In particular, the modification of the extent and the non-random spatial structure of the microscale variability observed here in terms of chlorophyll biomass and seawater excess viscosity over the course of a *Phaeocystis globosa* spring bloom imply that phytoplankton in general, and more specifically *P. globosa*, can induce high levels of spatial heterogeneity. The microscale adaptive processes discussed here are thus likely to cascade from the cell and/or colony scales up to larger scale and to influence critical processes such as biogeochemical fluxes through their impact on, e.g. sedimentation and remineralisation processes, and predator-prey and virus-host dynamics. It is finally stressed that our journey to elucidate the relationship between seawater viscous properties and the plankton components is only beginning, and that the future of plankton rheology should rely on interdisciplinary efforts focusing on seawater rheological properties (including viscosity and elasticity) as well as on phytoplankton taxonomy and standing stock, *P. globosa* colony size, turbulence intensity and the quality and quantity of exopolymeric materials.

Acknowledgements We thank D. Menu, who built the two-dimensional sampler, and D. Menu, D. Hilde, T. Caron, M. Priem, B. Thullier and D. Devreker for their assistance during the survey. Ian R. Jenkinson is acknowledged for his enlightening comments on topics related to the present work. This work has been financially and infrastructurally supported by a grant (Action Concertée Incitative ''Jeunes Chercheurs'' #3058) from the French Ministry of Research to L. Seuront, the CPER '*Phaeocystis*' (France), PNEC '*Chantier Manche Orientale-Sud Mer du Nord*' (France), Université des Sciences et Technologies de Lille (France), Australian Research Council (Australia) and Flinders University (Australia).

References

Al-Hasan RH, Ali AM, Radwan SS (1990) Lipids, and their constituent fatty acids, of *Phaeocystis* sp. From the Arabian Gulf. Mar Biol 105:9–14

Alderkamp AC, Buma AGJ, van Rijssel M (2007) The carbohydrates of Phaeocystis and their degradation in the microbial food web. Biogeochemistry, doi:10.1007/s10533-007-9078-2

Alderkamp AC, Nejstgaard JC, Verity PG, Zirbel MJ, Sazhin AF, van Rijssel M (2005) Dynamics in carbohydrate composition of Phaeocystis pouchetii colonies during spring blooms in mesocosm. J Sea Res 55:169–181

Andreassen IJ, Wassman P (1998) Vertical flux of phytoplankton and particulate biogenic matter in the marginal zone of the Barents Sea in May 1993. Mar Ecol Prog Ser 170:1–14

Azam F (1998) Microbial control of oceanic carbon flux: the plot thickens. Science 280:694–696

Azam F, Fenchel T, Field JG, Gray JS, Meyer-Reil LA, Thingstad F (1983) The ecological role of water-column microbes in the sea. Mar Ecol Prog Ser 10:257–263

Brylinski JM, Lagadeuc Y, Gentilhomme V, Dupont JP, Lafite R, Dupeuple PA, Huault MF, Auger Y, Puskaric E, Wartel M, Cabioch L (1991) Le 'fleuve côtier': un phénomène hydrologique important en Manche orientale (exemple du Pas de Calais). Oceanol Acta 11:197–203

Buskey EJ 1998. Components of mating behavior in planktonic copepods. J Mar Syst 15:13–21

Cartamil DP, Lowe CG (2004) Diel movement patterns of ocean sunfish *Mola mola* off southern California. Mar Ecol Prog Ser 266:245–253

Chin WC, Orellana MV, Verdugo P (1998) Spontaneous assembly of marine dissolved organic matter into polymer gels. Nature 391:568–572

Decho A, Fleeger J (1988) Microscale dispersion of meiobenthic copepods in response to food-resource patchiness. J Exp Mar Biol Ecol 118:229–243

Dekshenieks MM, Donaghay PL, Sullivan JM, Rines JEB, Osborn TR, Twardowski MS (2001) Temporal and spatial occurrence of thin phytoplankton layers in relation to physical processes. Mar Ecol Prog Ser 223:61–71

Dreyfuss R (1962) Note biologique à propos des eaux rouges. Cah Cent Rech Biol Océanogr Méd Nice 1:14–15

Geary RC (1954) The contiguity ratio and statistical mapping. Incop Statist 5:115–145

Janse I, van Rijssel M, Ottema A, Gottschal JC (1999) Microbial breakdown of *Phaeocystis* mucopolysaccharides. Limnol Oceanogr 44:1447–1457

Jenkinson IR (1986) Oceanographic implications of non-Newtonian properties found in phytoplankton cultures. Nature 323:435–437

Jenkinson IR (1993) Bulk-phase viscoelastic properties of seawater. Oceanol Acta, 16:317–334

Jenkinson IR, Biddanda BA (1995) Bulk-phase viscoelastic properties of seawater: relationship with plankton components. J Plankton Res 17:2251–2274

Jenkinson IR, Wyatt T, Malej T (1998) How viscoelastic effects of colloidal biopolymers modify rheological properties of seawater. In: Emri I, Cvelbar R (eds) Proceedings of the 5th European Rheology Conference, Portoroz, Slovenia, September 6–11, 1998, Progress and Trends in Rheology 5:57–58

Koike I, Shigemitsu H, Kazuki T, Kogure K (1990) Role of sub-micrometer particles in the ocean. Nature 345:242–244

Lancelot C, Billen G, Sournia A, Weisse T, Colijn F, Veldhuis MJW, Davies A, Wassman P (1987) *Phaeocystis* blooms and nutrient enrichment in the continental zones of the North Sea. Ambio 16:38–46

Legendre P, Legendre L (1998) Numerical ecology. Elsevier, New York

Levin SA (1992) The problem of patterns and scale in ecology. Ecology 73:1943–1967

Logan BE, Passow U, Alldredge AL, Grossard HP, Simon M (1995) Rapid formation and sedimentation of large aggregates is predictable from coagulation rates (half-lives) of transparent exopolymer particles (TEP). Deep-Sea Res 42:203–214

Mari X, Rassoulzadegan F, Brussaard CPD, Wassmann P (2005) Dynamics of transparent exopolymeric particles (TEP) production by *Phaeocystis globosa* under N- or P-limitation: a controlling factor of the retention/export balance. Harmful Algae 4:895–914

Martin AP (2003) Phytoplankton patchiness: the role of lateral stirring and mixing. Prog Oceanogr 57:125–174

Miyake Y, Koizumi M (1948). The measurements of the viscosity coefficient of sea water. J Mar Res 7:63–66

Myklestad S (1974) Production of carbohydrates by marine planktonic diatoms. I. Comparison of nine different species in culture. J Exp Mar Biol Ecol 15:261–274

Myklestad SM (1988) Production, chemical structure, metabolism, and biological function of the (1,3)-linked, beta-D-glucans in diatoms. BiolOceanogr 6:313–326

Moran PAP (1950) Notes on continuous stochastic phenomena. Biometrika 37:17–23

Peperzak L (2002) The wax and wane of *Phaeocystis globosa* blooms. Ph.D. Thesis, University of Groningen, The Netherland

Schoemann V, Becquevort S, Stefels J, Rousseau V, Lancelot C (2005) *Phaeocystis* blooms in the global ocean and their controlling mechanisms: a review. J Sea Res 53:43–66

Seuront L (2005) Hydrodynamic and tidal controls of small-scale phytoplankton patchiness. Mar Ecol Prog Ser 302:93–101

Seuront L, Spilmont N (2002) Self-organized criticality in intertidal microphytobenthos patch patterns. Physica A 313:513–539

Seuront L, Schmitt F, Lagadeuc Y (2001) Turbulence intermittency, small-scale phytoplankton patchiness and encounter rates in plankton: where do we go from here? Deep-Sea Res I 43:1199–1215

Seuront L, Vincent D, Mitchell JG (2006) Biologically-induced modification of seawater viscosity in the Eastern English Channel during a *Phaeocystis globosa* spring bloom. J Mar Syst 61:118–133

Seymour JR, Mitchell JG, Pearson L, Waters RL (2000) Heterogeneity in bacterioplankton abundance from 4.5 millimetre resolution sampling. Aquat Microb Ecol 22:143–153

Sokal R, Oden N (1978a) Spatial autocorrelation in biology. I Methodology Biol J Linn Soc London 10:199–228

Sokal R, Oden N (1978b) Spatial autocorrelation in biology. II. Some biological implications and four applications of evolutionary and ecological interest. Biol J Linn Soc London 10:229–249

Sokal RR, Rohlf FJ (1995) Biometry. The principles and practice of statistics in biological research. Freeman, San Francisco

Strickland JDH, Parsons TR (1972) A practical handbook of seawater analysis. Bull Fish Res Bd Canada 167:1–311

Suzuki R, Ishimaru T (1990) An improved method for the determination of phytoplankton chlorophyll using N, N-Dimethylformamide. J Oceanogr Soc Japan 46:190–194

Tiselius P (1992) Behavior of *Acartia tonsa* in patchy food environments. Limnol Oceanogr 8:1640–1651

Thingstad TF, Billen G (1994) Microbial degradation of *Phaeocystis* material in the water column. J Mar Syst 5:55–65

Turner JT, Ianora A, Esposito F, Carotenuto Y, Miralto A (2002) Zooplankton feeding ecology: does a diet of *Phaeocystis* support good copepod grazing, survival, egg production and egg hatching success? J. Plankton Res 24:1185–1195

van Rijssel M, Janse I, Noordkamp DJB, Gieskes WWC (2000) An inventory of factors that affect polysaccharide production by *Phaeocystis globosa*. J Sea Res 43:297–306

Verity PG, Medlin LK (2003) Observations on colony formation by the cosmopolitan phytoplankton genus *Phaeocystis*. J Mar Syst 43:153–164

Waters RL, Mitchell JG, Seymour JR (2003) Geostatistical characterisation of centimetre-scale spatial stucture of in vivo fluorescence. Mar Ecol Prog Ser 251:49–58

Zar JH (1996) Biostatistical analysis. Prentice Hall, NJ

Biogeochemistry (2007) 83:189–200
DOI 10.1007/s10533-007-9095-1

ORIGINAL PAPER

Haemolytic activity of live *Phaeocystis pouchetii* during mesocosm blooms

Marion van Rijssel · Anne-Carlijn Alderkamp · Jens C. Nejstgaard · Andrey F. Sazhin · Peter G. Verity

Received: 28 November 2005 / Accepted: 12 June 2006 / Published online: 10 March 2007

Abstract Chemical defence is a potential mechanism contributing to the success of *Phaeocystis* species that repeatedly dominate the phytoplankton in coastal areas. Species within the genus *Phaeocystis* have long been suspected of imposing negative effects on co-occurring organisms. Recently a number of toxins have been extracted and identified from *Phaeocystis* samples, but it is not clear if they do enhance the competitive advantage of *Phaeocystis* species.

In the present study the cytotoxic impact of live *Phaeocystis pouchetii* to human blood cells in close proximity, regardless of the nature of the responsible mechanism, was quantified using a bioassay. Haemolytic activity was measured during blooms of *P. pouchetii* in mesocosms. These environments were chosen to mimic natural conditions including chemically mediated interactions that could trigger defensive and/or allelopathic responses of *Phaeocystis*.

Haemolytic activity correlated with *P. pouchetii* numbers and was absent during the preceding diatom bloom. Samples containing live *P. pouchetii* cells showed the highest activity, while filtered sea water and cell extracts were less haemolytic or without effect. Dose-response curves were linear up to 70% lysis, and haemolysis in samples containing live *P. pouchetii* cells reached EC_{50} values comparable to known toxic prymnesiophytes ($1.9 * 10^7$ cells l^{-1}). Haemolytic activity was enhanced by increased temperature and light. The results indicate that unprotected and thus presumably vulnerable cells present in a *P. pouchetii* bloom may lyse within days.

Keywords Allelopathy · Chemical defence · Prymnesiophyte · PUFA · PUA

M. van Rijssel (✉) · A.-C. Alderkamp
Department of Marine Biology, Centre for Ecological and Evolutionary Studies, University of Groningen, P.O. Box 14, 9750AA Haren, The Netherlands
e-mail: m.van.rijssel@rug.nl

J. C. Nejstgaard
Department of Fisheries and Marine Biology, University of Bergen, P.O. Box 7800, 5020 Bergen, Norway

A. F. Sazhin
P. P. Shirshov Institute of Oceanology, Russian Academy of Science, 36 Nakhimovski Prospekt, 117997 Moscow, Russia

P. G. Verity
Skidaway Institute of Oceanography, 10 Oceans Science Circle, Savannah, GA, 31411, USA

Introduction

Phaeocystis is an algal genus which regularly dominates vernal blooms in coastal regions all over the world, especially in temperate and higher-latitude waters. These almost monospecific blooms (Lancelot et al. 1987) have a great impact on the local marine food web because they produce the bulk of the primary production in springtime (Arrigo et al. 1999; DiTullio et al. 2000; Schoemann et al. 2005, introduction of this issue). There is likely a combination of mechanisms behind the bloom forming capacity of this genus. For example, *Phaeocystis* is able to take advantage of eutrophication (Lancelot et al. 1987; Cadée and Hegeman 2002), resulting in high biomass production. Blooms are typically dominated by the colonial form of these species and adjustment of colony size could be a response to escape grazing pressure, thereby reducing population losses [Jacobsen and Tang 2002; Tang 2003, see however Nejstgaard et al. (this issue) for a discussion on this mechanism]. Envelopment of the cells by a colonial mucous layer could be another mechanism to reduce losses because only the motile cells seem to be susceptible to viral lysis (Brussaard et al. 2005, this issue).

Also, *Phaeocystis* has been suspected for a long time of having a negative effect on co-occurring organisms. Penguins died after consumption of krill that fed on *P. antarctica* (Sieburth 1960, 1961). Schools of herring seem to avoid *Phaeocystis* blooms (Savage 1930), and mass mortality of caged fish occurred during a *P. globosa* bloom in the China Sea (Huang et al. 1999). Cod larvae died in the presence of natural densities of *P. pouchetii* (Eilertsen and Raa 1995; Aanesen et al. 1998) and negative effects of *Phaeocystis* were recorded on the bryozoan *Electra pilosa* (Jebram 1980). Bacterial consumption rates of acrylate in field samples increased when the > 20 μm fraction containing *P. globosa* colonies was removed (Noordkamp et al. 2000). And then there is the so-called legend of *Phaeocystis* unpalatability (Huntley et al. 1987) that says that healthy colonies are not consumed due to some sort of chemical deterrence (Estep et al. 1990). The observed negative effects of *Phaeocystis* presence on other organisms may be a key to its bloom forming capacity; chemical deterrence could be a way to reduce grazing (Nejstgaard et al., this issue) as well as inhibiting competitors (allelopathy), thereby increasing fitness.

Up to now, three toxic components that could be involved in chemical deterrence have been identified in *Phaeocystis* species: acrylate, a polyunsaturated aldehyde, and a haemolytic glycolipid. (1) Acrylate is produced by *Phaeocystis* (Guillard and Hellebust 1971; Sieburth 1960) upon enzymatic cleavage of dimethylsulphoniopropionate (DMSP, Stefels and Dijkhuizen 1996) and accumulates in mM concentrations in the colonial mucous layer (Noordkamp et al. 1998). During growth, however, acrylate is unlikely to cause harmful effects on nearby living cells because it is not excreted from the colonies (Noordkamp et al. 2000). Additionally, the concentration of acrylate present in the water column is not expected to exceed the 4 μM observed in senescent cultures (Noordkamp et al. 2000). This is much lower than the mM range of $L(E)C_{50}$ values reported for marine organisms (Sverdrup et al. 2001). Therefore, acrylate is not a likely component to be involved in allelopathy. Acrylate produced by *Phaeocystis* could, however, have a negative impact on grazers (and their consumers) when *Phaeocystis* cells accumulate in their guts. In these acidic environments the high concentrations of acrylate will be in the protonated toxic form (below pH 4.25). A grazing-activated chemical defence system based on the conversion of DMSP into DMS and acrylate upon cell damage was already described for another prymnesiophyte, *Emiliania huxleyi* (Wolfe et al. 1997).

(2) The isolation and identification of an unsaturated aldehyde from *P. pouchetii* (Hansen et al. 2004) may indicate a line of defence that was recently revealed for diatoms (Paffenhöfer et al. 2005, and references therein). Membrane lipids are converted into mildly toxic polyunsaturated fatty acids (PUFAs) by a grazing-activated enzymatic conversion. In the presence of reactive oxygen species (ROS) PUFAs may be converted into highly toxic polyunsaturated aldehydes (PUAs). In laboratory tests these PUAs negatively affect copepod fecundity and egg-hatching, and induce apoptosis in sea urchin embryos and cytotoxicity in human cell lines (Pohnert and Boland 2002). Precursors for PUAs, such as the PUFA eicosapentaenoic acid (EPA), are abundantly present in *P. globosa* (Hamm and Rousseau 2003). Although these haemolytic PUFAs (Arzul et al. 1998; Fu et al. 2004) provide essential nutrition

for copepods high concentrations may be harmful (Jüttner 2001). Extracted culture fluid of *P. pouchetii* containing PUAs and corresponding with densities of 10^6 cells per ml completely blocked DNA replication in sea urchin embryos (Hansen et al. 2003; Hansen et al. 2004). Extraction of cells or sea water yielded fractions that were mildly haemolytic as well as anaesthetic upon injection in flies (Stabell et al. 1999).

(3) The massive *P. globosa* bloom of 1997 in the coastal waters of southeast China induced fish mortality and the economic losses were substantial (Huang et al. 1999). A haemolysin that was isolated and characterised as a glycolipid with a digalactose and a PUFA (heptadecadienoyl) group was found to be responsible for the fish mortality (He et al. 1999) by induction of pores in the cell membrane of target cells (Peng et al. 2005). Both the isolated toxin and supernatant of the *P. globosa* cultures inhibited cultures of other microalgae (J.-S. Liu pers. comm.).

Now that evidence is accumulating about toxic substances produced by *Phaeocystis* species, field studies are needed to assess if these toxins are used in chemical warfare against predators and/or competitors (allelopathy). In this study we tested the hypothesis that *P. pouchetii* excretes a lytic agent. The potential impact of live *P. pouchetii* to cells in close proximity was quantified in a bioassay during a *P. pouchetii* bloom in a mesocosm experiment. Red blood cells were selected as model targets representing unprotected cells, and lysis was quantified by monitoring dissolved haemoglobin. *Phaeocystis* blooms in mesocosms, with natural plankton communities containing all trophic levels up to macrozooplankton, were used to simulate in situ conditions including the chemically mediated interactions (Hay and Kubanek 2002, and references therein) that could trigger defensive responses of *P. pouchetii*.

Material and methods

Mesocosm set up

Haemolytic activity was studied in *P. pouchetii* blooms in experiments conducted from 27 February (experimental day 1) to 3 April 2003 at the marine biological field station in Raunefjorden, outside Bergen, Norway (60°16′N, 05°14′E). Three transparent floating polyethylene enclosures were used [4.5 m deep, 2 m diameter, ca 11 m^3, 0.12 mm thick walls with 90 % penetration of photosynthetically active radiation (PAR)]. Details of the location and general mesocosm design can be found in Svensen et al. (2001) and at the website of the University of Bergen (http://www.ifm.uib.no/LSF/inst2.html).

The mesocosms were filled on 27 February by pumping nutrient-poor fjord water from 5 m depth. The water column was well mixed with an airlift system, pumping 40 l min^{-1} (Jacobsen, 2000). In order to allow the introduction of new species, to avoid substantial pH changes due to primary production, and to replace water sampled during the mesocosm experiment, 10% of the water was renewed daily in each mesocosm from 3 March by pumping (peristaltic) fjord water from about 2.5 m depth outside the mesocosms. On 3 March, mesocosms M2 and M3 were enriched with nitrate ($NaNO_3$) and phosphate (NaH_2PO_4) to final concentrations of 16 μM and 1 μM, according to the Redfield ratio, to stimulate the development of a bloom of *P. pouchetii*. Mesocosm M1 received no nutrients and served as a control. Nutrient outflow with renewal of water was compensated for by daily additions of nutrients (1.6 μM nitrate and 0.1 μM phosphate per mesocosm in M2 and M3). On 20 March daily nutrient replacement in M2 was stopped to induce a fast decline of the *P. pouchetii* bloom.

Sampling

Sampling was performed at least every third day, in the morning, 4 h after sunrise, using 15 l carboys that were filled at the surface of the mesocosms. Samples were taken gently to avoid disruption of the colonies. Carboys were stored in the dark at ambient temperature (ca. 4°C). Cells were collected within 1 h from 250 ml subsamples filtered over GF/F filters (47 mm, Whatman) using gravity only. The GF/F filtrate was collected and stored in the dark at 4°C until analysis on the same day. The filters were transferred into test tubes containing 5 ml artificial seawater at 31.5 PSU (ambient salt concentration) which contained 23.4 g NaCl, 9.35 g $MgCl_2 \cdot 6H_2O$, 0.50 g $CaCl_2.2H_2O$, 3.05 g Na_2SO_4, and 0.81 g K_2SO_4 per litre (Admiraal and Werner 1983). To release water soluble cell content, cells on the filter were disrupted by three sonic bursts (5 s, amplitude 90, 28 Watts) with a probe (Sonics Vibratell^{TM}), followed by centrifugation (30 min, 8000 × *g*). The supernatant (hereafter referred to as extract) was ready to be tested for haemolysis. After experimental day 24 it was no longer possible to filter the 250 ml using gravity and sampling on filters was stopped.

Phytoplankton analysis

Phytoplankton abundance and species composition were determined in 60 ml samples that were fixed with glutaraldehyde (0.5% final concentration) and stored at 4°C. Samples were settled onto black-stained Nuclepore filters with 0.4 μm pore size and then frozen. Cells were counted at 200×, 400× and/or 600–1250× magnification on a light and epifluorescence Olympus or LUMAM-P8 microscope. Cellular carbon of the various phytoplankton species was determined using the conversion described by Menden-Deuer and Lessard (2000). Samples for chlorophyll *a* (Chl *a*) were filtered onto 0.45 μm cellulose-acetate filters (Sartorius AG, Germany), immediately extracted in 90% acetone overnight, and analyzed according to Parsons et al. (1984) on a Turner Designs 10-AU fluorometer.

Erythrocyte lysis analysis (ELA)

Whole mesocosm samples, GF/F filtrate and extracts (up to 23 March 2003) were diluted 1:1 with blood cell suspension. For whole-sample transfer, every pipette tip used was clipped with a pair of scissors to have a > 3 mm diameter inlet, making sure to include colonies while sampling. The blood cell suspension was prepared by adding five drops of fresh human blood to 30 ml buffer (Eschbach et al. 2001), centrifugation (5 min, 4500 × *g*) and addition of the same volume of buffer to the pellet. Standard incubations were done for 24 h in triplicate in test vials (2.5 ml, Eppendorf) at 15°C, 7 μmol photons m^{-2} s^{-1}. After incubation, intact blood cells and phytoplankton were removed by centrifugation (5 min, 4500 × *g*) and the supernatant was transferred into a cuvette and measured at 414 nm to quantify haemoglobin. Artificial seawater diluted 1:1 with ELA buffer was the 0% control, a sonified blood suspension diluted with artificial seawater was used as 100% lysis control. For comparison with other haemolytic studies a dose response curve with saponin (Sigma) was made to estimate the EC_{50} value for this control substance.

In addition to standard incubations, tests were performed. Dose-response curves were made with dilution series of whole mesocosm samples in artificial sea water. The effect of the incubation temperature (15°C) was also compared with a series at ambient temperature (4°C, all sample handling and incubation in cold room). Effects of light conditions were tested by incubations in the dark (wrapped in aluminium foil), 7 and at 40 μmol photons m^{-2} s^{-1}.

Data analysis

Results of regression analysis (Excel 2003, Microsoft Office) are presented as significant with the symbol *P* (<0.05); P^* (<0.01), P^{**} (<0.001), P^{***} (<0.0001), or by the actual *P* value.

Results

Bloom description

In the beginning of the experiment a diatom bloom dominated by *Chaetoceros socialis* developed in all mesocosms. Maximal densities of 4.5, 5.9 and 5.2×10^6 cells per litre were measured on day 12 of the experiment for M1, M2, and M3, respectively (ca. 4.5 ug l^{-1} Chl *a* in all mesocosms, Fig. 1, Table 1). After day 15, *P. pouchetii* blooms developed in all mesocosms, but were more pronounced in the two nutrient-enriched mesocosms (M2 and M3) compared to the control bag (M1). In M1 maximal *P. pouchetii* numbers were $9 \cdot 10^6$ cells per litre representing 55% of total carbon biomass at 2–3 μg total Chl *a* l^{-1}. In M2, that was fertilized until experimental day 21, maximal *P. pouchetii* numbers were 4.3×10^7 cells per litre (92% of total biomass at 23 μg Chl *a* l^{-1}), and in M3 5×10^7 cells

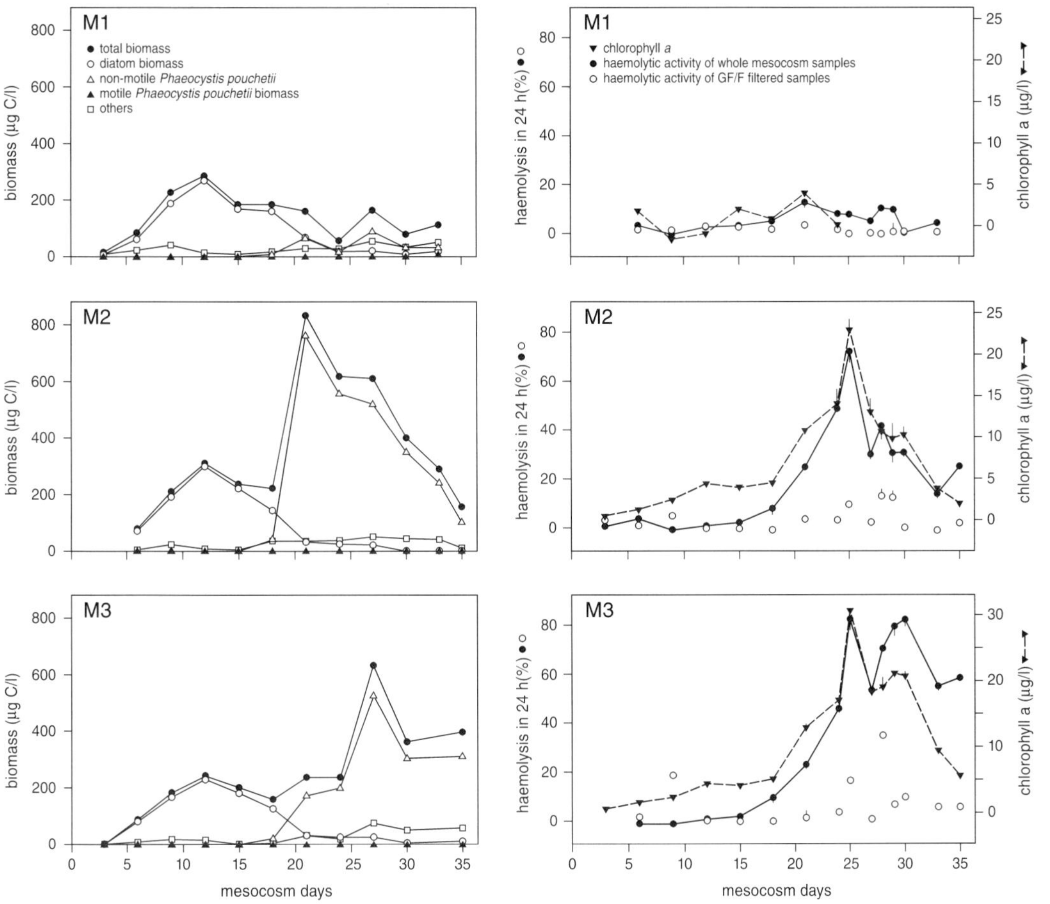

Fig. 1 Left-hand panel: development of biomass for different phytoplankton groups during the mesocosm incubations. M1, control bag; M2, fertilized until day 21; M3 continuously fertilized. Total biomass is the sum of carbon of diatoms, non-motile *Phaeocystis pouchetii*, motile *Phaeocystis* and others. Right-hand panel: development of chlorophyll *a* and haemolytic activity in the mesocosms of both whole mesocosm samples and GF/F filtered samples. Bars represent standard errors

per litre (84% of biomass at 30 µg Chl *a* l^{-1}). First the diatoms and then *P. pouchetii* completely dominated the phytoplankton community both in numbers (Table 1) and in biomass (Fig. 1).

Haemolytic activity

The haemolytic activity measured in whole-mesocosm samples was absent, or very low, during the diatom blooms (Fig. 1). In M1 haemolytic activity in whole mesocosm samples was low during the experiment, increasing from zero at the beginning of the mesocosm experiment to about 10% lysis in the period between day 21 and 29. In M2 and M3 the haemolytic activity was higher. In M2 it exceeded 10 % at day 21, reaching a peak of 70% lysis at day 25, followed by a slow decrease. Also in M3 values above 10% were observed from day 21 onwards, reaching a maximum of 81% lysis at day 30, followed by a slow decrease. The high value at day 25 seemed to be a high value within an increasing trend.

After day 15 the haemolytic activity followed the Chl *a* concentration in the nutrient enriched mesocosms (Figs. 1 and 2). The correlation was: lysis (%) = 3.40 × Chl *a* (ug l^{-1}) − 4.96, $n = 42$, $r^2 = 0.90$, P^{***}). Most of the haemolytic activity was associated with living cells, since only 14% of the whole mesocosm sample signal (total lysis) was measured in the GF/F filtrate: filtrate lysis (%) = 0.141 × [total lysis (%)] + 0.63, $n = 42$, $r^2 = 0.30$, P^{**}, Figs. 1 and 3).

Table 1 Cell densities (numbers × 1,000 per litre) enumerated by microscopy for the groups of plankton in three mesocosms in experiments conducted from 27 February (experimental day 1) to 3 April 2003 at the marine biological field station in Raunefjorden, outside Bergen, Norway (60°16′N, 05°14′E)

↓Categories of plankton in three mesocosms experimental day→	0	6	9	12	15	18	21	24	27	30	33	35
MESOCOSM 1 control, no NP additions												
Diatoms	80	1492	5103	7229	4557	4401	1700	465	511	306	467	
Motile *Phaeocystis*	239	134	120	90	120	157	225	120	949	1220	2842	
Non-motile *Phaeocystis*	7	15	45	0	195	358	6275	1210	8262	3119	2958	
Small phototrophic flagellates	1379	5314	18937	5294	709	2626	4300	1728	2603	1369	1457	
Large phototrophic flagellates and ciliates	2	2	1	1	1	2	2	1	26	0	103	
Small heterotrophic flagellates	596	356	909	121	741	801	1195	1141	2254	2617	5587	
Large heterotrophic flagellates and ciliates	3	5	26	35	13	11	15	31	98	39	96	
MESOCOSM 2 NP enriched until day 21												
Diatoms	159	1945	4996	8039	5916	3874	760	642	723	104	15	627
Motile *Phaeocystis*	60	15	45	60	60	390	990	420	2644	134	782	537
Non-motile *Phaeocystis*	0	0	30	60	105	2213	41810	30510	28606	19285	13216	5920
Small phototrophic flagellates	331	1486	7073	2436	1240	4678	4120	3685	2006	876	1114	455
Large phototrophic flagellates and ciliates	1	0	1	1	2	1	9	15	4	3	1	1
Small heterotrophic flagellates	105	241	242	114	48	840	302	241	3360	2640	2719	2834
Large heterotrophic flagellates and ciliates	25	3	40	5	6	26	53	27	80	51	133	6
MESOCOSM 3 NP continuously enriched												
Diatoms	108	2101	4389	6196	5174	3481	1022	684	724	207		474
Motile *Phaeocystis*	90	90	11	45	30	26	748	45	1763	353		1222
Non-motile *Phaeocystis*	0	0	7	30	165	1465	15687	18750	48654	28138		29017
Small phototrophic flagellates	561	5201	11018	8591	2810	1014	3150	2220	1830	1014		1216
Large phototrophic flagellates and ciliates	0	0	1	1	0	1	1	1	52	47		113
Small heterotrophic flagellates	39	76	63	16	19	19	50	138	1985	2016		1535
Large heterotrophic flagellates and ciliates	16	13	8	5	8	11	49	24	184	106		204

Diatoms, completely dominated by chain forming *Chaetoceros socialis*, 4–30 µm; motile *Phaeocystis* cells, about 3 µm; non-motile *Phaeocystis* cells, 5–6 µm; Small phototrophic flagellates (dominated by small flagellates other than motile *Phaeocystis* and only a few small *Scrippsiella* sp., 2–9 µm; large phototrophic flagellates and ciliate *Myrionecta rubra* (10–54 µm), small heterotrophic flagellates (2–9 µm), large heterotrophic flagellates and ciliates (10–69 µm)

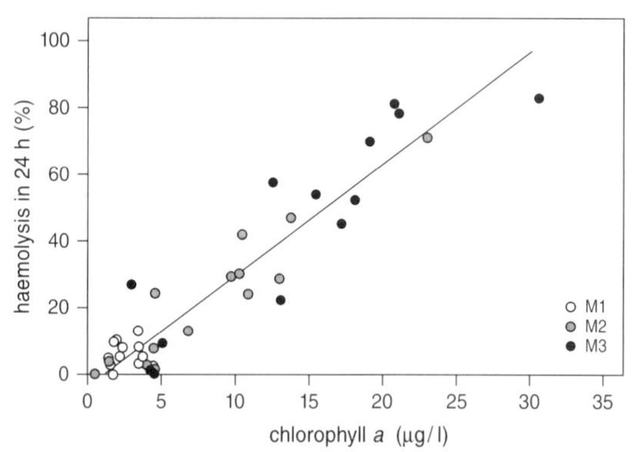

Fig. 2 Correlation between haemolytic activity of whole mesocosm samples and their chlorophyll *a* content for the three mesocosms

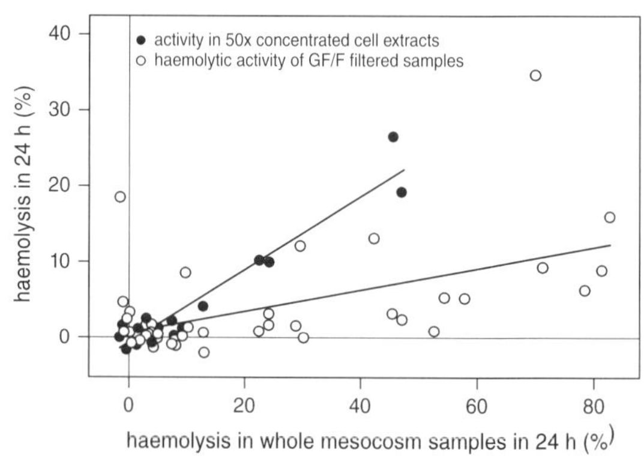

Fig. 3 The relation between the haemolytic activity measured in whole-mesocosm samples and the haemolytic activity in GF/F filtrate of the same samples, or the activity in the 50-times-concentrated cell extracts

In the period 17–23 March, haemolytic activity of the extracts was less than half of activity of whole-mesocosm samples (Fig. 3). Bearing in mind that the extracts are concentrated 50 times compared to the whole-mesocosm samples, the conclusion must be that the effect of the whole samples is not simply caused by components

Table 2 Densities of the different phytoplankton groups (see Table 1) of 31 mesocosm samples were related to the haemolytic activity of the samples as well as to the densities of non-motile *Phaeocystis pouchetii* using linear regression

Density of phytoplankton groups	Haemolytic activity		Non-motile *Phaeocystis*	
	Slope	*P*	Slope	*P*
Non-motile *Phaeocystis*	pos	<0.0005		
Large phototrophic flagellates and ciliates	pos	0.0056	pos	0.0230
Large heterotrophic flagellates and ciliates	pos	<0.0005	pos	<0.0005
Diatoms	neg	0.0011	neg	0.0008
Motile *Phaeocystis*	zero	0.1500	pos	0.0054
Small phototrophic flagellates	zero	0.0590	zero	0.1580
Small heterotrophic flagellates	zero	0.2490	zero	0.2120

released by lysing algae but rather being produced in the presence of intact cells. The activity of the soluble cellular contents was only 1% compared to the activity of the whole sample (extract lysis (%) = 0.010 × [total lysis (%)] −0.015, $n = 22$, $r^2 = 0.930$, P^{***}).

Correlation of haemolysis with phytoplankton groups

Correlations with haemolysis were determined for the most abundant phytoplankton groups. Haemolysis for whole mesocosms samples correlated with total *P. pouchetii* numbers (lysis (%) = 0.001203 * [cell density ml^{-1} sample] + 3.43, $n = 31$, $r^2 = 0.674$, P^{***}). For the non-motile *P. pouchetii* cell abundance, i.e., cells from colonies this regression equation was almost the same (% lysis = 0.001248 × [cell density ml^{-1} sample] + 3.63, $n = 31$, $r^2 = 0.677$, P^{***}, Fig. 4, Table 2). Interestingly, motile *P. pouchetii* numbers (Fig. 1) did not correlate with the haemolytic activity. This was most evident in M1 with low haemolytic activity and highest densities of motile cells (Table 1, Fig. 1). The relation between diatoms and haemolytic activity was negative because the haemolysis started just after the diatom bloom (P^{**}). There was no correlation between the haemolytic activity and heterotrophic flagellates or the small phototrophic flagellates (mainly unidentified flagellates and a few small developing cells of *Scrippsiella trochoidea* with their distinctive chloroplasts, Table 2). There was, however, a positive correlation between the haemolytic activity and the abundance of both the larger heterotophs (P^{*}) as well as the larger phototrophs (P^{*}). The abundance of these groups, however, were positively correlated with the non-motile *Phaeocystis* cells abundance (heterotrophs, P^{***} and phototrophs P, Table 2) which complicates assessment of which groups were responsible for the haemolytic activity. The phototrophs in M1 and M3 reached comparable densities while M1 had much lower haemolytic activity (Fig. 5a). In addition, in M2 the phototrophs displayed low densities in the second half of the mesocosm period while haemolytic activity was high. These observations suggest that the larger phototrophs were not responsible for haemolytic activity. In all bags the larger heterotrophs were present during the diatom blooms and increased during the second half of the experiment, while there was hardly any haemolytic activity present during the diatom blooms and the *P. pouchetii* period in M1 (Fig. 5b). Therefore, the larger heterotrophs are unlikely candidates for haemolysis as well.

Dose response curves at different temperatures

Dose-response curves from dilution series of samples taken at March 24 from the three mesocosms showed a linear relationship between 10 and 70% lysis. For the whole data set the activity at 4°C was approximately half the activity measured at 15°C (value at 4°C = 0.44 × [value at 15°C] + 3.24, $n = 21$, $r^2 = 0.963$, P^{***}, Fig. 6).

Light effects

Aanesen et al. 1998 measured a light dependent negative effect of *P. pouchetii* on fish larvae. To

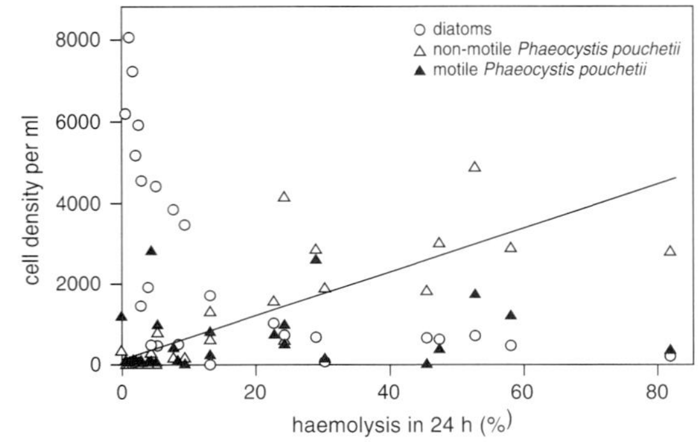

Fig. 4 Relation between the abundance of phytoplankton and the haemolytic activity measured in whole mesocosm samples. The regression line displayed is for non-motile *P. pouchetii* (i.e., cells present in colonies)

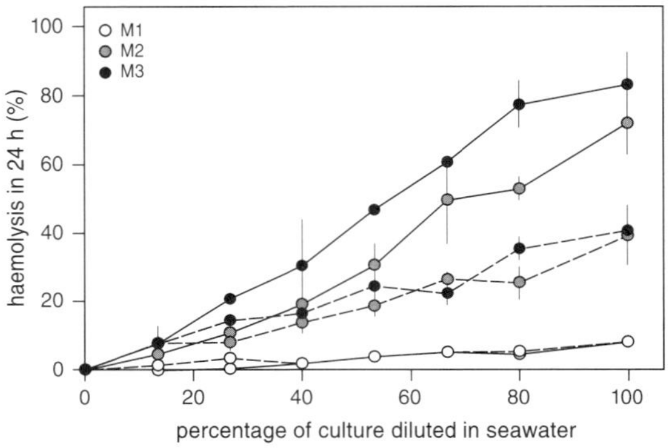

Fig. 6 Dose response curves for whole mesocosm samples taken at 24 March, 2003. Dilution series were measured at ambient temperature (4°C, dotted lines) and at standard experimental temperature (15°C, solid lines). Bars represent standard deviation

test possible influence of light intensity the haemolytic activity of whole mesocosms samples was analyzed not only at 7 but also at 40 μmol photons $m^{-2}\,s^{-1}$ on experimental day 21. For the three mesocosms the haemolytic activity increased twofold at higher light conditions (13 and 28% lysis for M1, 24 and 57% for M2; 23 and 55% for M3,). A similar experiment also including a dark incubation was performed on day 24 (Fig. 7). There was no difference between the dark and standard light conditions. Similar to the experiment on 20 March, at 40 μmol photons $m^{-2}\,s^{-1}$ the haemolytic activity was higher compared to 7 μmol photons $m^{-2}\,s^{-1}$, although not twice the amount observed in the previous experiment. The measurements at high light in this case could be an underestimation because they are higher than 70% where the response is no longer linear (Fig. 6, previous section). Extracts were also more active at the high-light conditions (data not shown).

Discussion

Haemolytic activity was observed in samples during *Phaeocystis pouchetii* blooms in mesocosms. The activity measured using incubations of whole mesocosm samples mixed with blood suspensions was proportional with the amount of sample added. The haemolytic activity in the mesocosms correlated best with the density of non-motile *P. pouchetii* but also with the abundance of the larger phototrophs and heterotrophs. The last two were also positively related to *P. pouchetii* numbers. Biomass during the *P. pouchetii* bloom was dominated by non-motile *P. pouchetii* cells and

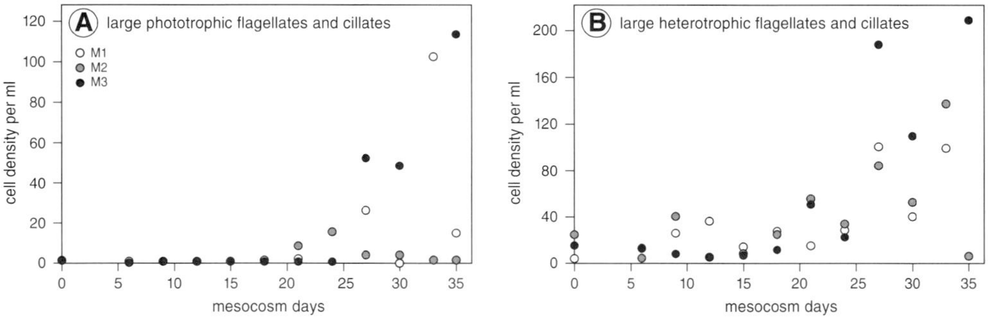

Fig. 5 Relation between haemolytic activity of whole mesocosm samples and abundance of (**A**) large phototrophic flagellates and ciliates (10–54 μm) and (**B**) large heterotrophic flagellates and ciliates (10–69 μm)

Table 3 Overview of haemolytic studies compared to the results obtained for non-motile *Phaeocystis pouchetii* using the EC_{50} saponin values of the tests used (adapted from de Boer et al. 2004)

Species	Microalgae EC_{50} density [10^7 cells l^{-1}]	Cell volume [μm^3]	Microalgae EC_{50} biovolume [10^9 μm^3 l^{-1}]	Reference
Alexandrium tamarense	0.0152	18700[a]	3	2
Phaeocystis pouchetii	1.9	65–110[a]	1–2	This study
Prymnesium parvum	0.3–24.7	280[b]	1–69	2,3,4,5,6
Fibrocapsa japonica	1.7–6.3	3800–4400[b]	64–277	1
Chrysochromulina polylepis	116–170	270[b]	310–460	5

[a] Volume calculated from cell dimensions or [b] Given cell volumes. References: (1) De Boer et al. 2004, (2) Eschbach et al. 2001, (3) Fistarol et al. 2003, (4) Johansson and Granéli 1999, (5) Simonsen and Moestrup 1997, (6) Stolte et al. 2002

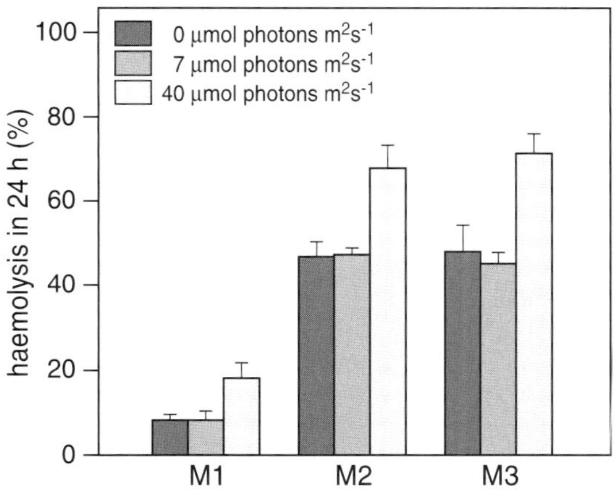

Fig. 7 Haemolytic activity observed in mesocosm samples (mesocosms day 24) incubated at different light intensities. Bars represent standard deviation

close examination of trends in densities of the other plankton groups revealed that these were unlikely to be the haemolytic agents. The non-motile *P. pouchetii* cells present in colonies were responsible for haemolytic activity, although it cannot be excluded that motile *P. pouchetii* cells (almost negligible in their biomass contribution) were also able to produce lytic substances. Haemolytic activity was absent during the diatom blooms that preceded the *P. pouchetii* blooms.

From the relationship between cell numbers and haemolytic activity, an EC_{50} cell density was estimated to 1.86×10^7 *P. pouchetii* cells l^{-1}. This value compares favourably with the EC_{50} densities of other algae that produce harmful algal blooms (Table 3). However, values for the other algae were measured with cell extracts where the EC_{50} values expressed on a biovolume basis indicate an internal toxicity level relevant for the assessment of grazing effects. The values in this mesocosms study were obtained with live cells, whereas *P. pouchetii* extracts contained only one percent of the total activity measured with live samples. Consistent with this, low haemolytic values were reported earlier for methanol extracts of *P. pouchetii* (Stabell et al. 1999) and extracts of *P. pouchetii* cells were not inhibitory to growth of yeast cells. Fractions derived from *P. pouchetii* culture water containing the PUA 2-*trans*-4-*trans*-decadienal, however, blocked sea urchin embryo development as well as the growth of yeast cells (Hansen et al. 2003, 2004).

Approximately 14% of the haemolytic activity in whole-mesocosm samples was present in GF/F filtrate, while almost no activity was extracted from the broken cells. It is unlikely that the bulk of the activity was bound to the cell membranes because a substantial part of the cell debris was still present in the extract (own observation). An alternative explanation would be that the live *P. pouchetii* cells produce an unstable component which greatly enhances haemolysis. In accordance with this is the observation that haemolytic activity did not accumulate over time in the mesocosms, but followed daily variations of chlorophyll *a*. A mechanism that involves physical contact between *Phaeocystis* colony and the blood cell could also explain the activity of live *Phaeocystis*.

During high-light conditions an increased mortality of cod larvae incubated in seawater from a *P. pouchetii* bloom has been observed (Eilertsen and Raa 1995; Aanesen et al. 1998). Similarly, light enhanced the effect of *P. pouchetii*-containing mesocosm samples on blood cells and may have been caused by the same toxic principle that killed the cod larvae with their exposed gills. It is tempting to speculate on the mechanism involved. Perhaps the haemolytic compound itself reacts in a light dependent manner. Light dependent haemolytic cytotoxins were recently identified in raphidophyte cultures (Kuroda et al. 2005). An alternative explanation is the following. During the mesocosms blooms the PUA that was identified earlier for *P. pouchetii* (Hansen et al. 2004) may have been converted into more toxic derivates by ROS produced by living cells: a cascade reaction sequence described earlier in diatoms (Jüttner 2001). This scenario provides an explanation for the action of living cells as well as the effect of light. The light-dependent production of liable ROS such as superoxide seems to be a common feature among microalgae, including prymnesiophytes (Marshall et al. 2005).

Haemolytic activity displayed by *P. pouchetii* was almost exclusively related to active cells and not to compounds within the cells. This situation seems to be different from the haemolytic glycolipids extracted from *P. globosa* isolated from ichthyotoxic blooms in Chinese coastal waters (He et al. 1999), the glycolipids from the prymesiophyte *Chrysochromulina polylepis* (Yasumoto et al. 1990), or the polyethers found for *Prymnesium* species (Legrand et al. 2003, and references therein). Because of the low haemolytic activity of the cell extracts, it is doubtful if predators on *P. pouchetii* in the mesocosm would experience negative effects after consumption of this prey, although healthy colonies are avoided by some copepods (Estep et al. 1990). Organisms such as bacteria and phytoplankton that co-occur with *Phaeocystis*, however, may be inhibited by this lytic action. If so, actively growing *P. pouchetii* colonies in this way further improve their competitive advantage, perhaps contributing to subsequent dominance.

Haemolytic activities observed in the nutrient-enriched mesocosms were higher than those measured in the control bag and values to be expected during blooms in the field. From the data it was possible to estimate the lysis rate at ambient temperature in a natural bloom, based on cell densities or chlorophyll *a* present. The 24 h of exposure to 7 μmol photons $m^2 s^{-1}$ was on the same order of magnitude as the average daily received illumination in the mesocosms and the field (cf. Nejstgaard et al. 2006). Cell densities reported during *P. pouchetii* blooms of 1–2 × 10^7 cells per litre (Schoemann et al. 2005) would lead to a daily lysis rate of 12–29%, whereas the reported Chl *a* values between 5–10 ug l^{-1} would lead to a 16–29% lysis rate. These rates indicate that unprotected cells like the blood cells used in this study, would lyse within days during a *P. pouchetii* bloom. Live *P. pouchetii* colonies are highly haemolytic and the mechanism seems to be fundamentally different from the haemolytic harmful algal bloom species studied so far.

Acknowledgments We would like to thank T. Sørlie, A. Aadnesen, and H. Gjertsen for their service at the Espegrend field station, S.R. Borrett and S.J. Whipple for light measurements, and M. Hordnes, E. F.Skjoldal, and S. Torkildsen for technical support. J.C. Nejstgaard was supported by the Norwegian Research Council (152714/120). The logistics and costs of the mesocosm study, and the contributions of P.G. Verity were supported by US National Science Foundation grant OPP-00-83381. M. van Rijssel and A.C. Alderkamp received funding for the fieldwork by the Dutch Schure-Beijerink-Popping Fund (SBP/JK/2003-14).

References

Aanesen RT, Eilertsen HC, Stabell OB (1998) Light-induced toxic properties of the marine alga *Phaeocystis pouchetii* towards cod larvae. Aquat Toxicol 40:109–121

Admiraal W, Werner D (1983) Utilisation of limiting concentrations of ortho-phosphate and production of extracellular organic phosphate in culture of marine diatoms. J Plankton Res 5:495–513

Arrigo KR, Robinson DH, Worthen DL, Dunbar RB, DiTullio GR, VanWoert M and, Lizotte MP (1999) Phytoplankton community structure and the drawdown of nutrients and CO_2 in the Southern Ocean. Science 283:365–367

Arzul G, Gentien P, Bodennec G (1998) Potential toxicity of microalgal polyunsaturated fatty acids (PUFAs). In: Baudimant G, Guezennec JH, Roy P, Samain JF (eds) Marine lipids. IFREMIER, Nantes, pp 53–62

Brussaard CPD, Kuipers B, Veldhuis MJW (2005) A mesocosm study of *Phaeocystis globosa* population dynamics - 1. Regulatory role of viruses in bloom. Harmful Algae 4:859–874

Brussaard CPD (2006) *Phaeocystis* and its interaction with viruses. Biogeochemistry (this issue)

Cadée GC, Hegeman J (2002) Phytoplankton in the Marsdiep at the end of the 20th century; 30 years monitoring biomass, primary production and *Phaeocystis* blooms. J Sea Res 48:97–110

de Boer MK, Tyl MR, Vrieling EG, van Rijssel M (2004) Effects of salinity and nutrient conditions on growth and haemolytic activity of *Fibrocapsa japonica* (Raphidophyceae). Aquat Microb Ecol 37:171–181

DiTullio GR, Grebmeier JM, Arrigo KR, Lizotte MP, Robinson DH, Leventer A, Barry JB, VanWoert ML, Dunbar RB (2000) Rapid and early export of *Phaeocystis antarctica* blooms in the Ross Sea, Antarctica. Nature 404:595–598

Eilertsen HC, Raa J (1995) Toxins in seawater produced by a common phytoplankter: *Phaeocystis pouchetii*. J Mar Biotechnol 3:115–119

Eschbach E, Scharsack JP, John U, Medlin LK (2001) Improved erythrocyte lysis assay in microtitre plates for sensitive detection and efficient measurement of haemolytic compounds from ichtyotoxic algae. J Appl Toxicol 21:513–519

Estep KW, Nejstgaard JC, Skjoldal HR, Rey F (1990) Predation by copepods upon natural populations of *Phaeocystis pouchetii* as a function of the physiological state of the prey. Mar Ecol Prog Ser 67:235–249

Fistarol GO, Legrand C, Granéli E (2003) Allelopathic effect of *Prymnesium parvum* on a natural plankton community. Mar Ecol Prog Ser 255:115–125

Fu M, Koulman A, Van Rijssel M, Lützen A, De Boer MK, Tyl MR, Liebezeit G (2004) Chemical characterisation of three haemolytic compounds from the microalgal species *Fibrocapsa japonica* (Raphidophyceae). Toxicon 43:355–363

Guillard RRL, Helleburst JA (1971) Growth and production of extracellular substances by two strain of *Phaeocystis pouchetii*. J Phycol 7:330–338

Hansen E, Eilertsen HC, Ernstsen A, Geneviere AM (2003) Anti-mitotic activity towards sea urchin embryos in extracts from the marine haptophycean *Phaeocystis pouchetii* (Hariot) Lagerheim collected along the coast of northern Norway. Toxicon 41:803–812

Hansen E, Ernstsen A, Eilertsen HC (2004) Isolation and characterisation of a cytotoxic polyunsaturated aldehyde from the marine phytoplankter *Phaeocystis pouchetii* (Hariot) Lagerheim. Toxicology 199:207–217

Hamm CE, Rousseau V (2003) Composition, assimilation and degradation of *Phaeocystis globosa*-derived fatty acids in the North Sea. J Sea Res 50:271–283

Hay ME, Kubanek J (2002) Community and ecosystem level consequences of chemical cues in the phytoplankton. J Chem Ecol 28:2001–2016

He J, Shi Z, Zhang Y, Liu Y, Jiang T, Yin Y, Qi Y (1999) Morphological characteristics and toxins of *Phaeocystis cf pouchetii* (Prymnesiophyceae). Oceanol Limnol Sin 30:76–83

Huang CJ, Dong QX, Zheng L (1999) Taxonomic and ecological studies on a large scale *Phaeocystis pouchetii* bloom in the Southeast Coast of China during late 1997. Oceanol Limnol Sin 30:581–590

Huntley M, Tande K, Eilertsen HC (1987) On the trophic fate of *Phaeocystis pouchetii* (Hariot). II. Grazing rates of *Calanus hyperboreus* (Kroyer) on diatoms and different size categories of *Phaeocystis pouchetii*. J Exp Mar Biol Ecol 110:197–212

Kuroda A, Nakashima T, Yamaguchi K, Oda T (2005) Isolation and characterization of light-dependent haemolytic cytotoxin from harmful red tide phytoplankton *Chattonella marina*. Comp Biochem Physiol C 141:297–305

Jacobsen A (2000) New aspects of bloom dynamics of *Phaeocystis pouchetii* (Haptophyta) in Norwegian waters. Ph. D. thesis, University of Bergen

Jakobsen HH, Tang KW (2002) Effects of protozoan grazing on colony formation in *Phaeocystis globosa* (Prymnesiophyceae) and the potential costs and benefits. Aquat Microb Ecol 27:261–273

Jebram D (1980) Prospection for a sufficient nutrition for the cosmopolitic marine bryozoan *Electra pilosa* (Linnaeus). Zoologische Jahrbücher Abteilung für Systematik, Ökologie und Geographie der Tiere 107:368–390

Johansson N, Granéli E (1999) Influence of different nutrient conditions on cell density, chemical composition and toxicity of *Prymnesium parvum* (haptophyta) in semi-continuous cultures. J Exp Mar Biol Ecol 239:243–258

Jüttner F (2001) Liberation of 5,8,11,14,17-eicosapantaenoic acid and other polyunsaturated fatty acids from lipids as a grazer defence reaction in epilithic diatom biofilms. J Phycol 37:744–755

Lancelot C, Billen G, Sournia A, Weisse T, Colijn F, Veldhuis MJW, Davies A, Wassmann P (1987) *Phaeocystis* blooms and nutrient enrichment in the continental coastal zones of the North Sea. Ambio 16:38–46

Legrand C, Rengefors K, Fistarol GO, Granéli E (2003) Allelopathy in phytoplankton – biochemical, ecological and evolutionary aspects. Phycologia 42:406–419

Marshall JA, Ross T, Pyecroft S, Hallegraeff G (2005) Superoxide production by marine microalgae. II. Towards understanding ecological consequences and possible functions. Mar Biol 147:541–549

Menden-Deuer S, Lessard EJ (2000) Carbon to volume relationships for dinoflagellates, diatoms, and other protist plankton. Limnol Oceanogr 45:569–579

Nejstgaard JC, Frischer ME, Verity PG, Anderson JT, Jacobsen A, Zirbel MJ, Larsen A, Martínez-Martínez J, Sazhin AF, Walters T, Bronk DA, Whipple SJ, Borrett SR, Patten BC, Long JD (2006) Plankton development and trophic transfer in sea water enclosures with nutrients and *Phaeocystis pouchetii* added. Mar Ecol Prog Ser 321:99–121

Nejstgaard JC, Tang KW, Steinke M, Dutz J, Koski M, Antajan E, Long JD (2006) Zooplankton grazing on *Phaeocystis*: a critical review and future challenges. Biogeochemistry. This volume

Noordkamp DJB, Schotten M, Gieskes WWC, Forney LJ, Gottschal JC, van Rijssel M (1998) High acrylate concentrations in the mucus of *Phaeocystis globosa* colonies. Aquat Microb Ecol.16:45–52

Noordkamp DJB, Gieskes WWC, Gottschal JC, Forney LJ, van Rijssel M (2000) Acrylate in *Phaeocystis* colonies does not affect the surrounding bacteria. J Sea Res 43:287–296

Paffenhöfer GA, Ianora A, Miralto A, Turner JT, Kleppel GS, d'Alcala MR, Casotti R, Caldwell GS, Pohnert G, Fontana A, Muller-Navarra D, Jonasdottir S, Armbrust V, Bamstedt U, Ban S, Bentley MG, Boersma M, Bundy M, Buttino I, Calbet A, Carlotti F, Carotenuto Y, d'Ippolito G, Frost B, Guisande C, Lampert W, Lee RF, Mazza S, Mazzocchi MG, Nejstgaard JC, Poulet SA, Romano G, Smetacek V, Uye S, Wakeham S, Watson S, Wichard T (2005) Colloquium on diatom-copepod interactions. Mar Ecol Prog Ser 286:293–305

Parsons T, Maita Y, Lalli C (1984) A manual of chemical and biological methods for seawater analysis. Pergamon Press, Oxford, UK, 173 pp

Peng XC, Yang WD, Liu JS, Peng ZY, Lu SH, Ding WZ (2005) Characterization of the hemolytic properties of an extract from *Phaeocystis globosa* Scherffel. J Integrative plant biol Y47:165–171

Pohnert G, Boland W (2002) The oxylipin chemistry of attraction and defence in brown algae and diatoms. Nat Prod Rep 19:108–122

Savage RE (1930) The influence of *Phaeocystis* on the migration of the herring. Fish Invest Lond (Ser II) 12:1–14

Schoemann V, Becquevort S, Stefels J, Rousseau V, Lancelot C (2005) *Phaeocystis* blooms in the global ocean and their controlling mechanisms: a review. J Sea Res 53:43–66

Sieburth JM (1960) Acrylic acid an "antibiotic" principle in *Phaeocystis* blooms in Antarctic waters. Science 132:676–677

Sieburth JM (1961) Antibiotic properties of acrylic acid, a factor in gastrointestinal antibiosis of polar marine animals. J Bacteriol 82:72–79

Simonsen S, Moestrup Ø (1997) Toxicity tests in eight species of *Chrysochromulina* (Haptophyta). Can J Bot 75:129–136

Stabell OB, Aanesen RT, Eilertsen HC (1999) Toxic peculiarities of the marine alga *Phaeocystis pouchetii* detected by in vivo and in vitro bioassay methods. Aquat Toxicol 44:279–288

Stefels J, Dijkhuizen L (1996) Characteristics of DMSP-lyase in *Phaeocystis* sp. (Prymnesiophyceae). Mar Ecol Prog Ser 131:307–313

Stolte W, Panosso R, Gisselson LA, Granéli E (2002) Utilization efficiency of nitrogen associated with riverine dissolved organic carbon (>1 kDa) by two toxin-producing phytoplankton species. Aquat Microb Ecol 29:97–105

Svensen C, Egge JK, Stiansen JE (2001) Can silicate and turbulence regulate the vertical flux of biogenic matter? A mesocosm study. Mar Ecol Prog Ser 217:67–80

Sverdrup LE, Källqvist T, Kelley AE, Fürst CS, Hagen SB (2001) Comparative toxicity of acrylic acid to marine and freshwater microalgae and the significance for environmental effects assessments. Chemosphere 45:653–658

Tang KW (2003) Grazing and colony size development in *Phaeocystis globosa* (Prymnesiophyceae): the role of a chemical signal. J of Plankton Res 25:831–842

Wolfe GV, Steinke M, Kirst GO (1997) Grazing-activated chemical defence in a unicellular marine alga. Nature 387:894–897

Yasumoto T, Underdal B, Aune T, Hormazabal V, Skulberg OM, Oshima Y (1990) Screening for hemolytic and ichthyotoxic components of *Chrysochromulina polylepis* and Gyrodinium aurolum from Norwegian coastal waters. In: Granèli E., Sundström B, Edler L, Anderson DM (eds) Toxic marine phytoplankton. Elsevier, New York, pp 436–440

Biogeochemistry (2007) 83:201–215
DOI 10.1007/s10533-007-9096-0

REVIEW

Phaeocystis and its interaction with viruses

**Corina P. D. Brussaard · Gunnar Bratbak ·
Anne-Claire Baudoux · Piet Ruardij**

Received: 10 November 2005 / Accepted: 14 February 2006 / Published online: 7 April 2007

Abstract Over the years, viruses have been shown to be mortality agents for a wide range of phytoplankton species, including species within the genus *Phaeocystis* (Prymnesiophyceae). With its polymorphic life cycle, its worldwide distribution, and the capacity of several of the *Phaeocystis* species to form dense blooms, this genus is a key player for our understanding of biogeochemical cycling of elements. This paper provides an overview of what is know to date about the ecological role of viruses in regulating *Phaeocystis* population dynamics. It explores which variables affect the algal host–virus interactions, and examines the impact of virally induced cell lysis of *Phaeocystis* on the function and structure of the pelagic food web as well as on the flow of organic carbon and nutrients.

Keywords Characteristics · Mortality · *Phaeocystis* · Phycodnaviridae · PgV · Viruses

C. P. D. Brussaard (✉) · A.-C. Baudoux · P. Ruardij
Department Biological Oceanography,
Royal Netherlands Institute for Sea Research,
P.O. Box 59, 1790AB Den Burg, The Netherlands
e-mail: corina.brussaard@nioz.nl

G. Bratbak
Department of Biology, University of Bergen,
Jahnebakken 5, 5020 Bergen, Norway

Abbreviations

PgV	Phaeocystis globosa virus
PpV	Phaeocysis pouchetii virus
TEP	Transparent exopolymeric particles
MPN	Most probable number
TEM	Transmission electron microscopy
DsDNA	Double-stranded DNA
HAB	Harmful algal bloom species
PFGE	Pulsed-field gel electrophoresis
DMS	Dimethyl sulfide
DMSP	Dimethylsulfoniopropionate
DOC	Dissolved organic carbon
DOM	Dissolved organic matter

Introduction

The presence of viruses in marine environments has been acknowledged for many years, and it is now well established that viruses are dynamic and important members of the microbial food web (Bergh et al. 1989; Proctor and Fuhrman 1990; Gobler et al. 1997; Fuhrman 1999; Wommack and Colwell 2000; Weinbauer 2004). Viruses infect not only the numerically dominant bacteria but also prokaryotic and eukaryotic primary producers. Unicellular photosynthetic organisms are a major group of organisms in natural aquatic communities and viruses have been recognized as mortality agents for phytoplankton (Van Etten et al. 1991; Reisser 1993; Brussaard 2004). Viruses or virus-like particles have been reported in many different taxa of eukaryotic algae, including harmful algal bloom (HAB) species (see review by Brussaard 2004; Nagasaki et al. 2004; Tomaru et al. 2004; Baudoux and Brussaard 2005).

The fate of phytoplankton biomass, whether through sinking, grazing or cell lysis, has major implications for carbon and energy cycles in marine ecosystems. Lysis-mediated release of the cellular content can greatly enhance bacterial activity and subsequently force the food web towards a more-regenerative system. Energy and nutrients released by cell lysis and excretion is transferred to higher trophic levels via the microbial loop (Azam et al. 1983). Lytic viral infection of phytoplankton causes rapid cell lysis, which may affect not only the energy and nutrient flow, but also the phytoplankton community composition and succession. Recognizing viral lysis of phytoplankton as a major process has emphasized the importance of the microbial loop including a viral shunt.

Theoretical models suggest that a 2–10% loss of phytoplankton due to viral infection increases the flow of organic carbon, bacterial production and respiration by more than 25% (Fuhrman 1999; Wilhelm and Suttle 1999). Especially during algal blooms, when high algal cell abundances enhance the virus-host encounter rates, virally mediated lysis can have a profound effect on population dynamics, community diversity and transfer of energy and matter within the pelagic food web.

The *Phaeocystis* genus, with its cosmopolitan distribution, includes several high-biomass-forming species (Cadée and Hegeman 2002; Verity and Medlin 2003; Schoemann et al. 2005). *Phaeocystis* has a life cycle dominated by single cells (with and without flagellae) and embedded colonial non-flagellated cells. Several species of *Phaeocystis*, e.g., *P. pouchetii* and *P. globosa*, regularly dominate the phytoplankton community and sequester huge amounts of nutrient resources, predominantly in the form of colonies. These blooms occur mostly in colder and temperate waters, such as the coastal zone of the North Atlantic and the North Sea. Because of the importance of these blooms for the pelagic ecosystem and the socioeconomic interest in these HAB species, substantial research has been conducted on factors controlling the wax and wane of these blooms. With light and nutrients as important factors initiating *Phaeocystis* blooms, grazing and viruses are considered the relevant loss factors. Field studies indicate that these viruses are a dynamic component, notably involved in the decline of the blooms. Laboratory and seminatural studies provided insight into host–virus interactions and revealed how environmental factors may influence viral infection. The scope of the present paper is to provide a summary and synthesis of available information and some unpublished data related to *Phaeocystis* and the viruses infecting it.

Isolation and characterization of viruses infecting *Phaeocystis*

Viruses that infect species of *Phaeocystis* have been isolated during and directly after natural blooms (Jacobsen et al. 1996; Brussaard et al. 2004; Baudoux and Brussaard 2005). *Phaeocystis pouchetii* viruses (PpV) were successfully isolated after 100-fold concentration by continuous centrifugation and exposure to ultraviolet (UV) for 15 and 30 s (Jacobsen et al. 1996). Exposure to UV light was intended to cause induction of virus production in algal cells containing lysogenic viruses, but as the virus isolated was lytic this treatment was most likely not necessary. So far, all viruses infecting eukaryotic microalgae are lytic and none have been reported to enter a lysogenic relationship with the host. *Phaeocystis globosa* viruses (PgV) were isolated from filtered (GF/F Whatman glass fiber filters) natural water that was added to exponentially growing *P. globosa* host cultures (Baudoux and Brussaard 2005). Incubation of the natural seawater with the addition of nutrients for a week at in situ temperature and irradiance (excluding UV) before adding a subsample to cultures of *P. globosa* occasionally advanced successful isolation of PgV. At the decline of the bloom, when most free viruses can be expected to occur, nutrients regularly become depleted. By adding nutrients more algal biomass was generated and the encounter rate between algal host and virus enhanced.

The *Phaeocystis* viruses isolated so far are species specific, i.e., they only infect one of the *Phaeocystis* species (Jacobsen et al. 1996; Baudoux and Brussaard 2005). Not all *P. globosa* or *P. pouchetii* host strains are infected by the same virus isolates, and not all viruses isolated infect the same algal host strains. There can be a high degree of specificity for these algal viruses.

All intracellular *Phaeocystis* virus particles observed microscopically to date have been located in the cytoplasm of the algal host cell, are hexagonal in their profile, are non-enveloped, and lack a tail. The particles are between 100 and 160 nm in diameter based on transmission electron microscopy (TEM) thin sectioning micrographs (particles tend to shrink when subjected to fixation and dehydration during preparation for thin sectioning). The hexagonal profiles of the viruses suggest that virions may contain an icosahedral capsid (Fig. 1). A recent investigation of the native morphology of PpV at a resolution of 3 nm using electron cryomicroscopy and three-dimensional image reconstruction methods revealed that the capsid had a maximum diameter of 220 nm between opposite vertices, and was composed of 2,192 capsomers that were organized in large triangular and pentagonal aggregates (Yan et al. 2005).

A common ancestor for *P. globosa* viruses and viruses infecting another prymnesiophyte (*Chrysochromulina brevifilum*) has also been suggested based on the phylogenetic analysis using the inferred amino acid sequences of a DNA polymerase gene fragment (Brussaard et al. 2004). This study also showed that seven PgV isolates formed a distinct monophyletic group with other eukaryotic algal viruses, despite differences in their genome size and other phenotypical characteristics. Wilson et al. (2006) recently isolated a virus infecting *P. globosa* from surface water in the English Channel, UK that did not cluster with the other PgVs. Instead, it was more closely related to *C. brevifilum*.

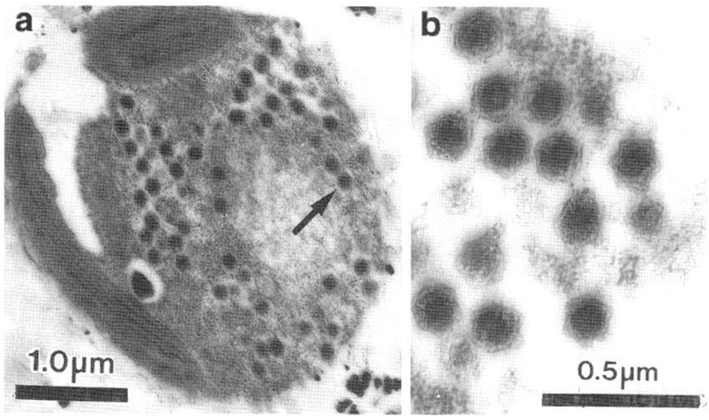

Fig. 1 Transmission electron micrographs of thin sections of infected cells of *Phaeocystis pouchetii*. (**a**) The virus-like particles (indicated by arrow) are found in the cytoplasm of the cells. (**b**) Detail of virus-like particles showing the hexagonal outline of the viruses

The fact that the DNA polymerase gene could be amplified using the algal virus-specific primers AVS1 and AVS2 (Chen and Suttle 1996) allows assignment of these viruses to the family Phycodnaviridae (Van Etten 1995). Many of the characterized phytoplankton viruses are indeed assigned to this family of large double-stranded (ds) DNA viruses that infect eukaryotic algae. The use of the highly conserved DNA polymerase gene turned out to be a good genetic marker for classification of dsDNA algal viruses. The *Phaeocystis* viruses have indeed large dsDNA genomes, about 485 kb in size for PpV (Castberg et al. 2002), and either 177 or 466 kb for PgV (Fig. 2; Baudoux and Brussaard 2005). After staining with a sensitive nucleic-acid-specific dye such as SYBR Green I, these large genome sized *Phaeocystis* viruses could be readily detected using epifluorescent microscopy or flow cytometry (Marie et al. 1999). Furthermore, the use of flow cytometry allowed the discrimination of these viruses from other viruses such as many other algal viruses (Brussaard et al. 2000) or bacteriophages in natural samples (Larsen et al. 2001; Baudoux and Brussaard 2005). The ability to detect, discriminate and enumerate samples containing *Phaeocystis* viruses in a rapid and objective manner promotes laboratory research on virus-host interactions, and ecological studies in the field.

Detailed laboratory studies showed that the total length of the lytic growth cycles of the *Phaeocystis* viruses infecting exponentially growing host cells ranged between 25 and 50 h (Jacobsen et al. 1996; Baudoux and Brussaard 2005). For PpV the latent period, the time period from infection until the first increase in the abundance of extracellular free viruses, was around 12–18 h (Jacobsen et al. 1996). The study by Baudoux and Brussaard (2005) showed three different latent periods for the various PgV isolates in culture, i.e., 10, 12 and 16 h (Fig. 3). These periods match the range of latent periods for all characterized phytoplankton viruses so far, and are somewhat

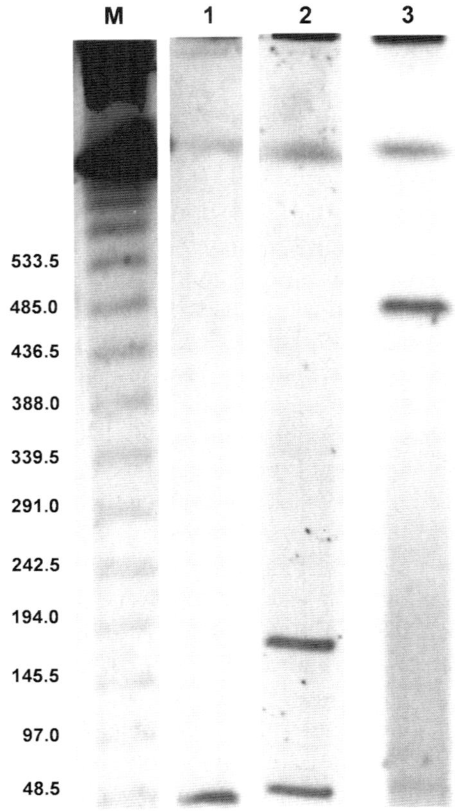

Fig. 2 Viral genome sizes of different *Phaeocystis globosa* virus isolates (PgV) determined by pulsed-field gel electrophoresis (PFGE). Lane M: Lambda concatamers ladder, Lane 1: uninfected culture of *P. globosa*, Lane 2: PgV-04 (genome size of 175 kb), Lane 3: PgV-12 T (genome size of 465 kb). The small-sized band (approximately 45 kb) as seen in lanes 1–3 correspond to bacteriophages since algal cultures were not axenic

shorter or comparable to the maximum growth rates of their host (Schoemann et al. 2005; Veldhuis et al. 2005). Based on the decline in the algal host population and the increase in extracellular virus particles, a conservative estimate of the burst size (number of viruses released per host cell that underwent lysis) can be estimated. For both *Phaeocystis* species, burst sizes ranged on average between 250 and 500, with the higher burst sizes for *P. pouchetii*. The variation in burst sizes between different PgV isolates was considerable, with values down to around 100 despite the exponential growth of the algal host cells (Baudoux and Brussaard 2005). Although the burst size strongly affects the chances for infection of the remaining cells in the host population, not all viruses are infectious. In exponentially growing cultures, the percentage of infective *Phaeocystis* viruses produced as determined by the most probable number (MPN) method is generally relatively high, ranging from 60 to about 100% (Bratbak et al. 1998; Brussaard, unpubl. data).

The devastating effect on *Phaeocystis* cells of an infection by lytic viruses is well illustrated by the morphological, physiological and viability status of the host population during infection (Jacobsen et al. 1996; Bratbak et al. 1998, Brussaard et al. 1999, 2001). Even though the photosynthetic apparatus of the infected algal cells seem to be active during the first hours after infection, sudden and sharp declines in the photosynthetic efficiency of the cells were observed at the end of the latent period (Fig. 4). In contrast to the above-mentioned assays, which reflect the status of the entire population, the use of flow cytometry allows the analysis of individual cells. Changes in the cell characteristics of the virally infected cells are dynamic in time, with the proportion of cells with increased cellular DNA increasing in the first hours after infection, followed by a decline in cellular scatter signals when virions are formed. Prior to cell lysis, the red autofluorescence declined concomitantly with the disruption of the organelles (as observed by transmission electron microscopy). By the time the first viruses are released from the host cells the portion of dead cells increased (Brussaard et al. 2001). Finally, during the period of cell lysis a subpopulation of cells showing reduced concentration of cellular DNA developed and increased with time.

Occurrence and dynamics of *Phaeocystis* viruses

Although the observations of *Phaeocystis* cells containing virus-like particles and the isolation of viruses infecting *Phaeocystis* species suggest that viruses may be potentially important, it is the succession of *Phaeocystis* algal cells and free viruses under seminatural conditions that implies a direct causative relationship and ecological significance. At present, two mesocosm studies and two field studies have been performed, all showing highly dynamic *Phaeocystis* virus abundances with time

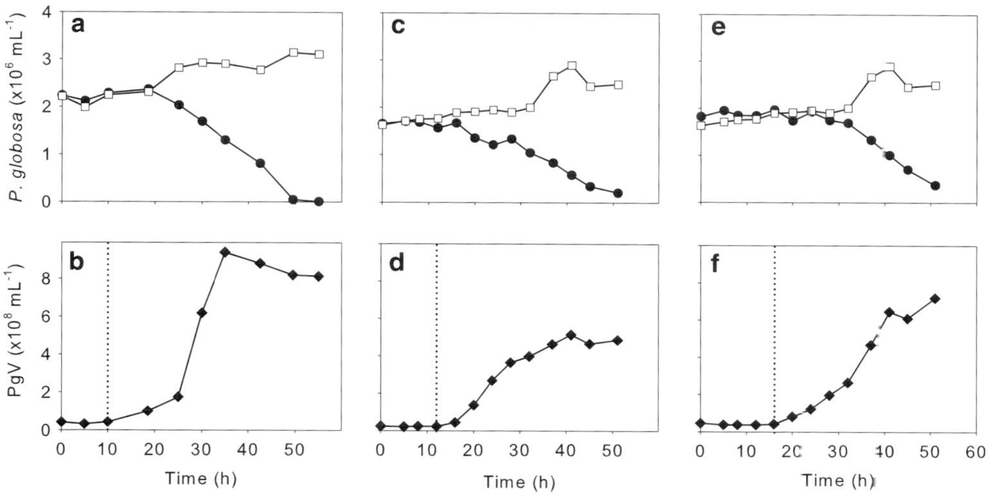

Fig. 3 Abundance of *Phaeocystis globosa* (**a**, **c**, **e**) and PgV (**b**, **d**, **f**) according to Baudoux and Brussaard (2005). Open square symbols represent uninfected cultures, while the filled circles represent virally infected *P. globosa*. (**a**) *P. globosa* infected with PgV-07T, (**c**) with PgV-05T, and (**e**) with PgV-04T. Filled diamond symbols represent the viral growth cycles of (**b**) PgV-07T, (**d**), PgV-05T, and (**f**) and PgV-04T. The length of the latent period (indicated by the dotted line) was 10 h for PgV-07T, 12 h for PgV-05T, and 16 h for PgV-04T

and closely linked to the abundance of their host (Larsen et al. 2001; Brussaard et al. 2004; Brussaard et al. 2005a; Baudoux et al. 2006). For all these studies the abundance of *Phaeocystis* viruses was 30- to 100-fold higher during bloom maxima than the abundance of their host, suggesting that viruses should indeed be regarded as important mortality agents for *P. globosa* and *P. pouchetii*. During the periods of study, the *Phaeocystis* viruses generally made up between 0 and 5% of the total virus population, independent of whether *Phaeocystis* was dominating the phytoplankton community or not. As bacteria are the numerically most abundant, the low share of *Phaeocystis* viruses is to be expected. However, under specific conditions favoring the single-cell morph as compared to the colonial form, the portion of PgV increased up to 30% of total virus abundance (Brussaard et al. 2005a). Under such conditions viruses were actually able to prevent a build-up of standing stock (i.e., bloom) of *P. globosa*.

A critical note here is that successful infection of *Phaeocystis* does not depend on the total abundance of *Phaeocystis* viruses, but on the number of infectious viruses. A recently published ecosystem model that was calibrated with a large data set from *P. globosa* mesocosm experiments (Ruardij et al. 2005) suggested that the fraction of infective PgV successfully infecting *P. globosa* cells increased steeply over the course of the bloom to a maximum of 0.035 (Fig. 5). The fraction of infective PgV that successfully infect *P. globosa* was highest when single cells dominate. The absorption of PgV to transparent exopolymer particles (TEP) that are formed upon disintegration of colonies reduces the available infective PgV and subsequently the fraction of PgV that successfully infects *P. globosa* (Fig. 5). The very low value prior to bloom formation (0.0005%) represents the situation at the start of the growing season when PgV standing stock was subjected for a long time (autumn till spring) to loss of infectivity and decline in actual virus particles. At the same time new virus production was insignificant for *P. globosa* when host cells were present in very low numbers and barely growing (due to light and temperature limitation during winter). A field study in the turbid coastal waters of the southern North Sea showed that the fraction of infectious PgV was around 0.04 (Baudoux et al., 2006), matching very well with the model situation. It is noteworthy that for *P. pouchetii* also a low host–virus adsorption efficiency was needed to produce reasonable good fit between simulation and experimental observations (Bratbak et al. 1998).

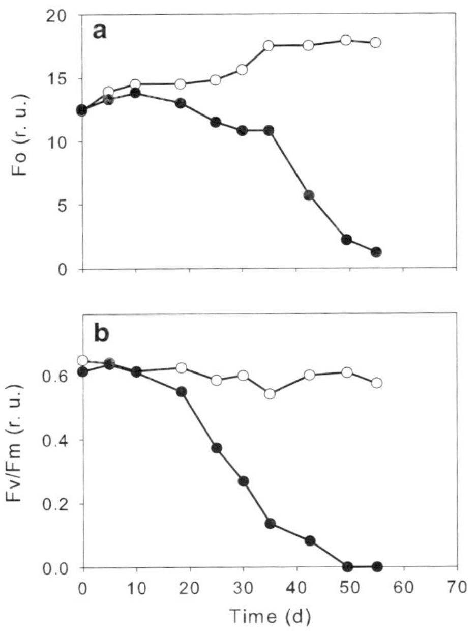

Fig. 4 Dynamics of in vivo chlorophyll fluorescence (Fo) and photosynthetic efficiency (Fv/Fm) of *Phaeocystis globosa* during viral infection as assessed by fluorometry. Open symbols represent uninfected cultures, while the filled symbols represent virally infected *P. globosa*. Maximum fluorescence (Fm) was obtained after addition of the photosystem II inhibitor DCMU (20 μM final concentration). Fv equals Fm-Fo. Data are expressed in relative units (r.u.)

The strong dynamics of the *Phaeocystis* viruses indicate that, besides substantial production, the viral particles are also lost. Ways of removal can include passive adsorption of viruses onto the abundant organic aggregates (TEP) that are grazed upon, or onto inorganic colloids (clay, sand) that are removed from the euphotic zone by sinking (Kapuscinski and Mitchell 1980; Brussaard et al. 2005b). Other factors affecting the loss of the virus particles or infectivity are grazing by protozoa, enzymatic hydrolysis, and UV radiation as it damages the viral nucleic acids (Kapuscinski and Mitchell 1980; Suttle and Chen 1992; González and Suttle 1993; Noble and Fuhrman 1997; Jacquet and Bratbak 2003). Despite the dynamic nature and the substantial losses of PgV, phenotypic characterization and molecular analysis of PgV isolates collected one year apart from the same area revealed identical sequences, indicating considerable stability of these PgV populations (Brussaard et al. 2004; Baudoux and Brussaard 2005). Isolation of PgV during periods with very low to undetectable *P. globosa* host abundance, furthermore, suggests this robust group of viruses has a constant presence in the water column in these coastal areas where *Phaeocystis* occurs.

Diversity of *Phaeocystis* viruses and its ecological role

To achieve successful infection a virus depends on the encounter rate and thus on the abundance of its host species. The *Phaeocystis* virus isolates have, however, not only a species-restricted host ranges but most often also a strain-specific spectrum of infection (Jacobsen et al. 1996; Baudoux and Brussaard 2005). Thus, not all strains of a *Phaeocystis* species (e.g., *P. globosa*) will become infected, even when coexisting in the same water mass. Factors influencing this are the ability of the virus to bind optimally to a proper host cell, as well as the sensitivity of the host to infection.

Based on the structural capsid protein composition, 12 PgV isolates that originate from the southern North Sea could be divided in two groups (Baudoux and Brussaard 2005). The proteins on the surface of free viruses serve as a means of virus-to-cells attachment and allow transfer of viral genomes into the host cells. Any changes in the composition of these structural proteins between viruses may affect the binding to the host cell's receptors, selecting for host range restriction.

Considering that some of these different PgV isolates and their algal host strains originate from the same water sample, the ecological impact of such virus–host diversity is intriguing. Infection by a specific PgV does not act merely at the total host species population level, but rather on the subpopulation level. Essential advantages of such virus diversity could be the promotion of coexisting *P. globosa* strains to guarantee the availability of algal hosts. If one (dominating) strain of *P. globosa* gets infected and undergoes lysis, another resistant *P. globosa* strain that otherwise might be less fit for competition for nutrients, for example, can fill the niche. The lysis of the infected *P. globosa* strain

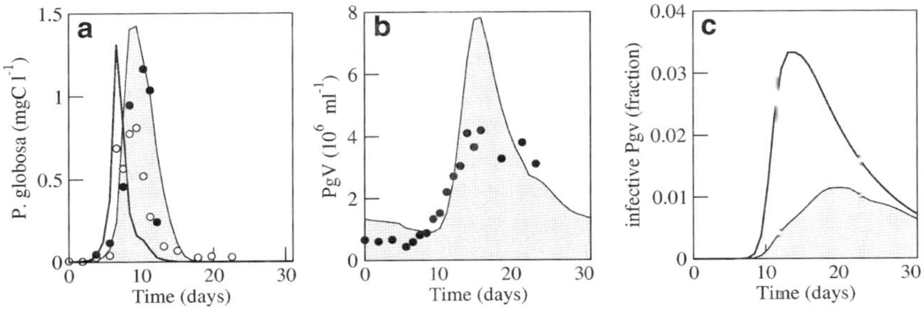

Fig. 5 Results of a mesocososm study on *Phaeocystis globosa* bloom dynamics and the ecological role of viruses. The measured data are represented by the symbols and the modeled results are represented by the lines. (**a**) measured and modeled biomass of *P. globosa* colonies (solid line and open symbol) and single cells (outline of grey area and filled symbol), (**b**) measured and modeled abundance of *P. globosa* viruses (PgV), and (**c**) modeled fraction of infective PgV without transparent exopolymer particles (TEP) present (white area) and in the presence of TEP (grey area). It was assumed that infectivity declined by 10% d^{-1}

might even promote the growth of the other strain(s) indirectly as regenerated nutrients become available. There is increasing evidence accumulating that specific geographical populations of an algal species can be morphologically and/or genetically heterogeneous (Barker et al. 1994; Rynearson and Armbrust 2000, 2004) and viruses might actually be an important regulating factor for host heterogeneity.

Besides different PgVs infecting different *P. globosa* strains, some of the different PgVs infected the same range of *P. globosa* strains. If present at the same time in the water, these different PgV species might have to compete for these host strains. To our knowledge, this is a completely unexplored field of research to date. Important factors affecting the outcome of competition between viruses for the same host cell include the total standing stock of infective viruses in order to enhance the contact rate, the ability to prevent co-infection by other virus types, the burst size, and the level of resistant to loss of infectivity. Differences in burst size and sensitivity to virucidal factors have been observed (Baudoux and Brussaard 2005), but to what extent this influence the actual competition is unknown.

Resistance to viral infection

As discussed above, some strains of the *Phaeocystis* species are resistant to viral infection. But even for one *Phaeocystis* strain the susceptibility to viral infection may not be constant. A laboratory study with *P. pouchetii* showed that some algal host cells escaped infection with the lytic PpV, and with time the algal population increased in abundance again (Thyrhaug et al. 2003). Since PpV originating from these cultures and the original stock affected sensitive *P. pouchetii* cultures equally, the resistance to viral infection must have come from changes in the algal phenotype. Such phenotypic plasticity of the host's susceptibility to infection may be an important stabilizing factor in host–virus interactions as it seems to sustain long-term (one-year) coexistence of host and virus.

The research related to PgV characterization and *Phaeocystis*–virus interactions performed to date generally involved single cells (flagellated and non-flagellated). *Phaeocystis* is, however, known to form colonies and most often the embedded colonial non-flagellated cells are the predominant cell morph during bloom events. A logical question is thus if embedded colonial cells are perhaps resistant to viral infection. This question was raised already by Jacobsen et al. in 1996 upon isolation of PpV, and it was speculated at the time that the gelatinous material the colonial cells are surrounded by serves as a protection against viral infection (Jacobsen et al. 1996; Bratbak et al. 1998). The presence of an outer thin, yet mechanically stable "skin", likely with pores <4.4 nm, has also been suggested as a defense against viral attack of the cells within the colony (Hamm et al. 1999). To test these hypotheses is nevertheless not easy as colonies will always shed at least some single cells, which become readily infected. The consequent production of viruses might thus incorrectly suggest that colonies can be infected. An attempt to infect colonial *P. pouchetii* cells with PpV have so far been unsuccessful (Jacobsen et al. 2005). Tests using 12 PgV isolates to infect strains of *P. globosa* that either formed distinct colonies or mucus aggregates were unsuccessful for all but one strain of PgV (Baudoux and Brussaard 2005). Thus, mucus formation might protect *Phaeocystis* cells from viral infection, but the positive infection of *P. globosa* strain Pg01MD-04 by PgV-01T indicates that the protection is not exclusive.

A mesocosm experiment studying the regulatory role of viruses on *P. globosa* population dynamics sheds more light on the topic (Brussaard et al. 2005a). Both single cells and colonies originating from the same clone were present at the start of the experiment in low concentrations. The results show that the morphology of the *P. globosa* cells (solitary vs. colonial) differently regulated viral control of *P. globosa*. Under non-limiting conditions, dense blooms of colonies and to a lesser extent single cells were formed. Viruses were found to be a significant loss factor but could not prevent bloom formation. Under conditions that restricted colony formation for the first 11 days but allowed single cells to grow, viruses were found to prevent bloom formation the moment conditions were no longer limiting colony formation. The maximum standing stock of PgV was also fivefold higher. The results suggest that the colonial form of *Phaeocystis* is indeed an excellent mechanism to prevent viral infection. Interestingly, a recently developed mathematical ecosystem model including a detailed virus module (Ruardij et al. 2005) suggests that the size of the colonies strongly reduces the chance of infection per cell (and not the gelatinous matrix). The physical principle of the spherical equivalent diameter determining the encounter rate was modeled earlier by Murray and Jackson (1992). With increasing diameter of the colonies, viral infection seems an insignificant loss factor (Fig. 6). The model implies that the single cells were still readily infected, which prevented a build-up of *P. globosa* biomass. By the time colony formation

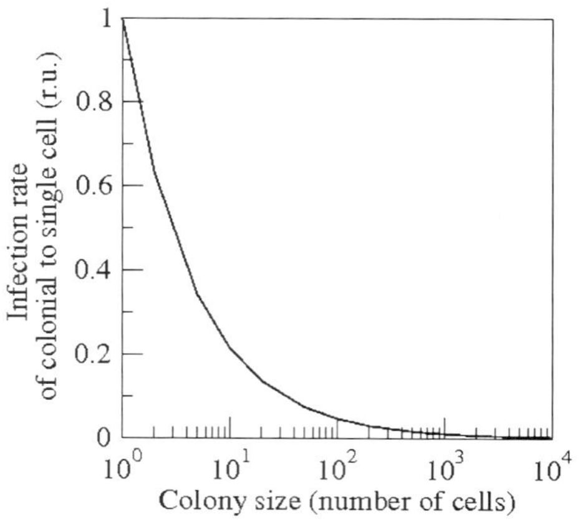

Fig. 6 Viral infection rate of *P. globosa* colonial cells embedded in a colonial matrix as compared to that of *P. globosa* single cells

was no longer limited there was not enough standing stock of single cells to form colonies.

An additional but important indirect defense mechanism against virus infection is the formation of TEP during colony disintegration. TEP was found to be a strong stabilizing negative-feedback mechanism (Brussaard et al. 2005b; Mari et al. 2005; Ruardij et al. 2005). Without colonies and thus without TEP, fewer viruses are needed to control the population of single cells.

Environmental factors influencing virus–host interactions

Keeping in mind that algal growth is regulated by environmental factors such as irradiance, nutrients and temperature, and the fact that viral replication is dependent on the metabolism of the host, it is of interest to know the impact of these factors on virus–host interactions. Under nutrient depletion, for example, the burst size of released *P. pouchetii* and *P. globosa* viruses per infected cell is lower than under nutrient-replete conditions, especially under P-depletion (Bratbak et al. 1998; Brussaard unpubl. data). The strongly reduced burst size found for PpV upon infection of *P. pouchetii* in the stationary phase of growth (15 compared to 240 $cell^{-1}$ for *P. pouchetii* in the exponential growth phase) is most likely the result of severe nutrient depletion inhibiting cell growth (Bratbak et al. 1998).

How nutrient depletion affects not only *Phaeocystis*' physiology and viral replication but also indirectly the relationship between the *Phaeocystis* host cell and the virus becomes clear when considering the different *Phaeocystis* morphotypes. Cells inside the colonial matrix are barely infected, but this changes when the colonies disintegrate due to nutrient depletion. Because of the presence of a diffusive boundary layer, colony formation will decrease nutrient uptake and, therefore, colonies experience nutrient depletion faster than single cells (Ploug et al. 1999). The liberation of large numbers of single cells promotes viral infection, resulting in high viral lysis rates (Ruardij et al. 2005). Nutrient deficiency also has another mode of controlling virus–host interactions, which is by the scavenging properties of TEP. The percentage of PgV attached to TEP was found to be higher under P- than under N-deficiency and consequently the number of PgV for successful infection will be lower (Brussaard et al. 2005b).

Another illustration that growth conditions can affect viral kinetics is the fourfold lower burst size for PpV-infected *P. pouchetii* when placed in the dark compared to those kept in the light. Interestingly, viral proliferation was not delayed or prevented in the dark, indicating that PpV was not dependent on host photosynthesis (Bratbak et al. 1998). Similar results have been found for the model system *P. globosa*–PgV (Brussaard, unpubl. data), which could indicate that this is a genus-wide feature. Since the level of irradiance strongly affects colony formation (Peperzak 1993), it also indirectly impacts on the level of viral control (Brussaard et al. 2005a).

UV radiation affected *P. pouchetii*–PpV interactions differently, depending on the type of UV radiation. UV-B radiation strongly inhibited viral infectivity, whereas UV-A radiation had no effect (Jacquet and Bratbak 2003). A fascinating additional finding was the reduced sensitivity to UV-B stress of *P. pouchetii* cells that previously escaped viral infection. Theoretically, these virus- and UV-resistant cells would have a huge advantage compared to sensitive cells. The fact that they are not dominating the population suggests that there must be a negative trade-off and it is tempting to speculate that these resistant cells may have lower growth rates due to inferior nutrient affinity. Anyway, it does stress the need for studies examining in more detail the effects of combined exposure to potentially regulating factors.

Virally induced mortality of *Phaeocystis*

Having established the presence, dynamics and diversity of *Phaeocystis* viruses, one may wonder what is the actual impact of those viruses on the loss rates of *Phaeocystis*. Earlier studies on *P. globosa* bloom dynamics showed that cell lysis was a relevant loss factor, with rates of up to 0.3 d^{-1} (Brussaard et al. 1995, 1996). Although it was not clear at that time whether viruses were causing the algal cells to undergo lysis, recent studies indicate that viruses are most likely the responsible lysis agents (Larsen et al. 2001; Brussaard et al. 2004; Baudoux and Brussaard 2005; Brussaard et al. 2005a). The use of live/dead assays indicated that viral lysis rates of infected *Phaeocystis* in cultures showed rates as high as 0.8 d^{-1} (Brussaard et al. 1999). Methods for specific and accurate determination of viral-mediated algal mortality in natural waters are however still lacking. The first attempt to estimate viral lysis rates of *P. globosa* cells during a bloom was performed for a set of mesocosm experiments (Brussaard et al. 2005a), and was based on the net production of PgV, a conservative viral loss rate of 0.07 d^{-1} used to correct viral production, and an assumed viral burst size of 300 (Baudoux and Brussaard 2005). The estimated viral lysis rates were around 0.2 d^{-1} during the bloom, and largely accounted for most of the total cell lysis, which was either obtained using the dissolved esterase activity assay or by subtracting the net algal growth rate (equals the change in net abundance) and the microzooplankton grazing rate (using the dilution method) from the gross algal growth rate (derived from DNA cell cycle analysis).

The only field study specifically estimating virally mediated mortality of *P. globosa* cells to date (Baudoux et al. 2006) was executed using a newly developed method, being an adaptation of the classical dilution method to determine microzooplankton grazing rates (Landry and Hassett

1982; Evans et al. 2003). Besides dilution of the natural sample in 0.2 μm pore-size filtered seawater at different ratios, parallel samples were diluted in seawater that was made virus-free by ultrafiltration through 30 kDa cartridges. As dilution results in a reduction in viral infection and/or grazing pressure, the subsequent increase in algal cell abundance after 24 h incubation is used to the determine the actual impact of grazing and viral lysis on the algal population. The usefulness of this adapted method for *Phaeocystis* was validated by testing a culture of *P. globosa* cells in the presence of viruses and the absence of grazers. During two consecutive years, natural viral lysis and microzooplankton rates were determined for *P. globosa* during the annual spring bloom events. Whereas during bloom development microzooplankton grazing seemed to be the dominant loss factor for *P. globosa* cells, viral lysis became increasingly important over the course of the bloom, with rates comparable to the grazing rates (max. $0.35\ d^{-1}$). At times, viral lysis made up more than 50% of the total losses of the single-cell population (Baudoux et al. 2006).

Despite the dependence on certain assumptions, the discussed studies clearly imply that viral lysis is an ecologically significant mortality factor for *Phaeocystis* cells. Modeling also backs up the evidence that viral lysis can be an important cause of *P. globosa* cell mortality, especially under conditions where single cells dominate (Ruardij et al. 2005). Comparing model situations that lack and contain colonies, grazing as well as viral lysis are strongly enhanced with sixfold higher maximum grazing rates and 40-fold higher maximum lysis rates (Fig. 7). Colonies thus seem to have the advantage of being largely protected against grazing and viral infection, explaining their build-up of biomass (bloom). The only significant loss factor for cells inside colonies seems to be automortality, which becomes important during nutrient deficiency as a result of strongly limited nutrient diffusion (Ploug et al. 1999; Ruardij et al. 2005).

Note that the model suggests that grazing indirectly affects the impact of viral infection on the population dynamics of *P. globosa* (assuming equal grazing on both uninfected and virally infected cells). Without grazing single-cell biomass would increase faster, but viral infection would take place earlier, resulting in an overall lower standing stock of single cells (Ruardij et al. 2005).

Impact of viral lysis of *Phaeocystis* on the microbial food web and element cycling

From earlier studies we know that the fate of *Phaeocystis* primary production is of importance for the distribution of energy and biogeochemical cycling within the pelagic and benthic ecosystems (Schoemann et al. 2005). Substantial viral lysis of *Phaeocystis* will provide a sizable source of dissolved organic matter, thereby promoting a retentive system that oxidizes organic matter and regenerates inorganic nutrients in the euphotic zone (Brussaard et al. 1996; Gobler et al. 1997; Wilhelm and Suttle 1999; Ruardij et al. 2005).

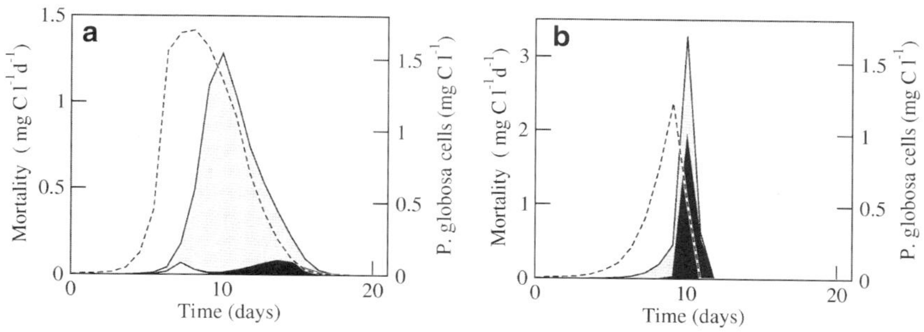

Fig. 7 Modeled abundance of *Phaeocystis globosa* cells and mortality rates of *P. globosa* during a mesocosm study (Ruardij et al. 2005). (**a**) A model run with colonies, and (**b**) a run without colonies present. Viral lysis is represented by the outline of the black area, microzooplankton grazing by the grey area, and automortality by the white area. Automortality affects cells with a net growth rate of <$0.002\ d^{-1}$, which was only found to be of importance for cells embedded in the colonies under nutrient depletion. The dotted line represents the total biomass of *P. globosa* cells

Bratbak et al. (1998) showed that viral infection of *P. pouchetii* resulted in a conversion of the entire algal biomass to dissolved organic carbon (DOC) within three days. In contrast, this was only a maximum of 20% in the uninfected cultures. In response to viral lysis of a *Phaeocystis* blooms, bacterial production has been shown to increase rapidly (Brussaard et al. 2005b). Assuming a bacterial conversion factor of 0.35 (intermediate of values reported for phytoplankton cell debris and lysis products; Biddanda 1988; Van Wambeke 1994), most to all of the bacterial C demand could be accounted for by *P. globosa* cellular C release upon viral lysis. Concomitantly, a shift in bacterial community composition was observed, which was most likely the result of the difference in the DOC pathway. Instead of the normally slow and steady release of small amounts of photosynthetic release of DOC (favoring K-strategists or equilibrium populations), there is the sudden virally induced release of large amounts of readily degradable and organic nutrient-rich DOC (favoring r-strategists or opportunistic populations).

Most of the field data that are available to date relate to *Phaeocystis* blooms. Although the virally mediated cell lysis of *Phaeocystis* during blooms is notably substantial, there is no proof of infection of single cells beyond the bloom period. This does not mean that there could not be significant viral lysis of *Phaeocystis* outside the blooming period. Production of single cells can also take place at high rates when colony formation is not possible, e.g., low irradiances and nutrient concentrations. This provides not only the potential of grazing by microzooplankton, but also of viral infection of the single cells. The relatively small fraction of dissolved organic matter (DOM) obtained in this way may constitute a significant portion of the cycling of rapidly degradable carbon in the pelagic zone.

The theoretical models considering the influence of (algal) viruses on the carbon cycle that exist to date are steady-state models assuming a fixed percentage of the algal population dying due to viral lysis. A bloom of *Phaeocystis* in, for example, temperate eutrophic coastal waters is, however, clearly not a steady-state situation. Based on the ecosystem model by Ruardij et al. (2005), we established a carbon budget for the main players during the wax and wane of a *P. globosa* bloom (Fig. 8 and Table 1). *P. globosa* dominated primary production only when colonies were present (68% of total), but was still one-third of the total primary production under conditions dominated by single cells. The averaged daily flux of viral lysis of *P. globosa* was tenfold higher for the model situation with only single cells present compared to the situation including colonies (115 vs. 11 μg C L^{-1}). Without colonies, viral lysis made up for 53% of the modeled daily *P. globosa* primary production (only 2% when colonies were present).

The reason that the averaged daily uptake of C by bacteria is not higher under the conditions that single cells dominate (compared to colony dominated conditions) and bacterial secondary production is stimulated due to *P. globosa* viral lysis, is due to the absence of TEP production upon disintegration of colonies. Because of the large influence colonies have on the C cycling as a result of their high content of intracellular carbon, the C budget for a model run without viruses resembles largely that of the run with viruses present (data not shown). When excluding both colonies and

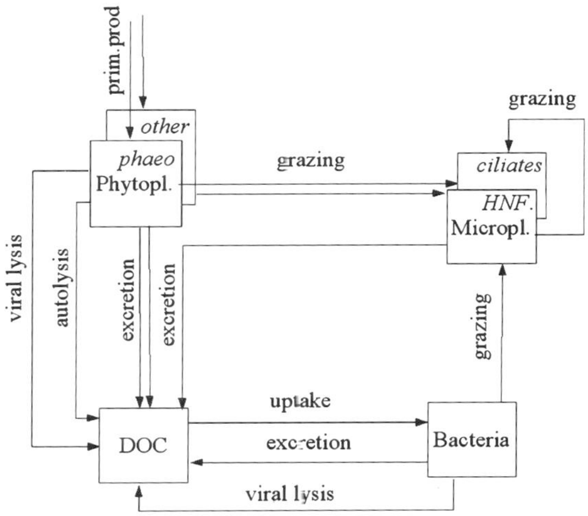

Fig. 8 Simplified representation of a pelagic food web used to calculate the C budget during a *Phaeocystis globosa* bloom as presented in Table 1. The release of cellular carbon due to viral lysis as sources of DOC, as well as automortality was only modeled for *P. globosa*. Excretion of DOC due to photosynthetic release was taken into account for all phytoplankton groups (Phytopl.: *Phaeocystis* and other algae). Micrograzers (Micropl.) include heterotrophic nanoflagellates (HNF) and ciliates. Export was considered a negligible loss

Table 1 Modeled C budget of the wax and wane of a *Phaeocystis globosa* bloom (Brussaard et al. 2005a)

	C-flux	(μg C L^{-1})
	With colonies (standard run)	No colonies
Phytoplankton		
Primary production *P. globosa*	605	215
Primary production other algae	290	435
Viral lysis *P. globosa*	11	115
Automortality *P. globosa*	4.2	0
Excretion *P. globosa*	189	35
Excretion other algae	104	117
Bacteria		
Bacterial uptake	614	364
Bacterial excretion	116	61
Bacterial viral lysis	225	115
Zooplankton		
Grazing on bacteria	64	53
Grazing on *P. globosa*	158	62
Grazing on other algae	54	143
Grazing on microzooplankton	143	115
Zooplankton excretion	125	113

The daily C fluxes (μg C L^{-1} d^{-1}) originate from an ecosystem model by Ruardij et al. (2005) and are averaged values over a period of 36 days. The standard run represents the situation as was observed during the mesocosm experiment, with both *P. globosa* single cells and colonies present. Viral lysis of phytoplankton is specific for P. globosa. Viral lysis of bacteria is a second order density-dependent mortality. Respiration and high refractory DOC were modeled but are not included in the table

viruses, grazing on *P. globosa* becomes logically more important (data not shown). This modeling exercise suggests that *Phaeocystis* viruses have a significant impact on organic C cycling, but the underlying processes and their interactions are complex.

The lysis of cells does not only release organic carbon into the surrounding waters, but also other geochemical important elements such as, N, P, S, Fe, Mn and Zn become available for biological cycling (Gobler et al. 1997; Ruardij et al. 2005). Nitrogen and phosphorus are well-known bottom-up controllers of phytoplankton productivity. Especially during periods that heterotrophic bacteria are carbon- instead of nutrient-deficient in their growth (common during *Phaeocystis* blooms) the shunting of lysis products through the viral shunt can result in a net remineralization of N and P. Since the available inorganic nutrients can nourish algal growth again, viral infection of *Phaeocystis* leads to death on one side and sustained primary production of uninfected *Phaeocystis* strains or different phytoplankton species on the other.

Iron is a nutritive trace element whose role as a limiting agent for algal growth has been demonstrated in areas where certain *Phaeocystis* species (e.g., *P. antarctica*) are also commonly found. Viral lysis will affect the absolute concentration of iron that is potentially available for biological requirement, but may also directly affect the speciation and bioavailability as iron is mostly complexed with organic ligands and colloids.

As a major dimethyl sulfide (DMS) and dimethylsulfoniopropionate (DMSP) producer, viral lysis of *Phaeocystis* cells has been reported an important mechanism for the release of these organic sulphur compounds (Malin et al. 1998). Viral infection of *P. pouchetii* resulted in an eight-fold increase of DMS concentration acompared to an uninfected culture. The high productivity associated with *Phaeocystis* blooms in combination with its world-wide distribution (Schoemann et al. 2005) makes the genus not only an important contributor to the marine carbon flux but also to the global sulphur cycle. It was estimated that the contribution of *Phaeocystis* to the global flux of DMS is 5–10% (Schoemann et al. 2005). As the atmospheric trace gas DMS affects cloud cover, viral lysis of *Phaeocystis* cells is a major link in the biogeochemical cycling of this climate-relevant element.

Future perspectives

To date viruses have been isolated and brought into culture for the bloom-forming species *P. pouchetii* and *P. globosa*. The genus includes, however, more species of which only one other is known to form a colony bloom (*P. antarctica*). Given the dependence of viral induced algal mortality on virus and host cell abundance, it would be of special interest to be able to bring into culture virus–host model systems of species that do not form such dense blooms and compare with those that do. Furthermore, the geographical distribution of the different *Phaeocystis* species is diverse which would provide an excellent opportunity to investigate the role of *Phaeocystis* viruses in the various ecosystems. A start has been by comparing the two different virus–host model systems that are available in culture, based on not only the more traditional basic virological characteristics but also on their genetic material. Full genome sequencing of PpV and PgV is currently in progress (University of Bergen, Norway and DOE-JGI). Once the sequences are available, exciting possibilities such as the development of methods for specific gene detection, function and activity assessment will be within reach.

As there are different virus types infecting the same *Phaeocystis* species, primer/probe development for the specific viruses will enhance our knowledge on virus diversity related issues. This type of research could also be of great help to study the ecological role of viruses for *Phaeocystis* during periods or in regions of low *Phaeocystis* abundance. Comparative genomics of the different virus types as planned for PgV is a challenge that is expected to provide insight into how these viruses coexist for the same host.

The newly developed data-based ecosystem model including a virus module by Ruardij et al. (2005) has proven to be very useful providing insight towards a comprehensive understanding of the role of viruses for *P. globosa* population dynamics and C cycling. It is, however, still far from complete. Aspects such as processes influencing PgV infectivity and losses are essentially unstudied. For example, is grazing of *Phaeocystis* viruses by protozoans an ecologically important loss factor? More detailed mechanistic-orientated experiments will help solve these unknowns.

Acknowledgements We acknowledge Runar Thyrhaug for providing TEM images of PpV viruses and PpV infected *Phaeocystis pouchetii* cells.

References

Azam F, Fenchel T, Field JG, Gray JS, Meyer-Reil LA, Thingstad F (1983) The ecological role of water-column microbes in the sea. Mar Ecol Prog Ser 10:257–263

Barker GLA, Green JC, Hayes PK, Medlin LK (1994) Preliminary results using the RAPD analysis to screen bloom populations of *Emiliania huxleyi* (Haptophyta). Sarsia 79:301–306

Baudoux A-C, Brussaard CPD (2005) Characterization of different viruses infecting the marine harmful algal bloom species *Phaeocystis globosa*. Virology 341:80–90

Baudoux A-C, Noordeloos AAM, Veldhuis MJW, Brussaard CPD (2006) Virally induced mortality of *Phaeocystis globosa* during a spring bloom in temperate coastal waters. Aquat Microb Ecol 44:201–217

Bergh Ø, Børsheim KY, Bratbak G, Heldal M (1989) High abundance of viruses found in aquatic environments. Nature 340:467–468

Biddanda BA (1988) Microbial aggregation and degradation of phytoplankton-derived detritus in seawater. II. Microbial metabolism. Mar Ecol Prog Ser 42:89–95

Bratbak G, Jacobsen A, Heldal M (1998) Viral lysis of *Phaeocystis pouchetii* and bacterial secondary production. Aquat Microb Ecol 16:11–16

Bratbak G, Jacobsen A, Heldal M, Nagasaki K, Thingstad F (1998) Virus production in *Phaeocystis pouchetii* and its relation to host cell growth and nutrition. Aquat Microb Ecol 16:1–9

Brussaard CP, Marie D, Bratbak G (2000) Flow cytometric detection of viruses. J Virol Methods 85:175–182

Brussaard CPD (2004) Viral Control of Phytoplankton Populations - a Review. J Eukaryot Microbiol 51:125–138

Brussaard CPD, Gast GJ, Van Duyl FC, Riegman R (1996) Impact of phytoplankton bloom magnitude on a pelagic microbial food web. Mar Ecol Prog Ser 144:211–221

Brussaard CPD, Kuipers B, Veldhuis MJW (2005a) A mesocosm study of *Phaeocystis globosa* population dynamics. I. Regulatory role of viruses in bloom control. Harmful Algae 4:859–874

Brussaard CPD, Mari X, Van Bleijswijk JDL, Veldhuis MJW (2005b) A mesocosm study of *Phaeocystis globosa* population dynamics. II. Significance for the microbial community. Harmful Algae 4:875–893

Brussaard CPD, Marie D, Thyrhaug R, Bratbak G (2001) Flow cytometric analysis of phytoplankton viability following viral infection. Aquat Microb Ecol 26:157–166

Brussaard CPD, Riegman R, Noordeloos AAM, Cadée GC, Witte HJ, Kop AJ, Nieuwland G, Van Duyl FC, Bak RPM (1995) Effects of grazing, sedimentation and phytoplankton cell lysis on the structure of a coastal pelagic food web. Mar Ecol Prog Ser 123:259–271

Brussaard CPD, Short SM, Frederickson CM, Suttle CA (2004) Isolation and phylogenetic analysis of novel viruses infecting the phytoplankter *Phaeocystis globosa* (Prymnesiophyceae). Appl Environ Microbiol 70:3700–3705

Brussaard CPD, Thyrhaug R, Marie D, Bratbak G (1999) Flow cytometric analyses of viral infection in two marine phytoplankton species, *Micromonas pusilla* (Prasinophyceae) and *Phaeocystis pouchetii* (Prymnesiophyceae). J Phycol 35:941–948

Cadée GC, Hegeman J (2002) Phytoplankton in the Marsdiep at the end of the 20th century; 30 years monitoring

biomass, primary production, and *Phaeocystis* blooms. J Sea Res 48:97–110
Castberg T, Thyrhaug R, Larsen A, Sandaa R-A, Heldal M, Van Etten JL, Bratbak G (2002) Isolation and characterization of a virus that infects *Emiliania huxleyi* (Haptophyta). J Phycol 38:767–774
Chen F, Suttle CA (1996) Evolutionary relationships among large double-stranded DNA viruses that infect microalgae and other organisms as inferred from DNA polymerase genes. Virology 219:170–178
Evans CGT, Archer SD, Jacquet S, Wilson AC (2003) Direct measurements of the contribution of viral lysis and microzooplankton grazing to the decline of a *Micromonas* spp. population. Aquat Microb Ecol 30:207–219
Fuhrman JA (1999) Marine viruses and their biogeochemical and ecological effects. Nature 399:541–548
Gobler CJ, Hutchins DA, Fisher NS, Cosper EM, Sanudo-Wilhelmy SA (1997) Release and bioavailability of C, N, P, Se, and Fe following viral lysis of a marine chrysophyte. Limnol Oceanogr 42:1492–1504
González JM, Suttle CA (1993) Grazing by marine nanoflagellates on viruses and virus-sized particles: ingestion and digestion. Mar Ecol Prog Ser 94:1–10
Hamm CE, Simson DA, Merkel R, Smetacek V (1999) Colonies of *Phaeocystis globosa* are protected by a thin but tough skin. Mar Ecol Prog Ser 187:101–111
Jacobsen A, Bratbak G, Heldal M (1996) Isolation and characterization of a virus infecting *Phaeocystis pouchetii* (Prymnesiophyceae). J Phycol 32:923–927
Jacobsen A, Martinez-Martinez J, Verity P, Frischer ME, Sandaa R-A, Larsen A (2005) Are colonies or colonial cells of *Phaeocystis pouchetii* (prymnesiophyceae) susceptible to virus infection? ASLO Summer Meeting, Santiago de Compostela, Spain, June 19–24
Jacquet S, Bratbak G (2003) Effects of ultraviolet radiation on marine virus-phytoplankton interactions. Fems Microbiol Ecol 44:279–289
Kapuscinski RB, Mitchell R (1980) Processes controlling virus inactivation in coastal waters. Wat Res 14:363–371
Landry MR, Hassett RP (1982) Estimating the grazing impact of marine micro-zooplankton. Mar Biol 67:283–288
Larsen A, Castberg T, Sandaa R-A, Brussaard CPD, Egge JK, Heldal M, Paulino A, Thyrhaug R, Van Hannen EJ, Bratbak G (2001) Population dynamics and diversity of phytoplankton, bacteria and viruses in a seawater enclosure. Mar Ecol Prog Ser 221:47–57
Malin G, Wilson WH, Bratbak G, Liss PS, Mann NH (1998) Elevated production of dimethylsulfide resulting from viral infection of cultures of *Phaeocystis pouchetii*. Limnol Oceanogr 43:1389–1393
Mari X, Rassoulzadegan F, Brussaard CPD, Wassmann P (2005) Dynamics of transparent exopolymeric particles (TEP) production by *Phaeocystis globosa* under N- or P-limitation: a controlling factor of the retention/export balance?. Harmful Algae 4:895–914
Marie D, Brussaard CPD, Thyrhaug R, Bratbak G, Vaulot D (1999) Enumeration of marine viruses in culture and natural samples by flow cytometry. Appl Environ Microbiol 65:45–52
Murray AG, Jackson GA (1992) Viral dynamics: a model of the effects of size, shape, motion and abundance of single-celled planktonic organisms and other particles. Mar Ecol Prog Ser 89:103–116
Nagasaki K, Tomaru Y, Katanozaka N, Shirai Y, Nishida K, Itakura S, Yamaguchi M (2004) Isolation and characterization of a novel single-stranded RNA virus infecting the bloom-forming diatom *Rhizosolenia setigera*. Appl Environ Microbiol 70:704–711
Noble RT, Fuhrman JA (1997) Virus decay and its causes in coastal waters. Appl Environ Microbiol 63:77–83
Peperzak L (1993) Daily irradiance governs growth rate and colony formation of *Phaeocystis* (Prymnesiophyceae). J Plankton Res 15:809–821
Ploug H, Stolte W, Jørgensen BB (1999) Diffusive boundary layers of the colony-forming plankton alga *Phaeocystis* sp.-implications for nutrient uptake and cellular growth. Limnol Oceanogr 44:1959–1967
Proctor LM, Fuhrman JA (1990) Viral mortality of marine bacteria and cyanobacteria. Nature 343:60–62
Reisser W (1993) Viruses and virus-like particles of freshwater and marine eukaryotic algae - a review. Arch Protistenkd 143:257–265
Ruardij P, Veldhuis MJW, Brussaard CPD (2005) Modelling bloom dynamics of the polymorphic phytoplankter *Phaeocystis globosa*: impact of grazers and viruses. Harmful Algae 4:941–963
Rynearson TA, Armbrust EV (2000) DNA fingerprinting reveals extensive diversity in a field population of the centric diatom *Ditylum brightwellii*. Limnol Oceanogr 45:1329–1340
Rynearson TA, Armbrust EV (2004) Genetic differentiation among populations of the planktonic marine diatom *Ditylum brightwellii* (Bacillariophyceae). J Phycol 40:34–43
Schoemann V, Becquevort S, Stefels J, Rousseau V, Lancelot C (2005) *Phaeocystis* blooms in the global ocean and their controlling mechanisms: a review. J Sea Res 53:43–66
Suttle CA, Chen F (1992) Mechanisms and rates of decay of marine viruses in seawater. Appl Environ Microbiol 58:3721–3729
Thyrhaug R, Larsen A, Thingstad FT, Bratbak G (2003) Stable coexistence in marine algal host-virus systems. Mar Ecol Prog Ser 254:27–35
Tomaru Y, Katanozaka N, Nishida K, Shirai Y, Tarutani K, Yamaguchi M, Nagasaki K (2004) Isolation and characterization of two distinct types of HcRNAV, a single-stranded RNA virus infecting the bivalve-killing microalga *Heterocapsa circularisquama*. Aquat Microb Ecol 34:207–218
Van Etten JL (1995) Phycodnaviridae. In: Murphy FA, Fauquet CM, Mayo MA, Jarvis AW, Ghabrial SA, Summers MD, Martelli GP, Bishop DHL (eds) The classification and nomenclature of viruses. *Sixth Report of the International Committee on Taxonomy of Viruses* Archives of Virology. Springer Verlag, Wien/New York

Van Etten JL, Lane LC, Meints RH (1991) Viruses and viruslike particles of eukaryotic algae. Microb Rev 55:586–620

Van Wambeke F (1994) Influence of phytoplankton lysis or grazing on bacterial metabolism and trophic relationships. Microb Ecol 27:143–158

Veldhuis MJW, Brussaard CPD, Noordeloos AAM (2005) Living in a *Phaeocystis* colony; a way to be a successful algal species. Harmful Algae 4:841–858

Verity PG, Medlin LK (2003) Observations on colony formation by the cosmopolitan phytoplankton genus *Phaeocystis*. J Mar Syst 43:153–164

Weinbauer MG (2004) Ecology of prokaryotic viruses. FEMS Microbiol Rev 28:127–181

Wilhelm SW, Suttle CA (1999) Viruses and nutrient cycles in the sea. BioScience 49:781–788

Wilson WH, Schroeder DC, Ho J Canty M (2006) Phylogenetic analysis of PfV-102P, a new virus from the English Channel that infects *Phaeocystis globosa*. J Mar Biol Ass UK 86:485–490

Wommack KE, Colwell RR (2000) Virioplankton: viruses in aquatic ecosystems. Microbiol Mol Biol Rev 64:69–114

Yan X, Chipman PR, Castberg T, Bratbak G, Baker TS (2005) The marine algal virus PpV01 has an icosahedral capsid wit hT = 219 quasisymmetry. J Virol 79:9236–9243

Biogeochemistry (2007) 83:217–234
DOI 10.1007/s10533-007-9093-3

ORIGINAL PAPER

Does *Phaeocystis* spp. contribute significantly to vertical export of organic carbon?

Marit Reigstad · Paul Wassmann

Received: 10 April 2006 / Accepted: 10 August 2006 / Published online: 23 March 2007

Abstract *Phaeocystis* spp. cell and colony mass fluxes and their contribution to the vertical particulate organic carbon (POC) export from a wide range of stations were quantified by short-term sediment traps. The compilation of available data, ranging from polar to sub-arctic and boreal regions, revealed that *Phaeocystis* colonial and single cells frequently are observed in shallow sediment traps at 30–50 m depth (average of 7 ± 11% of POC export). A strong vertical export decline between 40 m and 100 m diminished the contribution of *Phaeocystis* spp. cell carbon to vertical export of POC to only 3 ± 2% at 100 m depth, with two exceptions (deeper mixed stations). Estimates of potential corresponding mucus contribution increased the average *Phaeocystis* spp. contribution to <5% of POC export. The vertical flux attenuation efficiency is higher for *Phaeocystis* spp. than for diatoms. The overall contribution of *Phaeocystis* spp. to vertical carbon export based on direct investigations of vertical organic carbon export is small.

Keywords Carbon flux · Cells and mucus · *Phaeocystis* · Transparent exopolymer particles (TEP) · Vertical export

M. Reigstad (✉) · P. Wassmann
Norwegian College of Fishery Science, University of Tromsø, Tromsø 9037, Norway
e-mail: Marit.Reigstad@nfh.uit.no

Introduction

Only a few phytoplankton taxa have a significant biogeochemical impact on the global scale. One of these is the colony-forming haptophyte *Phaeocystis* spp. Most phytoplankton taxa, which have a significant biogeochemical impact on the carbon cycle, such as diatoms, coccolithophorids and dinoflagellates, leave behind cell structures and distinct, relatively stable biochemical signals in sedimented matter or the sediment record. This is not the case with the key taxon *Phaeocystis* spp. Its impact on the global C flux is thus easily disregarded since evidence often derived from long-term measurements of vertical export at depth (>1,000 m) or sediment cores where signals are scant or nonexistant.

The role of *Phaeocystis* as a globally distributed key genus for biogeochemical cycling, food web structure and impact on climate is acknowledged (Schoemann et al. 2005, and refs. therein; Verity and Smetacek 1996). However, the impact of *Phaeocystis*-derived carbon on the total organic carbon export to depth is not well known. To evaluate the role of *Phaeocystis* spp. in ecosystems and biogeochemical cycling, it is of utmost importance to evaluate our knowledge on the impact of *Phaeocystis*-derived material to the total vertical carbon export to depth. The importance of vertical export for the fate of *Phaeocystis* spp. and as contribution to the vertical organic carbon export is still a matter of controversy (e.g., Schoemann et al. 2005). Based upon visual examination and POC export, Wassmann et al. (1990) reported mass sedimentation of *P. pouchetii* from the Barents Sea. In the Antarctic region, indirect observations of vertical export through deep fluorescence signals together with elevated concentrations of DMSP and *P. antarctica* cells at >400 m in the Ross Sea (DiTullio et al. 2000), indicated that export of *P. antarctica*-derived material took place in this region. Formation of fast-sinking aggregates was suggested as a mechanism enhancing the vertical export in diatom- as well as *P. antarctica*-dominated regions of the Ross Sea (Asper and Smith 2003), although microscopic examination of the aggregate composition was not carried out. However, Gowing et al. (2001) and Pakomov et al. (2002) were unable to detect rapid and substantial export of *P. antarctica* using long- and short-term sediment traps in the Ross and Lazarev Seas, respectively. In the northern hemisphere, Smith et al. (1991) described significant blooms of *P. pouchetii* in the Norwegian and Greenland Sea, and suggested large-scale vertical export, in essence based upon the depth distribution of fluorescence signals. A *P. pouchetii* bloom in the Eastern Bering shelf, was followed by observations of cells close to the sediment surface at the shallow stations, but not in deeper regions (Sukhanova and Flint 2001). Despite significant blooms in the North Sea and observations of colonies within the benthic boundary layer (Riebesell 1993), no indications of fresh *Phaeocystis*-derived material (lipid composition) were found in surface sediments after a *P. globosa* bloom (Hamm and Rousseau 2003). Carbon budgets for a *Phaeocystis*-dominated ecosystem off Belgium, suggests sedimentation to be of minor importance (Rousseau et al. 2000).

The contradictory observations indicate that no universal rule can be applied regarding the vertical flux of *Phaeocystis* spp. and material derived from this genus. Episodic flux events that export *Phaeocystis* spp. material to deeper waters or the sediments obviously take place (based upon indirect evidence), but uncertainty exists regarding its role in biogeochemical cycles and if vertical export is important for the fate of the frequently observed, extensive *Phaeocystis* spp. biomass accumulations. What is needed is to directly measure the export of cells and colonies, evaluate the mechanisms involved and investigate the frequencies of events.

To quantify the vertical export of *Phaeocystis* spp. and its relative contribution to the particulate organic carbon (POC) export, direct measurements involving sediment traps with microscopic examination and biogeochemical analysis of the vertical flux is required. The catchment efficiency of sediment traps is at all times a matter of concern, but the assessment of Coppola et al. (2002) suggests that the majority of traps used in the current investigation measured particulate vertical export exactly. Even taxon-specific tracers such as high-performance liquid chromatography (HPLC) pigments and pigment ratios can confuse important groups such as diatoms, *Phaeocystis* spp. and dinoflagellates (Irigoien et al. 2004), highlighting the importance of time-consuming microscopic examination. Few studies are published where *Phaeocystis* spp. cells from sediment traps have been quantified through microscopic examination. As a consequence, attempts to estimate the relative contribution of *Phaeocystis* spp. to vertical carbon export are rare. Here, we bring together published and unpublished vertical export rates of *Phaeocystis* spp. for a more thorough evaluation of *Phaeocystis* export. As a final point we attempt to solve the controversy on the importance of *Phaeocystis* spp. for the contribution to the vertical organic carbon export.

Methods

Sampling sites and sediment trap measurements

Sediment trap investigations involving microscopic determination of *Phaeocystis* spp. cell fluxes have been conducted in North Norwegian fjords, the Barents Sea, the south-western part of the North Sea off the Dutch coast, and the Arctic Ocean (Amundsen Basin). Investigation sites, sampling periods, number of samplings and geographical positions are given with reference to data source in Table 1.

Table 1 Data from a wide range of localities and sampling dates relevant for *Phaeocystis* spp. blooms is included

Station	Lat.	Long.	Period	Sed. trap (m)	Depth (m)	Data source/ comment
Balsfjord	69.36	19.17	11.04.96	40, 100	190	This study
			15.04.96	40, 100		
			18.04.96	40, 100		
			22.04.96	40, 100		
			24.04.96	40, 100		
			30.04.96	40, 100		
			02.05.96	40, 100		
			07.05.96	40, 100		
			14.05.96	40, 100		
			21.05.96	40, 100		
			28.05.96	40, 100		
Balsfjord	69.37	19.12	08.04.97	40, 100	190	Reigstad et al. (2000) No SD 0–100 m
			11.04.97	40, 100		
			14.04.97	40, 100		
			18.04.97	40, 100		
			24.04.97	40, 100		
Balsfjord (n = 3)	69.37	19.12	19–25.04.01	100	190	This study
Malangen	69.50	18.35	09.04.97	40, 100	380	Reigstad et al. (2000) No SD 0–100 m
			12.04.97	40, 100		
			15.04.97	40, 100		
			17.04.97	40, 100		
			23.04.97	40, 100		
Malangen (n = 3)	69.5	18.35	18–24.04.01	100	380	This study
Ullsfjord	69.82	19.83	10.04.97	40, 100	279	Reigstad et al. (2000) No SD 0–100 m Susp. conc.
			13.04.97	40, 100		
			16.04.97	40, 100		
			22.04.97	40, 100		
			25.04.97	40, 100		
Ullsfjord (n = 3)	69.82	19.83	20–26.04.01	100	279	This study
North Sea off-shelf (n = 7)	53.93–54.11	4.35–4.71	09–29.04.02	30	45	This study, no SD
North Sea on shelf (n = 6)	52.94–53.14	4.42–4.57	07–28.04.02	15/20	25–30	This study, no SD
Barents Sea (ArW)	76.01	32.59	20.05–98	90	323	Olli et al. (2002)
Barents Sea (PF)	73.47	31.38	26.05.98	90	340	Olli et al. (2002)
Barents Sea (AW)	72.29	30.56	28.05.98	90	301	Olli et al. (2002)
Arctic Ocean (n = 6)	88–89	-9.5–3.7	02–22.08.01	90	4000	Olli et al. (2007)
Ross Sea/W Polynya (n = 5)	–76.5	–177.6–165	24.12.95–12.01.96	200		Gowing et al. (2001)
Lazarev Sea (n = 6)	–68.2–68.35	0.5–0.9	28.12.94–08.01.95	120		Pakomov et al. (2002)/this study

Each station is identified with number of samplings for averaged data (n=), and decimal positions as latitude (Lat. N) and longitude (Long. E). South and western positions are given as negative co-ordinates. Dates of sediment trap retrieval and depths of deployments (m) are given, as well as station depths (m) where available. Data sources are given with references for published data. Some stations lack suspended data (SD) from the 0–100 m depth interval

Vertical flux of particulate organic carbon (POC) and phytoplankton was measured with two parallel cylindrical sediment traps (aspect ratio 6.25), mounted in a gimballed frame equipped with a vane (KC maskiner og laboratorieudstyr, Denmark) (e.g., Reigstad et al. 2000; Olli et al. 2002). No baffles were used. The traps have been calibrated with the ^{234}Thorium technique with close to 100% trapping efficiency (Coppola et al. 2002; Lalande 2006). Sampling depths varied for the different investigations according to Table 1, and sampling time were in general about 24 h. The mooring was drifting in a Lagrangian mode, balanced with subsurface and surface buoys or anchored to ice (Barents Sea). During the seasonal study in Balsfjord, 1996, moored traps with duplicate set of cylinders were deployed for periods of 3–14 days, one set with Lugol–glutaraldehyde mix (Rousseau et al. 1990) for phytoplankton enumeration, and the other set with formaldehyde (all other analyses). Fixatives were injected with 400 ml NaCl-enriched water (40 psu) to the bottom of the sediment trap cylinders that were pre-filled with filtered seawater (2% final fixative concentration).

Analysis and calculations

Aliquots of 50–800 ml were taken for POC analyses and filtered onto pre-burned Whatman GF/F filters, frozen until analysis on a Leeman Lab 440 CHN analyser. For further details, see Reigstad et al. (2000). Total particulate matter (TPM) were filtered onto pre-weighed and pre-burned Whatman GF/F filters (50–500 ml) and analysed according to Wassmann (1991).

Phytoplankton was counted using a combination of methods (see further details in Wassmann et al. 2005). A standard light microscope, furnished with a counting stage (Semina 1978) was used. The most abundant nanoplankton algae (2–20 μm) were counted in Fuchs-Rosenthal chambers with magnification of 400×. After the smaller phytoplankton was enumerated, the sample was slowly decanted through a glass tube covered with two layers of fine-mesh nylon gauze to a 5–10 ml concentrated sample. After gentle mixing, a subsample was transferred to a 0.05 ml. chamber. Cells were counted under magnification of 200×. In order to count rare (usually larger) forms the whole sample (100–250 ml) was reduced to 1 ml by settling in a 1.0 ml chamber. These methods allowed also counting of non-sinking cells.

Biovolumes of individual cells were calculated from linear dimensions of measured cells applied to appropriate stereometric formulae (Smayda 1978). Carbon conversion factors established by Menden-Deuer and Lessard (2000) were used to estimate carbon content for diatoms (pg C $cell^{-1} = 0{,}288 \times vol^{0.811}$), and for non-diatoms (incl. dinoflagellates, flagellates, loricate ciliates) (pg C $cell^{-1} = 0{,}216 \times vol^{0.939}$). For *Phaeocystis pouchetii*, the estimated carbon content of 2.41 pg C (cell d = 3 μm) and 16.3 pg C (cell d = 6 μm) is comparable to measured rates of cell carbon for *P. globosa* and *P. antarctica* (10.8 ± 3.47 and 3.33 pg C $cell^{-1}$, respectively) as given by Schoemann et al. (2005, and references therein).

In the seasonal Balsfjord study, subsamples for transparent exopolymer particles (TEP) analysis (n = 5) were taken from the formaldehyde preserved cylinder (5–100 ml). TEP was gently filtered onto 0.4 μm membrane filters and stained with Alcian blue (Sigma A5268) according to Passow and Alldredge (1995). The samples were stored frozen, dissolved in 80% H_2SO_4 and measured spectrophotometrically at 787 nm. The amount of TEP is expressed as gum xanthan equivalents (mg gum xanthan equiv. m^{-2} d^{-1}) based on calibration with gum xanthan (Sigma 1253).

Daily loss rates were calculated according to the following equation:

$$\text{Loss rate } (\%d^{-1}) = (A/B) * 100$$

where B is the suspended concentration over an integrated depth (0–100 m, mg m^{-2}), and A is the flux at 100 m (mg m^{-2} d^{-1}). The sinking speed is estimated as

Sinking speed (m d^{-1}) = integrated depth (m) × loss rate (% d^{-1})/100

Results and discussion

How to define vertical export of *Phaeocystis*-derived matter?

An ambiguous use of the term export has lead to an impression of considerable deep vertical export following *Phaeocystis* blooms based on high export rates out of the euphotic zone in the Barents Sea (Wassmann et al. 1990; Wassmann 1994). However, high export out of the euphotic zone does not necessarily imply that the *Phaeocystis* spp.-derived material is exported to the sediments or deeper ocean regions as strong mesopelagic remineralisation of *Phaeocystis* spp. can take place (Wassmann 1994; Riebesell et al. 1995; Wassmann et al. 2003). Three stations from three distinct regions in the Barents Sea (e.g., the Atlantic-dominated region in southwest, the well-mixed Polar Front and Arctic water) illustrate how vertical fluxes of *Phaeocystis pouchetii* cells decline with increasing depth and in relation to the density profile (Fig. 1). The decline is not confined to any fixed depth, but strongly related to the depth of the upper mixed layer which is shallow and distinct in the strongly-stratified Arctic water, and deeper in the less-stratified Atlantic water more exposed to vertical mixing (Olli et al. 2002; Reigstad et al. 2002). The vertical flux attenuation is sharp and confined to depth between 30 m and 40 m in the Arctic waters (Fig. 1). In the less-stratified, deeper-mixed Polar Front and Atlantic water (Reigstad et al. 2002) algae are mixed deeper and the flux attenuation is weaker and continues down to 90 m depth (Fig. 1).

It is therefore important to differentiate between: (1) export of *Phaeocystis* spp. cells out of the euphotic zone as a part of a bloom termination, and (2) export of *Phaeocystis*-derived material injected into the deeper regions and benthic boundary layer (BNL) as a part of biogeochemical cycling. For the termination of a bloom it can be critical that the biomass is exported below the euphotic zone or upper-mixed layer preventing sufficient light for further growth. As a part of the larger-scale biogeochemical cycling and carbon sequestration the *Phaeocystis*-derived material must enter the deeper waters or BNL. Mass sedimentation of *Phaeocystis pouchetii* was observed at 50 m depth in the Barents Sea (Wassmann et al. 1990), but a strong decline in the vertical flux resulted in that only a fraction of this material was delivered to or below 100 m depth (Wassmann 1994). A similar pattern was encountered in a seasonal study in Balsfjord, northern Norway, in 1996, where <10% of the *P. pouchetii* cell flux at 40 m reached 100 m depth (Fig. 2). Similar observations of strong reductions in *P. pouchetii* cell flux in the 40–60 m depth interval have also been made in the Barents Sea by Andreassen and Wassmann (1998), and repeatedly in North Norwegian fjords (Lutter et al. 1989; Riebesell et al. 1995; Reigstad et al. 2000). A high vertical resolution of sediment traps in the upper 200 m of the pelagic zone is therefore indispensable. It is a precondition to

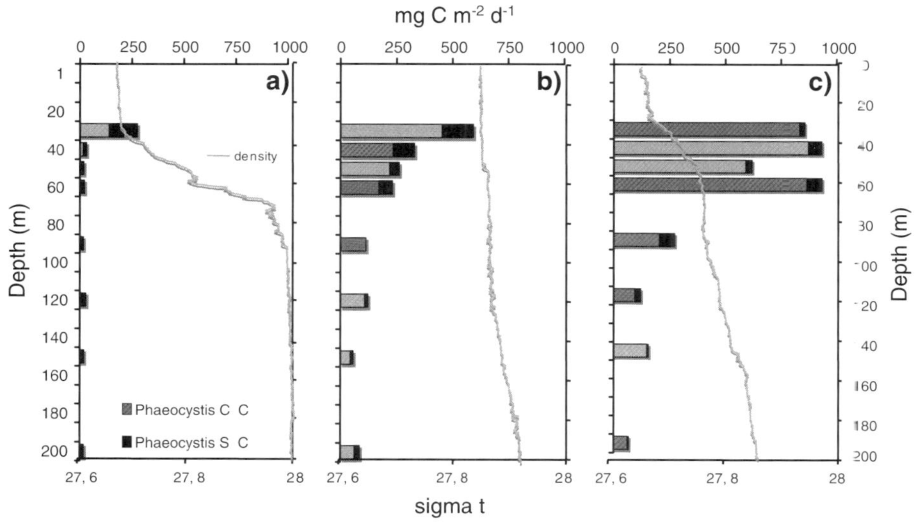

Fig. 1 Vertical pattern of *Phaeocystis pouchetii* export (mg C m^{-2} d^{-1}) in different water masses in the central Barents Sea (**a**) Arctic Water, (**b**) Polar Front and (**c**) Atlantic water measured by sediment traps at 30, 40, 50, 60, 90, 120, 150 and 200 m depth in May 1998 (modified from Olli et al. 2002). Sigma t for the water mass of the investigated stations is given *Phaeocystis* export is differentiated into colonical cells (CS) and single cells (SC)

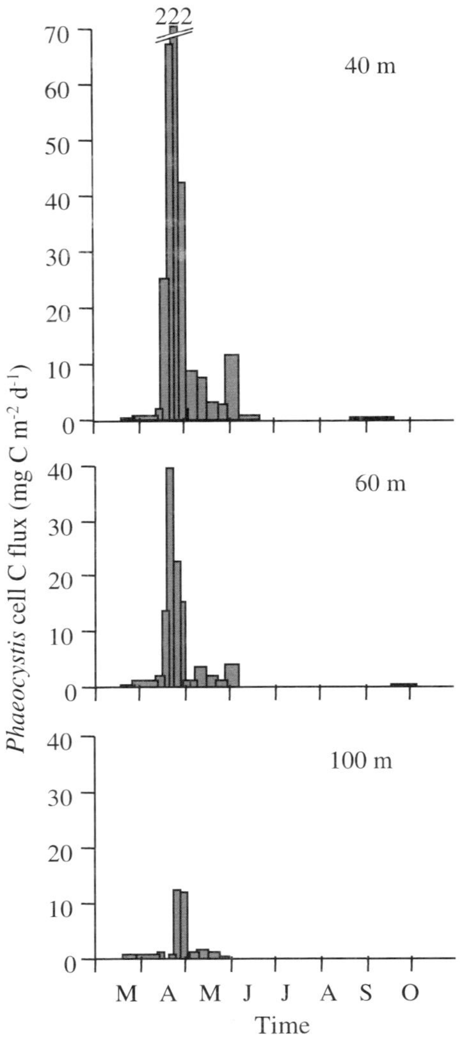

Fig. 2 Vertical flux of *Phaeocystis pouchetii* from March to October, 1996, in Balsfjord, northern Norway. Sediment traps were deployed at 40, 60 and 100 m depth, and the *P. pouchetii* cell carbon flux is given as mg C m^{-2} d^{-1}

evaluate how significantly phytoplankton and organic matter contributes to the recirculation and vertical export of bioelements.

The data compiled and presented below concern primarily *P. pouchetii* due to the abundance of sediment trap and microscopy data from the northern region, which is dominated by this species (Table 1). Limited numbers of data sets are available from regions dominated by *P. globosa* and *P. antarctica*. We will focus on the vertical export below the flux attenuation interval in the upper 50–100 m where retention, recycling, repackaging, grazing, remineralisation and aggregation shape the vertical flux profile according to ecosystem structure and environmental factors (Wassmann et al. 2003).

The morphological challenge: cells and colonies

There are several challenges in quantifying vertical flux of *Phaeocystis*-derived carbon. The quantification of exported cells and colonies and the conversion to carbon are two of the most prominent ones.

The presence of smaller and larger free-living flagellated stages (2–8 μm) as well as nonflagellated colonial cells within colonies of different size, morphology and fragility requires a dedicated microscopical approach combining several methods (Rousseau et al. 1990; Mathot et al. 2000; Gowing et al. 2001; Wassmann et al. 2005). Long-term deployments of sediment traps using formaldehyde, which does not adequately preserve flagellates, leads to an underestimation of *Phaeocystis* cells from colonies (T. Ratkova, pers. com.). Identification of single versus colonial cells can also be difficult due to loss of flagellae in formaldehyde-fixed samples. Use of short-term sediment traps without fixatives and the Lugol–glutaraldehyde cocktail recommended by Rousseau et al. (1990) improves cell and colony preservation. The fragility of *P. pouchetii* colonies combined with the handling of samples and the storage time prior to analysis does, however, result in an underestimation of size and number of colonies due to fragmentation and dissolution that is difficult to quantify exactly (Wassmann et al. 2005).

Quantification of *Phaeocystis*-derived carbon further requires appropriate carbon conversion factors for single cells and colonies in healthy as well as senescent conditions for all *Phaeocystis* species. Relationships between colony size, cell number and carbon have been established and published for *P. globosa* (Rousseau et al. 1990; Rijssel et al. 1997) and *P. antarctica* (Mathot et al. 2000) as described by Schoemann et al. (2005). Such relationships have not been established for *P. pouchetii*, and the variable colony morphology makes it problematic to apply carbon conversion factors established for other species.

At present, we are not able to meet all the requirements for a proper quantification of the carbon content of *P. pouchetii* colonies. Neither can we assure that the vertical fluxes of *P. pouchetii* colonies are reliable, as fragmentation of colonies can overestimate the colony vertical flux in terms of number. Similarly, dissolution of colonies may lead to an underestimation of the colony number and size. Through tedious microscopical analysis we do, however, have reliable measurements of depth-dependent vertical fluxes of *Phaeocystis* spp. cells (Figs. 1, 2). To provide a preliminary determination of the potential contribution of *Phaeocystis* spp. carbon to the vertical POC export at 100 m, we select available carbon conversion rates for other *Phaeocystis* spp. species.

The relative importance of *Phaeocystis* spp. cell carbon to vertical POC export

Few studies exist where mass fluxes of *Phaeocystis* cell carbon have been quantified along with POC export. We compiled the existing data sets to evaluate the contribution of *Phaeocystis* spp. cell carbon to vertical carbon flux. *Phaeocystis pouchetii* blooms are regularly observed in North Norwegian fjords (Heimdal 1974; Eilertsen et al. 1981; Riebesell et al. 1995; Reigstad et al. 2000; Wassmann et al. 2005). Due to the close vicinity of Tromsø, many years of observations from the water column as well as from sediment trap measurements (without fixatives) are available from these localities.

Three examples with focus on the vertical flux attenuation or geographical range are provided below. First we give a specific example on loss rates and the relative contribution of *P. pouchetii* to POC export in fjords. Second, the relative ratio of *P. pouchetii* cell carbon to diatom carbon is compared suspended and in the contemporary vertical export at 40 and 100 m depth, to point to the higher vertical flux attenuation of *P. pouchetii* compared to diatoms. As a third example, available published and own data quantifying the contribution of *Phaeocystis* spp. cells to vertical carbon flux at 100 m depth is presented and compared.

Although a *Phaeocystis* spp. colony bloom may be highly recognisable in surface waters, its contribution to the vertical export at 100 m in terms of cells carbon turns out to be minor. During a three-week field period in three North Norwegian fjords in spring 2001, nine 24 h sediment trap deployments (without fixatives) were completed (Tables 1, 2). The integrated suspended biomass of *Phaeocystis pouchetii* (presented as cell carbon) displayed an average that varied by a factor of three between the fjords, but cell carbon loss rates to 100 m depth were comparable and low, ranging from 0.2% to 2.1% (Table 2). The taxonomic composition of the exported phytoplankton carbon (PPC) showed a dominance of diatoms and diatom spores (Fig. 3a), despite the abundance of *P. pouchetii* in the suspended biomass in the Ullsfjord and Malangen fjords (Table 2, Wassmann et al. 2005). Among the *P. pouchetii* cells identified in the sediment trap at 100 m, ~97% were flagellated, about 6 μm in diameter, and categorised as single cells (Fig. 3b).

This feature of stronger attenuation of *P. pouchetii* cells compared to diatoms is prominent. The disappearance of *Phaeocystis* spp. in the upper aphotic zone flux attenuation interval is illustrated by the change in the *Phaeocystis*/diatom ratio (cell C/cell C) in the integrated suspended versus sediment trap material from 40 m to 100 m depth, in North Norwegian fjords, the Barents Sea and Arctic Ocean (Table 1, Fig. 4). Ratios of >1 reflect *Phaeocystis* spp. dominance, and ratios of <1 reflect diatom dominance. At 40 m depth, the exported particulate material reflects the integrated suspended *Phaeocystis*-diatom composition. A suspended *Phaeocystis* spp. dominance at 11 out of 26 stations examined, resulted in *Phaeocystis* spp. dominated vertical export at six stations at 40 m depth (Fig. 4a). At 100 m depth, *Phaeocystis* spp. dominated the integrated suspended composition at 12 out of 19 stations, while the *Phaeocystis* signal disappeared and the exported material at 100 m were dominated by diatom C at 17 stations (Fig. 4b). Only at two stations *Phaeocystis* spp. cell C dominated the vertical export (ratios > 1)

Table 2 Daily loss rates of *Phaeocystis pouchetii* cell carbon (C) from the integrated standing stock (0–100 m) to sediment trap at 100 m depth in three North Norwegian fjords, April, 2001

Phaeocystis pouchetii	Malangen	Balsfjord	Ullsfjord
Suspended biomass (mg C m^{-2})	2,897 ± 447	1,337 ± 494	3,727 ± 1,292
Vertical C flux (mg C m^{-2} d^{-1})	29 ± 9	13 ± 7	21 ± 7
Loss rates (%)	1.0 ± 0.4	1.2 ± 0.9	0.6 ± 0.4
Sinking rate (cm h^{-1})	4.3	5.0	2.7

Average ± standard deviation of three sampling events.

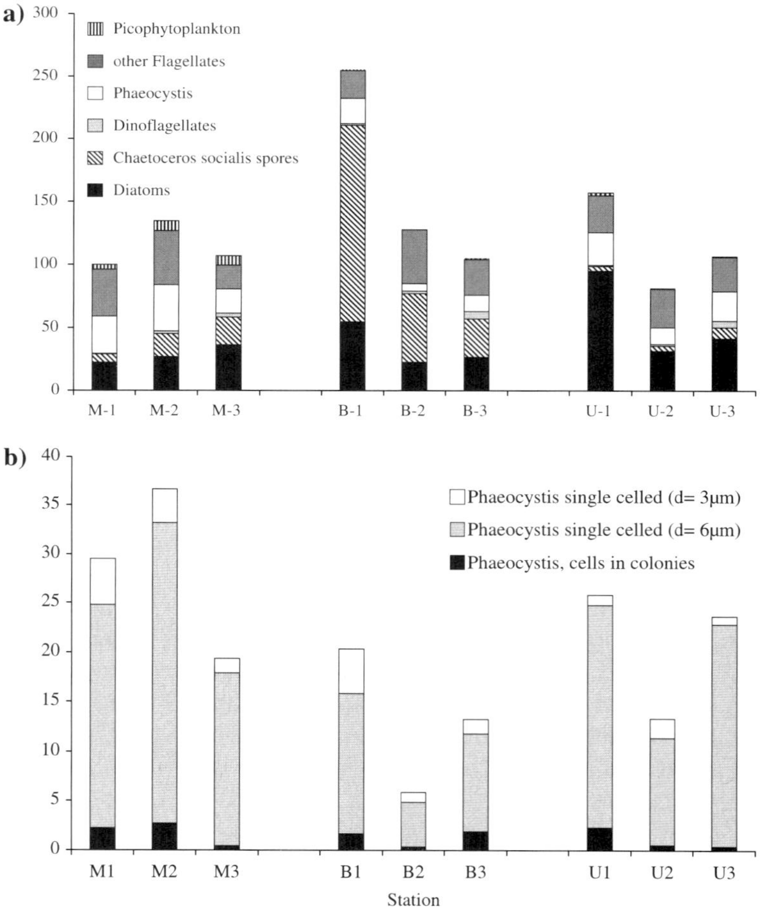

Fig. 3 (**a**) Carbon fluxes attributed to pico-, nano- and microplankton (e.g., diatoms, *Chaetoceros socialis* spores, dinoflagellates, *Phaeocystis pouchetii* cells, other flagellates and picophytoplankton) and (**b**) the carbon flux of *Phaeocystis pouchetii* cells shown as single cell (d = 3 and d = 6 μm) and colonial cells. Fluxes were measured using short-term sediment traps at 100 m depth in three North Norwegian fjords, Malangen (M), Balsfjord (B) and Ullsfjord (U), during three 24 h sampling events in each fjord in April, 2001

at 100 m depth. They were characterised by deep mixing >90 m depth in the Polar Front and the Atlantic region of the Barents Sea (Fig. 1b, c). Even if the integrated suspended biomass of *Phaeocystis* spp. was up to five times that of diatoms, the vertical export of diatoms exceeded the *Phaeocystis* export at all other stations.

A few published studies exist where *Phaeocystis* PPC as a fraction of POC has been quantified (Reigstad et al. 2000; Gowing et al. 2001; Olli et al. 2002; Pakhomov et al. 2002). Compiling these data with some previously unpublished data from investigations carried out by Wassmann and co-workers in Tromsø, Norway, provides

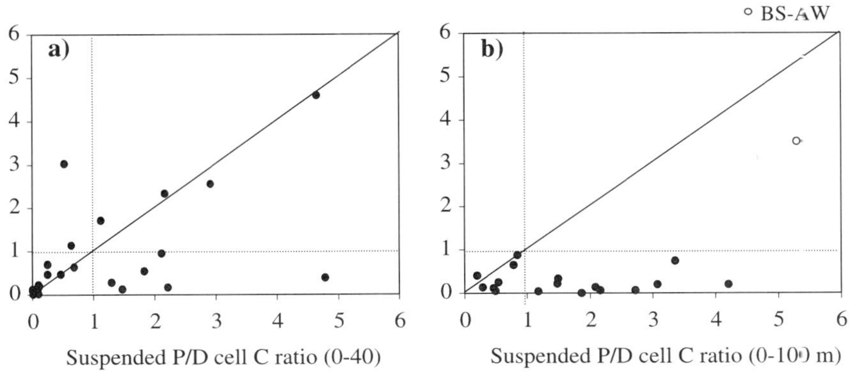

Fig. 4 *Phaeocystis pouchetii*/diatom (P/D) cell carbon ratio is compared suspended (integrated mg C m^{-2}, 0–40 and 0–100 m) and in the corresponding sediment trap material (mg C m^{-2} d^{-1}) at (**a**) 40 m and (**b**) 100 m depth during spring bloom events. The 40 m data includes measurements from North Norwegian fjords (Balsfjord in 1996 and Balsfjord, Ullsfjord and Malangen 1997, n = 26), and the 100 m data includes measurements from Balsfjord 1996, Ullsfjord 1997, Balsfjord, Malangen and Ullsfjord 2001, Barents Sea 1998 and Amundsen Basin, Arctic Ocean 2001 (n = 19) (for further details, see Table 1). A ratio >1 reflects *Phaeocystis* dominance, while a ratio <1 reflects diatom dominance. Open circles indicate stations were mixing exceeded 90 m at the Barents Sea Polar front and Atlantic water (BS-AW) region in May 1998 (for BS-AW; P/D cells C ratio = 43)

information on the contribution from *Phaeocystis* cell carbon from 15 different ecologically settings in five regions (Antarctica, coastal North Sea, North Norwegian fjords, the Barents Sea and the central Arctic Ocean; Fig. 5). Except for Balsfjord –96, all rates were obtained by short-term (24 h) sediment traps, deployed at or close to 100 m depth (± 10 m). In the shallow North Sea, sediment trap sampling was only possible at 20 m and 30 m. The vertical fluxes presented are averages of several measurements (68 observations in total) covering periods during and closely after *Phaeocystis* spp. blooms (see Table 1 and Fig. 5 legend for details).

The averaged, daily POC fluxes are highly variable between the different regions, with rates on the North Sea shelf and the central Arctic Ocean representing the extremes of 1,500 and 50 mg C m^{-2} d^{-1}, respectively. The highest and lowest PPC fluxes were also measured in these two regions (430 and 12 mg C m^{-2} d^{-1}, respectively). The polar region exhibited POC fluxes that were a factor of 10 or more below the North Sea. Within the Barents Sea, there was a distinct decrease in POC fluxes from the station in the deeper-mixed and more-productive Atlantic region, to the Polar Front, and the more-stratified station in the ice-covered Arctic region. In the North Sea, the <40 m depth obviously played an important role for the delivery of high POC fluxes 10 m above the seafloor. The flux gradient in the Barents Sea is influenced by vertical mixing in frontal zones and less-stratified waters that fertilises primary production and increase sinking speed through active vertical transport of organic carbon and phytoplankton (Olli et al. 2002; Reigstad et al. 2002). The vertical export of POC can therefore be relatively higher in shallow regions like the North Sea and vertically well-mixed regions such as the central Barents Sea, where retention time and time available for degradation is shortened.

Phaeocystis spp. was present at all stations (Fig. 5), with a *Phaeocystis*/diatom ratio (cell C/cell C) of 0.2–5.3. Despite the differences, two aspects were common for the investigated sites. First, PPC never exceeded 45% of POC flux at ~100 m depth, with an average of 26% (7–42%). This implies that a large phytoplankton-derived carbon is partly degraded before it reaches depths of 100 m. Second, the contribution from exported *Phaeocystis* spp. cell C to the POC export was small, ranging from 0.7–7%, with an average of 3% for all stations. The two deep-mixed stations made an exception with 15 and 35% *P. pouchetii* cell C of POC flux at 90 m depth, increasing the overall average to 6%. Comparatively, the diatom carbon contributed 1–27% to the 100 m measured POC flux, with an average of 11%.

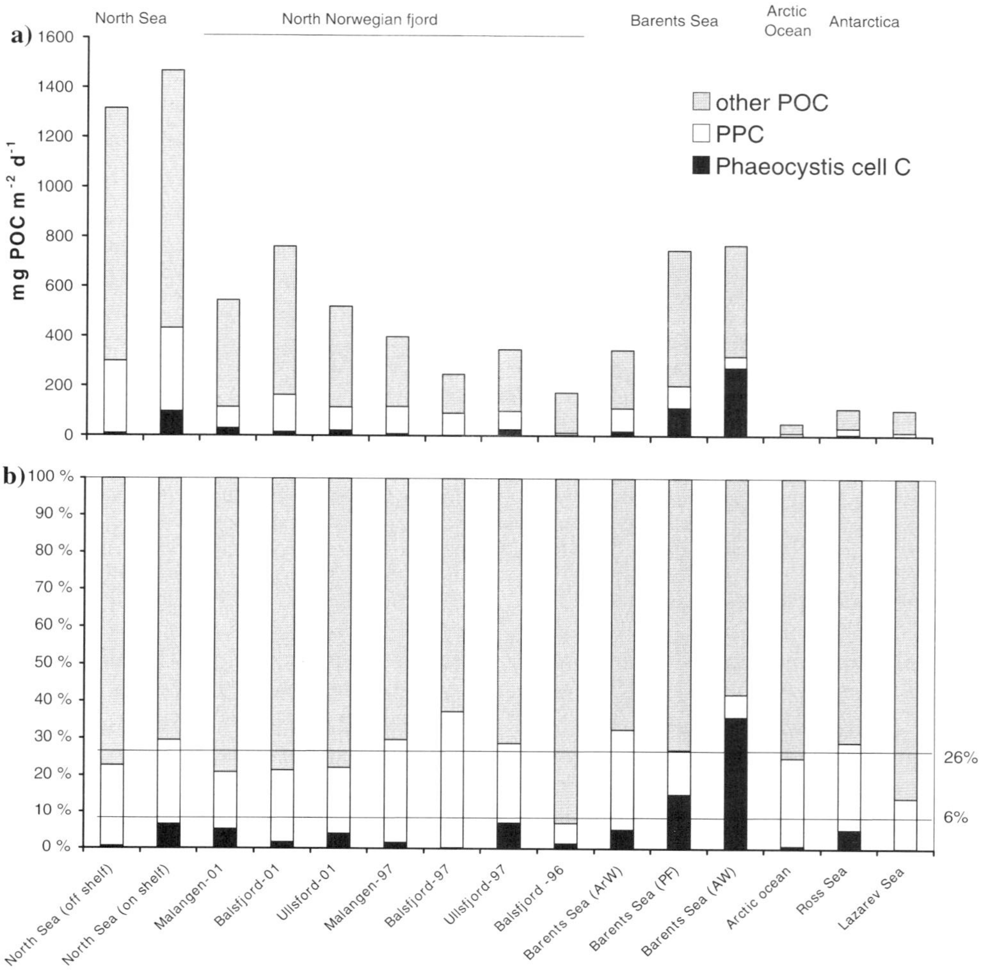

Fig. 5 (**a**) Vertical export of particulate organic carbon (POC) and (**b**) the percent contribution of phytoplankton carbon (PPC) and *Phaeocystis* spp. cell carbon at 100 m depth in different regions; on and off the shelf of the North Sea, April-98 ($n = 6$ and $n = 7$), Malangen, Balsfjord and Ullsfjord, April-01 ($n = 9$), Malangen, Balsfjord and Ullsfjord, April-97 ($n = 15$, Reigstad et al. 2000), Balsfjord 1996 ($n = 11$, Reigstad 2000), Atlantic water (AW), Polar Front (PF) and Arctic water (ArW) regions of the Barents Sea, May-98 ($n = 1$ for each region, Olli et al. 2002), Arctic Ocean, Aug-01 ($n = 6$, Olli et al. 2007), Ross Sea, December-95/January-96 ($n = 5$, Gowing et al. 2001), Lazarev Sea, December-95/January-95 ($n = 6$, Pakhomov et al. 2002). The number of samplings averaged in each region is given in parenthesis after each station name. For further information on geographical position and sampling dates, see Table 1. Data are obtained through microscopic examination of short-term sediment trap material, and conversion to carbon

Despite occasionally high concentrations and dominance of *Phaeocystis* spp. in the water column, calculations based on sediment-trap measurements suggest low daily loss rates, implying high retention of *Phaeocystis* cell carbon in the upper 50–100 m. Unless deep mixing accelerates vertical export, the contribution to the vertical carbon flux at 100 m is on average 3%. The conclusion is that *Phaeocystis* cell carbon does not contribute significantly to vertical carbon export.

Does mucus contribute significantly to the vertical export of *Phaeocystis* spp-derived C?

Mucus carbon estimates

The above calculations were provided without taking the contribution from *Phaeocystis* spp. colony mucus into account. Rousseau et al. (1990) established a relationship for the number of cells per colony of *P. globosa*, and estimated a carbon content per volume of mucilaginous matter of 335 ng C mm^{-3}. The authors concluded that, when the colony size exceeded 400 μm, mucus carbon dominated the biomass, and at colony sizes of 1 mm, mucus contributed 90% of the *Phaeocystis* colony biomass. Their assumption was a colony filled with mucus structures. Rijssel et al. (1997) challenged this assumption, discovering that the total amounts of sugars and carbon were correlated with the colony surface area for *P. globosa*, indicating a hollow colony structure with a mucus layer of fixed thickness. The authors argued that the estimates by Rousseau et al. (1990) over- and underestimated the mucus carbon pool for larger and smaller colonies, respectively. Rijssel et al. (1997) measured C content including mucus of 57 pg C cell^{-1} for field samples and 122 pg C cell^{-1} for cultures. For *P. antarctica*, another cell-per-colony ratio and a carbon-per-unit-colony volume of 213 ng C mm^{-3} was suggested by Mathot et al. (2000) based on investigations from the Ross Sea, Antarctica.

A relationship between cells per colony or colony carbon content has so far not been established for *Phaeocystis pouchetii* (Schoemann et al. 2005). This species has lobed-formed colonies where the cells are grouped, contrasting the more spherical or elongated colonies characterising *P. globosa* and *P. antarctica*. There are also indications that *P. pouchetii* colonies are more fragile (Wassmann et al. 2005) compared to the tough structure demonstrated for *P. globosa* (Hamm et al. 1999). It is therefore likely that a difference in the cell per colony ratio as well as carbon per unit of colony exists between *P. pouchetii* and the other two species. In an attempt to suggest a possible range of contributions from mucus to vertical carbon export based on vertical export rates of *P. pouchetii* when the mucus sinks as intact colonies with cells we have chosen to apply the relationships for *P. globosa* and *P. antarctica*.

First, the range of carbon content in a colony is estimated for the smallest and largest observed colonies in the dataset from the central Barents Sea, and two different cell sizes using the available relationships (Table 3). Estimates of the relative contribution from mucus C to the total *Phaeocystis* colony carbon based on the relationships given by Rousseau et al. (1990) and Mathot et al. (2000), show that they indicate very similar mucus contributions, adding a maximum of 43–48%, for the largest colonies with small cells (d = 3 μm) (Table 3) and 10–12% for similar-sized colonies but larger cells (d = 6 μm). As colonies often contain larger cells, the addition of mucus carbon is most likely in the order of 10%, and of minor importance compared to the cell carbon for the total colony carbon content. The relationship suggested by Rijssel et al. (1997) does not allow a separation of cell and mucus carbon, but for the entire colony including cells, the estimated carbon content ranges from the upper levels suggested by the two others, to three times their highest suggested value (Table 3). The reason for this high discrepancy is likely our use of the relatively small colonies (65–115 μm) present in the preserved samples examined from the Barents Sea. Rijssel et al. (1997) based their relationship on larger colonies ranging >250–5,000 μm in diameter.

Estimated mucus contribution to carbon export

To approximate the potential impact of the estimated mucus contribution to vertical flux, maximum values of colony flux and observed colony size were applied to give an upper range (Table 4). The *P. pouchetii* fluxes are from the Arctic region of the marginal ice zone in the Barents Sea during spring, 40 m depth and above the vertical mass flux attenuation interval. The estimate of *Phaeocystis* spp. carbon export based on relationships from Mathot et al. (2000) is 25% lower compared to Rousseau et al. (1990) with 168 mg C m^{-2} d^{-1} and 225 mg C m^{-2} d^{-1}, respectively (Table 4). The mucus contribution to these estimates is however only 10% and 12%, indicating that even with very high vertical fluxes of *Phaeocystis* spp. colonies, mucus will not add more than about 10% C to estimates based on cell carbon (given the observed colony diameter of 115 μm). The total POC export measured at the Barents Sea Arctic water station was about 900 mg C m^{-2} d^{-1} (Olli et al. 2002). The *Phaeocystis* spp. carbon export estimates based on Rijssel et al. (1997) exceeded this, and were 2–3 times higher compared to the others, using

Table 3 Estimated carbon content (C) of *Phaeocystis pouchetii* colonies separated as cells and mucus carbon. Estimates are based on relationships established for living *P. globosa* (Rijssel et al. 1997; Rousseau et al. 1990) and *P. antarctica* (Mathot et al. 2000), but using colony size, cell sizes and vertical fluxes measured for *P. pouchetii*. Maximum and minimum values are examplified including smaller and larger cells and colonies (diameter = *d*). Cell carbon is calculated according to Menden-Deuer and Lessard (2000)[a,b]

	Cell size (μm)	Calculations according to		
		Rousseau et al. (1990) (*P. globosa*)	Mathot et al. (2000) (*P. antarctica*)	Rijssel et al. (1997) (*P. globosa*)
Colony size (μm)		65–115	65–115	
# cells col^{-1}		51–122	33–93	33–122
P. pouchetii cell C (pg col^{-1})	$d = 3$ [a]	122–294	79–224	
	$d = 6$ [b]	831–1,988	538–1,516	
P. pouchetii mucus C (pg C col^{-1})		48–267	31–170	
Total *P. pouchetii* C	$d = 3$	170–561	110–394	
(cell+mucus) pg C col^{-1}	$d = 6$	879–2,255	569–1,686	1,881–6,954
% mucus C col^{-1}	$d = 3$	24–48%	28–43%	
	$d = 6$	6–12%	5–10%	

Calculations based on Rijssel et al. (1997) assume a hollow colony structure and give C content per cell including colony mucus, where the given field data value of 57 pg C $cell^{-1}$ is applied. Number of cells col^{-1} used is the highest and lowest from the relationships given by Rousseau et al. (1990) and Mathot et al. (2000)

Notes: [a] Cell $d = 3$ μm gives 2.41 pg C $cell^{-1}$

[b] Cell $d = 6$ μm gives 16.3 pg C $cell^{-1}$

Table 4 Estimated vertical fluxes of *Phaeocystis pouchetii* cell carbon and mucus contribution based on maximum observed colony fluxes (40 m depth in the central Barents Sea, May 1998)

Vertical flux, Barents Sea, May, 40 m	Calculations according to		
	Rousseau et al. (1990) (*P. globosa*)	Mathot et al. (2000) (*P. antarctica*)	Rijssel et al. (1997)
100,000,000 col m^{-2} d^{-1}, Max colony size $d = 115$ μm, colony cell $d = 6$ μm			(93–122 cells col^{-1})
Vertical flux *P. pouchetii* colony incl. mucus	225 mg C m^{-2} d^{-1}	168 mg C m^{-2} d^{-1}	530–695 mg C m^{-2} d^{-1}
% Mucus	12%	10%	
Assuming larger colonies ($d = 400$ μm, 170 cells col^{-1}), (average from fresh samples, Balsfjord 2004)			(170 cells col^{-1})
Total *P. pouchetii* C vertical flux	390 mg C m^{-2} d^{-1}	349 mg C m^{-2} d^{-1}	969 mg C m^{-2} d^{-1}
Underestimate of *Phaeocystis* contribution without mucus C	40%	25%	

Calculation of mucus according to relationships suggested for *P. globosa* (Rijssel et al. 1997; Rousseau et al. 1990) and *P. antarctica* (Mathot et al. 2000). The observed colonies were small, and likely fragmented larger ones. The shallow depth and probably overestimated settling colony number makes this a maximum estimate to illustrate the potential extra carbon contribution from *Phaeocystis* colony mucus. An example using colony size and cell col^{-1} measured from Balsfjord on fresh samples (T. Ratkova. unpublished) is included to illustrate the difference of larger colonies (similar vertical flux used in both examples). The carbon content of cells, including mucus, given by Rijssel et al. (1997) does not allow an estimate of the mucus fraction the total *Phaeocystis* carbon

the maximum cell $colony^{-1}$ estimate based on Rousseau et al. (1990) and Mathot et al. (2000) of 122 and 93, respectively. Again, the small colony size is probably the reason.

Since there is a possibility of colony fragmentation (leading to colony size underestimation) caused by preservation and transportation prior to microscopic analyses, an example based on average colony diameter of 400 μm and 170 cells col^{-1} from fresh *P. pouchetii* samples in Balsfjord, Northern Norway, is included (Table 4). Larger colonies increased the estimated contribution from colony mucus, and the assumed underestimation of *Phaeocystis* carbon export using only cell carbon is estimated to be 40% and 25% using the relationships by Rousseau et al. (1990) and Mathot et al. (2000), respectively. The comparative estimate based on Rijssel et al. (1997) is again almost tripled compared to the other estimates (Table 4), exceeding the total measured POC fluxes. We therefore assume that for smaller colonies and/or for *P. pouchetii*, this relationship overestimates the carbon values. Gowing et al. (2001) made calculations on the contribution from *P. antarctica* mucus C in the Ross Sea using the number of flagellated cells in the sediment trap, and cell per colony ratios and carbon per colony volume, according to relationships established by Mathot et al. (2000). They concluded that, while the *P. antarctica* cell carbon contributed 1–13% of the POC export, mucus contributed only 0.5–4%.

The potential underestimation of *Phaeocystis* spp. carbon by only including cell carbon is thus not likely to be >40%. Based on measurements of vertical flux around 100 m depth from Antarctica, the coastal North Sea, North Norwegian fjords, the Barents Sea and the Arctic Ocean (Fig. 5), the average carbon contribution from *Phaeocystis* spp. to the POC flux will increase from 3% to <5% by including mucus carbon. Despite the lack of appropriate species-specific carbon conversion rates for all *Phaeocystis* species, the evidence clearly indicates that *Phaeocystis* spp. mucus, sinking as intact colonies with cells, is not significantly contributing to the total POC export.

Fate of Phaeocystis colonies visualised through TEP

Aggregation is hypothesised to be important for vertical export of *Phaeocystis* spp.-derived material (Asper and Smith 2003). During experiments testing the aggregation potential of senescent *P. pouchetii* colonies, increased stickiness was not observed, but rather a lower aggregation potential was found compared to diatoms (Passow and Wassmann 1994). They hypothesised that mucus flocks derived from the dissolved colonial matrix could represent a secondary pathway for *Phaeocystis* spp.-derived carbon export. This mechanism was investigated in Balsfjord, northern Norway through a vertical flux study of transparent exopolymer particles (TEP).

TEP are formed from acid polysaccharides and stainable with Alcian blue (Alldredge et al. 1993). TEP can be formed from disintegrating senescent *Phaeocystis* spp. colonies (Passow and Wassmann 1994; Hong et al. 1997; Passow 2000; Passow 2002), and are suggested to induce mass flux events. Riebesell et al. (1995) observed continued high vertical carbon export associated with increased C/N and C/Chl *a* ratios and decreased surface-water TEP concentrations after a *P. pouchetii* bloom in Balsfjord, Northern Norway. The vertical fluxes of *P. pouchetii* cells were low, but export of *Phaeocystis*-derived carbon in the form of mucilageneous matter was suggested. Our seasonal study in Balsfjord, 1996, revealed that TEP could be closely associated with vertical export events. The maximum suspended distribution of TEP (1,415 mg xanthan equiv. m^{-3}) coincided with the termination of the *P. pouchetii* bloom. TEP were measured in sediment traps along with total particulate matter (TPM) and POC, and a sudden increase in vertical fluxes of TPM and POC (93–440 mg POC m^{-2} d^{-1}) in April, were closely associated with an increase in the flux of TEP from 95 to 400 mg xanthan equiv. m^{-2} d^{-1}, at 100 m depth (Fig. 6).

A conversion of TEP to carbon using the conversion factor established by Engel and Passow (2001; Passow 2002) reveals a TEP-C export maximum of 390 mg C m^{-2} d^{-1} at 100 m, exceeding the POC mass flux estimate of 350 mg POC m^{-2} d^{-1}. This is not impossible as TEP are retained on 0.4 μm filters, and will include a carbon size fraction that will pass through the 0.7 μm GF/F filters used for POC determination. The observations emphasize that, during transient particle flux events, a significant part of the exported carbon signal could be *Phaeocystis* spp.-derived carbon. This contribution is not detected as pigments or cells in the microscopy, and perhaps not even picked up

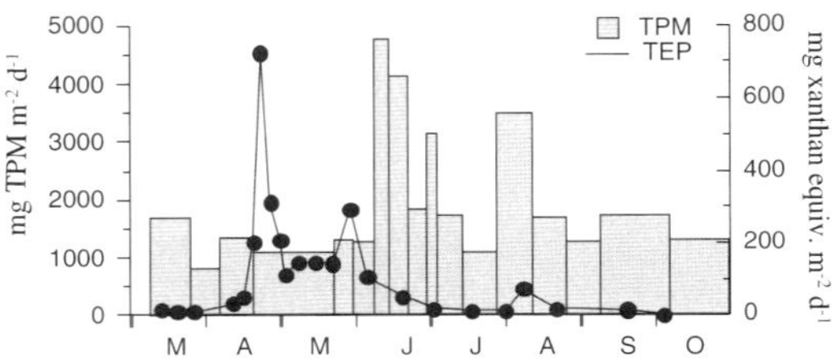

Fig. 6 Episodic vertical flux events of total particulate material (TPM) and particulate organic carbon (POC) at 100 m depth induced by *P. pouchetii*-derived material quantified as transparent polymer particles (TEP, in terms of gum xanthan equivalents) in Balsfjord, northern Norway, from March to October, 1996. Vertical flux is given as mg m^{-2} d^{-1}.

correctly by conventional POC measurements. Microscopic inspection of stained TEP from our sediment traps in Balsfjord showed that they were associated with phytoplankton cells and particles (data not shown). A second, but smaller TEP export event in late May, was not associated with any increased POC or TPM export (Fig. 6), but induced increased vertical flux of flagellates (data not shown).

Mari et al. (2005) found that nutrient limitation played a critical role for the TEP characteristics during a *P. globosa* bloom in a mesocosm experiment. While N-limitation induced a slow degradation of colonies to non-sticky mucus particles (exposed to microbial degradation and recycling), P-limitation induced rapid colony disruption resulting in heavy and sticky mucus aggregates (suggested to sink rapidly). We do not know whether the *P. pouchetii* blooms in Balsfjord were N or P limited, and we are not able to link the different fate of TEP observed in Balsfjord to the mechanisms suggested by Mari et al. (2005). Still, the fate of a *Phaeocystis* spp. bloom and the *Phaeocystis*-derived mucus carbon could strongly depend on the nutrient regime. A mass export event, like the one observed in April in Balsfjord, is thus a possible result of bloom termination induced by sticky TEP. Export of *Phaeocystis*-derived carbon and associated chemical properties like dimethylsulfide (DMS) can thus be exported in the form of TEP without being retained on the filters or visible under microscope without staining.

Mechanisms for vertical export

Having concluded that sinking colonies are not a major driver of vertical carbon export to deeper waters despite the major blooms observed, and that mucus measured as TEP only episodically induces flux events, we wish to focus also on additional mechanisms inducing vertical export of *Phaeocystis* spp.-derived carbon. Zooplankton grazing can repack cells and colonies to larger and faster-sinking faecal pellets (FP). Controversies exist regarding the impact of grazing on *Phaeocystis* spp. cells and colonies (Turner et al. 2002; Schoemann et al. 2005). With the wide size range of *Phaeocystis* morphotypes as well as potential grazers, and recent observations of cell-type-dependent grazing (Dutz and Koski 2006), grazing on *Phaeocystis* spp. becomes a complex task to investigate. In terms of vertical flux, it is mainly larger grazers that contribute through FP production, as smaller zooplankton FP are generally recycled (Turner 2002). Comparing FP production with FP fluxes suggests that 50–95% of the FP from larger copepods such as *Calanus* spp. are recycled in the upper layers rather than being exported (Urban-Rich et al. 1999; Wexels Riser et al. 2002). The resulting fraction of *Phaeocystis* spp.-derived carbon exported after mesozooplankton grazing and in the form of detritus can thus not be extensive.

Macrozooplankton producing larger and fast-sinking FP such as krill can be efficient FP exporters. Using combined fingerprint techniques (lipids and stereols), Hamm et al. (2001) argues that krill grazed *P. pouchetii* in Balsfjord, and contributed episodically to a considerably increase in the POC flux though *Phaeocystis*-containing FP export. The frequency of such episodes in northern regions is related to the distribution and patchiness of krill, which is variable and thus difficult to evaluate.

Downwelling of suspended biogenic matter that may contribute to vertical export is not a well-described process, although it may be of significant importance in certain regions and periods (Head and Horne 1993; Wassmann et al. 2000; Belviso et al. 2006). Mixing of surface water and denser coastal water at a fjord sill resulted in denser water containing fresh *P. pouchetii* cells being advected into the fjord basin (Wassmann et al. 2000). Increased chlorophyll *a* concentrations, fresh cells and DMS at the bottom of the fjord (180 m depth) did not result from vertical export, but rather from horizontal advection combined with mixing processes in bloom-containing surface water (Wassmann et al. 2000; Belviso et al. 2006). In a similar manner, fresh biogenic material and nutrient-depleted water were present at >100 m depth in the Polar Front and Atlantic water region of the Barents Sea, resulting from extensive mixing processes of non-stratified surface waters (Reigstad et al. 2002). This process was effective in bringing *P. pouchetii* cells well below the euphotic zone in the deeper-mixed Barents Sea stations (Olli et al. 2002; Ratkova and Wassmann 2002; Reigstad et al. 2002). Locally such processes might be of importance for vertical export of *Phaeocystis* spp. especially as it has been suggested that *Phaeocystis* spp. is favoured in regions with deeper and well-mixed waters (60–80 m) (Goffart et al. 2000; Lancelot et al. 1998; Sakshaug 2004).

Concluding remarks

The main body of evidence based upon direct observations suggests that *Phaeocystis* spp. blooms are to a large extent recycled in the upper layers. Contradictory evidence derives from indirect evidence, suggesting that sediment traps may underestimate vertical export of *Phaeocystis* spp. An alternative to the deployment of sediment traps, the ^{234}Th technique, has recently been applied in concert with sediment traps in the Barents Sea during the presence of *P. pouchetii* (Lalande 2006). Sediment-trap POC fluxes were much lower than large-volume POC fluxes at almost every station in summer 2005 (Lalande 2006). This may reflect either an under-collection by the drifting sediment traps or an over-collection by the large-volume ^{234}Th sampling. The offset between the two methods may be attributed to the prominent presence of *P. pouchetii in summer 2005*, which is potentially causing the large variation observed in POC/^{234}Th ratios (Lalande 2006). Due to the large proportion of carbon released by *P. pouchetii* and because *P. pouchetii* cell C does not contribute significantly to the vertical export of biogenic matter, the utilization of large-volume sampling of ^{234}Th may yield relatively high, and possibly incorrect, POC/^{234}Th ratios and hence POC fluxes in regions where *P. pouchetii* is dominant. Large-volume ^{234}Th sampling appears to be an inadequate alternative for vertical flux estimates in *Phaeocystis* spp.-dominated ecosystems. The use of sediment traps may be more reliable for the accurately measurement of the vertical export of biogenic matter in the Barents Sea (Lalande 2006).

The fate of *Phaeocystis* spp. blooms, and the possible contribution to carbon export have been investigated and discussed for several decades (Wassmann 1994; Lancelot et al. 1998; Schoemann et al. 2005; Whipple et al. 2005). Clearly, there is extraordinary flexibility and plasticity in the ecology, life cycle and fate of this genus, related to species, morphotypes and environmental conditions (Veldhuis and Wassmann 2005). Beside the many mechanisms and processes promoting retention and recycling of *Phaeocystis* spp.-derived carbon, there are also several pathways for vertical carbon export. An attempt to summarize the different pathways is presented in a conceptual diagram (Fig. 7). The evidence presented in this publication attempts to quantify or at least examplify these export mechanisms, and illustrate their regional and episodic significance. Considering the biomass and dominance of *Phaeocystis* in pelagic systems, the vertical export of cell carbon is of minor importance. This is based on the following findings:

- A strong decline in the vertical export of *P. pouchetii* cells between 40 m and 100 m has been shown in the Barents Sea and North Norwegian fjords.
- Despite potential *Phaeocystis* spp. cell C dominance in the water column, vertical export of diatom cell C was always larger than that *Phaeocystis* spp. cell C at 100 m depth, unless deep mixing short-circuits retention.
- *Phaeocystis* spp. cell carbon contributed on average only 3% to the vertical POC flux at 100 m depth.

Fig. 7 Conceptual diagram indicating possible pathways for *P. pouchetii*-derived carbon

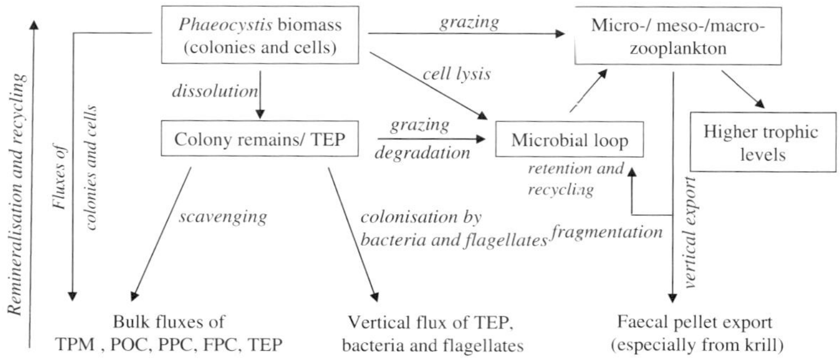

- The estimated contribution from mucus to the colony sizes observed in the north-eastern North Atlantic suggest that mucus may contribute an additional 40% (maximum) to the estimated cell C export. In concert, *Phaeocystis* spp. cell and colony C may thus contribute <5% of the vertical POC export.
- Possible grazing from krill inducing FP export, downwelling of *Phaeocystis* cell/colonies/remains through deep physical mixing processes or nutrient-related flocculation and export of senescent *Phaeocystis* spp. mucus may induce episodic *Phaeocystis* flux events.
- TEP measurements provides a method to estimate vertical export of *Phaeocystis*-derived mucus-remains that otherwise are easy to overlook based on their semiparticulate and transparent nature.

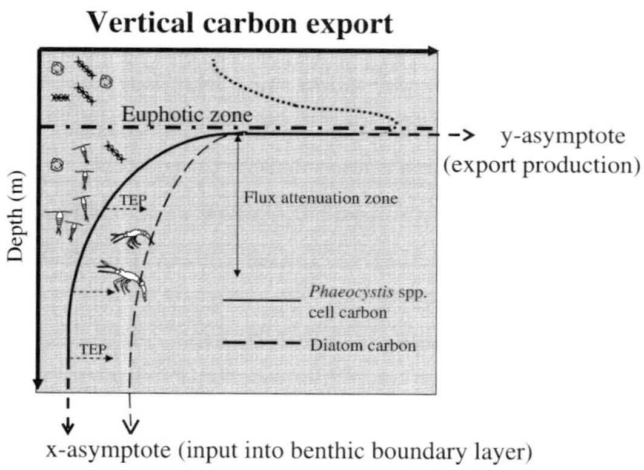

Fig. 8 Attenuation of vertical flux below the euphotic zone is stronger for *Phaeocystis* spp. cells compared to diatoms despite similar biomass in the upper mixed layer (modified from Wassmann et al. 2003). Export of transparent exopolymer particles (TEP) can reduce the attenuation of *Phaeocystis*-derived carbon

The increased vertical flux attenuation efficiency in *Phaeocystis* spp.-dominated regions has thus to be taken into account and the channelling of the *Phaeocystis* spp. biomass through the pelagic has to be carefully considered (Fig. 8). We suggest that there must be significant differences between the vertical export of *Phaeocystis* spp. and diatoms. The potential loss of TEP and mucilaginous matter on filters prior to analysis may result in an underestimation of the *Phaeocystis* spp.-associated C flux, and Fig. 8 indicates that the vertical flux attenuation efficiency during *Phaeocystis* spp. blooms may be smaller then the vertical export of cell carbon may suggest.

Acknowledgement The tedious microscopic effort by Tatjana Ratkova made this contribution possible, and is highly acknowledged. Contributions from colleagues through discussions and fieldwork over many years during the ESCAPE (EU-MAS3-CT96-0050) and BIOHAB (EU-EVK3-CT99-00015) projects, the fjord projects within North Norwegian Coastal ecology (NFR), ALV (NFR 121521/720), CABANERA (NFR 155936/700), MACESIZ (NFR 155945/700) and the Arctic Ocean Expedition-2001, are highly acknowledged as well as the economic support received through these projects. This publication is a contribution to all those projects, and to the SCOR#120 working group. We appreciate constructive comments from two anonymous referees, and thank the organisers of the symposium "*Phaeocystis, major link in the biogeochemical cycling of climate-relevant elements*" for the opportunity to publish these data.

References

Alldredge AL, Passow U, Logan BE (1993) The abundance and significance of a class of large, transparent organic particles in the ocean. Deep-Sea Res 40:1131–1140

Andreassen IJ, Wassmann P (1998) Vertical flux of phytoplankton and particulate biogenic matter in the marginal ice zone of the Barents Sea in May 1993. Mar Ecol Prog Ser 170:1–14

Asper VL, Smith WOJ (2003) Abundance, distribution and sinking rates of of aggregates in the Ross Sea, Antarctica. Deep-Sea Res I 50:131–150

Belviso S, Thouzeau G, Schmidt S, Reigstad M, Wassmann P, Arashkevich E, Stefels J (2006) Significance of vertical flux as a sink for surface water DMPS and as a source for the sediment surface in coastal zones of northern Europe. Estuar Coast Shelf 68:473–488

Coppola L, Roy-Barman M, Wassmann P, Mulsow S, Jeandel C (2002) Calibration of sediment traps and particulate organic carbon export using ^{234}Th in the Barents Sea. Mar Chem 80:11–26

DiTullio GR, Grebmeier JM, Arrigo KR, Lizotte MP, Robinson DH, Leventer A, Barry JP, VanWoert ML, Dunbar RB (2000) Rapid and early export of *Phaeocystis antarctica* blooms in the Ross Sea Antarctica. Nature 404:595–598

Dutz J, Koski M (2006) Trophic significance of solitary cells of the prymnesiophyte *Phaeocystis globosa* depends on cell type. Limnol Oceanogr 51:1230–1238

Eilertsen HC, Schei B, Taasen JP (1981) Investigations on the plankton community of Balsfjorden, northern Norway: the phytoplankton 1976–1978. Abundance, species composition and succession. Sarsia 66:129–141

Engel A, Passow U (2001) Carbon and nitrogen content of transparent exopolymer particles (TEP) in relation to their Alcian Blue adsorption. Mar Ecol Prog Ser 219:1–10

Goffart A, Catalano G, Hecq JH (2000) Factors controlling the distribution of diatoms and *Phaeocystis* in the Ross Sea. J Mar Syst 27:161–175

Gowing MM, Garrison DL, Kunze HB, Winchell CJ (2001) Biological components of Ross Sea short-term particle fluxes in the austral summer of 1995–1996. Deep-Sea Res I 48:2645–2671

Hamm C, Reigstad M, Wexels Riser C, Mühlebach A, Wassmann P (2001) On the trophic fate of *Phaeocystis pouchetii*: VII. Sterols and fatty acids reveal sedimentation of *Phaeocystis*-derived organic matter via krill fecal strings. Mar Ecol Prog Ser 209:55–69

Hamm C, Simson DA, Merkel R, Smetacek V (1999) Colonies of *Phaeocystis globosa* are protected by a thin but tough skin. Mar Ecol Prog Ser 187:101–111

Hamm CE, Rousseau V (2003) Composition, assimilation and degradation of *Phaeocystis globosa*-derived fatty acids in the North Sea. J Sea Res 50:271–283

Head EJH, Horne EPW (1993) Pigment Transformation and Vertical Flux in an Area of Convergence in the North-Atlantic. Deep-Sea Res II 40:329–346

Heimdal BR (1974) Composition and abundance of phytoplankton in the Ullsfjord area, North Norway. Astarte 7:17–42

Hong Y, Smith WO, White A-M (1997) Studies on transparent exopolymer particles, (TEP) produced in the Ross Sea (Antarctica) and by *Phaeocystis antarctica* (Prymnesiophyceae). J Phycol 33:368–376

Irigoien X, Meyer B, Harris R, Harbour D (2004) Using HPLC pigment analysis to investigate phytoplankton taxonomy: the importance of knowing your species. Helgol Mar Res 58:77–82

Lalande C (2006) Vertical export of biogenic carbon in the Barents and Chukchi Seas. PhD thesis. University of Knoxville, Tennessee

Lancelot C, Keller MD, Rousseau V, Smith WOJ, Mathot S (1998) Autecology of the marine haptophyte *Phaeocystis* sp. In: Anderson DM, Cembella AD, Hallegraeff GM (eds) Physiolcgical ecology of harmful algal blooms. Springer-Verlag, Berlin Heidelberg, pp 209–224

Lutter S, Taasen JP, Hopkins CCE, Smetacek V (1989) Phytoplankton dynamics and sedimentation prosesses during spring and summer in Balsfjord, northern Norway. Polar Biol 10:113–124

Mari X, Rassoulzadegan F, Brussard CPD, Wassmann P (2005) Dynamics of transparent exopolymeric particles (TEP) production by *Phaeocystis globosa* under N- or P- limitation: A controlling factor of the retention/export balance. Harmful Algae 4:895–914

Mathot S, Smith WOJ, Carlson CA, Garrison DL, Gowing MM, Vickers CL (2000) Carbon partitioning within *Phaeocystis antarctica* (Prymnesiophyceae) colonies in the Ross Sea. J Phycol 36:1049–1056

Menden-Deuer S, Lessard EJ (2000) Carbon to volume relationships for dinoflagellates, diatoms and other protist plankton. Limnol Oceanogr 45:569–579

Olli K, Wassmann P, Reigstad M, Ratkova TN, Arashkevich E, Pasternak A, Matrai P, Knulst J, Tranvik L, Haarakalju R, Jakobsen A (2007) The fate of production in the central Arctic Ocean-top-down regulation by zooplankton expatriates? Prog. Oceanogr 72(1):84–113

Olli K, Wexels Riser C, Wassmann P, Ratkova TN, Arashkevich E, Pasternak A (2002) Seasonal variation in vertical flux of biogenic matter in the marginal ice zone and the central Barents Sea. J Mar Syst 38:189–204

Pakhomov EA, Froneman PW, Wassmann P, Ratkova T, Arashkevich E (2002) Contribution of algal sinking and zooplankton grazing to downward flux in the Lazarev Sea (Southern Ocean) during the onset of phytoplankton bloom: a lagrangian study. Mar Ecol Prog Ser 233:73–88

Passow U (2000) Formation of transparent expopolymer particles, TEP, from dissolved precursor material. Mar Ecol Prog Ser 192:1–11

Passow U (2002) Transparent exopolymer particles (TEP) in aquatic environments. Prog Oceanogr 55:287–333

Passow U, Alldredge AL (1995) A dye-binding assay for the spectrophotometric measurement of transparent exopolymer particles (TEP). Limnol Oceanogr 40:1326–1335

Passow U, Wassmann P (1994) On the trophic fate of *Phaeocystis pouchetii* (Hariot): IV. The formation of marine snow by *P. pouchetii*. Mar Ecol Prog Ser 104:153–161

Ratkova TN, Wassmann P (2002) Seasonal variation and spatial distribution of phyto- and protozooplankton in the central Barents Sea. J Mar Syst 38:47–75

Reigstad M (2000) Plankton community and vertical flux of biogenic matter in north Norwegian fjords: Regulating factors, temporal and spatial variations. PhD thesis. Norwegian College of Fishery Science. University of Tromsø, Tromsø

Reigstad M, Wassmann P, Ratkova T, Arashkevich E, Pasternak A, Øygarden S (2000) Comparison of the springtime vertical export of biogenic matter in three northern Norwegian fjords. Mar Ecol Prog Ser 201:73–89

Reigstad M, Wassmann P, Wexels Riser C, Øygarden S, Rey F (2002) Variations in hydrography, nutrients and chlorophyll *a* in the marginal ice zone and the central Barents Sea. J Mar Syst 38:9–29

Riebesell U (1993) Aggregation of *Phaeocystis* during phytoplankton spring blooms in the southern North Sea. Mar Ecol Prog Ser 96:281–289

Riebesell U, Reigstad M, Wassmann P, Noji T, Passow U (1995) On the trophic fate of *Phaeocystis pouchetii* (Hariot): VI. Significance of *Phaeocystis*-derived mucus for vertical flux. Netherland J Sea Res 33:193–203

Rijssel Mv, Hamm CE, Gieskes WWC (1997) *Phaeocystis globosa* (Prymnesiophyceae) colonies: hollow structures built with small amounts of polysaccharides. Eur J Phycol 32:185–192

Rousseau V, Becquevort S, Parent JY, Gasparini S, Daro MH, Tackx M, Lancelot C (2000) Trophic efficiency of the planktonic food web in a coastal ecosystem dominated by *Phaeocystis* colonies. J Sea Res 43:357–372

Rousseau V, Mathot S, Lancelot C (1990) Calculating carbon biomass of *Phaeocystis* sp. from microscopic observations. Mar Biol 107:305–314

Sakshaug E (2004) Primary and secondary production in the Arctic Seas. In: Stein R, Macdonald RW (eds) The organic carbon cycle in the Arctic Ocean. Springer-Verlag, Berlin Heidelberg, pp 57–81

Schoemann V, Becquevort S, Stefels J, Rousseau V, Lancelot C (2005) *Phaeocystis* blooms in the global ocean and their controlling mechanisms: a review. J Sea Res 53:43–66

Semina HJ (1978) Treatment of an aliquot sample. In: Sournia A (ed) Phytoplankton manual. Unesco publications, Paris, p 181

Smayda TJ (1978) From phytoplankters to biovolume. In: Sournia A (ed) Phytoplankton manual. Unesco publications, Paris, pp 273–279

Smith WO, Jr. Codispoti LA, Nelson DM, Manley T, Buskey EJ, Niebauer HJ, Cota GF (1991) Importance of *Phaeocystis* blooms in the high-latitude ocean carbon cycle. Nature 352:514–516

Sukhanova IN, Flint MV (2001) *Phaeocystis pouchetii* at the Eastern Bering Sea Shelf. Oceanology 41:75–85

Turner JT (2002) Zooplankton fecal pellets, marine snow and sinking phytoplankton blooms. Aquat Microb Ecol 27:57–102

Turner JT, Ianora A, Esposito F, Carotenuto Y, Miralto A (2002) Zooplankton feeding ecology: does a diet of *Phaeocystis* support good copepod grazing, survival, egg production and egg hatching success? J Plankton Res 24:1185–1195

Urban-Rich J, Nordby E, Andreassen IJ, Wassmann P (1999) Contribution by mesozooplankton faecal pellets to the carbon flux on Nordvestbanken, North Norwegian shelf 1994. Sarsia 84:253–264

Veldhuis MJW, Wassmann P (2005) Bloom dynamics and biological control of a high biomass HAB species in European coastal waters: a *Phaeocystis* case study. Harmful Algae 4:805–809

Verity PG, Smetacek V (1996) Organism life cycles, predation, and the structure of marine pelagic ecosystems. Mar Ecol Prog Ser 130:277–293

Wassmann P (1991) Sampling and analysis of marine particles with PEBENOCO (PElagic-BEntic coupling in the NOrwegian COastal zone), University of Tromsø, Norway. Geophys Monogr 63:97–99

Wassmann P (1994) Significance of sedimentation for the termination of *Phaeocystis* blooms. J Mar Syst 5:81–100

Wassmann P, Olli K, Wexels Riser C, Svensen C (2003) Ecosystem function, biodiversity and vertical flux regulation in the twilight zone. In: Wefer G, Lamy F, Mantoura F (eds) Marine science frontiers for Europe. Springer-Verlag, Berlin Heidelberg New York Tokyo, pp 277–285

Wassmann P, Ratkova TN, Reigstad M (2005) The contribution of single and colonial cells of *Phaeocystis pouchetii* to spring and summer blooms in the north-eastern North Atlantic. Harmful Algae 4:823–840

Wassmann P, Reigstad M, Øygarden S, Rey F (2000) Seasonal variation in hydrography, nutrients and suspended biomass in a subarctic fjord: applying hydrographic features and biological markers to trace watermasses and circulation significant for phytoplankton production. Sarsia 85:237–249

Wassmann P, Vernet M, Mitchell BG, Ray F (1990) Mass sedimentation of *Phaeocystis pouchetii* in the Barents Sea. Mar Ecol Prog Ser 66:183–195

Wexels Riser C, Wassmann P, Olli K, Pasternak A, Arashkevich E (2002) Seasonal variation in production, retention and export of zooplankton faecal pellets in the marginal ice zone and the central Barents Sea. J Mar Syst 38:175–188

Whipple SJ, Patten BC, Verity PG (2005) Colony growth and evidence for colony multiplication in *Phaeocystis pouchetii* (Prymnesiophyceae) isolated from mesocosm blooms. J Plankton Res 27:495–501

Biogeochemistry (2007) 83:235–244
DOI 10.1007/s10533-007-9094-2

ORIGINAL PAPER

Vernal sedimentation trends in north Norwegian fjords: temporary anomaly in ^{234}Th particulate fluxes related to *Phaeocystis pouchetii* proliferation

Sabine Schmidt · Sauveur Belviso · Paul Wassmann · Gérard Thouzeau · Jacqueline Stefels

Received: 23 May 2006 / Accepted: 5 October 2006 / Published online: 8 March 2007

Abstract We report data of a naturally occurring radionuclide, ^{234}Th, an in situ tracer, to investigate vertical export of biogenic matter during a vernal bloom of *Phaeocystis pouchetii* in the fjords of northern Norway. To optimise sampling of different stages of the bloom, three fjords with increasing oceanic influence (Balsfjord, Malangen fjord and Ullsfjord, respectively) were investigated in April 1997. Contrasting situations were encountered between the three fjords: the proliferation of *P. pouchetii* in Ullsfjord surface waters coincided with a drastic reduction of particulate ^{234}Th fluxes in traps, although particulate organic carbon (POC) and dimethylsulphoniopropionate (DMSP) were exported and ^{234}Th was available in surface waters. When large colonies make up a significant fraction of the vertical flux, as observed in Ullsfjord in April 1997, there may be a large and rapid change in the POC/^{234}Th ratio, further complicating the use of ^{234}Th as a tracer for POC export. The results suggest that the proliferation of *Phaeocystis pouchetii* during vernal bloom could temporary increase OC/^{234}Th ratio of particles and delay the particulate export of ^{234}Th, and probably of other particle-reactive species, from surface waters.

Keywords fjord · Northern Norwegian shelf waters · Particle flux · *Phaeocystis* · Sediment trap · ^{234}Th

S. Schmidt (✉)
EPOC, UMR 5805, Avenue des facultés, Talence Cedex 33405, France
e-mail: s.schmidt@epoc.u-bordeaux1.fr

S. Belviso
LSCE, UMR 1572, Gif sur Yvette Cedex 91 198, France

P. Wassmann
Institute for Aquatic Resources and Environmental Biology, The Norwegian College of Fishery Science, University of Tromso, Tromso, Norway

G. Thouzeau
IUEM, UMR 6539, Technopôle Brest-Iroise, Plouzane 29280, France

J. Stefels
Laboratory of Plant Physiology, University of Groningen, P.O. Box 14, 9750 AA Haren, The Netherlands

Introduction

The microalga genus *Phaeocystis* has a worldwide distribution and is known to develop massive blooms in polar and temperate waters (Riebesell et al. 1995). Spring blooms of *Phaeocystis* often follow diatom blooms, after the decline of dissolved silicate (Lancelot et al. 2005; Wassman et al. 2005). This colony-forming alga is an important source of the volatile organic sulphur compound dimethyl sulphide (DMS) and its dense blooms can act as a carbon sink (Rousseau et al. 2000; Belviso et al. 2006). Both sulfur and carbon cycles are relevant to climate change studies, sulphur as an important source of cloud condensation nuclei, and carbon being the main contributor to the greenhouse effect. Quantifying the role of *Phaeocystis* blooms in particle fluxes and export is therefore important if we want to understand their potential control on the escape of carbon from the photic zone of coastal waters. Indeed, the community structure determines how much of the primary production settles via the classical food web, via large phytoplankton and their grazers, and how much nutrients are regenerated via the microbial food web (Eppley and Peterson 1979; Verity 2000; Stelfox-Widdicombe et al. 2004). Within the EU funded program Escape (Entangled Sulphur and Carbon cycles in *Phaeocystis* dominated Ecosystems), this study focused upon the vertical export of biogenic material in relation to the plankton community dynamics and structures in north Norwegian fjords.

Thorium-234 (^{234}Th, $t_{1/2}$ = 24.1 days), a naturally occurring radionuclide, was used in this work to explore particle dynamics during the development of a *Phaeocystis* bloom. ^{234}Th is produced in seawater via the decay of its long-lived and highly soluble parent, uranium-238 (^{238}U). Since ^{234}Th is highly particle-reactive, and hence sticks to all particle surfaces, the clearance of ^{234}Th from surface waters is a direct indication of the removal rate of material on sinking particles from the upper ocean. Review papers clearly show the interest in this tracer as a tool for estimating particulate organic carbon export (Buesseler et al. 2006 and references herein). In surface waters, biological activity is the main source of particles and can vary considerably on short time scales. Previous studies have shown the close coupling between dissolved ^{234}Th scavenging and new production in the ocean (Coale and Bruland 1985). The particulate ^{234}Th residence time in oceanic surface waters is of the order of a few days to a few weeks (Moran and Buesseler 1992) and appears to be mainly governed by the classical food web, via the export of detritus (marine snow, aggregates, faecal pellets) (Schmidt et al. 1992; Buesseler et al. 2006).

The present paper discusses ^{234}Th data in the water column and in settling particles, together with the determination of the stocks of *Phaeocystis pouchetii* and DMSP in surface waters of north Norwegian fjords. The aims of this study are: (i) to describe temporal variability of ^{234}Th activities and fluxes during the progress of a vernal bloom, (ii) to calibrate trap collection efficiencies using ^{234}Th data, and (iii) to assess the impact of *Phaeocystis* proliferation on vertical export of particles.

Material and methods

Study area

To optimize sampling of different stages of the recurrent vernal *Phaeocystis pouchetii* bloom in northern Norway, three fjords with decreasing oceanic influence were investigated in the north Norwegian coastal zone (Fig. 1). The westernmost fjord, Malangen fjord, is relatively open and exposed to shelf waters. Ullsfjord, the northernmost is the most open to coastal waters with a wide entrance. Both fjords are wide and have deep sills. Balsfjord is the least exposed: it is a tapered inland fjord, which is separated from the Malangen fjord and Ullsfjord by three narrow straits. More details on sampling areas including their physical, chemical and biological characteristics can be found in Wassman et al. (1996), Keck and Wassmann (1996), Reigstad et al. (2000) and Archer et al. (2000). The stations were visited sequentially five times, between April 7 and April 25, 1997, aboard the RV Johan Ruud.

Measurements

To sample the settling flux directly, free-floating multiple-sample programmable sediment traps (Pro-Trap) were deployed at all three fjords for about 16 h at each time (trap depth: 60 m). The Pro-Trap system consists of four polyvinyl chloride (PVC) sediment tubes, each of 0.018 square meter exposed area (cylinder height of 80 cm, trap aspect ratio A ¼ 5.3), mounted on a stainless-steel frame. Depth and angle sensors allowed defining the position of the traps in the water

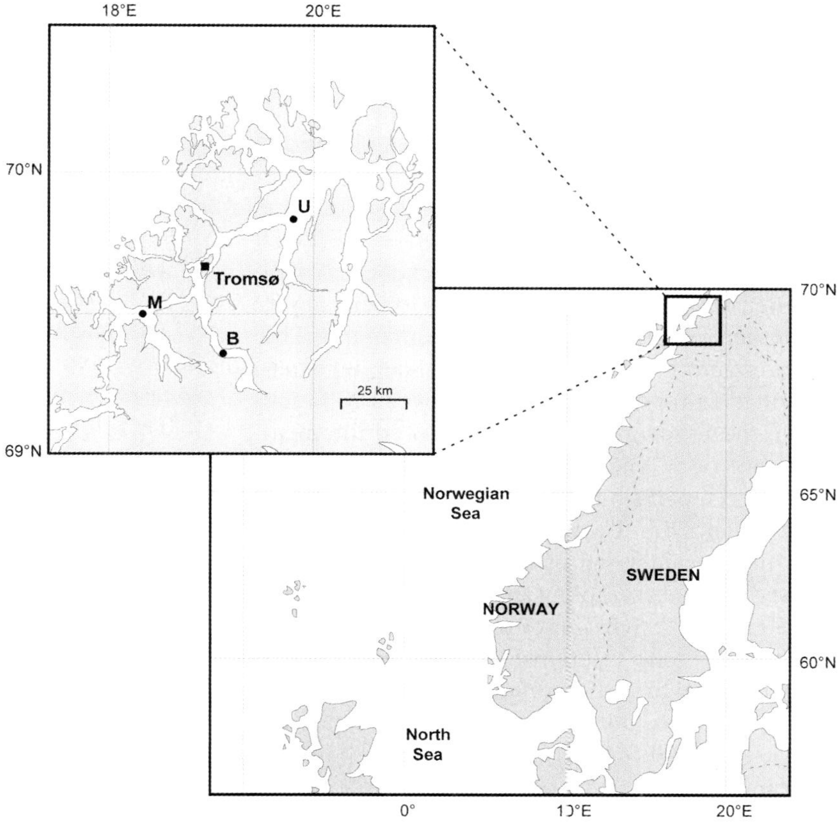

Fig. 1 Map of the investigated fjords

column, while light scattering was measured by a Sea Tech LS sensor. A VALEPORT 800-0 series electromagnetic current meter provided 2-axis flow velocity measurements at the trap aperture (see Belviso et al. 2006 for details). Upon recovery, swimmers (living zooplankton) were removed from the samples prior to all analyses, in order to assess properly the passive particulate fluxes. Trap samples were immediately filtered on precombusted and preweighed Whatman GF/F filters of 25 or 47 mm diameter and stored at –20°C until analysis (within six months). All carbon and nitrogen analyses were performed on a Carlo Erba NC 2500 gas chromatography analyzer. The difference-on-ignition (DOI) method was used to separate organic and inorganic carbon (Majeed 1987). Details on the determination of DMSP stocks and fluxes are given in Belviso et al. (2006).

One sediment tube, devoted to the determination of ^{234}Th, took a single sample for the entire duration of each deployment. On recovery of the traps, trapped particles were filtered on GF/F; ^{234}Th activities were measured with a low-background high-efficiency γ detector. ^{234}Th activities were measured from its 63.2 and 92.4–92.8 keV -gamma rays, and decay corrected to the time of sample collection (Schmidt and Reyss, 2000). The standards used for the calibration of the γ detector are: (1) a mock-up of sediment and U and Th US standards from National Bureau of Standards (NBS) at 1,000 ppm, and (2) a known amount of ^{238}U deposited on an aluminum disk checked by α-counting using a grid chamber of a known efficiency. Counting efficiencies and backgrounds for ^{234}Th are detailed in Schmidt and Reyss (2000) and Schmidt (2006).

During the fifth (and last) sampling date, a profile of ^{234}Th was sampled between the surface and the trap depth (60 m) at each station. Immediately after sampling, the 20 l of seawater was passed through a 0.45 μm pore size filter to separate dissolved from particulate phases. Within one month after the collection, particulate

^{234}Th ($^{234}Th^P$) was directly measured on the filter as trapped particles. After acidification to pH 2, pre-weighed 50 mBq ^{229}Th chemistry yield tracer and 120 mg Fe (as $FeCl_3$) were added to the dissolved sample. After spike equilibration, $Fe(OH)_3$ was precipitated by adding NH_4OH to pH 7. After recovery of the precipitate, the purification of ^{234}Th was obtained by anionic exchange. After elution, Th was extracted with 1-(2thenoyl)-3,3,3-trifluoracetone in toluene at pH 3 and then evaporated onto an aluminum foil. The first α counting of this foil allowed the determination of ^{229}Th for chemical yield (between 20% and 60%); the following γ-counting allowed the measurement of ^{234}Th. Due to the short half-life of ^{234}Th, the separation of ^{234}Th from its parent was done within 24 h. Calculations of dissolved ^{234}Th activities consider ingrowth corrections for time elapsed between sampling and chemistry (Schmidt and Reyss 2000; Schmidt et al. 2002).

The U–salinity relationship (Chen et al. 1986) is appropriate for estimating dissolved ^{238}U in the open ocean; in other regimes, i.e., continental shelves, estuaries or marginal seas, the U concentration must be measured (Schmidt and Reyss 1991; Rutgers van der Loeff et al. 2006). Seawater samples were collected at each station for ^{238}U determination. Uranium was concentrated from 2 l seawater, as previously described for ^{234}Th, in the presence of a known amount of ^{232}U spike and 20 mg Fe. Then uranium activities were determined by α-counting (Schmidt 2004).

Phytoplankton abundance was determined in samples from depth profiles taken at dawn as described by Archer et al. (2000). Data on *Phaeocystis* abundance provided here are means from the 1, 4 and 8 m depth samples. Phytoplankton carbon is calculated from biovolume measurements as described by Archer et al. (2000).

Irreversible scavenging model of ^{234}Th

^{234}Th as a tracer is widely used and critical for two tasks: to quantify fluxes and residences time of particles, and to calibrate trap efficiency by comparing estimated water-column ^{234}Th fluxes with those measured by traps (Cochran et al. 2000). In surface waters, ^{234}Th activities are the result of a balance between its continuous production from ^{238}U, its decay, removal onto rapidly sinking particles, and advection/diffusion. The temporal change in total ^{234}Th is expressed by the classical transport equation:

$$\delta A^{Th}/\delta t = \lambda A^{U} - \lambda A^{Th} - P + V \qquad (1)$$

where A^U is the ^{238}U activity, A^{Th} is the total ^{234}Th activity, λ is the decay constant of ^{234}Th (=0.0288 day^{-1}), P is the net removal flux of $^{234}Th^P$, and V is the sum of the advective/diffusive fluxes (Savoye et al. 2006 and references therein). Measurements of both dissolved and particulate ^{234}Th allow us to calculate rates of exchange between dissolved and particulate phases, removal fluxes and particle residence times. By assuming steady state and negligible advective/diffusive terms, particulate ^{234}Th flux, P, must balance the measured deficit relative to A^U. Particle residence times with respect to removal from surface layer are given by the ratio $^{234}Th^P/P$.

Results

Hydrography, *Phaeocystis* and DMSP distribution in the three fjords

The differences in temperature, salinity and density indicated no major exchange of water masses throughout the cruise. Salinity was lowest in Balsfjord (33.5–33.7), and highest in Ullsfjord (34.2–34.5) (Reigstad et al. 2000).

The development of the *Phaeocystis* bloom was different between the three fjords (Fig. 2). The lowest cell numbers were observed in Balsfjord, with contributions of *Phaeocystis* cells to total phytoplankton cells of less than 12% throughout the study period; diatoms dominated in Balsfjord. On the other hand, Ullsfjord, with open access to coastal waters, presented a well-developed vernal bloom, with the highest *Phaeocystis* abundances; this alga dominated the phytoplankton community (in numbers of cells) in April. The abundance of *Phaeocystis* in Malangen fjord waters was relatively low and intermediate between the two other fjords.

DMSP in surface waters, produced primarily by *Phaeocystis*, paralleled the distribution of this

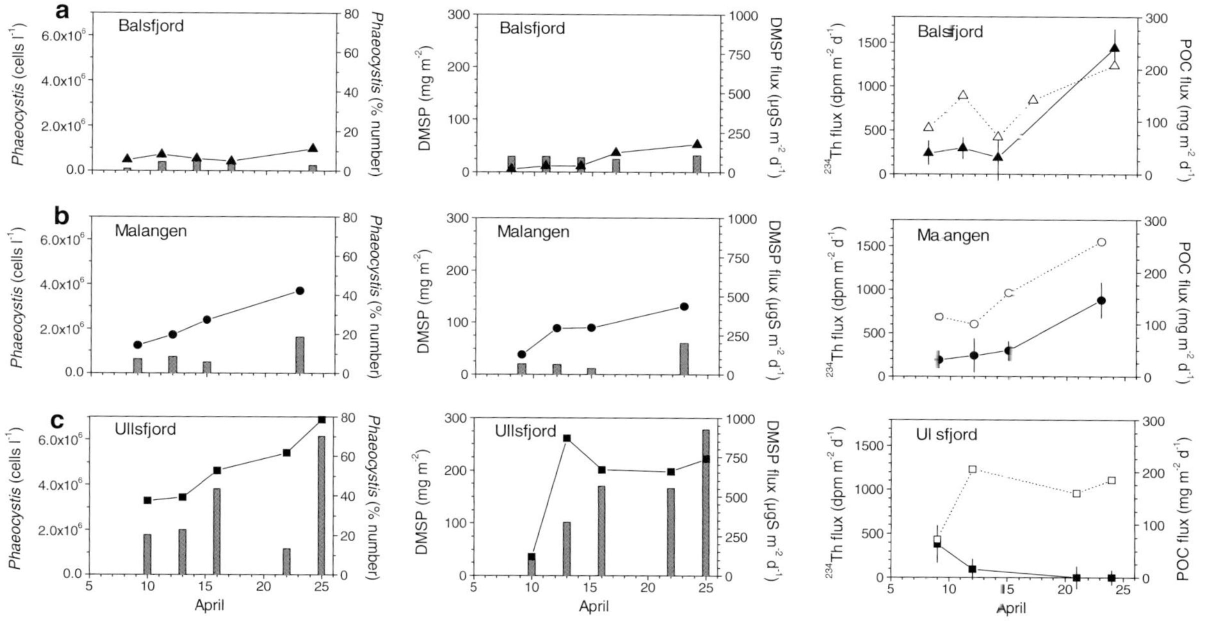

Fig. 2 Summary of changes in the stocks of *Phaeocystis* (left; bars: number of cells; closed symbols: % of *Phaeocystis* cells to total cell number) and DMSP (middle; bars: total water column standing stock; closed symbols: export fluxes) and particulate ^{234}Th flux (right, closed symbols) and POC flux (right, open symbols) for the three fjords studied in April 1997. The measured DMSP, POC and ^{234}Th fluxes are from sediment trap samples collected at 60 m depth on the same cruises as the profiles

alga, with increasing stocks from Balsfjord Malangen to Ullsfjord (Fig. 2). Fluxes and standing stock of DMSP were positively correlated in the three fjords, but the slope of the relationship was different from one fjord to the other. Measured POC fluxes in the same trap mimic temporal evolution in DMSP fluxes, although DMSP to POC ratios in the settled material differed between fjords. This is related to the difference in phytoplankton community composition between the three fjords: *Phaeocystis*-dominated phytoplankton in Ullsfjord being richer in DMSP than diatom-dominated phytoplankton in Balsfjord.

Vertical distribution of ^{234}Th

One profile of particulate and dissolved ^{234}Th between surface and 60 m depth was taken at each station in late April (Table 1). ^{238}U activities increased with oceanic influence, in association with the salinity, from 2.10 dpm l^{-1} (Balsfjord) to 2.38 dpm l^{-1} (Ullsfjord; Table 2). Particulate ^{234}Th was extremely variable from negligible to 0.64 dpm l^{-1} and represented up to 62% of the total ^{234}Th. The mean 0–40 m ^{234}Th^{P} to total ^{234}Th ratio was lowest in Ullsfjord (20%). Total ^{234}Th activities varied both with depth and with fjords, ranging from 0.73 dpm l^{-1} to 1.96 dpm l^{-1}. These values were always lower than ^{238}U activities (2.10–2.38 dpm l^{-1}). Marked deficits (33–65%) were observed in all three fjords at the end of April (33–65%). Equilibrium state (^{234}Th/^{238}U = 1) was only observed at 60 m depth in Balsfjord. These trends are consistent with a common pattern of increasing scavenging in shallow waters (Gustafsson et al. 1998). As the ^{234}Th deficit is due to exported particles, these large deficits may indicate that efficient particle export had occurred during the sampling period or few weeks before. Such deficits are comparable to the range of values reported in other Arctic regions (Coppola et al. 2002; Trimble and Baskaran, 2005).

The calculated mean particulate ^{234}Th residence times for the upper 60 m were about

Table 1 Particulate, dissolved and total ^{234}Th activities

Site	Depth (m)	$^{234}Th^{p}$ dpm l^{-1}	$^{234}Th^{d}$ dpm l^{-1}	$^{234}Th^{T}$ dpm l^{-1}
Balsfjord April, 24, 1997	0	0.32 ± 0.06	0.41 ± 0.06	0.73 ± 0.08
	20	0.40 ± 0.05	0.88 ± 0.09	1.28 ± 0.10
	40	0.64 ± 0.09	0.77 ± 0.17	1.40 ± 0.19
	60	0.61 ± 0.09	1.35 ± 0.28	1.96 ± 0.29
Malangen April, 23, 1997	0	0.57 ± 0.14	0.34 ± 0.34	0.92 ± 0.37
	20	0.38 ± 0.08		
	60	0.10 ± 0.10	1.21 ± 0.36	1.31 ± 0.27
Ullsfjord April, 25, 1997	0	0.24 ± 0.08	0.77 ± 0.14	1.00 ± 0.17
	40	0.22 ± 0.10	1.11 ± 0.13	1.33 ± 0.17
	60	0.55 ± 0.10		

Table 2 Mean 0–60 m activities of particulate and total ^{234} Th, and ^{238}U; residence time of particulate ^{234}Th ($T(P)$); and ^{234}Th fluxes

Site	$^{234}Th^{p}$ dpm l^{-1}	$^{234}Th^{T}$ dpm l^{-1}	^{238}U dpm l^{-1}	$T(P)$ days	Calculated ^{234}Th fluxes dpm m^2 d^{-1}	Trapped ^{234}Th fluxes dpm m^2 d^{-1}
Balsfjord April, 24, 1997	0.49 ± 0.15	1.34 ± 0.37	2.10 ± 0.15	22.6	1307 ± 93	1444 ± 184
Malangen April, 23, 1997	0.35 ± 0.19	1.06 ± 0.45	2.29 ± 0.12	10.0	2122 ± 127	880 ± 162
Ullsfjord April, 25, 1997	0.34 ± 0.16	1.17 ± 0.23	2.38 ± 0.10	9.6	2097 ± 37	<82

Calculated ^{234}Th fluxes are based on ^{234}Th data using the model described in the method section; trap ^{234}Th fluxes correspond to measurements on trap samples for the corresponding period

10 days for the Ullsfjord and Malangen stations and 23 days for the Balsfjord station. These residence times are comparable to the mean values calculated for the Barents Sea (15 days, Coppola et al. 2002).

Particulate fluxes of ^{234}Th

^{234}Th fluxes presented a more confusing signal (Fig. 2). Like the DMSP fluxes, the ^{234}Th fluxes increased throughout April in both the Balsfjord and Malangen fjords. In contrast, ^{234}Th fluxes dropped dramatically (from moderate to negligible values) in Ullsfjord after mid April. This trend was rather unexpected in spring when fluxes are usually increasing during the development of the spring bloom (Buesseler et al. 1992). Profiles in the upper 60 m of the three fjords showed that total ^{234}Th activities ranged from 0.7 to 1.9 dpm l^{-1}. There were no large differences in ^{234}Th availability between the fjords. This result would signify that the rapid decrease in ^{234}Th fluxes at Ullsfjord was not related to a complete depletion of this radionuclide. In such coastal systems, it is not possible to exclude exchange with adjacent coastal waters, especially for Ullsfjord, the most open fjord. Nevertheless hydrological data do not indicate major exchange of water masses during the sampling period (Reigstad et al. 2000). In addition such exchanges, if occurring, would be likely to convey ^{234}Th, as coastal waters have higher salinity and consequently higher ^{234}Th.

Discussion

^{234}Th fluxes and trap collection efficiencies

During ESCAPE, integrating ^{234}Th profiles down to 60 m (to get a particulate ^{234}Th flux) provided a meaningful comparison with the sediment trap, which collected particulate ^{234}Th flux passing the depth horizon of 60 m. Steady-state fluxes ranged between 1300 and 2100 dpm m^{-2} d^{-1} (Table 2). For Balsfjord, measured and calculated fluxes compared rather well, which indicates a high efficiency of the trap.

In contrast, measured particulate ^{234}Th fluxes in Malangen and Ullsfjord were far below the calculated ones (Table 2). A rapid interpretation of these results would lead to the conclusion that recovery efficiencies of the sediment trap were low, even negligible, for Ullsfjord. This is, however, unlikely since temporal evolution of biological parameters in trap collections (DMSP, POC) were in good correlation to their standing stocks (Fig. 2). Another explanation may be related to the ^{234}Th calibration method. In the present work, ^{234}Th fluxes are calculated from a single profile of ^{234}Th assuming steady state and neglecting advection. Although the use of non-steady-state modeling is more accurate (Buesseler et al. 1992; van der Loeff et al. 1997), steady-state models have been shown to provide adequate estimations of ^{234}Th fluxes (Benitez-Nelson et al. 2000), except when sampling occurs just after a high export event that depleted the ^{234}Th stock (Tanaka et al. 1983). To test the pertinence of the steady-state model used in the present work, we reconstructed the depletion levels of ^{234}Th during the sampling period using measured data and fluxes from the traps (with the assumption of optimal recovery; Fig. 3). The results clearly show that the situation is different between the three fjords. The reconstructed ^{234}Th activities are always present at significant levels in Balsfjord, but in the two other fjords, high deficiencies occurred in early April. The deficits of ^{234}Th measured in Malangen fjord and Ullsfjord indicate a memory effect of an export event in early April. Consequently, ^{234}Th cannot be used to calculate trap efficiencies in theses cases.

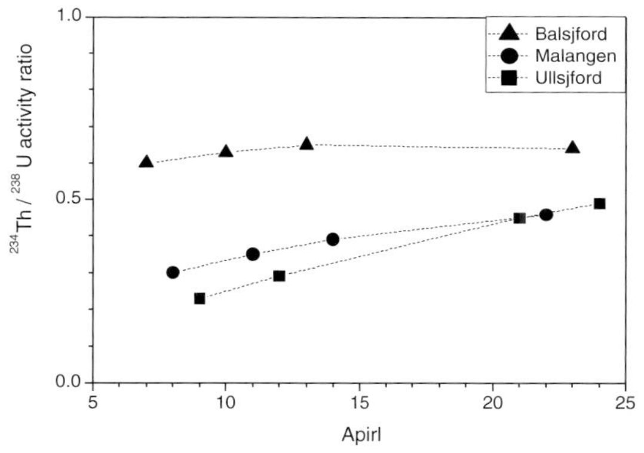

Fig. 3 Reconstruction of the temporal evolution of ^{234}Th:^{238}U activity ratio in the upper 0–60 m. Total ^{234}Th for the layer 0–60 m was calculated from the values of late April, taking into account the production rate from ^{238}U, the measured fluxes in trap and the decay of ^{234}Th

Impact of *Phaeocystis* proliferation on settling flux

The most striking feature in the temporal evolution of particulate fluxes throughout April 1997 was the differences between its components (Fig. 2). The relationship between particulate organic carbon, DMSP and ^{234}Th were clearly different between fjords. POC fluxes showed a wide range (70–260 mg m^{-2} d^{-1}), with the highest values observed after mid April. DMSP fluxes exhibited the same trend, although DMSP:POC ratios varied between the fjords (Fig. 2) while ^{234}Th fluxes presented a very unusual pattern in Ullsfjord, decreasing to negligible levels at the end of April. This cannot be fully explained with a ^{234}Th deficiency due to an earlier export event, since ^{234}Th is always available in surface waters and activities were increasing during April (Fig. 3).

The major difference among the three fjords was the timing of the vernal bloom. In April 1997, Balsfjord and Malangen were diatom-dominated ecosystem, whereas Ullsfjord was *Phaeocystis*-dominated (Reigstad et al. 2000). DMSP fluxes were positively related to the proliferation of *Phaeocystis*, while ^{234}Th fluxes showed an inverse pattern (Fig. 4). Hence the question was raised what mechanism could lead to a system exporting DMSP and POC but not ^{234}Th.

We hypothesize therefore that the observed low ^{234}Th flux in traps is associated to the composition of settling material. Faecal pellets did not dominate the vertical export in April 1997: in fact phytoplankton, mainly diatoms and *Phaeocystis*, comprised a major fraction of the sinking particles (Reigstad et al. 2000). The high PPC (total phytoplankton carbon)/POC ratios

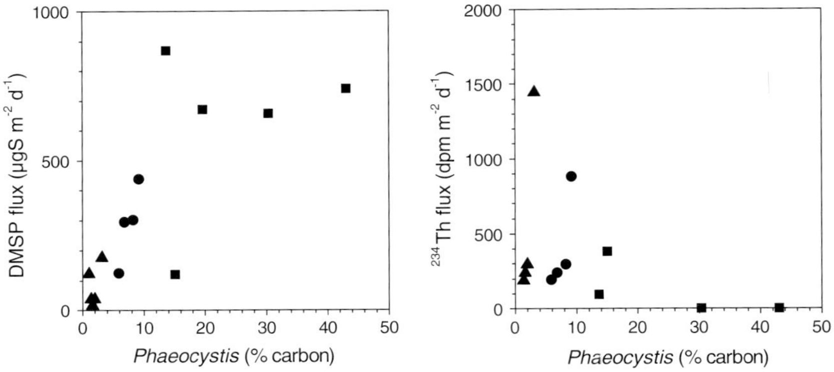

Fig. 4 DMSP (left) and particulate ^{234}Th fluxes (right) as a function of *Phaeocystis* content of the surface waters (expressed as a percentage of total phytoplankton carbon; same symbols as in Fig. 3)

(0.37, 0.39 and 0.29 for Ullsfjord, Balsfjord and Malangen, respectively) suggest that much of the spring bloom in these three fjords in April 1997 was vertically exported as ungrazed phytoplankton. Moreover, during this experiment, it was observed that the vertical export of *Phaeocystis pouchetii* cells (free and colonial) was negligible in Balsfjord, rather low in Malangen Fjord and highest in Ullsfjord (1.2, 8.0 and 25.3% of phytoplankton-carbon export; Reigstad et al. 2000).

The application of the ^{234}Th method for investigating particle dynamics is based on the assumption that ^{234}Th is scavenged onto fine particles that can be aggregated into larger particles able to settle, either due to grazing or bloom senescence and aggregate formation (Buesseler et al. 2006; Savoye et al. 2006). However several recent works point out intricate relationship between ^{234}Th and particles depending on their nature (phytoplankton cells, flocs, lithogenic material) and the occurrence of transparent exopolymer particles (TEP) (Passow et al. 2006; Waite and Hill, 2006). Passow et al. (2006) investigated the affinity of TEP and inorganic clays to ^{234}Th and concluded that POC/^{234}Th ratios depend largely on the types of particles presents (Passow et al. 2006). Buesseler et al. (1995) already postulated that fresh biological particles are likely to have a higher POC:^{234}Th ratio. In Ullsfjord after mid April $^{234}Th^{P}$ activities are the lowest and trap POC/^{234}Th ratios (>120 µmol dpm^{-1}) far exceeded values in Balsfjord and Malangen (12–30 and 25–49 µmol dpm^{-1}, respectively) and all published values (<70 µmol dpm^{-1}) in the review of Buesseler et al. (2006), except a data set from the low-salinity Baltic (up to 800 µmol dpm^{-1}).

In Ullfjord, *Phaeocystis* comprised a major fraction of the sinking particles (Reigstad et al. 2000). Sedimentation of these cells may account for the high DMSP and POC fluxes and low POC/^{234}Th ratio measured in the traps with the highest rates in Ullsfjord.

Conclusions

The main objective of this study was to assess the impact of *Phaeocystis* proliferation on vertical export of particles based on ^{234}Th data in water column and in sediment traps. A specific task was to calibrate the trap collection efficiencies. From our present investigation, we conclude the following:

(1) Special care must be taken when using ^{234}Th measurements for estimating particle fluxes and trap efficiencies in coastal systems. During the early stage of large cells bloom, like *Phaeocystis* proliferation, a complete data set will be require: high-frequency three-dimensional (3D) sampling, determination of $^{234}Th^{P}$ and POC/^{234}Th ratios on suspended and sinking particles, properties of flocs, and the use of a non-steady-state model.

(2) We hypothesize that the proliferation of large, fresh and unflocculated cells and their direct settling temporally affect the partitioning of ^{234}Th within the in situ particles (high POC-to-^{234}Th ratios) and delay, even prevent, the export, via grazing or aggregate

formation, of ^{234}Th and probably of similar particle-reactive elements.

Acknowledgments This work was partly supported by the European Union MAST programme ESCAPE (contract MAS3-CT96-0050) and the Centre National de la Recherche Scientifique. We are grateful to the crew of RV Johan Ruud for their strong support in field work.

References

Archer SD, Verity PG, Stefels J (2000) Impact of microzooplankton on the progression and fate of the spring bloom in fjords of northern Norway. Aquat Microb Ecol 22:27–41

Belviso S, Thouzeau G, Schmidt S, Reigstad M, Wassmann P, Arashkevich E, Stefels J (2006) Significance of vertical flux as a sink for surface water DMSP and as a source for the sediment surface in coastal zones of northern Europe. Estuar Coast Shelf Sci 68:473–488

Benitez-Nelson C, Buesseler KO, Crossin G (2000) Upper ocean carbon export, horizontal transport, and vertical eddy diffusivity in the southwestern Gulf of Maine. Cont Shelf Res 20:707–736

Buesseler KO, Bacon MP, Cochran JK, Livingston HD (1992) Carbon and nitrogen export during the JGOFS North Atlantic Bloom Experiment estimated from Th-234/U-238 disequilibria. Deep-Sea Res 39:1115–1137

Buesseler KO, Andrews JA, Hartman MC, Belastock R, Chai F (1995) Regional estimates of the export flux of particulate organic carbon derived from thorium-234 during the JGOFS EqPac program. Deep-Sea Res II 42:777–804

Buesseler KO, Benitez-Nelson CR, Moran SB, Burd A, Charette M, Cochran JK, Coppola L, Fisher NS, Fowler SW, Gardner WD, Guo LD, Gustafsson O, Lamborg C, Masque P, Miquel JC, Passow U, Santschi PH, Savoye N, Stewart G, Trull T (2006) An assessment of particulate organic carbon to thorium-234 ratios in the ocean and their impact on the application of ^{234}Th as a POC flux proxy. Mar Chem 100:213–233

Chen JH, Edwards LR, Wasserburg GJ (1986) ^{238}U, ^{234}U and ^{232}Th in seawater. Earth Planet Sci Lett 80:241–251

Coale KH, Bruland KW (1985) Th-234/U-238 disequilibria within the California Current. Limnol Oceanogr 32:22–33

Cochran JK, Buesseler KO, Bacon MP, Wang HW, Hirschberg DJ, Ball, Andrews J, Crossin G, Fleer A (2000) Short-lived thorium isotopes (^{234}Th, ^{228}Th) as indicators of POC export and particle cycling in the Ross Sea, Southern Ocean. Deep-Sea Res II 47:3451–3490

Coppola L, Roy-Barman M, Wassmann P, Mulsow S, Jeandel C (2002) Calibration of sediment traps and particulate organic carbon export using Th-234 in the Barents Sea. Mar Chem 80:11–26

Eppley RW, Peterson BJ (1979) Particulate organic matter flux and planktonic new production in the deep ocean. Nature 282:677–680

Gustafsson O, Buesseler KO, Geyer W, Rockwell, Moran B, Gschwend PM (1998) On the relative significance of horizontal and vertical transport of chemicals in the coastal ocean: application of a two-dimensional Th-234 cycling model. Cont Shelf Res 18:805–829

Keck A, Wassmann P (1996) Temporal and spatial patterns of sedimentation in the subarctic fjord Malangen, northern Norway. Sarsia 80:259–276

Lancelot C, Spitz Y, Gypens N, Ruddick K, Becquevort S, Rousseau V, Lacroix G, Billen G (2005) Modelling diatom and *Phaeocystis* blooms and nutrient cycles in the Southern Bight of the North Sea: the MIRO model. Mar Ecol Prog Ser 289:63–78

Majeed S (1987) Organic matter and biotic indices on the beaches of North Brittany. Mar Pollution Bulletin 18:490–495

Moran SB, Buesseler KO (1992) Short residence time of colloids in the upper ocean estimated from U-238–Th-234 disequilibria. Nature 359:221–223

Passow U, Dunne J, Murray JW, Balistrieri L, Alldredge AL (2006) Organic carbon to 234Th ratios of marine organic matter. Mar Chem 100:323–336

Reigstad M, Wassmann P, Ratkova T, Arashkevich E, Pasternak A, Oygarden S (2000) Comparison of the springtime vertical export of biogenic matter in three northern Norwegian fjords. Mar Ecol Prog Ser 201:73–89

Riebesell U, Reigstad M, Wassmann P, Noji Y, Passow U (1995) On the trophic fate of *Phaeocystis pouchetii* (Hariot): VI. Significance of *Phaeocystis*-derived mucus for vertical flux. Netherl J Sea Res 33:193–203

Rousseau V, Becquevort S, Parent JY, Gasparini S, Daro MH, Tackx M, Lancelot C (2000) Trophic efficiency of the planktonic food web in a coastal ecosystem dominated by *Phaeocystis* colonies. J Sea Res 43:357–372

Rutgers van der Loeff M, Friedrich J, Bathmann UV (1997) Carbon export during the Spring Bloom at the Antarctic Polar Front, determined with the natural tracer ^{234}Th. Deep-Sea Res II 44:457–478

Rutgers van der Loeff M, Sarin MM, Baskaran M, Benitez-Nelson C, Buesseler KO, Charette M, Dai M, Gustafsson Ö, Masque P, Morris P, Rodriguez y Baena A, Savoye N, Schmidt S, Vöge I, Waples J (2006) A review of present techniques and methodological advances in analyzing 234Th in aquatic systems. Mar Chem 100:190–212

Savoye N, Benitez-Nelson C, Burd A, Cochran K, Charette M, Buesseler KO, Jackson G, Roy-Barman M, Schmidt S, Elskens M (2006) ^{234}Th sorption and export models in the water column: a review. Mar Chem 100:213–233

Schmidt S (2004) Investigation of dissolved uranium content in the watershed of the Seine River (France). J Environ Radioactivity 78:1–10

Schmidt S (2006) Impact of the Mediterranean Outflow Water on particle dynamics in intermediate waters of the North-East Atlantic, as revealed by ^{234}Th and ^{228}Th. Mar Chem 100:289–298

Schmidt S, Reyss JL (1991) Uranium concentrations of Mediterranean seawaters with high salinities. C R Acad Sci Paris 312:379–384

Schmidt S, Reyss JL (2000) Improvement in radiochemical and γ-counting procedures for the determination of ^{234}Th in seawater. Radiochim Acta 88:459–463

Schmidt S, Nival P, Reyss JL, Baker M, Buat-Ménard P (1992) Relation between Th-234 scavenging and zooplankton biomass in Mediterranean surface waters. Oceanol Acta 15:227–231

Schmidt S, Andersen V, Belviso S, Marty JC (2002) Strong seasonality in particle dynamics of north-western Mediterranean surface waters (JGOFS France DYFAMED time-series station) as revealed by ^{234}Th/^{238}U. Deep-Sea Res I 49:1507–1518

Stelfox-Widdicombe CE, Archer SD, Burkill PH, Stefels J (2004) Microzooplankton grazing in *Phaeocystis* and diatom-dominated waters in the southern North Sea in spring. J Sea Res 5:37–51

Tanaka N, Takeda Y, Tsunogai S (1983) Biological effect on removal of Th-234, Po-210 and Pb-210 from surface water in Funkai Bay, Japan. Geochim Cosmochim Acta 47:1783–1790

Trimble SM, Baskaran M (2005) The role of suspended particulate matter in ^{234}Th scavenging and ^{234}Th-derived export fluxes of POC in the Canada Basin of the Arctic Ocean. Mar Chem 96:1–19

Verity PG (2000) Grazing experiments and model simulations of the role of zooplankton in *Phaeocystis* food webs. J Sea Res 43:317–343

Waite AM, Hill PS (2006) Flocculation and phytoplankton cell size can alter 234Th-based estimates of the vertical flux of particulate organic carbon in the sea. Mar Chem 100:366–375

Wassmann P, Svendsen H, Keck A, Reigstad M (1996) Selected aspects of the physical oceanography and particle fluxes in fjords of northern Norway. J Mar Syst 8:53–71

Wassmann P, Ratkova T, Reigstad M (2005) The contribution of single and colonial cells of *Phaeocystis pouchetii* to spring and summer blooms in the north-eastern North Atlantic. Harmful Algae 4:823–840

Biogeochemistry (2007) 83:245–275
DOI 10.1007/s10533-007-9091-5

REVIEW

Environmental constraints on the production and removal of the climatically active gas dimethylsulphide (DMS) and implications for ecosystem modelling

Jacqueline Stefels · Michael Steinke · Suzanne Turner · Gill Malin · Sauveur Belviso

Received: 14 June 2006 / Accepted: 5 August 2006 / Published online: 27 April 2007

Abstract Seawater concentrations of the climate-cooling, volatile sulphur compound dimethylsulphide (DMS) are the result of numerous production and consumption processes within the marine ecosystem. Due to this complex nature, it is difficult to predict temporal and geographical distribution patterns of DMS concentrations and the inclusion of DMS into global ocean climate models has only been attempted recently. Comparisons between individual model predictions, and ground-truthing exercises revealed that information on the functional relationships between physical and chemical ecosystem parameters, biological productivity and the production and consumption of DMS and its precursor dimethylsulphoniopropionate (DMSP) is necessary to further refine future climate models. In this review an attempt is made to quantify these functional relationships. The description of processes includes: (1) parameters controlling DMSP production such as species composition and abiotic factors; (2) the conversion of DMSP to DMS by algal and bacterial enzymes; (3) the fate of DMSP-sulphur due to, e.g., grazing, microbial consumption and sedimentation and (4) factors controlling DMS removal from the water column such as microbial consumption, photo-oxidation and emission to the atmosphere. We recommend the differentiation of six phytoplankton groups for inclusion in future models: eukaryotic and prokaryotic picoplankton, diatoms, dinoflagellates, and other phytoflagellates with and without DMSP-lyase activity. These functional groups are characterised by their cell size, DMSP content, DMSP-lyase activity and interactions with herbivorous grazers. In this review, emphasis is given to ecosystems dominated by the globally relevant haptophytes *Emiliania huxleyi* and *Phaeocystis* sp., which are important DMS and DMSP producers.

Keywords Climate modelling · DMS · DMSP · *Emiliania huxleyi* · Functional groups · *Phaeocystis*

J. Stefels (✉)
Laboratory of Plant Physiology, University of Groningen, P.O. Box 14, 9750 AA Haren, The Netherlands
e-mail: j.stefels@rug.nl

M. Steinke · S. Turner · G. Malin
Laboratory for Global Marine and Atmospheric Chemistry, School of Environmental Sciences, University of East Anglia, Norwich NR4 7TJ, UK

S. Belviso
Laboratoire des Sciences du Climat et de l'Environnement, UMR CEA-CNRS 1572, CEN/Saclay, Bât 709, L'Orme des Merisiers, 91191 Gif-sur-Yvette, France

Present Address:
M. Steinke
Department of Biological Sciences, University of Essex, Wivenhoe Park, Colchester CO4 3SQ, UK

Introduction

Dimethylsulphide (DMS) is a semivolatile organic sulphur compound that accounts for 50–60% of the total natural reduced sulphur flux to the atmosphere, including emissions from volcanoes and from vegetation (Andreae 1990; Bates et al. 1992; Spiro et al. 1992). By providing 95% of the flux to the atmosphere, the oceans are the main source for DMS, with estimates of its emission ranging between 15 and 33 Tg S y^{-1} (Kettle and Andreae 2000). In the late 1980s, the hypothesis that DMS is involved in the biological regulation of global climate was put forward (Bates et al. 1987; Charlson et al. 1987). It is, however, only recently that DMS has been incorporated in global climate models (e.g., Aumont et al. 2002; Bopp et al. 2003, 2004; Gabric et al. 2004; Kettle and Andreae 2000; Simo and Dachs 2002). After emission to the atmosphere, this volatile sulphur compound is oxidised to sulphur dioxide (SO_2) and other products. From SO_2, non-sea-salt (nss) sulphate is produced, which can form sulphate (SO_4^{2-}) particles that act as condensation nuclei for water vapour. These nuclei affect the radiative properties of the atmosphere and clouds, with implications for climate. Higher numbers of condensation nuclei will deflect more incoming solar radiation back into space and thereby reduce the temperature on earth. The hypothesis that this process may modulate the greenhouse effect of increased anthropogenic CO_2 input to the atmosphere, was indirectly supported by the modelling results of the effect of anthropogenic SO_2 input to the atmosphere (Andreae et al. 2005; Mitchell et al. 1995). Although this study gave a rough indication of the counteracting effects of atmospheric SO_2 and CO_2, a quantitative understanding of all sources and sinks of atmospheric aerosols is still lacking (Andreae and Crutzen 1997).

Currently, anthropogenic SO_2 production exceeds natural SO_2 production by a factor of 3 (Bates et al. 1992), but the impact of the former on aerosol production is largely confined to industrialised areas of the Northern Hemisphere. The oceans, on the other hand, cover approximately 70% of the Earth's surface and much of this area is remote from man-made atmospheric contaminants. Consequently, the exchange of marine DMS is of high regional importance and may affect climate globally. Since the publication of a global inventory of DMS data by Kettle and co-authors (Kettle et al. 1999), it has become possible to include DMS in global climate models. Indeed, recent model calculations have shown that in the Southern Hemisphere, where anthropogenic sulphate emission is low, DMS plays a major role in the production of atmospheric nss-sulphate (Gondwe et al. 2003). Gondwe et al. calculated that the contribution of DMS to the total (global) atmospheric nss-sulphate burden is 18% and that it shows significant regional and temporal differences; e.g., in the Southern Hemisphere its annual contribution is 43% and over the Southern Ocean it is in excess of 80% during summer.

In one of the early attempts to add DMS in a global ocean climate model, Bopp et al. (2003) showed that a doubling of the atmospheric CO_2 concentration resulted in a reduction of the DMS flux at low latitudes and in enrichment at mid-latitudes. Thus, depending on the sign of the chance in DMS flux, the subsequent climate forcing by sulphur products could either alleviate or amplify the greenhouse effect (Bopp et al. 2004). The inclusion of DMS in the model was achieved by coupling the production of dimethylsulphoniopropionate (DMSP) and its conversion to DMS with the trophic status of the ecosystem, which in turn was based on the silica ratio, defined as the local simulated production of biogenic silica relative to the maximum production and is related to the proliferation of diatoms. It is computed from local silica concentrations and a variable Si:C ratio, which depends on the silica concentration. Although simplistic in its ecological approach, this study showed that an increased CO_2 concentration doesnot necessarily result in increased DMS production that may counterbalance the greenhouse effect, as suggested by the Charlson-Lovelock-Andreae-Warren (CLAW) hypothesis (Charlson et al. 1987). Clearly, an improved understanding of the biological processes is necessary to address the role of DMS in climate feedback mechanisms.

The production of DMS is almost exclusively through biogenic processes and shows strong seasonal and latitudinal variation (Kettle et al. 1999). DMS mainly results from the enzymatic cleavage of DMSP, a compound that is produced in several groups of marine phytoplankton. A complex network of production and consumption pathways of both DMSP and DMS involves most of the microbial food web (Fig. 1) and determines the concentration of

DMS in surface water and consequently its flux to the atmosphere (Malin and Kirst 1997). Physical and chemical ecosystem parameters all affect this network, potentially resulting in dramatic shifts in the DMS flux to the atmosphere. Although our knowledge on the qualitative aspects of the marine sulphur cycle has improved considerably during the past two decades, it is still difficult to quantify the effects of controlling factors on the various pathways.

Ecosystem modelling provides a tool for investigating how the DMS concentration and subsequently its flux to the atmosphere are regulated and what the most critical processes are. In a recent review on DMS and DMSP ecosystem models, Vezina (2004) concluded that although all current models will greatly benefit from improvements to the underlying ecosystem model, the quantitative understanding of the processes that drive variations in DMS and DMSP

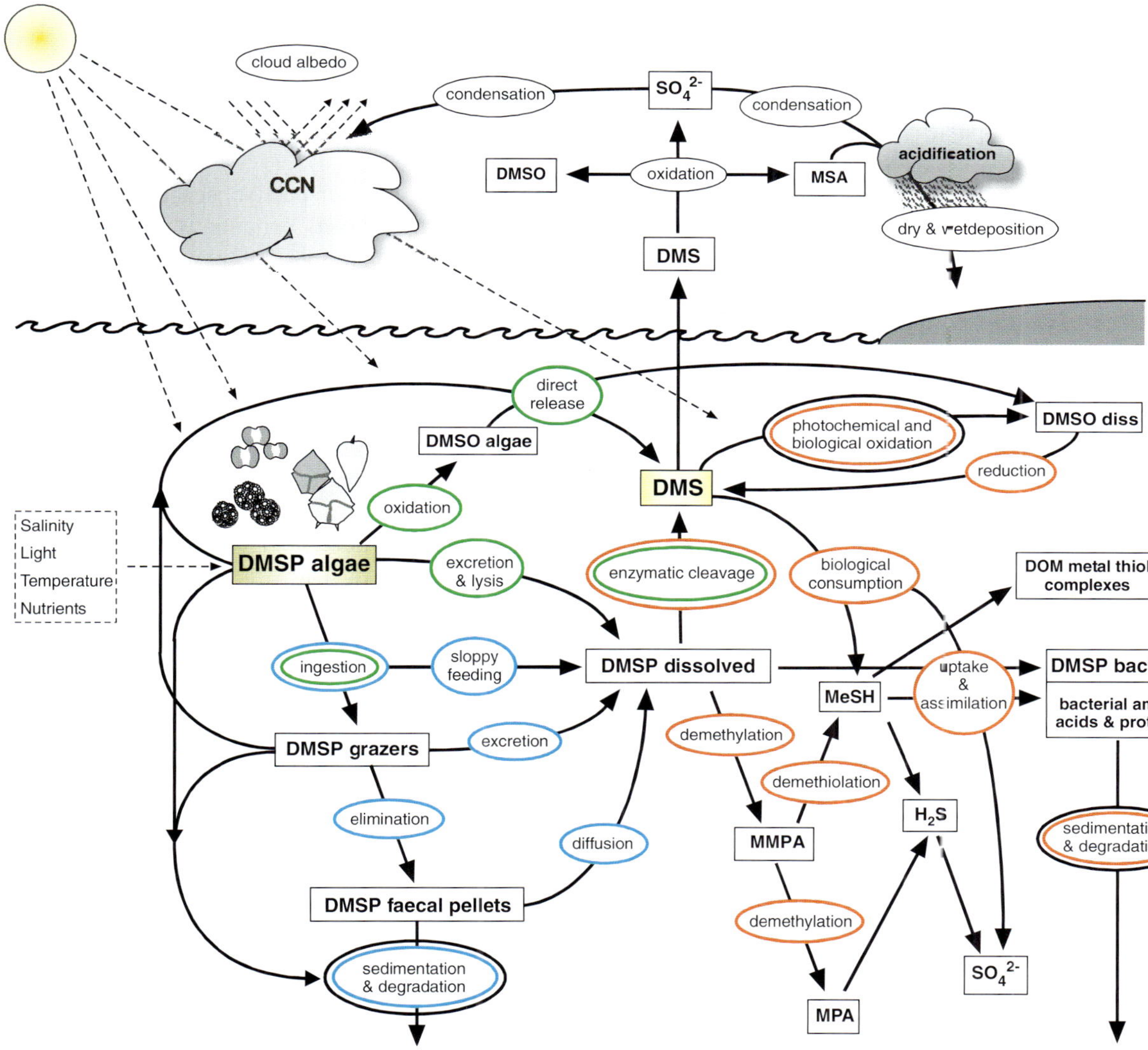

Fig. 1 Schematic representation of the processes and pools involved in the marine biogeochemical cycling of DMSP and DMS. Dominant role of functional groups in the different processes is indicated by coloured ellipses: green, phytoplankton; blue, zooplankton; red, bacteria; black, abiotic factors. CCN, cloud-condensation nuclei; DOM, dissolved organic material; DMSO, dimethyl sulphoxide; MeSH, methanethiol; MPA, mercaptopropionate; MMPA, methylmercaptopropionate; MSA, methanesulphonic acid

[DMS(P)] quotas and microbial yields is still too limited. Such insights are needed to inform laboratory and field studies and aid us in the development of more robust DMS(P)-modules within ecosystem models. During the past decade, many excellent reviews have been written on several aspects of the marine sulphur cycle. One of the emerging pictures is that this cycle is not only of interest for global climate, but that DMS and DMSP are compounds which are central to the microbial food web in their own right. The purpose of this review is not so much to reiterate these reviews, but rather to use pertinent information from them in an attempt to assist the development of parameterisations for DMSP and DMS modelling.

In this review, much attention has been given to two specific algal haptophyte taxa: *Phaeocystis* sp. and *Emiliania huxleyi*. These algae are well known as prolific producers of DMS and DMSP and their blooms can cover extensive areas in neritic and open ocean waters, respectively. Due to an increased interest to define the phytoplankton realm in models in more detail, we have tried to find unifying processes, but most published information is from *Phaeocystis* and *E. huxleyi*. Since our main goal is to provide an understanding of the complexity of the system, we have chosen to describe the various processes independently, even though this may have resulted in some repetition of observations. The level of detail may not be equal throughout the paper, but is a reflection of the current state of knowledge, our judgment of the potential impact of a specific process on the marine sulphur cycle and the assignment of different processes and functional groups in recent (complex) ecosystem models (e.g., Archer et al. 2004). We have tried to be as concise as possible, without losing information necessary for a holistic description. In order to evaluate the relative importance of individual pathways, we have provided the reader with an educated guess of the quantitative aspects, whenever possible. Obviously, when describing the different processes in detail, one comes across many gaps in knowledge. We have therefore taken the opportunity to highlight these gaps and make recommendations for future research.

Factors controlling DMSP production

A direct coupling of DMSP production with primary production would be ideal for modelling. However, there is no straightforward relationship since DMSP production is confined to a limited number of algal taxa. A further complicating factor is that the physiological conditions of the algal cells affect DMSP production. As a result, there is no definitive global relationship between algal biomass parameters such as chlorophyll-*a* and algal DMSP. Many ecosystem models are expressed in pools of carbon or nitrogen, whereas many global climate models are more often expressed in units of carbon. These models would benefit from conversion factors to describe the particulate DMSP pool, hence we choose to estimate DMSP:C ratios (on a molar basis) from literature data whenever possible.

Species composition

From an ecosystem perspective, species composition is the factor that affects community-DMSP production the most. Keller et al. (1989) made an extensive inventory of the DMSP content of 123 clones of marine phytoplankton, analysed during mid-exponential growth in nutrient replete media, and concluded that the major production of DMSP is found in a limited number of species, which mainly belong to the classes of Haptophyceae (=Prymnesiophyceae) and Dinophyceae (dinoflagellates). However, some members of the Chrysophyceae and Bacillariophyceae (diatoms) can also produce significant amounts of DMSP. In order to use this knowledge in models, we have recalculated published data of cellular DMSP concentrations to provide DMSP-to-carbon (DMSP:C) ratios, that can be used to estimate DMSP production in blooms of different taxonomic groups (Table 1). In addition, we have provided DMSP-to-chlorophyll-*a* (DMSP:chl-*a*) ratios, since several global models use satellite-derived chlorophyll-*a* data multiplied with a trophic status factor, as a proxy for particulate DMSP (Anderson et al. 2001; Aumont et al. 2002; Bopp et al. 2003; Simo and Dachs 2002). The carbon-to-chlorophyll-*a* conversion factor we used for this calculation (60 g/g) is typical for cultures that grow under nutrient-replete conditions and saturating light intensities (Geider 1987). Variations in abiotic factors in the field will, however, have a strong impact on this ratio and, as will be discussed in the following sections, on the DMSP:C and DMSP:chl-*a* ratios.

Table 1 Mean DMSP:C ratios, proportion of cell carbon composed of DMSP and DMSP:chlorophyll-*a* ratios in species groups, with the standard deviation in brackets. Data are recalculated from published data. Carbon per cell was calculated from cell volumes, according to the formula given by Menden-Deuer and Lessard (2000): diatoms: pgC/cell = 0.288 × cell volume (μm^3)$^{0.811}$; all other algae: pgC/cell = 0.216 × cell volume (μm^3)$^{0.939}$. Chlorophyll-*a* per cell was calculated assuming a C:chl-*a* ratio of 60 g/g (see comments in the text)

Species group	DMSP:C (mol:mol)	Proportion DMSP-carbon	DMSP:chl-*a* (mmol:g)	Data from
Diatoms (*n* = 22)	0.00086 (0.00126)	0.0043 (0.0063)	4 (6)	2
Chrysophytes (*n* = 6)	0.019 (0.015)	0.094 (0.075)	94 (75)	2
Dinoflagellates (*n* = 32)	0.022 (0.032)	0.111 (0.158)	111 (168)	2, 3, 5
Prasinophytes (*n* = 18)	0.005 (0.007)	0.025 (0.034)	25 (34)	1, 2
Haptophytes (*n* = 32)	0.011 (0.007)	0.053 (0.037)	52 (37)	1, 2, 3, 4
Prochlorophytes/Cyanophytes (*n* = 17)	0.0000015 (0.000004)	0.000008 (0.00002)	0.008 (0.02)	1, 2

1. Corn et al. 1996; 2. Keller et al. 1989; 3. Niki et al. 2000; 4. Stefels and van Leeuwe 1998; 5. Wolfe et al. 2002

As is evident from the standard deviations in Table 1, the variability within the groups is high. In fact, the haptophytes are the only group where all the species tested were observed to produce DMSP. In all other groups, one or several clones did not produce DMSP. Of the groups depicted in Table 1, the prochlorophytes/cyanophytes and diatoms produced the least DMSP. Some exceptions have been observed, but these are not usually found in open ocean areas, but rather at the fringes: in estuaries (e.g., *Melosira nummuloides* in Keller et al. 1989), in ice algal communities (Baumann et al. 1994; Kirst et al. 1991; Levasseur et al. 1994) and in benthic microbial mats (van Bergeijk et al. 2002). Such data have been excluded from the calculations summarised in Table 1.

The nano/picoplankton are a very diverse group of species that belong to several taxonomic groups and, hence, vary considerably in their DMSP production. For instance, almost all cyanophytes (e.g., *Synechococcus* species, *Trichodesmium*), prochlorophytes and cryptophytes tested did not produce DMSP, whereas other species that belong to the chrysophytes and prasinophytes do produce DMSP (Table 1). There are only a few field studies that report on the production of DMSP in picophytoplankton (<2 μm diameter). Corn et al. (1996) showed that picoplankton can contribute up to 25% of depth-integrated total DMSP in oligotrophic waters of the subtropical Atlantic. Within this fraction, the picoeukaryotes were the main DMSP producers, whereas the prokaryotes contributed less than 1% of picoplanktonic DMSP. In contrast, Wilson et al. (1998) investigated DMSP and DMS production in mesocosm enclosures and suggested that *Synechococcus* can be a significant producer of DMSP under nutrient-replete conditions.

Another aspect of the contribution of different species to global DMSP production is related to their ability to form blooms. Although Table 1 shows that the haptophytes do not have unusually high concentrations of DMSP per cell, the ability of *Phaeocystis* sp. and *Emiliania huxleyi* to dominate phytoplankton biomass during the formation of extensive coastal and oceanic blooms can result in elevated levels of DMS(P) (Table 2). These species therefore have received a great deal of research attention. *Phaeocystis* particularly is known to form almost monospecific spring blooms. Within such blooms, DMSP production is constrained by abiotic factors that affect cell physiology and the ability to produce and retain DMS(P). This topic is explored in the next section.

Physiological condition

The effect of abiotic parameters (light, nutrients, temperature and salinity) on the physiological condition of algal cells and subsequently on the production of DMSP, has been reviewed by Stefels (2000). DMSP is a multifunctional compound and there is no doubt that it has a role as a compatible solute in cell metabolism, but the regulation of its internal concentration is still unresolved. Because many abiotic parameters appear to have an effect to some extent, Stefels (2000) hypothesised that the production of

Table 2 DMS, dissolved DMSP ($DMSP_d$) and particulate DMSP ($DMSP_p$) concentrations and ratios to carbon and chlorophyll-*a* in blooms of *Phaeocystis* sp. and *Emiliania huxleyi*; within brackets are mean concentrations over the period of observation. When available, contribution to the bloom composition is given. See for a discussion on artificially elevated $DMSP_d$ concentrations the introductory paragraphs of the section 'Mechanisms of release to the dissolved fraction' and Kiene and Slezak (2006)

Area	DMS (nM)	$DMSP_d$ (nM)	$DMSP_p$ (nM)	DMSP:C (mol/mol)	DMSP:chla (mmol/g)	Period of observation	Data and comments
Phaeocystis							
S. Ocean—Davis Station			20–290			2 months	1
S. Ocean—Davis station			94–590			1 month	2
S. Ocean—Ross Sea	20–110				15*	Transect	3
S. Ocean—Bellingshausen Sea		8–300 (77)				1 month	4
	27			0.03*	18*	Single obs.	5
Bering Sea	3–19					3 days	6
Barents Sea—''station II''	9*	40*	30*	0.004	17*	Transect	7
N. Norway—Ullsfjord	5–41 (17)		25–213 (90)	0.006	31	15 days	8
NE Atlantic—Faerøe	20–85		20–220	0.02*	50*	Transect	9
West English Channel	24					Single obs.	10
S. North Sea	3–48 (25)				43	1 month	11
S. North Sea—coastal zone	7–47	6–22	110–470	0.004	15	1 month	12
Wadden Sea inlet	8–18	20–60	100–1,500		22	1 month	13
	2–40	3–35	200–1,650			1 month	14
Emiliania huxleyi							
UK nearshore waters	1–9					Periodic	15
Northeast Atlantic	3–51*		85–285	0.02	50–140	Transect	16
Northeast Atlantic	2–16#					Transect	17

DMSP might also serve as an overflow mechanism for excess reduced sulphur under conditions of unbalanced growth, when carbon and nitrogen flows are out of tune. The continued production and possible loss of DMSP would serve as a sink for excess carbon and at the same time regenerate intracellular nitrogen from methionine, which can then be used for synthesis of other amino acids. Although such a mechanism seems wasteful, the benefits are the continuation of the metabolic machinery. In this respect, it is comparable to the commonly observed exudation of carbohydrates by cells at high light and low nutrient concentrations. If indeed DMSP production is connected to overflow metabolism, this requires that DMSP is mainly located in the cytosol and that the intracellular equilibrium concentration is regulated by its degradation or loss from the cell rather than by its production. Transport of DMSP out of the cell would then be facilitated by the extracellular cleavage of DMSP (see also 'Maintenance of intracellular DMSP concentration: algal DMSP-lyase activity'). Although this hypothesis can explain many of the observed changes in DMSP content presented in the literature, direct evidence is still lacking.

Another hypothesis on the physiological function of DMSP was presented by Sunda et al. (2002). It suggests that DMSP and its breakdown products DMS, acrylate, dimethylsulphoxide (DMSO) and methane sulphinic acid (MSNA) together form a cascade of radical scavengers that may serve as an efficient anti-oxidant system that would need to be regulated in part by the enzymatic cleavage of DMSP. If true, one would expect the production of DMSP and its enzymatic cleavage to take place in the chloroplast, where most reactive oxygen species (ROS) are produced. This indeed seems to be the case with respect to the production of DMSP in

Table 2 Continued

Area	DMS (nM)	$DMSP_d$ (nM)	$DMSP_p$ (nM)	DMSP:C (mol/mol)	DMSP:chla (mmol/g)	Period of observation	Data and comments
Northeast Atlantic	4–8		29–52		44–112	5 days	18
Northern North Sea	3 ± 0.6	7.2 ± 3.9	60.9 ± 19.9		100 ± 20.2	5 days	19
Gulf of Maine	0.8–8	60–175	60–260			Single profile	20

1. Gibson et al. 1990. Original data published as DMS concentrations, but due to the use of $HgCl_2$, most probably an estimation of total DMS + DMSP (see also Curran et al. 1998)

2. Yang et al. (1994): data as for Gibson et al. (1990)

3. DiTullio and Smith 1995. Data from samples with 19′-hexanoyloxyfucoxanthin as the dominant pigment. Calculated from $DMSP_p$-data plus 11% of the DMS-data; the latter was estimated to be the methodological bias due to diluting and not-filtering the samples

4. Crocker et al. 1995. Data as for Gibson et al. (1990), but only dissolved DMS + DMSP

5. Turner et al. 1995

6. Barnard et al. 1984. Mean of samples in which *Phaeocystis* makes up >25% of cell density

7. Matrai and Vernet 1997. Mean of samples in which *Phaeocystis* makes up 43% of phytoplankton-C

8. Data from April 1997 field campaign of the EU-funded ESCAPE project: Belviso et al. 2006 (DMS and DMSP data) and Stefels (POC and chlorophyll data). Ratios are from the *Phaeocystis* maximum on April 25

9. Malin et al. 1993

10. Holligan et al. 1987

11. Liss et al. 1994 and Turner et al. 1996

12. Unpublished data from April 1998 field campaign of the EU-funded ESCAPE-project: Belviso (DMS and DMSP data) and Stefels (POC and chlorophyll data). Ratio's are derived from regression coefficients of the respective parameters. Carbon represents phytoplankton carbon as derived from the regression of POC versus chlorophyll-*a*

13. Kwint and Kramer 1996

14. van Duyl et al. 1998

15. Turner et al. 1988. Samples with >20% coccolithophores; winter and summer 1985

16. Malin et al. 1993. June-July 1987. Samples with >50% of total carbon biomass as coccolithophores

17. Holligan et al. 1993. *E. huxleyi* bloom, June 1991

18. Jickells et al. unpublished data from ACSOE cruise in the NE Atlantic, June 1998 (eddy—Lagrangian), with *E. huxleyi* dominating

19. Archer et al. 2002. DISCO Lagrangian experiment, June 1999; *E. huxleyi* contributed 16% of the $DMSP_p$ standing stock in surface waters

20. Matrai and Keller 1993. Centre of an *E. huxleyi* bloom, July 1990; data range in top 10 m of one depth profile

* Estimated from published figures

\# Surface concentrations along a transect at 20°W

higher plants (Trossat et al. 1996), but there is no conclusive evidence for this in marine algae, which use a different biochemical pathway for DMSP production (Gage et al. 1997; Summers et al. 1998). Another complicating factor is that with the common techniques for DMS(P) analysis it is impossible to measure the fluxes through this cascade of compounds. Sunda and co-workers suggested the anti-oxidant hypothesis on the basis of elevated concentrations of intracellular DMSP under stress conditions. In the process of radical scavenging, however, DMSP would be converted into one of its breakdown products. Therefore, a loss of DMSP would be expected, unless the stress reaction results in increased de novo synthesis (up-regulation) of DMSP. Only in those cases, a subsequent overshoot production may lead to increased intracellular concentrations of DMSP and/or one of the downstream products. A method for the measurement of de novo synthesis of DMSP is clearly warranted.

Since unbalanced growth and the production of ROS often co-occur under high irradiance and/or nutrient-limited conditions, it is difficult to test the two hypotheses individually without detailed investigation of the physiological condition of the cells and fluxes through the relevant biochemical pathways. Moreover, the two hypotheses do not necessarily need to be mutually exclusive, since a function in oxidative stress management does not exclude additional functions in cell metabolism. However, for a better understanding, one should be aware of the fundamental differences in operating principles and value the published data accordingly. Here, we present only an update of the current knowledge and quantify the relationships whenever possible.

Salinity

There is little doubt that an increase in salinity will result in an increase of the equilibrium concentration of intracellular DMSP, but an active up- or down regulation of its concentration upon short-term salinity changes (minutes to hours) has not been observed (reviewed by Stefels 2000). Indeed, phytoplankton from high-salinity environments such as coastal rock pools, hypersaline lakes or sea-ice pore waters are known to accumulate relatively high concentrations of DMSP. In *Phaeocystis* cultures, an increase of salinity resulted in an exponential increase of intracellular DMSP (Stefels 2000). Therefore, salinity effects need to be taken into account when modelling DMSP production in extreme environments, e.g. when modelling the production of DMSP in sea ice.

Light

Most experiments on the effect of light on DMSP production by algae have been carried out with *Emiliania huxleyi*, a species that is unusual in its ability to grow at very high light (>1,000 μmol m^{-2} s^{-1}) intensities (Paasche 2001). In order to make the data from the various publications comparable, we have converted intracellular DMSP quota to DMSP:C ratios. From these data, a good correlation between increasing DMSP:C-ratio

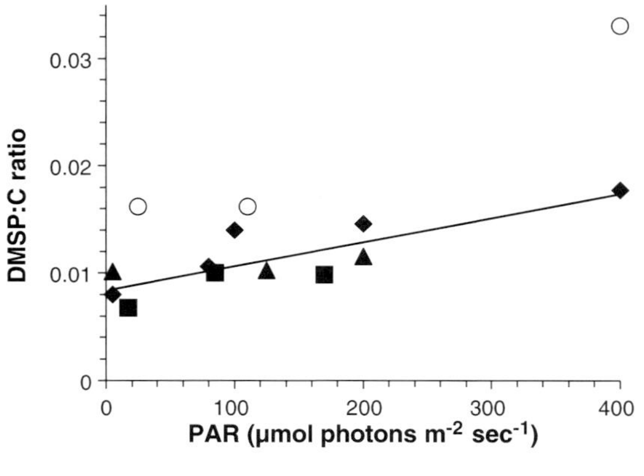

Fig. 2 Effect of irradiance on the cellular DMSP:C-ratio (mol:mol) in *Emiliania huxleyi* (closed symbols). Data are recalculated from: Slezak and Herndl (2003) (diamonds); Keller and Korjeff-Bellows (1996) (squares); van Rijssel and Buma (2002) and van Rijssel and Gieskes (2002) (triangles). For comparison, data from experiments with *Phaocystis antarctica* are included (open symbols; Stefels and van Leeuwe 1998; Stefels unpublished). Equation of the linear regression fit (*E. huxleyi* data only): DMSP:C = 0.00002 × PAR + 0.0084 (r^2 = 0.693)

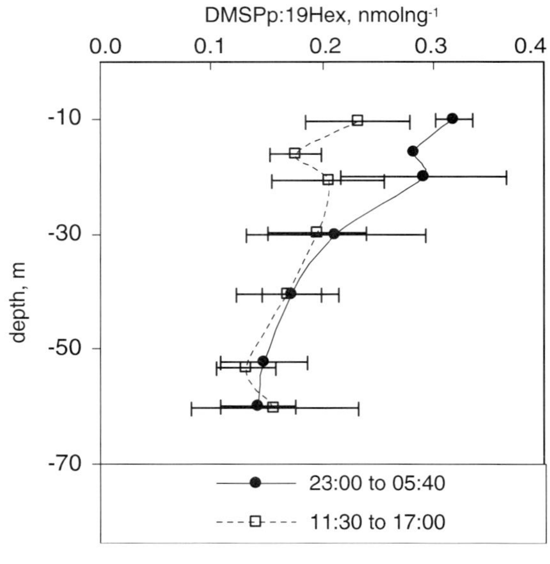

Fig. 3 $DMSP_p$: 19′-hexanoyloxyfucoxanthin (19′Hex) ratios with depth in the Southern Ocean during light and dark periods. Daylight hours (open symbols), night hours (filled symbols). The data were collected over an 18-day period during which daylight and dark profiles were evenly distributed and the mixed layer depths were between 50 m and 100 m. The data are binned by depth and shown as averages with range bars of one standard deviation. There are 4-10 data values in each average except for the single dark 15 m depth and the average depth of the 1% light level was 55 m

and increasing irradiance can be observed (Fig. 2). For comparison, data for *Phaeocystis antarctica* are included in the figure. A comparable increase with irradiance can be observed, but with an overall higher offset. This offset might be the result of additional temperature effects.

Getting information from field data on the controls on cellular DMSP is not straightforward but there is some evidence from a Lagrangian experiment in the Southern Ocean, which tends to suggest that light is an important factor at the daily scale (Fig. 3; S. Turner and I. Peeken, unpublished data). Figure 3 shows vertical profiles of $DMSP_p$, normalised to 19′-hexanoyloxyfucoxanthin (19′Hex), an accessory pigment which is indicative of Prymnesiophytes. Since the changes in the ratios are dominated by variation in DMSP, the results suggest that there is a diel cycle, in which $DMSP_p$ is consumed during the day. This may be supportive of an antioxidant function, but also shows that short-term effects can be opposite to long-term effects as presented by Sunda et al. (2002) and as shown in Fig. 2. Clearly, better methods are warranted to measure fluxes through the $DMSP_p$ pool in field samples.

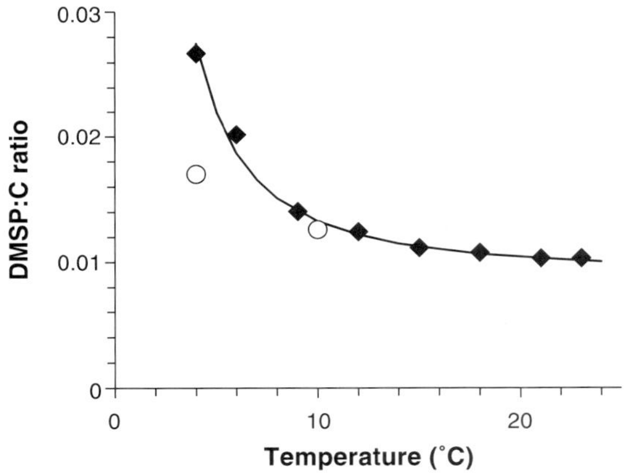

Fig. 4 Effect of temperature on the cellular DMSP:C-ratio (mol:mol) in *Emiliania huxleyi* (closed symbols; recalculated from van Rijssel and Gieskes (2002). For comparison, data from experiments with *Phaeocystis antarctica* (4°C; Stefels and van Leeuwe 1998) and *P. globosa* (10°C; Stefels and van Boekel 1993) are included. Except for *P. antarctica*, carbon data are calculated from cell volume data (see Table 1). Equation of the power fit to the *E. huxleyi* data is: $DMSP{:}C = 0.009 + 0.17\ T^{-1.6}$

Temperature

DMSP has been found to be a compatible solute for cell metabolism under cold conditions (Karsten et al. 1996; Nishiguchi and Somero 1992, reviewed in Stefels 2000). The observation that DMSP is present in many ice algae, including the diatoms, and in many pelagic algae from polar regions (Matrai and Vernet 1997) may be indicative of its functionality under cold conditions. Surprisingly, only two studies report on the acclimatisation of the intracellular DMSP concentration at various temperatures (Sheets and Rhodes 1996; van Rijssel and Gieskes 2002). Converting the Van Rijssel and Gieskes (2002) data on *Emiliania huxleyi* to DMSP:C ratios gives a correlation with temperature as shown in Fig. 4. The added data points for *Phaeocystis* are for *P. globosa* at 10°C and for *P. antarctica* at 4°C (J. Stefels unpublished data) and fall close to the *E. huxleyi* relationship. However, further data are needed to establish whether the same relationship holds for DMSP-producing species that do not belong to the haptophytes.

Nutrients

After Challenger (1951) had noticed the structural analogy between DMSP and glycine betaine (GBT), many have suggested that DMSP could replace GBT as an osmoregulator under nitrogen-limited conditions. Indeed, there are several reports of increased cellular DMSP content under N-limited growth, but there are also reports of the contrary (Stefels 2000 and refs. therein). As a whole, the effects of nutrient limitations on DMSP production are still enigmatic.

From a physiological point of view, the intracellular concentrations of organic solutes are most relevant, as it is this concentration that affects enzymatic processes. Shifting from unlimited towards limited growth in a batch culture, cell volume often reduces under nitrogen and iron limitation and stays constant under phosphate limitation. Therefore, an increased intracellular DMSP concentration under N or Fe limitation is at least partly due to a reduction in cell volume (Bucciarelli and Sunda 2003; Keller et al. 1999a; Stefels and van Leeuwe 1998).

From a modeller's point of view, it is more relevant whether or not the DMSP production can be related to primary production. A complicating factor is that under nutrient-limited conditions the concen-

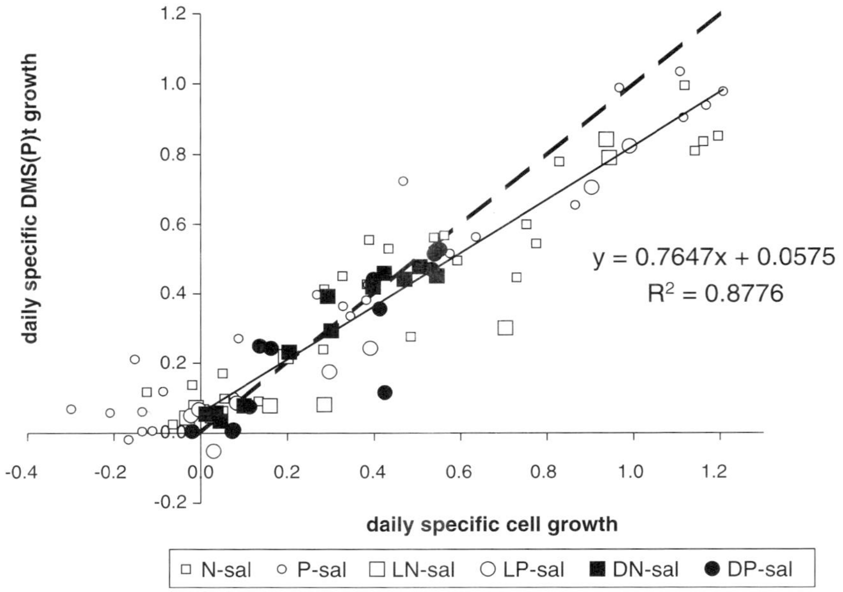

Fig. 5 Daily specific growth rates of total DMS + $DMSP_d$ + $DMSP_p$ versus specific cell growth in a variety of axenic *Phaeocystis globosa* batch cultures under different conditions. Cultures were grown at the same temperature (11°C), but with different salinities, nutrient ratios and/or light conditions. Specific growth rates were calculated per day. Culture growth details are as follows: N-sal: salinity range from 25 to 50 PSU, nitrogen limited; P-sal: salinity range from 25 to 50 PSU, phosphate limited; LN-sal: high irradiance (120 μmol PFD), 30 and 40 PSU, nitrogen limited; LP-sal: high irradiance (120 μmol PFD), 30 and 40 PSU, phosphate limited; DN-sal: low irradiance (10 μmol PFD), 30 and 40 PSU, nitrogen limited; DP-sal: low irradiance (10 μmol PFD), 30 and 40 PSU, phosphate limited. Regression line is computed from data with positive specific cell growth only. Dashed line indicates a 1:1 relationship

trations of DMS and dissolved DMSP often increase, either due to cell lysis or to active exudation (Laroche et al. 1999). The question is how the total production of DMS and DMSP is related to algal growth. Unfortunately, total pools of DMS(P) are rarely presented in the literature. In Fig. 5, a compilation of experiments with *Phaeocystis globosa* is presented (unpublished data, J. Stefels), in which it is clearly shown that the specific DMS + DMSP production is coupled to cell growth and compares well under a variety of conditions: a range of salinities, low or high light conditions and either nitrogen or phosphorus limitation. The fact that the regression coefficient deviates from 1 reflects the observation that under unlimited (high) growth rates, cells tend to divide faster than they grow in terms of carbon, which results in a cell-size reduction during exponential growth. The positive *Y*-intercept indicates that at limited (low) cell growth, DMSP production continues under all conditions, even when cell numbers decline (negative growth). Whether this production is due to a few healthy cells that are still growing amidst a majority of inactive or dead cells, or because of a reaction to stress is unknown. This compilation shows that there is no increased production under stress conditions as suggested in the anti-oxidant hypothesis and that a modeller's practice of coupling DMSP production to cell growth is an appropriate approach.

A potential pitfall in carbon-based models is, however, the decoupling between cell growth and carbon growth, which would result in shifts of the DMSP:C ratios. Direct evidence for such a decoupling between the carbon and sulphur cycles is still lacking, but indirect evidence can be sought in calculated DMSP:C ratios relative to growth rates. In Fig. 6, a compilation of available data is given. In those cases where only cell volume data were available, cell carbon is calculated according to Menden-Deuer and Lessard (2000). This may result in an underestimation of cell carbon under N or Fe limitation (and thus an overestimation of the DMSP:C ratio), since, under those conditions, cells often become carbon denser, i.e., an increase in cell carbon per cell volume (Stefels and van Leeuwe

1998). In the low-DMSP producing diatom *Thalassiosira pseudonana* a clear effect of nutrient limitation on DMSP:C ratios could be found, with highest ratios under N limitation and all other conditions comparable. However, in all other high-DMSP-producing species, changes of the ratio are negligible (see figure legend for description and references). This suggests that cells with a high DMSP content do not respond to nutrient limitation—or at least that a response in the DMSP production is too low to affect the DMSP:C ratio—whereas cells with low DMSP content do react. Modellers may implement this by assigning different behaviour towards nutrient limitation to different plankton groups in their model.

Maintenance of intracellular DMSP concentration: algal DMSP-lyase activity

Algae can adjust the intracellular concentration of DMSP through the biosynthetic (anabolic) or the degradation (catabolic) pathways. DMSP-lyase enzymes facilitate the degradation pathway, in which DMSP is cleaved to DMS, acrylate and a proton. What controls the activity of DMSP-lyases in phytoplankton is still unknown. Stefels (2000) suggested

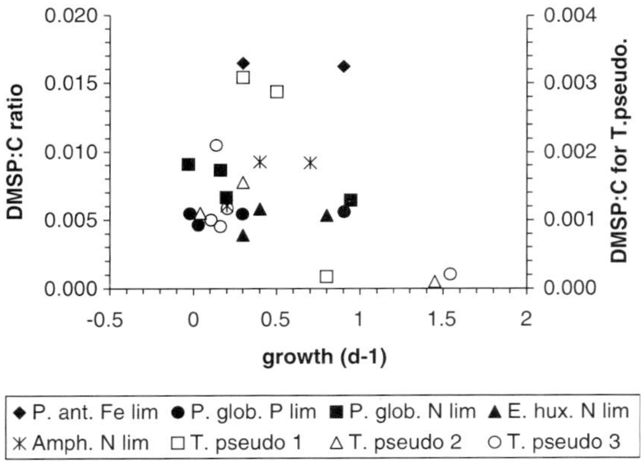

Fig. 6 Cellular DMSP-to-carbon ratios (mol:mol) versus specific growth rates under various nutrient limited conditions. *Phaeocystis antarctica*: Fe limited (Stefels and van Leeuwe 1998); *P. globosa*: P and N limited (Stefels unpublished; carbon is calculated from cell volume; see Table 1); *Emiliania huxleyi*, *Amphidinium carterae* and *Thalassiosira pseudonana* 1: N-limited chemostats (Keller et al. 1999b); *T. pseudonana* 2: Fe and CO_2 limited (Sunda et al. 2002; carbon is calculated from cell volume; see Table 1); *T. pseudonana* 3: N, P, Si and CO_2 limited (data are taken from the exponential and early stationary phase of growth, Figs. 2–7 in Bucciarelli and Sunda 2003; carbon is calculated from cell volumes, see Table 1)

that in the case of the production of DMSP as an overflow mechanism, the intracellular equilibrium concentration has to be regulated by its degradation rather than by its production. Since DMSP is a zwitter ion, this could be achieved by actively transporting DMSP out of the cell. Subsequent removal of DMSP from the transporter site by extracellular cleavage would facilitate the release, since concentration gradients are kept maximal. Such a mechanism would necessitate a membrane-bound extracellularly located DMSP-lyase. This is particularly relevant for organisms with a thick boundary layer such as *Phaeocystis* colonies. The idea of such a role for DMSP-lyase was instigated by the observations that *Phaeocystis* exhibits high in vivo lyase activities when dissolved DMSP is added to the culture medium and that extracellular inhibitors can repress this activity (Stefels and Dijkhuizen 1996). In addition, it has been observed that acrylate, one of the products of DMSP degradation, accumulates in the mucus layer of *Phaeocystis* colonies (Noordkamp et al. 2000), which is in agreement with an extracellular location of the enzyme. Sunda et al. (2002) suggested that DMSP-lyase is involved in the scavenging cascade that detoxifies cells of harmful oxygen radicals, as it facilitates the production of DMS and acrylate, two products that are highly efficient ROS scavengers. This hypothesis would favour an intracellular and possibly chloroplastic location of the lyase.

Despite being so central for the release of DMS, relatively little is known about DMSP-lyase and its physiological regulation. Molecular information on the genetic sequences of DMSP-lyases is lacking and it still is possible that nonspecific house-keeping enzymes that accept a wide range of substrates perform this reaction. In the studies done so far, both intra- and extracellular isozymes of DMSP-lyase have been found in several *Emiliania huxleyi* and *Phaeocystis* strains, apparently with different kinetic characteristics (Steinke et al. 1998; Stefels et al. in prep). Both studies conclude that DMSP-lyases are constitutively present and there is as yet no indication of up- or down-regulation by abiotic factors, although only few dedicated experiments have been carried out to date. Shifts in enzyme affinity during growth have been observed in batch cultures, but are as yet unexplained (Table 3; Stefels and van Boekel 1993; Steinke et al. unpublished data). In addition, several

Table 3 Kinetic parameters of DMSP-lyase in whole-cell samples (in vivo activity) and in crude extracts (in vitro activity) of cultures or field samples. Data are usually published on a per-cell basis, but the data here have been recalculated to cell-carbon units (see also legend Table 1). In many cases, activities were measured at non-saturating substrate concentrations. Therefore, if no V_{max} or K_m values are available, the apparent enzyme efficiency (i.e., the initial slope of a *V*/*S* plot) was calculated, assuming Michaelis Menten kinetics with linearly increasing activities in the non-saturating substrate range and a K_m in the millimolar ranges

	Culture conditions	Assay conditions	V_{max} (μmol DMS gC^{-1} min^{-1})	K_m (mM DMSP)	Efficiency (L gC^{-1} min^{-1})	Data from
Laboratory—in vivo activity						
Phaeocystis globosa	10°C, 85 μE, early exp. growth	20°C, dark	167	0.012	7.65	1
	10°C, 85 μE, late exp. growth	20°C, dark			1.37	1
Phaeocystis globosa (5 strains)	12–23°C, 40 μE, late exp. growth	Growth temp, dim light			0.06 (0.03)	8
Phaeocystis antarctica	4°C, variable light, late exp. growth	4°C, dim light			0.03	8
Scrippsiella trochoidea	20°C, exp. growth	20°C, dim light			9.81	5
Heterocapsa triquetra	20°C, exp. growth	20°C, dim light			19.55	5
Laboratory—in vitro activity						
Phaeocystis globosa	11°C, 85 μE, late exp. growth	pH 8.3, 3 mM DTT, 20°C, dark	670	2.25	0.15	3
Emiliania huxleyi CCMP370 (low lyase act.)	15°C, 80 μE, late exp. growth	pH 6, 1 mM DTT, 30°C	1.3	5.5	0.0001	4
Emiliania huxleyi CCMP373 (high lyase act.)	15°C, 80 μE, late exp. growth	pH 6, 1 mM DTT, 30°C	968	1.4	0.35	4
Alexandrium tamarense (3 strains)	15°C, 80 μE, exp. growth	pH 4–6, 30°C	46			7
Alexandrium fundyense	15°C, 80 μE, exp. growth	pH 4–6, 30°C	225			7
Field—in vitro activity						
Phaeocystis	Dutch coast *Phaeocystis* bloom	pH 8.5, 5 mM DTT, 20°C, dark			0.08	2
Mixed population	North Atlantic coccolithophorid bloom	pH 6, 22–27°C	25*			6
Field—in vivo activity						
Mixed phytoplankton mesocosm:	Peak of *E. hux.* Numbers	pH 8.2, in situ temp. = 12°C, dim light	22	0.5	0.02	9
Emiliania huxleyi dominated	After *E. hux.* collapse		34	0.01	1.75	

1. Stefels and van Boekel 1993; 2. Stefels et al. 1995; 3. Stefels and Dijkhuizen 1996; 4. Steinke et al. 1998; 5. Niki et al. 2000; 6. Steinke et al. 2002a; 7. Wolfe et al. 2002; 8. Stefels, unpublished; 9. Steinke, unpublished

*Data taken from Fig 4b of Steinke et al. 2002a, assuming that dinoflagellates were responsible for the lyase activity

studies suggest that not all DMSP-producing algae have DMSP-lyase activity. Niki et al. (2000) tested five species of phytoplankton for their in vivo lyase activity, three of which did not show any activity. These were a non-DMSP-producing raphidophyte and two DMSP-producing haptophytes. A lack of DMSP-lyase activity was also found in a suite of DMSP-producing coccolithophorid species (Franklin et al. unpublished data). Steinke et al. (1996) did not find any in vitro lyase activity in the DMSP-producing prasinophyte *Tetraselmis subcordiformis*. Non-optimised assay conditions may contribute to these findings but this could also indicate that the ability to synthesise DMSP is not necessarily related to the presence of a DMSP-lyase, which makes the role of the enzyme all the more puzzling.

For modelling purposes two aspects of algal DMSP-lyases are relevant. Firstly, the activity of an extracellularly located enzyme may contribute to the conversion of the (ambient) dissolved DMSP pool. In this case, the apparent enzyme efficiency (α; i.e., the initial slope of a *V/S* plot showing DMSP-lyase activity versus DMSP concentration) at ambient conditions obtained during in vivo assays is relevant. Secondly, the activity of intra- or extracellular enzymes may be crucial for the direct conversion of intracellular DMSP, after cells break up due to autolysis, viral attack or grazing (see below for a description of these processes). In these cases, one can envisage that the lyase, formerly separated from its substrate, is exposed to elevated DMSP concentrations, resulting in high production rates of DMS. The conversion rate of this DMSP might then be related to the V_{max} of the enzyme. In Table 3, we tried to make the available published data comparable by recalculating them on a cell-carbon basis.

Another important question is how algal lyase activity compares to bacterial activity. Although the latter will be discussed in the section 'Microbial consumption of DMSP', several studies on field samples indicate that algal lyase activity can be as important, if not more, as bacterial lyase activity. Stefels et al. (1995) found a strong correlation between *Phaeocystis* abundance and in vitro lyase activity in Dutch coastal waters. No lyase activity was found in size fractions <10 µm. During sampling of a bloom of *E. huxleyi* in the North Atlantic, Steinke et al. (2002a) found >74% of the in vitro lyase activity associated with particles >10 µm and suggested that dinoflagellates were responsible for this activity. Niki et al. (2000) calculated that the algal lyase pathway is as important as the bacterial lyase pathway, in samples from Tokyo Bay. Also other size fractionation experiments have shown that lyase activity is often associated with the large size fractions (Cantin et al. 1999; Scarratt et al. 2000), although it cannot be excluded that attached bacteria are partly responsible for this activity. In a modelling study of the seasonal evolution of DMS in the Southern Bight of the North Sea, van den Berg et al. (1996) showed that the presence of an algal DMSP-lyase associated with the occurrence of *Phaeocystis* was essential to properly describe the DMS spring peak.

These results suggest that for modelling purposes it might be beneficial to distinguish between phytoplankton groups with and without DMSP-lyase. The limited data we have so far suggest that the dinoflagellates and part of the haptophytes have DMSP-lyase. This has been incorporated in our recommendation on plankton groups and their characteristics (section 'Conclusions'; Table 6).

Fate of DMSP-sulphur

Table 1 shows that the proportion of DMSP-carbon to the total cell-carbon in the different phytoplankton groups varies between 0% and 11%. This suggests that this single component can be an important fraction of the labile dissolved organic carbon pool and therefore highly relevant as a carbon source for bacteria. It also indicates that the coupling of the fate of dissolved DMSP and that of DOC (Archer et al. 2004) might be promising in advancing future models. There is considerable evidence that in many marine areas the DMS that ultimately reaches the atmosphere accounts for only a small percentage of the DMSP-sulphur originally produced by the algae (e.g., Bates et al. 1994; Kiene and Bates 1990; Kiene and Service 1991). More recently, it was shown that this percentage yield can vary considerably (between 1% and 40%) and can be correlated to the mixed layer depth (Simo and Pedros-Alio 1999a). Both biotic and abiotic processes will affect the yield of DMS from DMSP. We will first focus on those factors involved in the release and conversion of DMSP and then discuss processes affecting the consumption of DMS.

Mechanisms of release to the dissolved fraction

Pathways through which particulate DMSP is released into the dissolved phase are active exudation, cell lysis due to senescence or to viral attack and grazing by zooplankton. In the many publications that describe these processes, this release is presented either in the form of dissolved DMSP ($DMSP_d$) or in one of its degradation products. For models, however, an important question is whether this release goes through the dissolved DMSP pool, with subsequent conversion by bacterial and algal enzymes, or that release and conversion of the particulate DMSP pool is intrinsic to the release process. In the vicinity of or within a bursting cell, high-affinity enzyme systems will result in the rapid conversion of DMSP to DMS, whereas in the case of no or only low-affinity enzyme systems the same process may result in dissolved DMSP alone. The subsequent fate of $DMSP_d$ will then be consumption by bacteria with a DMS yield depending on the development of the bacterial community. The same release process may thus result in completely different DMS yields and therefore needs to be modelled differently for the different plankton groups.

A complicating factor in the evaluation of published data on the fractionation between dissolved and particulate DMSP is the potential for overestimations of $DMSP_d$, due to release from the cells during filtration (Kiene and Slezak 2006). $DMSP_d$ is taken to be the filter fraction of the less than 1 μm fraction, but for fragile, DMSP-containing flagellates such as *Phaeocystis*, filtration can easily result in operational release of dissolved organic material from the cells. Any use of published data should, therefore, involve a thorough check of the analytical methods used for the quantification of $DMSP_d$. The following sections address the different mechanisms of release of DMSP-sulphur from algal cells.

Exudation

Information on active exudation of DMSP is limited. Laroche et al. (1999), using a modelling study, indicated that DMSP exudation in the dinoflagellate *Prorocentrum minimum* is independent of growth phase at the level of 1% of its DMSP quota per day. In contrast, they showed that *Phaeocystis* sp. exudation rates, measured as cumulative DMS concentration in the cultures, are highest at the end of logarithmic growth and account for 3–11% per day. DMSP in ice diatoms can be an important source of $DMSP_d$ and DMS, following ice break-up in polar waters when interstitial plankton communities are released from hypersaline porewaters into seawater of much lower salinity (Levasseur et al. 1994). The ability to remove and rapidly cleave DMSP to DMS may be advantageous under such circumstances but as yet ice algae have not been tested for the presence of DMSP lyases. Hence, the exudation of DMSP is species-specific and can be affected by abiotic parameters such as salinity and temperature shifts and nutrient limitation. It thus can be a significant source of $DMSP_d$ and, where active DMSP-lyase is present, DMS. Especially in stress situations, however, it will be difficult to distinguish active exudation as a physiological response from cell lysis.

Autolysis

Recent findings indicate a distinction between programmed cell death (PCD) or apoptosis—a form of autocatalytic cell suicide in which morphological and biochemical changes lead to cell dissolution—and necrosis, which is a passive form of cell death and leads to immediate cell rupture e.g., after injury (Bidle and Falkowski 2004). In multicellular organisms, PCD results in phagocytosis without releasing the cellular contents, whereas necrosis does result in release of cell content. Whether a comparable distinction can be made for phytoplankton cells is unknown. It has been suggested that in algae, initiation of PCD occurs under nutrient limited or otherwise stressful conditions and that PCD can be regarded as a refinement of the term autolysis (Bidle and Falkowski 2004). Autolysis has long been assigned as the prime pathway for release of particulate DMSP (Nguyen et al. 1988), since, at the end of a bloom, cells disintegrate and DMSP is released into the water. This process thus parallels possible active exudation by nutrient-limited cells that are alive.

Viral lysis

To date only a few studies have examined how viral infection and lysis of DMSP-containing cells affect the production of DMS. Malin et al. (1998) found that viral infection of a *Phaeocystis pouchetii* culture resulted in cell lysis and four fold increase of DMS after 20 h. However, this strain was not axenic so it was not possible to ascertain whether the measured DMS derived directly from the algal cells or from bacterial conversion of DMSP to DMS. In contrast a parallel study on *Micromonas pusilla* showed DMSP release, but DMS production was only present in bacterised cultures (Hill et al. 1998). More recently Evans et al. (2007) have found that infected axenic cultures of *Emiliania huxleyi* do produce elevated levels of DMS (see Wilson et al. 2005 and references therein for more details of this virus/host system). Whether the quantity of DMS produced relates to the DMSP lyase activity of the host has yet to be proven. As far as we are aware there are no data concerning the activity of this enzyme in *M. pusilla* and viral pathogens of high-lyase *E. huxleyi* have only been isolated very recently (Matrai et al., pers. com.).

Whilst these culture studies suggest that viral lysis could be a significant DMS production pathway, quantification of the contribution of this process versus grazing or autolysis in natural plankton assemblages is challenging. Evans et al. (2003) used a modified dilution protocol on three successive days to estimate viral and grazing related mortality in a mesocosm population of *Micromonas* spp. They found turnover rates of 10%, 25% and 9% of the standing stock per day compared to a microzooplankton grazing turnover of 48%, 26% and 23% per day. This technique shows promise but has yet to be applied successfully in natural blooms of phytoplankton that contain high levels of DMSP, so at the present time we lack quantitative data for this process. Nevertheless, it is interesting to consider more qualitative data from field-based studies. Bratbak et al. (1995) conducted a mesocosm study where DMS and $DMSP_d$ levels did not respond to virus-induced termination of a bloom of *E. huxleyi* and reasoned that this was due to bacterial degradation which prevented the accumulation of these compounds. In a Lagrangian study of an *E. huxleyi* bloom in the Northern North Sea, Wilson et al. (2002) concluded that microzooplankton grazing out-competed viral infection. In contrast, in mesocosm studies in 2000 and 2003 (University of East Anglia group, unpublished data) we found substantial increase in DMS concentration concomitant with the virus-induced demise of *E. huxleyi* blooms. There is a clear need for reliable new tools that would allow for the quantification and differentiation of DMS(P) production by virus-induced mortality, mortality due to autolysis or grazing

Grazing by microzooplankton

A first indication of an important role for the microbial foodweb in the conversion of DMSP to DMS was provided by size-fractionation experiments of field samples (Belviso et al. 1990; Christaki et al. 1996). A mechanistic explanation for the release of DMS during grazing was later provided by laboratory experiments. Wolfe and Steinke (1996) found that grazing by the herbivorous dinoflagellate *Oxyrrhis marina*, which engulf their prey by phagotrophy, could greatly increase the production of DMS and that the DMS yield of the ingested prey-DMSP was correlated to the in vitro lyase activity of the prey (Table 4). This led to the assumption that DMSP and DMSP-lyases are present in *E. huxleyi* but are segregated into different cell compartments and react with each other once they are mixed in the food vacuoles of microzooplankton after ingestion. Strom et al. (2003a) also found proportionality between DMS production and prey-DMSP lyase activity in grazing experiments with *O. marina* and four *E. huxleyi* strains. The relationship between DMS yield and lyase activity of the prey, suggests that the predator merely accelerates the conversion of DMSP into DMS and does not affect the conversion pathways. Due to a lack of measurements, the fate of the unconverted prey DMSP in these experiments is unknown. In grazing experiments with another heterotrophic dinoflagellate, *Gyrodinium dominans*, Tang and Simo (2003) calculated that 32% and 44% of ingested prey DMSP was retained in the grazer, irrespective of the lyase activity of the prey (Table 4). This would roughly comply with a growth efficiency of 0.3 as has also been suggested by Archer et al. (2001b).

Since DMS production from DMSP results in equimolar concentrations of acrylate, a compound that has been attributed with antimicrobial properties (Sieburth 1960), grazing-induced DMSP cleavage could affect microbial grazers. Wolfe et al. (1997) tested the grazing behaviour of *O. marina* with low- and high-lyase *E. huxleyi* and mixed-prey experiments with the non-DMSP-producing chlorophyte *Dunaliella tertiolecta*. On the basis of the results, the authors suggested the existence of a grazing-activated

Table 4 Grazing mediated production of DMS and $DMSP_d$ as percentage of algal $DMSP_p$ stock and as a percentage of ingested $DMSP_p$, by micro- and macrozooplankton. In the column with DMS + $DMSP_d$ production of ingested DMSP, notations in brackets refer to release of DMSP repacked into faecal pellets

Area	Grazer	Dominant prey species	DMSP stock grazed (% d^{-1})	DMS + $DMSP_d$ production as % of stock (% d^{-1})	DMS + $DMSP_d$ production of ingested DMSP (%)	DMS yield of ingested DMSP (%)	DMSP retained by grazers (%)	Data and comments
Laboratory—micrograzers								
	Gyrodinium dominans	*Phaeocystis*			nd	nd	44	1
		Isochrysis galbana					32	
	Oxyrrhis marina	*Emiliania. huxleyi 370*				8.4	nd	2
		Emiliania. huxleyi 373				60.6		
	Oxyrrhis marina	*Isochrysis galbana*	19		110	63	nd	3
	Strombidium sulcatum	*Isochrysis galbana*			>65		nd	4
Field								
Mediterranean	Ciliates	Mixed			40–50		50–60	5
S. North Sea—coastal	Ciliates, heterotr. dinofl.	*Phaeocystis*	31	10	37	23		8
S. North Sea—offshore	Ciliates, heterotr. dinofl.	Mixed with minor *Phaeocystis*	12	3	23	75		8
S. North Sea	Ciliates, heterotr. dinofl.	*Phaeocystis*	20	14	69	9		6
Iceland Basin	Ciliates, heterotr. dinofl.	Nano-, dinoflagellates	17	12	67	32		6
N. North Sea	Ciliates, heterotr. dinofl.	*E.huxleyi*, *P.minimum*, var. dinoflagellates	36	25				7
North Atlantic	Ciliates, heterotr. dinofl.	*E.huxleyi* bloom	44	30	67			9

Table 4 continued

Area	Grazer	Dominant prey species	DMSP stock grazed (% d^{-1})	DMS + $DMSP_d$ production as % of stock (% d^{-1})	DMS + $DMSP_d$ production of ingested DMSP (%)	DMS yield of ingested DMSP (%)	DMSP retained by grazers (%)	Data and comments
Laboratory—macrograzers								
	Labidocera aestiva	*Gymnodinium nelsonii*				37	0	10
	Centropages hamatus	=				35	0	10
	Eurytemora affinis	*Phaeodactylum tricornutum*			20 (+80)	0	0	11
		Thalassiosira weisflogii			20 (+80)	0	0	
	Temora longicornis	*Tetraselmis impellucida*			nd	nd	+	12
	Acartia tonsa	*Phaeocystis* sp.			nd	nd	4	13
		Isochrysis galbana					5	
	Krill	*Phaeocystis antarctica*			(+)	3–16x control	nd	14

1. Tang and Simo 2003
2. Wolfe and Steinke 1996
3. Archer et al. 2001a [assumed that DMSP content of *I. Galbana* is 4 fmol/cell (Keller et al. 1989; Niki et al. 2000; Tang and Simo 2003), which means that only one third of the DMSP stock was in *I. Galbana*; Archer et al. assumed that all DMSP was in the prey]
4. Christaki et al. 1996
5. Belviso et al. 1990
6. Archer et al. 2001a (dilution approach; Iceland Basin data is mean of two experiments; DMS yield from DMSP includes consumed DMS)
7. Archer et al. 2001b [mean of eight dilution experiments; DMS(P) release calculated from the assumption that microzooplankton graze with a growth efficiency of 30%]
8. Archer et al. 2003 (coastal: mean of three experiments; offshore: mean of four experiments; DMS yield from DMSP includes consumed DMS)
9. Simo et al. 2002 (mean of four experiments, DMSP-release was calculated, not measured)
10. Dacey and Wakeham 1986
11. Kwint et al. 1996 (20% of ingested DMSP was released as DMSPd; 80% as $DMSP_p$ in faecal pellets)
12. Tang et al. 1999
13. Tang and Simo 2003
14. Daly and DiTullio 1996

(+ = undefined part of ingested $DMSP_p$ was repacked into faecal pellets). nd = not determined chemical defence mechanism based on acrylate. Such a mechanism could also be partly responsible for the significantly reduced feeding levels on high-lyase phytoplankton by other microzooplankton taxa (Strom et al. 2003a; Wolfe 2000). However, it was recently suggested that DMSP rather than acrylate is the active deterrent of microzooplankton grazing (Strom et al. 2003b) and readers are directed to the paper presented by Nejstgard et al. (this issue) for a discussion on this subject.

A few field experiments have combined the determination of microzooplankton grazing rates (the Landry-Hassett dilution technique) with DMS(P) analyses and have shown that DMSP-containing algal species may be proportionally less grazed than total algal biomass (Archer et al. 2001b; Olson and Strom 2002; Wolfe et al. 2000). However, several problems are associated with this technique and we refer to Archer et al. (2001b) for a discussion on the potential pitfalls. Nevertheless, the emerging picture is that generally 20% to 70% of the ingested $DMSP_p$ is released to the dissolved phase (Table 4). This is based on the assumption that no selective grazing takes place and all $DMSP_p$ is in algal cells. Since part of the $DMSP_p$ stock is in the micrograzers themselves, ingested $DMSP_p$ as calculated from grazing rates multiplied with the DMSP concentration may be an overestimation and hence the percentage converted to DMS and $DMSP_d$ an underestimation. In addition, possible selective grazing on none-DMSP-containing algae and microbial consumption of DMS and $DMSP_d$ not resulting in DMS, will further increase the calculated percentage of ingested $DMSP_p$ being transferred to the dissolved pool. Archer et al. (2001b) therefore suggested that the transformation of algal DMSP through microzooplankton could simply be calculated from grazing rates by assuming a growth efficiency factor of 0.3, which implies that 70% of the grazed DMSP is released to the dissolved pool or as faecal pellets. Tang and Simo (2003) showed that intracellular DMSP concentrations of a heterotrophic dinoflagellate vary passively with the DMSP content of the food source. If indeed microzooplankton do not regulate internal DMSP concentration for physiological purposes, the amount of DMSP retained by grazers and released to solution is a fixed percentage of DMSP consumption. The variation in DMS yield from exuded $DMSP_p$ is the result of a multitude of microbial processes, but seems to be at least partly related to the DMSP-lyase activity of the prey. Further refinement of methods is needed to fully understand the competing pathways.

Grazing by meso- and macrozooplankton

Grazing by larger zooplankton such as copepods and krill can increase the release of dissolved organic material by mechanical disruption of intact cells (sloppy feeding). This mechanism can result in the production of DMS (Dacey and Wakeham 1986; Daly and DiTullio 1996; Kasamatsu et al. 2004; Malin et al. 1994). In those cases, the release of $DMSP_d$ during sloppy feeding may stimulate the conversion by algal or bacterial lyase enzymes, but microbial activity in the intestinal tract of the zooplankton and in faecal pellets may also be responsible for the high DMS production observed. Unfortunately, remarkably few dedicated grazing studies have published DMS and DMSP in both the dissolved and particulate phase necessary to draw a complete budget of the effect of grazing. When data are available, comparing the amount of DMSP ingested with the amount of DMS produced shows that a considerable part is lost (Table 4). Possible explanations are demethylation by bacteria in- or outside the grazer, storage of DMSP in the body tissues of some grazers (Tang et al. 1999) or repackaging of prey DMSP into faecal pellets (Kwint et al. 1996). The latter will result in high concentrations of DMSP in suspended material that could rapidly sink out of the euphotic zone and result in export of DMSP into deeper waters. The DMSP in this material appears to be readily available for biological degradation (see also 'Significance of vertical flux as a sink for particulate DMSP'), which does not necessarily result in the production of DMS (Kwint et al. 1996).

Phaeocystis colonies are considered to be less palatable to copepods than diatoms. This has mainly been attributed to the large size of these colonies. In addition it has been suggested that acrylate may accumulate to high concentrations in the mucus surrounding *Phaeocystis* colonies (Davidson and Marchant 1987; Guillard and Hellebust 1971). Since acrylate may have antimicrobial properties at elevated concentrations (Sieburth 1960; Slezak et al. 1994), DMSP-lyase could be involved in the accumulation of this compound to prevent microbial attack and grazing. In intact and growing *Phaeocystis* colonies, acrylate indeed appears to accumulate in the mucus to concentrations between 1 mM and 7 mM (Noordkamp et al. 1998 ,2000), but a direct impact on bacterial and grazing activities has not been established yet. Moreover, *Phaeocystis* can be readily grazed by krill, which results in elevated production of DMS (Daly and DiTullio 1996).

Evidently, the effect of macro-grazers on the sulphur budget depends on both the grazer and on the algae that are grazed. It appears that the most important role for copepods and krill is merely to be mediators in the release of DMSP through sloppy feeding or through repackaging into faecal pellets. Although there is limited evidence, we speculate that the amount of DMS produced from ingested DMSP is related to the DMSP-lyase activity of the prey, since DMS production was only observed in grazing experiments with *Phaeocystis* and a dinoflagellate, but not with diatoms (Table 4).

Microbial consumption of DMSP

The chemical half-life of DMSP in seawater is >8 years (Dacey and Blough 1987), which results in high abiotic stability under natural conditions (moderate temperatures and pH). Therefore, most of the DMSP removal is through enzymatic processes. In the microbial food web, dissolved DMSP has many fates and several recent reviews on the microbial pathways and involved mechanisms have been published (Bentley and Chasteen 2004; Kiene et al. 2000; Lomans et al. 2002; Yoch 2002). They all show that DMSP can be readily used in a complex network of enzymatic conversions. This versatility indicates that this single compound is of major importance for the nutrition of the bacterial community. Indeed, several studies have shown that DMSP alone can contribute 1 to 15% of the total bacterial carbon demand in surface waters. Moreover, DMSP assimilation can satisfy most, if not all the, sulphur demand of marine bacteria (Kiene and Linn 2000; Simo et al. 2002; Zubkov et al. 2001). Since the focal point of this section is the quantification of DMSP removal, only the overall effects of the main pathways originating from DMSP (Fig. 1) will be discussed here.

The two major pathways of bacterial DMSP degradation are cleavage to DMS and acrylate and demethylation/demethiolation, with or without prior uptake into the cells. Quantitatively, the most important degradation pathway of DMSP, is demethylation. This pathway does not yield DMS, but 3-methiolpropionate (MMPA) and, after a second demethylation, 3-mercaptopropionate (MPA). MMPA can also yield methanethiol (MeSH) after a demethiolation reaction. Next to DMS, MeSH is another important volatile sulphur compound. It is metabolised rapidly and appears to be the major sulphur source for the production of sulphur-containing amino acids and proteins in bacterioplankton (Kiene et al. 1999). Members of the *Roseobacter* group, a ubiquitous group of marine bacteria, are well known MeSH producers (Zubkov et al. 2001) and in waters rich in these bacteria, consumption of DMSP via demethylation can be tenfold higher than DMS production (Kiene et al. 2000).

The cleavage of DMSP in aerobic DMS-producing bacteria is thought to be similar to the cleavage by algae and differentiating between the two in field samples is thus a challenge. As far as this has been investigated, it appears that acrylate is further used as a carbon source, leaving DMS untouched (Yoch 2002 and references therein). The current three conceptual models for the uptake and metabolism of DMSP (Yoch 2002) suggest considerable diversity of DMSP-lyases amongst the bacteria. The presence of DMSP and DMSP-cleaving enzymes is a prerequisite but cannot be used as proxies for DMS production. Measurements in the Northeast Atlantic Ocean showed maximum in vitro DMSP-lyase activity (which indicates the potential for DMS production from both bacterial and algal enzymes) on the order of 6–185 nM DMS h^{-1} (Steinke et al. 2002b), whereas in vivo DMS production (from short-term (8 h) bottle incubations) resulted in only 1.2–14.4 nM DMS h−1 (Simo and Pedros-Alio 1999b).

A more extensive screening of 15 *Roseobacter* strains showed that the cleavage pathway was present in all strains and that the demethylation pathway co-occurred within five of these strains (Gonzalez et al. 1999). In addition, several strains were able to degrade DMS and MeSH. This versatility of sulphur metabolism within a single genus of bacteria indicates that identification of the community structure alone is insufficient to determine the dominating degradation pathways. Furthermore, the fact that several isolates were also able to consume DMS means that the difference between DMSP loss and DMS production in natural waters not necessarily indicates a dominance of the demethylation pathway. While considering the total myriad of conversion pathways, Kiene et al. (2000) proposed a hypothetical model that helps us to understand the relative importance of these pathways. These authors suggested that bacterioplankton will prefer the demethylation/demethiolation over the lyase pathway at low $DMSP_d$ concentrations. This is because this pathway provides more energetic benefits and it is a relatively economic way to assimilate reduced sulphur. In fact, it was proposed that the total sulphur demand of bacteria can be derived in this way and as such, this pathway can be directly linked to bacterial production. At higher $DMSP_d$ concentrations, the DMSP that is not assimilated is then available to the cleavage pathway. In other words, the fraction that is converted to DMS depends on the biomass and growth of the bacterial community. Kiene et al. (2000) also stipulated that it is not necessary for the $DMSP_d$ concentration to be high, but that it is the bacterial sulphur demand relative to the $DMSP_d$ availability that is critical. For instance, if sulphur demand is low due to nutrient limitation or UV stress, the demethylation pathway is minor and DMS yield may increase. Such a DMSP-availability hypothesis is consistent with the shifts in the relative contribution of the demethylation- and cleavage pathway, as observed by Simo and Pedros-Alio (1999a), who found DMS yields from consumed DMSP ranging between 5% and 100%. These shifts were correlated to the depth of the mixed water layer, with higher DMS yields in shallow mixed layers, and could be related to UV stress on the bacterial population. In addition to this bacteria-oriented hypothesis, it appears that at high DMSPd concentrations, either in blooms or in microenvironments around algal cells or aggregates, the algal cleavage pathway may overrule the bacterial cleavage. For instance, the peak of the DMS concentration in *Phaeocystis* blooms is often associated with the younger parts of the bloom and not with the senescent part (van Duyl et al. 1998). This may be related to the fact that the bacterial community of younger blooms is often still developing and therefore their sulphur demand low. A relatively large proportion of $DMSP_p$ may then be rapidly converted to DMS by algal DMSP-lyases. During the senescent stages of the bloom, the mature bacterial community may divert DMSP into the demethylation pathway and/or consume a large proportion of the DMS. In terms of enzyme kinetics, one can envision that a plankton bloom diverts from a low affinity but high capacity cleavage system to a high-affinity but low-capacity demethylation/demethiolation system (Kiene et al. 2000).

The particulate DMSO pathway

In comparison to DMS and DMSP, the nonvolatile dimethyl sulphoxide (DMSO) is a poorly understood component of the marine sulphur cycle. In addition to DMSO being ubiquitous in seawater in the dissolved (or filterable $DMSO_d$) phase, evidence is also gathering for the direct production of particulate DMSO (GF/F retained $DMSO_p$) by cultures and natural assemblages of marine microalgae. DMSO has been the subject of three reviews (Hatton et al. 2004; Lee and de Mora 1999a; Lee and de Mora 1999b) and whilst we aim not to be overly repetitive here, it is nonetheless important to underline some pertinent information regarding particulate DMSO. This compound is remarkable in its ability to permeate intact biological membranes and it is well known as an effective radical scavenger. Indeed, it is widely used medically to deliver drugs through the skin and as a cryoprotectant for the storage of cells at freezer temperatures. This ability to transfer across membranes and the DMSP antioxidant cascade hypothesis put forward by Sunda et al. (2002) could explain the marked increase in $DMSO_d$ seen in dinoflagellate cultures as they approach and enter stationary phase (Simo et al. 1998). $DMSO_p$ production has been noted for cultures of the dinoflagellate *Amphidinium carterae* and the coccolithophorids *Pleurochrysis carterae* and *E. huxleyi* (Simo et al. 1998), suggesting that a wide range of phytoplankton may produce $DMSO_p$. Exponential phase cultures of *A. carterae* and *E. huxleyi* gave mean $DMSP_p$:$DMSO_p$ molar ratios of 25 and 8, respectively. In a study of the Peruvian upwelling system Riseman and DiTullio (2004) found a strong positive correlation though with ratios of 0.7 to 4.2, and they noted that concentrations of both sulphur pools and the antioxidant β-carotene increased under low-iron conditions which might be consistent with the production of free radicals due to iron deficiency. These authors provide a table that includes $DMSP_p$:$DMSO_p$ values from five other studies that span the range 0.02–100. More recently Simo and Vila-Costa (2006) discussed geographic and temporal patterns of $DMSO_p$ distribution using the largest data set for $DMSO_p$ and $DMSP_p$ compiled to date. They report that the ratio of $DMSP_p$:$DMSO_p$ varied from 1 to 13 with an average of 5.2, and noted a trend towards higher proportions of $DMSO_p$ in phytoplankton in warmer seas. Further research is needed to better evaluate whether these variable ratios result from differences in phytoplankton speciation, bloom stages or prevailing growth conditions.

Significance of vertical flux as a sink for particulate DMSP

The significance of vertical flux as a sink for particulate DMSP was recently reviewed and reassessed using new data gained in coastal waters of northern Europe (Belviso et al. 2006 and references therein). Special attention was paid to the many biases that can affect estimates of downward fluxes of DMSP (catchment efficiency of traps, releases of particulate material to the dissolved pool, etc). A method was also suggested to correct DMSP fluxes from biological losses during the sedimentation process. Indeed, the lability of DMSP during the sedimentation process has been shown to be comparable to that of chlorophyll-*a* (Cailliau et al. 1999). The chlorophyll-*a* degradation products resulting from grazing and senescence are, in order of least to most degraded: phaeophytin-*a*, phaeophorbide-*a* and pyrophaeophorbide-*a*. These phaeopigments can all be detected by high-performance liquid chromatography (HPLC). Assuming that phaeopigments trace the degradation products of DMSP, the following equation can be used to restore DMSP fluxes and to revise the daily vertical loss rates for DMSP:

$$F_{\text{DMSP,revised}} = F_{\text{DMSP,observed}} \times [(\text{Chl } a + \text{Phaeo } a)/(\text{chl } a)]$$

Since grazing also provides products, not detected by HPLC and fluorometry, and since some of the DMSP is also lost in the form of DMSO (Hatton 2002b) the revised export rates may still be underestimated. Nevertheless, Table 5 shows an updated picture of fluxes and export rates of DMSP for a number of different sites in open-ocean and coastal waters. Export rates of DMSP range between 0.1% and 16.6% d^{-1} in coastal waters and between 0.03% and 0.74 % d^{-1} in the open ocean.

Factors controlling DMS removal

Bacterial consumption

In incubation studies with water samples from the equatorial Pacific, Kiene and Bates (1990) demonstrated that bacterial degradation of DMS dominated over sea-to-air gas exchange. The major microbial degradation pathways of DMS are consumption via DMS monooxygenases and methyltransferases and oxidation via DMS dehydrogenase. DMS can be oxidised to DMSO by a variety of sulphur and ammonia oxidisers, methylotrophs and phototrophs, which suggests that this is a versatile worldwide process (reviewed by Bentley and Chasteen 2004). The reverse reaction—reduction of DMSO to DMS via DMSO reductases under aerobic and anaerobic conditions is well known from laboratory studies, but as far as we are aware, this pathway has not been documented for any natural assemblage. The enzyme responsible for the conversion of DMS to MeSH and formaldehyde has long been thought to be DMS monooxygenase. However, both the enzyme and the metabolic pathway are enigmatic (Bentley and Chasteen 2004). Detailed biochemical and molecular level studies are needed to improve understanding of these important microbial processes. DMS can also be assimilated as a sulphur source although Zubkov et al. (2002) found this to be a minor pathway in the northern North Sea. However, it is interesting to note that Fuse et al. (2000) isolated a γ-proteobacterium *Marinobacterium* sp. that assimilated DMS at wavelengths of 380–480 nm via the production of heat-stable photosensitisers. A light-requiring mechanism like this would not operate under the dark incubation conditions that are often used for incubation experiments.

It is clear that bacterial activity is an important factor in reducing the quantity of DMS that is emitted to the atmosphere. Simo (2004) compiled data from several studies in different marine areas, which illustrates the linear relationship between biological DMS consumption and DMS production, and sug-

Table 5 Values for the calculated correction factor for potential losses based on the phaeopigment to chlorophyll-*a* ratio, the corrected daily downward flux of DMSPt, the seawater $DMSP_p$ standing stocks and the revised daily vertical loss rates for DMSP at different sites

Area and coordinates		(Chl-*a* + Phaeo-*a*) /Chl-*a*	DMSP fluxes (μmol m^{-2} d^{-1})	$DMSP_p$ standing stock (mmol m^{-2})	Export rates (% d^{-1})	Data*
Southern Ocean						
52°S–62°E		2.85[a]	2.05	3.8	0.05	1
63°S–70°20′ E		3.46	0.68	2.0	0.03	1
61°S–62°E		2.97	0.45	1.6	0.03	2
64°S–62°E		5.41	1.04	2.4	0.04	2
66°41′S–61°50′ E		9.07	24.3	3.3	0.74	2
Ligurian Sea						
43°25′N–7°51′ E		2.81[a]	2.08	2.5	0.08	2
=		3.10	1.46	2.3	0.06	2
Northern Norway						
Balsfjord	69°22′ N–19°5′ E	1.23–17[b]	1.05–20.5	0.76–0.94	0.1–0.7	3
Malangen	69°31′ N–18°10′ E	2.33–3.66	9.1–50.1	0.64–1.89	1.4–8.0	3
Ullsfjord	69°49′ N–19°45′ E	1.76–2.23	9.8–47.7	1.0–8.67	0.5–1.5	3
North Sea						
S. Bight	53°30′ N–3°30′ E	1.44–2.78[c]	36.5–106	0.25–0.69	11.2–16.6	3

[a] Pigments measured by HPLC

[b] Pigments measured by fluorometry. Data digitised from Reigstad et al. (2000), Fig. 4

[c] Pigments measured by fluorometry

* Data are from: 1. Cailliau et al. 1999; 2. This study; 3. Belviso et al. 2006

gests that consumption accounts for between 50% and 80% of the production. However, the significance of bacterial DMS consumption also alters depending on the strength of other, competing loss processes. Since all of these can alter rapidly according to meteorological forcing it is difficult to derive a range of widely applicable quantification terms. This is best illustrated by a study in the subpolar North Atlantic which showed that: (1) photochemical degradation dominated under clear skies and shallow mixing conditions, (2) bacterial consumption was most important when skies were cloudy and/or when the water column mixed to a greater depth and (3) loss of DMS due to sea-to-air transfer was roughly equivalent to bacterial consumption during a storm (Simo and Pedros-Alio 1999a).

Photochemical oxidation of DMS

Brimblecombe and Shooter (1986) established that photochemical oxidation is an important loss process for DMS, that the reaction rate varies with DMS concentration and that photolysis can occur at visible wavelengths via photosensitisers. It is important to emphasise that DMS photolysis does not always result in DMSO production. However, when DMSO is produced there remains a possibility that a reduction pathway could operate to return DMS to the seawater pool. Kieber et al. (1996) showed that maximum photolysis occurred at wavelengths between 380 nm and 460 nm and in Pacific waters only 14% of the DMS was converted to DMSO. Hatton (2002a) found that DMS was removed from seawater by UVA/visible light (>315 nm) and UVB (<315 nm) though only UVA/visible wavelengths resulted in DMSO production and in northern North Sea waters ~37% of total DMS was lost by this route under full natural sunlight. Toole and Siegel (2004) investigated the factors that influenced DMS cycling using a DMS time series for the Sargasso Sea and found that the UV radiation dose explained 77% of the variability in DMS concentrations. Brugger et al. (1998) demonstrated that the initial rate of DMS removal is proportional to initial DMS concentration (5–

100 nM tested), irradiance intensity and DOC concentration. However, we note that photolysis is most likely to be mediated by the portion of DOC that is known as coloured dissolved organic matter (CDOM). The apparent quantum yield for DMS also doubles with a temperature increase of 20°C (Toole et al. 2003). Interestingly nitrate concentrations are also relevant, in that Toole et al. (2004) found that in high-nitrate Antarctic waters 35% of the observed DMS photolysis was related to nitrate photochemistry and photolysis rates increased linearly with added nitrate. This nitrate dependence was also observed by Bouillon and Miller (2004). Moreover, Bouillon and Miller (2005) found that nitrate-induced photolysis of DMS was strongly enhanced by the presence of bromide ions and to a lesser extent by bicarbonate/carbonate ions. Contrary to what was expected bicarbonate/carbonate-induced photolysis reduced with increasing pH.

By pooling data from different oceanic regions, Hatton et al. (2004) found a highly significant correlation between DMS and DMSO in (near-)surface waters, with the DMSO concentration comparing to 1.5 times the DMS concentration. This clearly shows that photolysis is an important DMS loss process and that accurate photolysis parameterisations are needed to improve our understanding of DMS cycling and DMS models. Some of the studies mentioned above give DMS photolysis rates that are highly variable: 0.03–0.07 h^{-1} for the northern North Sea (Hatton 2002a), 0.12 h^{-1} for the coastal Adriatic (Brugger et al. 1998), 0.04 h^{-1} for the Pacific (Kieber et al. 1996), 0.026–0.086 h^{-1} for the western Atlantic (Toole et al. 2006) and 0.16–0.23 h^{-1} for the Antarctic (Toole et al. 2004). Given the previous discussion of CDOM, nitrate concentration, wavelength and temperature, all of which can vary with depth, we might conclude that an ideal parameterisation could be rather complex. For now modelling studies appear to use a variety of relatively simple approaches such that the DMS photolysis rate varies with depth and season (e.g., Archer et al. 2004; Lefevre et al. 2002).

Flux of DMS to the atmosphere

Until very recently there was no direct method for measuring the sea-air flux of DMS. Micrometeorological techniques are considered to be the best approach for determining fluxes (Businger and Delany 1990) but are a considerable technological challenge. Huebert et al. (2004) made the first measurements of DMS flux using the eddy correlation technique, on board ship. Their breakthrough was due to the development of the atmospheric pressure ionization mass spectrometer-isotopically labelled standard (APIMS-ILS) technique, which can measure DMS at high frequency. Another micrometeorological technique used recently for DMS is relaxed eddy accumulation (Zemmelink et al. 2004). Both methods require highly sophisticated equipment and specialist knowledge, so it is likely that the emission rates of DMS will continue to be determined by semi-empirical methods for some time to come.

The most widely used method for calculations of gas fluxes requires the concentration gradient between surface water and the atmosphere and a kinetic parameter known as the transfer or piston velocity (Liss and Slater 1974):

$$F_{\mathrm{DMS}} = k_{(\mathrm{T})} \times \Delta C$$

where F_{DMS} = net flux; $k_{(\mathrm{T})}$ = transfer velocity; ΔC = concentration gradient across the air-sea interface i.e., $C_w - C_a H^{-1}$, where C_w = concentration in water, C_a = concentration in air and H = Henrys law constant. Since C_a is very low relative to C_w, C_a is often taken to be zero which generally may lead to an overestimation of about 6% or less which is not significant, relative to other uncertainties (Turner et al. 1996).

Over the last decade the results of oceanic multiple deliberate tracer experiments have refined existing estimates of k (Nightingale et al. 2000) and, including the findings of the micrometeorological studies, there is still disagreement. Estimates vary by factors of about two and there is generally greater uncertainty for k at higher wind speeds. What is agreed, however, is that wind speed alone is not adequate for the parameterization of k and that other factors, such as surfactants and breaking wave bubble generation are important (e.g., Frew et al. 2004). Quantification of these parameters in the field is not straightforward and future measurement and modelling efforts are required before improvements in the quantification of k can be achieved.

The concentration term of the flux calculation also has major uncertainties, mainly associated with the spatial and temporal scales of measurements. These include bias towards Spring and Summer measurements with many areas of the world oceans still sparsely represented (Kettle et al. 1999). More important, in the context of ecosystem modelling and validation, is the uncertainty in measured DMS concentrations arising from the depth at which the samples are taken: operationally defined as surface water, but can be as deep as 11 m. Vertical profiles of DMS concentration through the oceanic mixed layer are rarely homogenous and transient stratification (surface heating and freshwater input) can have marked affects (see Ward et al. 2004). At the smaller vertical scale of the sea-surface microlayer ($\sim$60 μm thickness) lies further uncertainty for the DMS concentration term. Study of the microlayer is technologically and logistically difficult and the few reports suggest a wide range of enrichments of DMS, relative to subsurface concentrations (e.g., 0.38–2.94, mean 1.1; Yang and Tsunogai 2005). It is not possible to assess the uncertainty of the DMS concentrations in the literature and further work is required to determine how important the depth of sampling is for a variety of biological and physical drivers and hence for the concentration term in the air-sea flux equation.

In summary, there is still a great deal of uncertainty in the calculation of sea-to-air fluxes and, for the time being, we recommend use of the transfer velocity parameterisation of Nightingale et al. (2000), which appears to be intermediate between all other estimations.

Vertical and horizontal mixing of DMS

Although oceanic mixing and stirring processes do not constitute loss of DMS per se, in dynamic oceanographic systems, dispersion and homogenisation can have significant control on the amount of DMS measured in the surface layer over timescales of hours and days. Vertical profiles of DMS in the mixed layer generally show some structure, often with a subsurface maximum. However, during high-wind events increased mixing causes homogenisation of the upper water column and deepening of the mixed layer. Thus, in the course of a few hours the concentration of DMS at the surface can increase or decrease depending on the prior vertical profile structure and degree of enhanced turbulence. Post storms and during warm quiescent periods, when turbulence decreases, the mixed layer tends to shallow which causes DMS to be trapped below the new depth of the picnocline. This, in effect, reduces the size of the reservoir of DMS that is available for sea-to-air flux, inter alia.

Winds, currents and tides all contribute to horizontal mixing, which can also affect DMS concentrations. As an example, during a Lagrangian iron-addition experiment in the Southern Ocean a sulphur-hexafluoride-labelled patch of water spread from about 70–1,000 km^2 in 18 days which led to rapid dilution of the bloom with water containing low biomass and DMS (Turner et al., in prep). Vertical and horizontal mixing processes can thus be overruling loss processes after a local built-up of the DMS concentration, which warrants inclusion of these processes in models.

Conclusions

As we have shown in the previous sections, the transformation of DMSP to DMS and the accumulation of DMS in surface waters are intricately linked to food-web dynamics and physico-chemical processes, including photochemical degradation, vertical mixing, and sea to air flux. By recalculating and compiling current literature data we have provided an insight in the relative magnitude of different DMS-related processes. Hopefully, this will assist future modelling efforts to choose the most relevant processes and establish correct parameterisations. To our knowledge, Archer et al. (2004) has published the most complex ecosystem/DMS model to date, following on from the pioneering work done by Gabric et al. (1993). This model generates DMS(P) concentrations that are generally comparable with field data. It appears to be most sensitive to changes in DMSP as a proportion of total phytoplankton carbon and to the proportion of $DMSP_p$ that is transformed to $DMSP_d$ or DMS during excretion, lysis and grazing. The importance of correct intracellular DMSP quota was also shown by the model of Lefevre et al. (2002) for the subtropical North Atlantic, in which intracellular DMSP:N ratios varying with season and depth by a factor of five were needed to reproduce the field data.

In the section 'Factors controlling DMSP production', we have given an update of the DMSP:C ratios in various phytoplankton groups and explained how abiotic factors can affect these ratios. Species composition has the largest impact on the DMSP:C ratio. Light and temperature appear to affect the DMSP:C ratio by a factor of 2-3 and nutrients may only have an effect on species with low levels of intracellular DMSP.

Another important feature recently recognised is the variation in DMS yield from consumed $DMSP_d$, ranging between 1% and 100%. This appears not only to be related to the bacterial community structure and its productivity as presented in the DMSP-availability hypothesis of Kiene et al. (2000), but also to the ability of phytoplankton cells to cleave DMSP into DMS. There is ample evidence that within aggregates, during viral lysis, autolysis and grazing the DMS yield from released DMSP is related to the DMSP-lyase activity of the phytoplankton involved. This argues for the implementation of group-specific DMSP-lyase activity for algae in modelling studies. As a result, the DMS yield during grazing and lysis, but also the conversion of the dissolved DMSP pool will be affected by the dominant algal group. Van den Berg et al. (1996) already showed that this was vital to her model in order to reproduce the DMS peak during a *Phaeocystis* bloom. Nonetheless, group-specific DMSP-lyase activity has not been implemented in recent models.

We recommend distinguishing six phytoplankton groups, based on their size, DMSP content, DMSP-lyase activity and interactions with grazers (Table 6). The diatoms and dinoflagellates are taxonomically well defined and have for reasons of simplification been confined to the larger size class. The dinoflagellates are characterised by having DMSP-lyase. We propose that the diatoms are the only group in which nutrients have a distinct effect on the DMSP:C ratio. Although they have only low levels of DMSP, their global importance may make this nutrient dependency an important controlling factor. More dedicated experiments are needed to be able to quantify the effect of nutrient limitations on the DMSP content of other groups. The phytoflagellates are a highly diverse group that contains the haptophytes, including *Phaeocystis* and coccolithophorids, several larger prasinophytes, such as *Tetraselmis*, but possibly also the smaller dinoflagellates. Although this group also

Table 6 Recommended phytoplankton groups

Phytoplankton group	Size class	% DMSP-C of cell carbon	DMSP-lyase activity	Grazed by microzooplankton	Grazed by mesozooplankton	Nutrient effect on % DMSP-C
Diatoms	20–200 μm	0.004	–	–	+	+
Dinoflagellates	20–200 μm	0.1	+	–	+	–
Phytoflagellates A	2–20 μm	0.05	–	+	+	–
Phytoflagellates B*	2–20 μm	0.08	+	–/+	+	–
Eukaryotic picoplankton	0.2–2 μm	0.05	–	+	–	–
Prokaryotic picoplankton (incl N2-fixers)	0.2–2 μm	0	–	+	–	–

* Includes *Phaeocystis* and high-lyase *Emiliania huxleyi* strains

holds non-DMSP-containing algae, it seems justifiable from Table 1 to assign this group as DMSP producers. Given the importance of algal DMSP-lyase activity, we propose the division of this group into two subgroups, based on this property: The phytoflagellates B contain DMSP-lyase and are represented by species such as *Phaeocystis* and *E. huxleyi* and possibly small dinoflagellates. They are potentially less grazed by microzooplankton than the phytoflagellates A, due to their DMSP-lyase activity. The picoplankton is a very heterogeneous group that is difficult to identify and describe for modelling purposes, but nonetheless very important for biogeochemical fluxes in open ocean areas. In order to address the heterogeneous nature of this group, we recommend separation between eukaryotic picoplankton, which do contribute to $DMSP_p$ and prokaryotic picoplankton, with no intracellular $DMSP_p$.

With the current overview, we hope to have given better insight into the various pathways and constraints; to have provided the modelling community with the means to parameterise these pathways in order to improve their models and to have highlighted areas where knowledge is limited and requires future laboratory and field studies.

Acknowledgements This is a contribution to the European Union IRONAGES project funded under contract N8 ENV4-CT97-0472 by the Key Action Global Change of the Environment and Sustainable Development Programme of the European Commission and to the Scientific Committee on Oceanic Research (SCOR) working group 120 'Marine Phytoplankton and Global Climate Regulation: The *Phaeocystis* spp. Cluster as a Model'. GM thanks the Natural Environmental Research Council of the United Kingdom (NERC) for the award of a NERC advanced research fellowship (NERC NE/B501039/1) and additional grants for research on DMS (NER/T/S/2000/00638, NER/O/S/2001/01244, NER/A/S/2002/00917). MS was supported by a postdoctoral fellowship from the UK Natural Environment Research Council (NER/I/S/2000/00897 and NE/B500282/1).

References

Anderson TR, Spall SA, Yool A, Cipollini P, Challenor PG, Fasham MJR (2001) Global fields of sea surface dimethylsulfide predicted from chlorophyll, nutrients and light. J Mar Syst 30:1–20

Andreae MO (1990) Ocean-atmosphere interactions in the global biogeochemical sulfur cycle. Mar Chem 30:1–29

Andreae MO, Crutzen PJ (1997) Atmospheric aerosols: biogeochemical sources and role in atmospheric chemistry. Science 276:1052–1058

Andreae MO, Jones CD, Cox PM (2005) Strong present-day aerosol cooling implies a hot future. Nature 435:1187–1190

Archer SD, Gilbert FJ, Allen JI, Blackford J, Nightingale PD (2004) Modelling of the seasonal patterns of dimethylsulphide production and fate during 1989 at a site in the North Sea. Can J Fish Aquat Sci 61:765–787

Archer SD, Smith GC, Nightingale PD, Widdicombe CE, Tarran GA, Rees AP, Burkill PH (2002) Dynamics of particulate dimethylsulphoniopropionate during a Lagrangian experiment in the northern North Sea. Deep-Sea Res Part II 49:2979–2999

Archer SD, Stelfox-Widdicombe CE, Burkill PH, Malin G (2001a) A dilution approach to quantify the production of dissolved dimethylsulphoniopropionate and dimethyl sulphide due to microzooplankton herbivory. Aquat Microb Ecol 23:131–145

Archer SD, Stelfox-Widdicombe CE, Malin G, Burkill PH (2003) Is dimethyl sulphide production related to microzooplankton herbivory in the southern North Sea? J Plankton Res 25:235–242

Archer SD, Widdicombe CE, Tarran GA, Rees AP, Burkill PH (2001b) Production and turnover of particulate dimethylsulphoniopropionate during a coccolithophore bloom in the northern North Sea. Aquat Microb Ecol 24:225–241

Aumont O, Belviso S, Monfray P (2002) Dimethylsulfoniopropionate (DMSP) and dimethylsulfide (DMS) sea surface distributions simulated from a global three-dimensional ocean carbon cycle model. J Geophys Res-Oceans 107:Article no. 3029

Barnard WR, Andreae MO, Iverson RL (1984) Dimethylsulfide and Phaeocystis poucheti in the southeastern Bering sea. Cont Shelf Res 3:103–113

Bates TS, Charlson RJ, Gammon RH (1987) Evidence for the climatic role of marine biogenic sulphur. Nature 329:319–321

Bates TS, Kiene RP, Wolfe GV, Matrai PA, Chavez FP, Buck KR, Blomquist BW, Cuhel RL (1994) The cycling of sulfur in surface seawater of the Northeast Pacific. J Geophys Res-Oceans 99:7835–7843

Bates TS, Lamb BK, Guenther A, Dignon J, Stoiber RE (1992) Sulfur emissions to the atmosphere from natural sources. J Atmos Chem 14:315–337

Baumann MEM, Brandini FP, Staubes R (1994) The influence of light and temperature on carbon-specific DMS release by cultures of Phaeocystis antarctica and three antarctic diatoms. Mar Chem 45:129–136

Belviso S, Kim SK, Rassoulzadegan F, Krajka B, Nguyen BC, Mihalopoulos N, Buatmenard P (1990) Production of Dimethylsulfonium Propionate (DMSP) and Dimethylsulfide (DMS) by a Microbial Food Web. Limnol Oceanogr 35:1810–1821

Belviso S, Thouzeau G, Schmidt S, Reigstad M, Wassmann P, Arashkevich E, Stefels J (2006) Significance of vertical flux as a sink for surface water DMSP and as a source for the sediment surface in coastal zones of northern Europe. Estuar Coast Shelf Sci 68:473–488

Bentley R, Chasteen TG (2004) Environmental VOSCs-formation and degradation of dimethyl sulfide, methanethiol and related materials. Chemosphere 55:291–317

Bidle KD, Falkowski PG (2004) Cell death in planktonic, photosynthetic microorganisms. Nat Rev Microbiol 2:643–655

Bopp L, Aumont O, Belviso S, Monfray P (2003) Potential impact of climate change on marine dimethyl sulfide emissions. Tellus B 55:11–22

Bopp L, Boucher O, Aumont O, Belviso S, Dufresne JL, Pham M, Monfray P (2004) Will marine dimethylsulfide emissions amplify or alleviate global warming? A model study. Can J Fish Aquat Sci 61:826–835

Bouillon RC, Miller WL (2004) Determination of apparent quantum yield spectra of DMS photo-degradation in an in situ iron-induced Northeast Pacific Ocean bloom. Geophys. Res. Lett. 31:Article no. L06310

Bouillon RC, Miller WL (2005) Photodegradation of dimethyl sulfide (DMS) in natural waters: Laboratory assessment of the nitrate-photolysis-induced DMS oxidation. Environ Sci Technol 39:9471–9477

Bratbak G, Levasseur M, Michaud S, Cantin G, Fernandez E, Heimdal BR, Heldal M (1995) Viral activity in relation to Emiliania huxleyi blooms: A mechanism of DMSP release? Mar Ecol Prog Ser 128:133–142

Brimblecombe P, Shooter D (1986) Photo-oxidation of dimethylsulfide in aqueous solution. Mar Chem 19:343–353

Brugger A, Slezak D, Obernosterer I, Herndl GJ (1998) Photolysis of dimethylsulfide in the northern Adriatic Sea: Dependence on substrate concentration, irradiance and DOC concentration. Mar Chem 59:321–331

Bucciarelli E, Sunda WG (2003) Influence of CO_2, nitrate, phosphate, and silicate limitation on intracellular dimethylsulfoniopropionate in batch cultures of the coastal diatom Thalassiosira pseudonana. Limnol Oceanogr 48:2256–2265

Businger JA, Delany AC (1990) Chemical Sensor Resolution Required For Measuring Surface Fluxes By 3 Common Micrometeorological Techniques. J Atmos Chem 10:399–410

Cailliau C, Belviso S, Goutx M, Bedo A, Park Y, Charriaud E (1999) Sedimentation pathways in the Indian sector of the Southern Ocean during a production regime dominated by regeneration. Mar Ecol Prog Ser 190:53–67

Cantin G, Levasseur M, Schultes S, Michaud S (1999) Dimethylsulfide (DMS) production by size-fractionated particles in the Labrador Sea. Aquat Microb Ecol 19:307–312

Challenger F (1951) Biological methylation. Adv Enzymol 12:429–491

Charlson RJ, Lovelock JE, Andreae MO, Warren SG (1987) Oceanic phytoplankton, atmospheric sulphur, cloud albedo and climate. Nature 326:655–661

Christaki U, Belviso S, Dolan JR, Corn M (1996) Assessment of the role of copepods and ciliates in the release to solution of particulate DMSP. Mar Ecol Prog Ser 141:119–127

Corn M, Belviso S, Partensky F, Simon N, Christaki U (1996) Origin and importance of picoplanktonic DMSP. In: Kiene RP, Visscher PT, Keller MD, Kirst GO (eds) Biological and environmental chemistry of DMSP and related sulfonium compounds. Plenum Press, New York, pp 191–201

Crocker KM, Ondrusek ME, Petty RL, Smith RC (1995) Dimethylsulfide, algal pigments and light in an Antarctic Phaeocystis sp. bloom. Marine Biol 124:335–340

Curran MAJ, Jones GB, Burton H (1998) Spatial distribution of dimethylsulfide and dimethylsulfoniopropionate in the Australasian sector of the Southern Ocean. J Geophys Res-Atmos 103:16677–16689

Dacey JWH, Blough N (1987) Hydroxide decomposition of dimethylsulfoniopropionate to form dimethylsulfide. Geophys Res Lett 14:1246–1249

Dacey JWH, Wakeham SG (1986) Oceanic dimethylsulfide: production during zooplankton grazing. Science 233:1314–1316

Daly KL, DiTullio GR (1996) Particulate dimethylsulfoniopropionate removal and dimethylsulfide production by zooplankton in the Southern Ocean. In: Kiene RP, Visscher PT, Keller MD, Kirst GO (eds) Biological and environmental chemistry of DMSP and related sulfonium compounds. Plenum Press, New York, pp 223–238

Davidson AT, Marchant HJ (1987) Binding of manganese by Antarctic Phaeocystis pouchetii and the role of bacteria in its release. Marine Biol 95:481–487

DiTullio GR, Smith WO (1995) Relationship Between Dimethylsulfide And Phytoplankton Pigment Concentrations In The Ross Sea, Antarctica. Deep-Sea Res Part I 42:873–892

Evans C, Archer SD, Jacquet S, Wilson WH (2003) Direct estimates of the contribution of viral lysis and microzooplankton grazing to the decline of a Micromonas spp. population. Aquat Microb Ecol 30:207–219

Evans C, Kadner SV, Darroch LJ, Wilson WH, Liss PS, Malin G (2007) The relative significance of viral lysis and microzooplankton grazing as pathways of dimethylsulfoniopropionate (DMSP) cleavage: An Emiliania huxleyi culture study. Limnol Oceanogr (in press)

Frew NM, Bock EJ, Schimpf U, Hara T, Haussecker H, Edson JB, McGillis WR, Nelson RK, McKenna SP, Uz BM, Jahne B (2004) Air-sea gas transfer: Its dependence on wind stress, small-scale roughness, and surface films. J. Geophys. Res.-Oceans 109:Article no. C08S17

Fuse H, Takimura O, Murakami K, Yamaoka Y, Omori T (2000) Utilization of dimethyl sulfide as a sulfur source with the aid of light by Marinobacterium sp strain DMS-S1. Appl Environ Microbiol 65:5527–5532

Gabric A, Murray N, Stone L, Kohl M (1993) Modeling the production of dimethylsulfide during a phytoplankton bloom. J Geophys Res-Oceans 98:22805–22816

Gabric AJ, Simo R, Cropp RA, Hirst AC, Dachs J (2004) Modeling estimates of the global emission of dimethylsulfide under enhanced greenhouse conditions. Global Biogeochem Cycles 18:Article no. GB2014

Gage DA, Rhodes D, Nolte KD, Hicks WA, Leustek T, Cooper AJL, Hanson AD (1997) A new route for synthesis of dimethylsulphoniopropionate in marine algae. Nature 387:891–894

Geider RJ (1987) Light and temperature-dependence of the carbon to chlorophyll-*a* ratio in microalgae and Cyanobacteria—Implications for physiology and growth of phytoplankton. New Phytol 106:1–34

Gibson JAE, Garrick RC, Burton HR, McTaggart AR (1990) Dimethylsulfide and the alga Phaeocystis pouchetii in antarctic coastal waters. Mar Chem 104:339–346

Gondwe M, Krol M, Gieskes W, Klaassen W, de Baar H (2003) The contribution of ocean-leaving DMS to the global atmospheric burdens of DMS, MSA, SO2, and NSS SO4=. Global Biogeochem. Cycles 17:Article no. 1056

Gonzalez JM, Kiene RP, Moran MA (1999) Transformation of sulfur compounds by an abundant lineage of marine bacteria in the alpha-subclass of the class Proteobacteria. Appl Environ Microbiol 65:3810–3819

Guillard RRL, Hellebust JA (1971) Growth and the production of extracellular substances by two strains of Phaeocystis pouchetii. J Phycol 7:330–338

Hatton AD (2002a) Influence of photochemistry on the marine biogeochemical cycle of dimethylsulphide in the northern North Sea. Deep-Sea Res Part II 49:3039–3052

Hatton AD (2002b) DMSP removal and DMSO production in sedimenting particulate matter in the northern North Sea. Deep-Sea Res Part II 49:3053–3065

Hatton AD, Darroch L, Malin G (2004) The role of dimethylsulphoxide in the marine biogeochemical cycle of dimethylsulphide. Oceanogr Mar Biol Ann Rev 42:29–55

Hill RW, White BA, Cottrell MT, Dacey JWH (1998) Virus-mediated total release of dimethylsulfoniopropionate from marine phytoplankton: a potential climate process. Aquat Microb Ecol 14:1–6

Holligan PM, Fernandez E, Aiken J, Balch WM, Boyd P, Burkill PH, Finch M, Groom SB, Malin G, Muller K, Purdie DA, Robinson C, Trees CC, Turner SM, Vanderwal P (1993) A Biogeochemical Study Of The Coccolithophore, Emiliania-Huxleyi, In The North-Atlantic. Global Biogeochem Cycles 7:879–900

Holligan PM, Turner SM, Liss PS (1987) Measurements of dimethyl sulphide in frontal regions. Cont Shelf Res 7:213–224

Huebert BJ, Blomquist BW, Hare JE, Fairall CW, Johnson JE, Bates TS (2004) Measurement of the sea-air DMS flux and transfer velocity using eddy correlation. Geophys Res Lett 31:Article no. L23113

Karsten U, Kück K, Vogt C, Kirst GO (1996) Dimethylsulfoniopropionate production in phototrophic organisms and its physiological function as a cryoprotectant. In: Kiene RP, Visscher PT, Keller MD, Kirst GO (eds) Biological and environmental chemistry of DMSP and related sulfonium compounds. Plenum Press, New York, pp 143–153

Kasamatsu N, Kawaguchi S, Watanabe S, Odate T, Fukuchi M (2004) Possible impacts of zooplankton grazing on dimethylsulfide production in the Antarctic Ocean. Can J Fish Aquat Sci 61:736–743

Keller MD, Bellows WK, Guillard RRL (1989) Dimethyl sulfide production in marine phytoplankton. In: Saltzman ES, Cooper WJ (eds) Biogenic sulfur in the environment. American Chemical Society, Washington DC, pp 167–182

Keller MD, Kiene RP, Matrai PA, Bellows WK (1999a) Production of glycine betaine and dimethylsulfoniopropionate in marine phytoplankton. I Batch cultures. Marine Biol 135:237–248

Keller MD, Kiene RP, Matrai PA, Bellows WK (1999b) Production of glycine betaine and dimethylsulfoniopropionate in marine phytoplankton. II. N-limited chemostat cultures. Marine Biol 135:249–257

Keller MD, Korjeff-Bellows W (1996) Physiological aspects of the production of dimethylsulfoniopropionate (DMSP) by marine phytoplankton. In: Kiene RP, Visscher PT, Keller MD, Kirst GO (eds) Biological and environmental chemistry of DMSP and related sulfonium compounds. Plenum Press, New York, pp 131–142

Kettle AJ, Andreae MO (2000) Flux of dimethylsulfide from the oceans: A comparison of updated data seas and flux models. J Geophys Res-Atmos 105:26793–26808

Kettle AJ, Andreae MO, Amouroux D, Andreae TW, Bates TS, Berresheim H, Bingemer H, Boniforti R, Curran MAJ, DiTullio GR, Helas G, Jones GB, Keller MD, Kiene RP, Leck C, Levasseur M, Malin G, Maspero M, Matrai P, McTaggart AR, Mihalopoulos N, Nguyen BC, Novo A, Putaud JP, Rapsomanikis S, Roberts G, Schebeske G, Sharma S, Simo R, Staubes R, Turner S, Uher G (1999) A global database of sea surface dimethylsulfide (DMS) measurements and a procedure to predict sea surface DMS as a function of latitude, longitude, and month. Global Biogeochem Cycles 13:399–444

Kieber DJ, Jiao JF, Kiene RP, Bates TS (1996) Impact of dimethylsulfide photochemistry on methyl sulfur cycling in the equatorial Pacific Ocean. J Geophys Res-Oceans 101:3715–3722

Kiene RP, Bates TS (1990) Biological removal of dimethyl sulfide from sea water. Nature 345:702–705

Kiene RP, Linn LJ (2000) Distribution and turnover of dissolved DMSP and its relationship with bacterial production and dimethylsulfide in the Gulf of Mexico. Limnol Oceanogr 45:849–861

Kiene RP, Linn LJ, Bruton JA (2000) New and important roles for DMSP in marine microbial communities. J Sea Res 43:209–224

Kiene RP, Linn LJ, Gonzalez J, Moran MA, Bruton JA (1999) Dimethylsulfoniopropionate and methanethiol are important precursors of methionine and protein-sulfur in marine bacterioplankton. Appl Environ Microbiol 65:4549–4558

Kiene RP, Service SK (1991) Decomposition of dissolved DMSP and DMS in estuarine waters—Dependence on temperature and substrate concentration. Mar Ecol Prog Ser 76:1–11

Kiene RP, Slezak D (2006) Low dissolved DMSP concentrations in seawater revealed by small volume gravity filtration and dialysis sampling. Limnol Oceanogr Methods 4:80–95

Kirst GO, Thiel C, Wolff H, Nothnagel J, Wanzek M, Ulmke R (1991) Dimethylsulfoniopropionate (DMSP) in ice-algae and its possible biological role. Mar Chem 35:381–388

Kwint RLJ, Irigoien X, Kramer KJM (1996) Copepods and DMSP. In: Kiene RP, Visscher PT, Keller MD, Kirst GO (eds) Biological and environmental chemistry of DMSP and related sulfonium compounds. Plenum Press, New York, pp 239–252

Kwint RLJ, Kramer KJM (1996) The annual cycle of the production and fate of DMS and DMSP in a marine coastal system. Mar Ecol Prog Ser 134:217–224

Laroche D, Vezina AF, Levasseur M, Gosselin M, Stefels J, Keller MD, Matrai PA, Kwint RLJ (1999) DMSP synthesis and exudation in phytoplankton: a modeling approach. Mar Ecol Prog Ser 180:37–49

Lee PA, de Mora SJ (1999a) A review of dimethylsulfoxide in aquatic environments. Atmosphere-Ocean 37:439–456

Lee PA, de Mora SJ (1999b) Intracellular dimethylsulfoxide (DMSO) in unicellular marine algae: Speculations on its origin and possible biological, role. J Phycol 35:8–18

Lefevre M, Vezina A, Levasseur M, Dacey JWH (2002) A model of dimethylsulfide dynamics for the subtropical North Atlantic. Deep-Sea Res Part I 49:2221–2239

Levasseur M, Gosselin M, Michaud S (1994) A New Source of Dimethylsulfide (DMS) for the Arctic Atmosphere - Ice Diatoms. Marine Biol 121:381–387

Liss PS, Malin G, Turner SM, Holligan PM (1994) Dimethyl Sulfide And Phaeocystis - A Review. J Mar Syst 5:41–53

Liss PS, Slater PG (1974) Flux of gases across the air-sea interface. Nature 247:181–184

Lomans BP, van der Drift C, Pol A, Op den Camp HJM (2002) Microbial cycling of volatile organic sulfur compounds. Cell Mol Life Sci 59:575–588

Malin G, Kirst GO (1997) Algal production of dimethyl sulfide and its atmospheric role. J Phycol 33:889–896

Malin G, Liss PS, Turner SM (1994) Dimethyl sulfide: production and atmospheric consequences. In: Green JC, Leadbeater BSC (eds) The Haptophyte Algae. Clarendon Press, Oxford, pp 303–320

Malin G, Turner S, Liss P, Holligan P, Harbour D (1993) Dimethylsulfide and Dimethylsulphoniopropionate in the Northeast Atlantic During the Summer Coccolithophore Bloom. Deep-Sea Res Part I 40:1487–1508

Malin G, Wilson WH, Bratbak G, Liss PS, Mann NH (1998) Elevated production of dimethylsulfide resulting from viral infection of cultures of Phaeocystis pouchetii. Limnol Oceanogr 43:1389–1393

Matrai PA, Keller MD (1993) Dimethylsulfide In A Large-Scale Coccolithophore Bloom In The Gulf Of Maine. Cont Shelf Res 13:831–843

Matrai PA, Vernet M (1997) Dynamics of the vernal bloom in the marginal ice zone of the Barents Sea: Dimethyl sulfide and dimethylsulfoniopropionate budgets. J Geophys Res-Oceans 102:22965–22979

Menden-Deuer S, Lessard EJ (2000) Carbon to volume relationships for dinoflagellates, diatoms, and other protist plankton. Limnol Oceanogr 45:569–579

Mitchell JFB, Johns TC, Gregory JM, Tett SFB (1995) Climate response to increasing levels of greenhouse gases and sulphate aerosols. Nature 376:501–504

Nguyen BC, Belviso S, Mihalopoulos N, Gostan J, Nival P (1988) Dimethyl Sulfide Production During Natural Phytoplanktonic Blooms. Mar Chem 24:133–141

Nightingale PD, Malin G, Law CS, Watson AJ, Liss PS, Liddicoat MI, Boutin J, Upstill-Goddard RC (2000) In situ evaluation of air-sea gas exchange parameterizations using novel conservative and volatile tracers. Global Biogeochem Cycles 14:373–387

Niki T, Kunugi M, Otsuki A (2000) DMSP-lyase activity in five marine phytoplankton species: its potential importance in DMS production. Marine Biol 136:759–764

Nishiguchi MK, Somero GN (1992) Temperature- and concentration-dependence of compatibility of the organic osmolyte ß-dimethylsulfoniopropionate. Cryobiol 29:118–124

Nejstgaard JC, Tang KW, Steinke M, Dutz J, Koski M, Antajan E, Long JD (2007) Zooplankton grazing on *Phaeocystis*: a quantitative review and future challenges. Biogeochemistry doi:10.1007/s10533-007-9098-y

Noordkamp DJB, Gieskes WWC, Gottschal JC, Forney LJ, van Rijssel M (2000) Acrylate in Phaeocystis colonies does not affect the surrounding bacteria. J Sea Res 43:287–296

Noordkamp DJB, Schotten M, Gieskes WWC, Forney LJ, Gottschal JC, van Rijssel M (1998) High acrylate concentrations in the mucus of Phaeocystis globosa colonies. Aquat Microb Ecol 16:45–52

Olson MB, Strom SL (2002) Phytoplankton growth, microzooplankton herbivory and community structure in the southeast Bering Sea: insight into the formation and temporal persistence of an Emiliania huxleyi bloom. Deep-Sea Res Part II 49:5969–5990

Paasche E (2001) A review of the coccolithophorid Emiliania huxleyi (Prymnesiophyceae), with particular reference to growth, coccolith formation, and calcification-photosynthesis interactions. Phycologia 40:503–529

Reigstad M, Wassmann P, Ratkova T, Arashkevich E, Pasternak A, Oygarden S (2000) Comparison of the springtime vertical export of biogenic matter in three northern Norwegian fjords. Mar Ecol Prog Ser 201:73–89

Riseman SF, DiTullio GR (2004) Particulate dimethylsulfoniopropionate and dimethylsulfoxide in relation to iron availability and algal community structure in the Peru Upwelling System. Can J Fish Aquat Sci 61:721–735

Scarratt M, Cantin G, Levasseur M, Michaud S (2000) Particle size-fractionated kinetics of DMS production: where does DMSP cleavage occur at the microscale? J Sea Res 43:245–252

Sheets EB, Rhodes D (1996) Determination of DMSP and other onium compounds in Tetraselmis subcordiformis by plasma desorption mass spectrometry. In: Kiene RP, Visscher PT, Keller MD, Kirst GO (eds) Biological and environmental chemistry of DMSP and related sulfonium compounds. Plenum Press, New York, pp 55–63

Sieburth JM (1960) Acrylic acid, an ''antibiotic'' principle in Phaeocystis blooms in Antarctic waters. Science 132:676–677

Simo R (2004) From cells to globe: approaching the dynamics of DMS(P) in the ocean at multiple scales. Can J Fish Aquat Sci 61:673–684

Simo R, Archer SD, Pedros-Alio C, Gilpin L, Stelfox-Widdicombe CE (2002) Coupled dynamics of dimethylsulfoniopropionate and dimethylsulfide cycling and the microbial food web in surface waters of the North Atlantic. Limnol Oceanogr 47:53–61

Simo R, Dachs J (2002) Global ocean emission of dimethylsulfide predicted from biogeophysical data. Global Biogeochem. Cycles 16:Article no. 1078

Simo R, Hatton AD, Malin G, Liss PS (1998) Particulate dimethyl sulphoxide in seawater: production by microplankton. Mar Ecol Prog Ser 167:291–296

Simo R, Pedros-Alio C (1999a) Role of vertical mixing in controlling the oceanic production of dimethyl sulphide. Nature 402:396–399

Simo R, Pedros-Alio C (1999b) Short-term variability in the open ocean cycle of dimethylsulfide. Global Biogeochem Cycles 13:1173–1181

Simo R, Vila-Costa M (2006) Ubiquity of algal dimethylsulfoxide in the surface ocean: Geographic and temporal distribution patterns. Mar Chem 100:136–146

Slezak D, Herndl GJ (2003) Effects of ultraviolet and visible radiation on the cellular concentrations of dimethylsulfoniopropionate (DMSP) in Emiliania huxleyi (strain L). Mar Ecol Prog Ser 246:61–71

Slezak DM, Puskaric S, Herndl GJ (1994) Potential role of acrylic acid in bacterioplankton communities in the sea. Mar Ecol Prog Ser 105:191–197

Spiro PA, Jacob DJ, Logan JA (1992) Global inventory of sulfur emissions with 1°x1° resolution. J Geophys Res 97:6023–6036

Stefels J (2000) Physiological aspects of the production and conversion of DMSP in marine algae and higher plants. J Sea Res 43:183–197

Stefels J, Dijkhuizen L (1996) Characteristics of DMSP-lyase in *Phaeocystis* sp (Prymnesiophyceae). Mar Ecol Prog Ser 131:307–313

Stefels J, Dijkhuizen L, Gieskes WWC (1995) DMSP-lyase activity in a spring phytoplankton bloom off the Dutch coast, related to *Phaeocystis* sp. abundance. Mar Ecol Prog Ser 123:235–243

Stefels J, van Boekel WHM (1993) Production of DMS from dissolved DMSP in axenic cultures of the marine phytoplankton species Phaeocystis sp. Mar Ecol Prog Ser 97:11–18

Stefels J, van Leeuwe MA (1998) Effects of iron and light stress on the biochemical composition of Antarctic *Phaeocystis* sp. (Prymnesiophyceae). I. Intracellular DMSP concentrations. J Phycol 34:486–495

Steinke M, Daniel C, Kirst GO (1996) DMSP lyase in marine macro- and microalgae: intraspecific differences in cleavage activity. In: Kiene RP, Visscher PT, Keller MD, Kirst GO (eds) Biological and environmental chemistry of DMSP and related sulfonium compounds. Plenum Press, New York, pp 317–324

Steinke M, Malin G, Archer SD, Burkill PH, Liss PS (2002a) DMS production in a coccolithophorid bloom: evidence for the importance of dinoflagellate DMSP lyases. Aquat Microb Ecol 26:259–270

Steinke M, Malin G, Gibb SW, Burkill PH (2002b) Vertical and temporal variability of DMSP lyase activity in a coccolithophorid bloom in the northern North Sea. Deep-Sea Res Part II 49:3001–3016

Steinke M, Wolfe GV, Kirst GO (1998) Partial characterisation of dimethylsulfoniopropionate (DMSP) lyase isozymes in 6 strains of Emiliania huxleyi. Mar Ecol Prog Ser 175:215–225

Strom S, Wolfe G, Holmes J, Stecher H, Shimeneck C, Lambert S, Moreno E (2003a) Chemical defense in the microplankton I: Feeding and growth rates of heterotrophic protists on the DMS-producing phytoplankter Emiliania huxleyi. Limnol Oceanogr 48:217–229

Strom S, Wolfe G, Slajer A, Lambert S, Clough J (2003b) Chemical defense in the microplankton II: Inhibition of protist feeding by beta-dimethylsulfoniopropionate (DMSP). Limnol Oceanogr 48:230–237

Summers PS, Nolte KD, Cooper AJL, Borgeas H, Leustek T, Rhodes D, Hanson AD (1998) Identification and stereospecificity of the first three enzymes of 3-dimethylsulfoniopropionate biosynthesis in a chlorophyte alga. Plant Physiol 116:369–378

Sunda W, Kieber DJ, Kiene RP, Huntsman S (2002) An antioxidant function for DMSP and DMS in marine algae. Nature 418:317–320

Tang KW, Dam HG, Visscher PT, Fenn TD (1999) Dimethylsulfoniopropionate (DMSP) in marine copepods and its relation with diets and salinity. Mar Ecol Prog Ser 179:71–79

Tang KW, Simo R (2003) Trophic uptake and transfer of DMSP in simple planktonic food chains. Aquat Microb Ecol 31:193–202

Toole DA, Kieber DJ, Kiene RP, Siegel DA, Nelson NB (2003) Photolysis and the dimethylsulfide (DMS) summer paradox in the Sargasso Sea. Limnol Oceanogr 48:1088–1100

Toole DA, Kieber DJ, Kiene RP, White EM, Bisgrove J, del Valle DA, Slezak D (2004) High dimethylsulfide photolysis rates in nitrate-rich Antarctic waters. Geophys Res Lett 31:Article no. L11307

Toole DA, Siegel DA (2004) Light-driven cycling of dimethylsulfide (DMS) in the Sargasso Sea: Closing the loop. Geophys Res Lett 31:Article no. L09308

Toole DA, Slezak D, Kiene RP, Kieber DJ, Siegel DA (2006) Effects of solar radiation on dimethylsulfide cycling in the western Atlantic Ocean. Deep-Sea Res Part I 53:136–153

Trossat C, Nolte KD, Hanson AD (1996) Evidence that the pathway of dimethylsulfoniopropionate biosynthesis begins in the cytosol and ends in the chloroplast. Plant Physiology 111:965–973

Turner SM, Bakker DCE, Goldson LE, Messias M-J, Nightingale PD, Riebesell U (in prep) Considerations on the use of sulfur hexafluoride as a deliberate tracer in oceanic biogeochemical lagrangian studies: lessons from EisenEx

Turner SM, Malin G, Liss PS, Harbour DS, Holligan PM (1988) The seasonal variation of dimethylsulfoniopropionate concentrations in near-shore waters. Limnol Oceanogr 33:364–375

Turner SM, Malin G, Nightingale PD, Liss PS (1996) Seasonal variation of dimethyl sulphide in the North Sea and an assessment of fluxes to the atmosphere. Mar Chem 54:245–262

Turner SM, Nightingale PD, Broadgate W, Liss PS (1995) The Distribution of Dimethyl Sulfide and Dimethylsulphoniopropionate in Antarctic Waters and Sea-Ice. Deep-Sea Res Part II 42:1059–1080

van Bergeijk SA, Schonefeldt K, Stal LJ, Huisman J (2002) Production and consumption of dimethylsulfide (DMS) and dimethylsulfoniopropionate (DMSP) in a diatom-dominated intertidal sediment. Mar Ecol Prog Ser 231:37–46

van den Berg AJ, Turner SM, van Duyl FC, Ruardij P (1996) Model structure and analysis of dimethylsulphide (DMS) production in the southern North Sea, considering phytoplankton dimethylsulphoniopropionate- (DMSP) lyase and eutrophication effects. Mar Ecol Prog Ser 145:233–244

van Duyl FC, Gieskes WWC, Kop AJ, Lewis WE (1998) Biological control of short-term variations in the concentration of DMSP and DMS during a Phaeocystis spring bloom. J Sea Res 40:221–231

van Rijssel M, Buma AGJ (2002) UV radiation induced stress does not affect DMSP synthesis in the marine prymnesiophyte Emiliania huxleyi. Aquat Microb Ecol 28:167–174

van Rijssel M, Gieskes WWC (2002) Temperature, light, and the dimethylsulfoniopropionate (DMSP) content of Emiliania huxleyi (Prymnesiophyceae). J Sea Res 48:17–27

Vezina AF (2004) Ecosystem modelling of the cycling of marine dimethylsulfide: a review of current approaches and of the potential for extrapolation to global scales. Can J Fish Aquat Sci 61:845–856

Ward B, Wanninkhof R, McGillis WR, Jessup AT, DeGrandpre MD, Hare JE, Edson JB (2004) Biases in the air-sea flux of CO_2 resulting from ocean surface temperature gradients. J Geophys Res-Oceans 109:Article no. C08S08

Wilson WH, Schroeder DC, Allen MJ, Holden MTG, Parkhill J, Barrell BG, Churcher C, Harnlin N, Mungall K, Norbertczak H, Quail MA, Price C, Rabbinowitsch E, Walker D, Craigon M, Roy D, Ghazal P (2005) Complete genome sequence and lytic phase transcription profile of a Coccolithovirus. Science 309:1090–1092

Wilson WH, Tarran G, Zubkov MV (2002) Virus dynamics in a coccolithophore-dominated bloom in the North Sea. Deep-Sea Res Part II 49:2951–2963

Wilson WH, Turner S, Mann NH (1998) Population dynamics of phytoplankton and viruses in a phosphate-limited mesocosm and their effect on DMSP and DMS production. Estuar Coast Shelf Sci 46:49–59

Wolfe GV (2000) The chemical defense ecology of marine unicellular plankton: Constraints, mechanisms, and impacts. Biol Bull 198:225–244

Wolfe GV, Levasseur M, Cantin G, Michaud S (2000) DMSP and DMS dynamics and microzooplankton grazing in the Labrador Sea: application of the dilution technique. Deep-Sea Res Part I 47:2243–2264

Wolfe GV, Steinke M (1996) Grazing-activated production of dimethyl sulfide (DMS) by two clones of Emiliania huxleyi. Limnol Oceanogr 41:1151–1160

Wolfe GV, Steinke M, Kirst GO (1997) Grazing-activated chemical defence in a unicellular marine alga. Nature 387:894–897

Wolfe GV, Strom SL, Holmes JL, Radzio T, Olson MB (2002) Dimethylsulfoniopropionate cleavage by marine phytoplankton in response to mechanical, chemical, or dark stress. J Phycol 38:948–960

Yang GP, Tsunogai S (2005) Biogeochemistry of dimethylsulfide (DMS) and dimethylsulfoniopropionate (DMSP) in the surface microlayer of the western North Pacific. Deep-Sea Res Part I 52:553–567

Yang H, McTaggart AR, Davidson AT, Burton H (1994) Measurement of acrylic acid and dimethyl sulfide in Antarctic coastal water during a summer bloom of Phaeocystis pouchetii. the NIPR Symposium on Polar Biology, no. 7. National Institute of Polar Research, Tokyo, pp 43–52

Yoch DC (2002) Dimethylsulfoniopropionate: Its sources, role in the marine food web, and biological degradation to dimethylsulfide. Appl Environ Microbiol 68:5804–5815

Zemmelink HJ, Gieskes WWC, Klaassen W, Beukema WJ, de Groot HW, de Baar HJW, Hintsa EJ, McGillis WR, Dacey JWH (2004) Relaxed eddy accumulation measurements of the sea-to-air transfer of dimethylsulfide over the northeastern Pacific. J Geophys Res-Oceans 109: C01025, doi: 1029/2002JC001616

Zubkov MV, Fuchs BM, Archer SD, Kiene RP, Amann R, Burkill PH (2001) Linking the composition of bacterioplankton to rapid turnover of dissolved dimethylsulphoniopropionate in an algal bloom in the North Sea. Environ Microb 3:304–311

Zubkov MV, Fuchs BM, Archer SD, Kiene RP, Amann R, Burkill PH (2002) Rapid turnover of dissolved DMS and DMSP by defined bacterioplankton communities in the stratified euphotic zone of the North Sea. Deep-Sea Res Part II 49:3017–3038

Biogeochemistry (2007) 83:277–292
DOI 10.1007/s10533-007-9092-4

ORIGINAL PAPER

Variability in abundance and fluxes of dimethyl sulphide in the Indian Ocean

D. M. Shenoy · M. Dileep Kumar

Received: 16 February 2006 / Accepted: 16 November 2006 / Published online: 8 March 2007

Abstract Dimethyl sulphide (DMS) is a biogenic gas of climatic significance on which limited information is available from the Indian Ocean. To fill this gap, we collected data on DMS and total dimethylsulphoniopropionate ($DMSP_t$) by participating in a dozen cruises. Here, we discuss the variability in DMS and $DMSP_t$ in the north and central Indian Ocean in terms of their spatial and temporal variation. DMS and $DMSP_t$ exhibited significant spatial and temporal variability. Apart from the concentration gradients in DMS within the Arabian Sea, Bay of Bengal and Central Indian Ocean basins, differences in average abundances were conspicuous between these basins. The Arabian Sea contained more DMS (mixed layer average was 7.8 nM) followed by the Bay of Bengal (2.8 nM) and the Central Indian Ocean (2.7 nM). The highest concentrations of DMS and $DMSP_t$ (525 nM and 916 nM, respectively) were found in upwelling regimes along the west coast of India during the Southwest monsoon and fall intermonsoon seasons. Average surface DMS was the highest in the Arabian Sea. On the other hand observed sea-to-air fluxes of DMS were higher in the Bay of Bengal due to the prevalence of turbulent conditions. In the Arabian Sea wind speeds were low and hence the sea-to-air fluxes. The total diffusive flux of DMS from the study area to atmosphere is estimated to be about 1.02×10^{12} g S y^{-1}, which contributes to 4.1–6.3% of the global DMS emission

Keywords DMS · $DMSP_t$ · Fluxes · Indian Ocean · Monsoon

D. M. Shenoy (✉) · M. D. Kumar
Chemical Oceanography Division, National Institute of Oceanography, Dona Paula, Goa 403 004, India

Present Address:
D. M. Shenoy
SAMS, Dunstaffnage Marine Laboratory, Oban, Argyll PA37 1QA, Scotland, UK
e-mail: damodar.shenoy@sams.ac.uk

Introduction

Dimethyl sulphide is the most dominant of the reduced sulphur gas found in surface layers of the ocean (Lovelock et al. 1972). The emission of dimethyl sulphide from seawater is expected to balance the excess sulphur deposition over the remote oceans (Charlson et al. 1992). Charlson et al. (1987) proposed a hypothesis, known as the CLAW (after the authors Charlson, Lovelock, Andreae and Warren) hypothesis connecting biogenic DMS emissions to changes in albedo, in which increased production of DMS due to global warming is expected to lead to more sulphate aerosols and subsequently to more cloud condensation nuclei (CCN) that in turn enhances back radiation.

In seawater, DMS is produced from dimethylsulphoniopropionate (DMSP), which in turn is produced by phytoplankton. Specific species of phytoplankton are found to be responsible for most of the DMSP production in seawater. According to Liss et al. (1993) the potential for production of DMS in various taxonomic groups follows the order:

Coccolithophores > *Phaeocystis*
> Dinoflagellates > Diatoms

DMSP in phytoplankton was initially proposed to function as an osmolyte (Vairavamurty et al. 1985). Over the years we now have evidence for its role in other processes such as for cryopreservation (Kirst et al. 1991), as a deterrent against grazing (Dacey and Wakeham 1986; Wolfe and Steinke 1996) and against bacteria (Wolfe et al. 1997). In the recent past the anti-oxidant role of DMSP is becoming more popular (Sunda et al. 2002). Factors controlling DMSP production, its conversion to DMS, fate of DMSP and DMS sulphur in the marine environment are well documented in the review by Stefels et al. (2000) and in the present issue.

Extensive measurements on DMS and DMSP have been done in different parts of the oceans (Kettle et al. 1999 and references therein) including estuarine and lagoon waters (Iverson et al. 1989; Moret et al. 2000). A few studies have been made in the Indian Ocean, which include Hatton et al. (1999), Shenoy et al. (2000, 2002), Kumar et al. (2002) and Shenoy and Patil (2003) and some unpublished data in eastern Indian Ocean (of T.S. Bates, as mentioned by Kettle et al. 1999) and for Amsterdam station (Nguyen et al. 1990, 1992).

Hatton et al. (1999) found elevated concentrations of DMS, DMSP and dimethyl sulphoxide (DMSO) in the eutrophic zone (Arabian Sea) immediately after the SW monsoon. In addition DMSO concentrations were found to correlate with near surface DMS and DMSP. Shenoy et al. (2000) reported the first ever measurements of DMS from the Bay of Bengal where salinity appeared to play an important role in the DMSP production. Shenoy et al. (2002) documented five times higher concentrations of DMS and DMSP in winter season of 1999 than that in 1998, which has been attributed to differences in physical forcing and associated biological processes. In a time series experiment carried out in estuarine waters of Goa, on the west coast of India, Shenoy and Patil (2003) recorded elevated concentrations of DMS and DMSP during the monsoon season (maximum DMS 15.4 nM and $DMSP_t$ 419.5 nM) due to the possible prevalence of mixed population of diatoms and dinoflagellates. Covariations among oceanic and atmospheric DMS and atmospheric SO_2, wet deposition of methane sulphonic acid (MSA), non-sea salt (NSS) SO_4^{-2}, and rain acidity have been shown by Nguyen et al. (1990, 1992) near Amsterdam island (southern Indian Ocean).

While the biogeochemistry of carbon dioxide and nitrogen species are reasonably well known, particularly in the northern Indian Ocean, our knowledge of sulphur species is limited. DMS and associated sulphur compounds are climatically important, particularly in the monsoon-dominated tropical Northern Indian Ocean, in regard to air–sea interactions and feedbacks. Significant diversity in physical and associated biological regimes in the Indian Ocean together with dynamic climate offers a potentially interesting region to understand the DMS cycling and its controlling factors. Therefore, this study has been undertaken to study spatial (especially inshore–offshore gradients off the coast of India and surface variation) and temporal variations of DMS and $DMSP_t$ in the Indian Ocean and to evaluate its atmospheric emissions in regional and global context.

Material and methods

Sampling for this study was attempted keeping in view the large spatial and temporal variability of the biogeochemical processes in the Indian Ocean. Data were collected for temperature, salinity, dissolved oxygen, nitrate, chlorophyll *a*, DMS and $DMSP_t$ in seawater. Besides, data were collected for wind speed and atmospheric temperature. Figure 1a depicts the study area in the Indian Ocean, where a dozen oceanic expeditions were undertaken on board ORV Sagar Kanya (SK) and FORV Sagar Sampada (SS) covering a total of 320 stations. Data were mainly collected

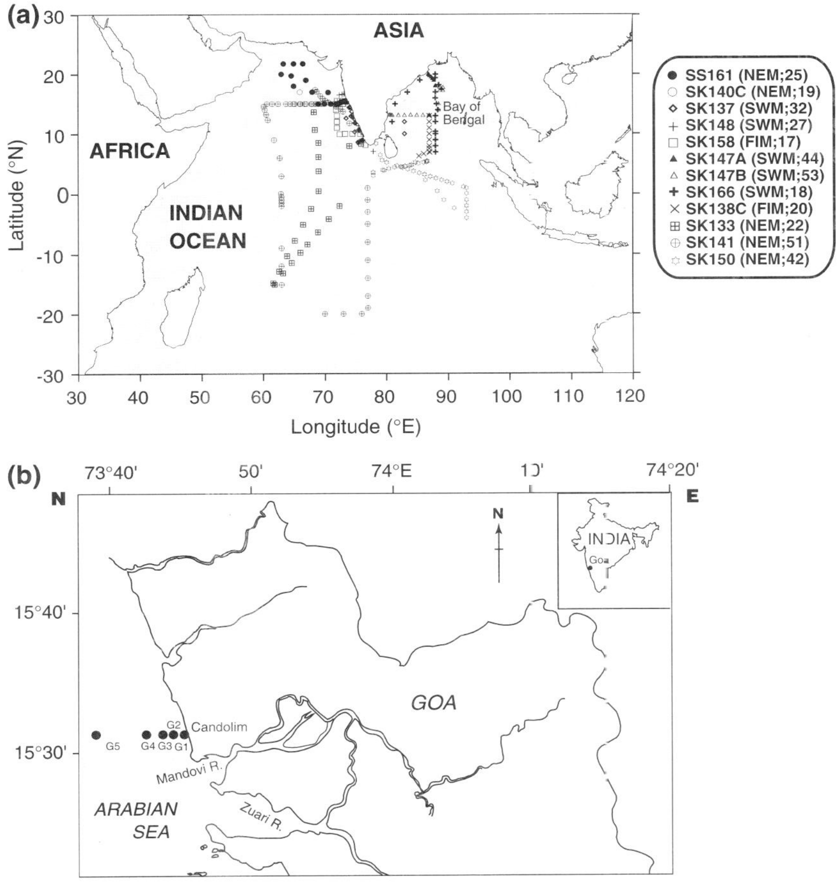

Fig. 1 (**a**) Station locations occupied for dimethyl sulphide studies in the central and north Indian Ocean. Symbol index indicates cruise number, season and number of stations. (**b**) Station locations off Candolim, Goa

from the upper 200 m of the water column as the initial study revealed DMS and $DMSP_t$ to be below detectable levels at depths greater than 200 m. Surface waters were sampled at many locations besides regular stations. Figure 1b depicts the stations locations along a section off Candolim, Goa used for collection of DMS data particularly during and after the southwest monsoon. The data presented in this paper were collected under different projects in the Indian Ocean: land-ocean interaction in the coastal zone (LOICZ), the Bay of Bengal monsoon experiment (BOBMEX), the Bay of Bengal process studies (BOBPS), ocean observing systems (OOS) and the Indian Ocean experiment (INDOEX).

Hydrographic and biological properties

Seawater samples were collected using a rosette attached to the conductivity–temperature–depth (CTD) system, fitted with 12 Niskin bottles. Subsampling of water was made in the order: dissolved oxygen (DO), DMS, $DMSP_t$, nutrients, salinity and chlorophyll *a*. Care was taken while sampling dissolved gases to avoid trapping of bubbles. Temperature and conductivity were measured using the sensors fitted to the Sea-bird CTD system. Temperature sensed by the probe was periodically checked against the values obtained from reversing thermometers. Conductivity measurements of CTD, and hence salinity were calibrated against data obtained using an Autosal salinometer (model 8400). Dissolved oxygen was measured by the classical Winkler titration method. Nutrients (nitrate) were measured using a Skalar autoanalyser (SA 4000).

For chlorophyll *a* analysis, a known volume of seawater sample (0.5–1 l for coastal and estuarine samples or 2–3 l for open ocean waters) was filtered through Whatman GF/F filters under low vacuum. Chlorophyll pigments on the filter paper were then extracted into 10 ml of 90% acetone under dark and refrigerated conditions for 24 h. Fluorescence was measured using a Hitachi spectrofluorometer (UNESCO, 1994).

DMS and DMSP measurements

Water samples for DMS and $DMSP_t$ analyses were collected in two separate 60 ml dark ground-glass bottles. Water samples were not filtered because even with extreme care DMS losses cannot be prevented during the filtrations and this process can alter the contents of $DMSP_t$. Thus the concentrations reported here refer to total dissolved DMS and DMSP. Following the collection DMS samples were immediately kept in the dark at 4°C. DMS was measured using a Hewlett Packard 5890 series II plus gas chromatograph fitted with a flame photometric detector (FPD). The details of the analysis are described in Shenoy et al. (2002). The detection limit for DMS measurements, found as twice the standard deviation from regression analysis for values <0.1 nmol, was 0.012 nmol. Precision in DMS measurements was found to be 6% for standards. Reproducibility of DMS and $DMSP_t$ in seawater samples was periodically checked in all the cruises for paucity of time and availability of a dedicated gas chromatograph.

Immediate repeat analysis of a sample showed no detectable DMS indicating its negligible production from $DMSP_t$, if any, during the stripping process. $DMSP_t$ was measured after hydrolyzing the unfiltered water sample for 6 h using 10 M NaOH (1 ml), which was added immediately after sample collection (Turner et al. 1990; Shenoy et al. 2002). Alkali hydrolysis resulted in the cleavage of $DMSP_t$ into DMS and acrylic acid. $DMSP_t$ concentrations were read from the DMS calibration curve. Tests revealed about 95% conversion of $DMSP_t$ to DMS during the alkali hydrolysis for 6 h. The precision of seawater DMS analysis is found to be 8–10%.

Wind Speed

Wind speed data used in DMS flux calculations were collected using an automatic weather station (AWS) on board the research vessels or on top of the institute building. Wind speed was measured using Young's wind monitor (model 05103). The wind speeds were normalized for 10 m height and corrected for direction.

DMS Flux

DMS fluxes were calculated using the formulations proposed by Turner et al. (1996) and the correction factors given by Saltzman et al. (1993). The flux of DMS gas between sea and air is proportional to the concentration gradient across the air-sea interface and is calculated according to the following equation.

$$F_{\mathrm{DMS}} = k \cdot \Delta C$$

where F_{DMS} = net flux of DMS
k = transfer (or piston) velocity and
ΔC = concentration gradient across the air–sea interface.

$$\Delta C = C_{\mathrm{w}} - C_{\mathrm{a}} \cdot h^{-1}$$

where C_{w} = concentration of DMS is seawater
C_{a} = concentration of DMS in air
h = Henry's law constant, expressed as the ratio of air to water concentrations at equilibrium.

As the concentration of DMS in air is very low (nearly three orders of magnitude less than that in water) C_{a} is generally considered to be zero and thus C_{w} equals ΔC.

Mixed layer depth (MLD)

Mixed layer depths (MLD) were calculated based on density criteria where MLD was defined as the depth at which density increased by 0.125 kg dm^3 with reference to that at the sea surface. ML-DMS is the DMS averaged in the water column above the MLD.

Results

Hydrography in brief

The southwest and northeast monsoons of the Indian Ocean region play a major role in controlling the biogeochemical processes. Between November and February, the intertropical convergence zone (ITCZ) lies in the southern hemisphere (~10°S) and the region experiences the northeast monsoon (NEM). During this period cold dry winds from the north (north-easterlies) blow over the northern Indian Ocean. These cool dry winds facilitate enhanced evaporation in the northern Arabian Sea and thus set in winter convection (Fig. 2). The occurrence of convection pumps in nutrients into the surface layers. During the present study the average nitrate concentration in surface waters of the Arabian Sea where convection occurred was found to be ~4.9 μM. Such high nitrate levels promote primary production. During NEM the surface circulation in the Arabian Sea is quite similar to the circulation in the North Pacific and Atlantic Oceans.

The extensive heating of the Indo-Gangetic plain and Himalayan terrain from March to May (spring intermonsoon) leads to the development of low pressure which makes the ITCZ move northward. This transition changes the direction of the winds, which now move in the southwesterly direction during the southwest monsoon (June to September, SWM). The surface circulation in the Arabian Sea in SWM is totally opposite to that of the NEM. The circulation along the coasts of Oman and Somali is towards the pole and away from the coast. This causes intense upwelling in these regions. Similarly upwelling also occurs along the south west coast of India, where the flow of water is towards the equator.

One of the major characteristics of the northern Indian Ocean is the differential riverine discharges in to the Arabian Sea and the Bay of Bengal. The Bay of Bengal receives 1.6×10^{12} m^3 y^{-1}, which is five times more than the run off received by the Arabian Sea (0.3×10^{12} m^3 y^{-1}). This reduces the surface salinity in the Bay and the lowest salinities are mostly confined to areas of river mouths. The salinity increases from north to south in the Bay. The low salinity lens results in a strong surface stratification and prevents surfacing of subsurface nutrients through vertical mixing.

The month of October is referred to as the fall intermonsoon (FIM) period. During the FIM the temperature contours show a stratified structure unlike in the SWM. Low wind speeds during this period lead to shallow MLD's. There are no detectable nutrients in the surface layers. One of the prominent features of the fall-intermonsoon is the occurrence of high nitrate levels near the base of MLD, which occurs within the euphotic zone. This leads to basin wide subsurface chlorophyll maximum (SCM, Bhattathiri et al. 1996) as shown in Fig. 3. All these make the Northern Indian Ocean one of the most productive basins in comparison to the rest of the world in general (Behrenfeld and Falkowski 1997).

Central Indian Ocean characteristics have been discussed in detail in Shenoy et al. (2002). In order to get an overview on the variation of DMS and DMSP in the Indian Ocean, their variability is discussed in the following sections.

Vertical distributions

Vertical distributions of DMS and $DMSP_t$ in the coastal and open ocean waters of the Arabian Sea (AS), Bay of Bengal (BoB) and in the Central Indian Ocean (CIO) are shown in Fig. 4. Both DMS and $DMSP_t$ showed very distinct trends in their vertical distributions. Concentrations of DMS and $DMSP_t$ were very low (less than or around the detection limits) in waters deeper than 200 m in the Indian Ocean. Therefore, DMS sampling was restricted to the top 200 m of water column in most cruises. In coastal waters of the Arabian Sea (76°E, 10.19°N) maximal concentrations of DMS and $DMSP_t$ were found at shallow depths. In the present study the highest values of 525 nM of DMS and 916 nM of $DMSP_t$ were found, respectively, at a station off Candolim and at a location off Mangalore (10.19°N, 76°E, SK148). In the open Arabian Sea (65°E, 18°N), a $DMSP_t$ maximum of 15.5 nM occurred at 10 m whereas the peak DMS of 1.7 nM was found at the sea surface. Therefore, depths of occurrence of maximal levels of DMS and $DMSP_t$ did not necessarily coincide at all stations but the higher abundances mostly occurred closer to the surface. The DMS and $DMSP_t$ concentrations rapidly decreased to undetectable levels below 120 m. Similar behaviours were observed in their vertical distributions in the coastal and open ocean waters of the Bay of Bengal. In the central Indian Ocean,

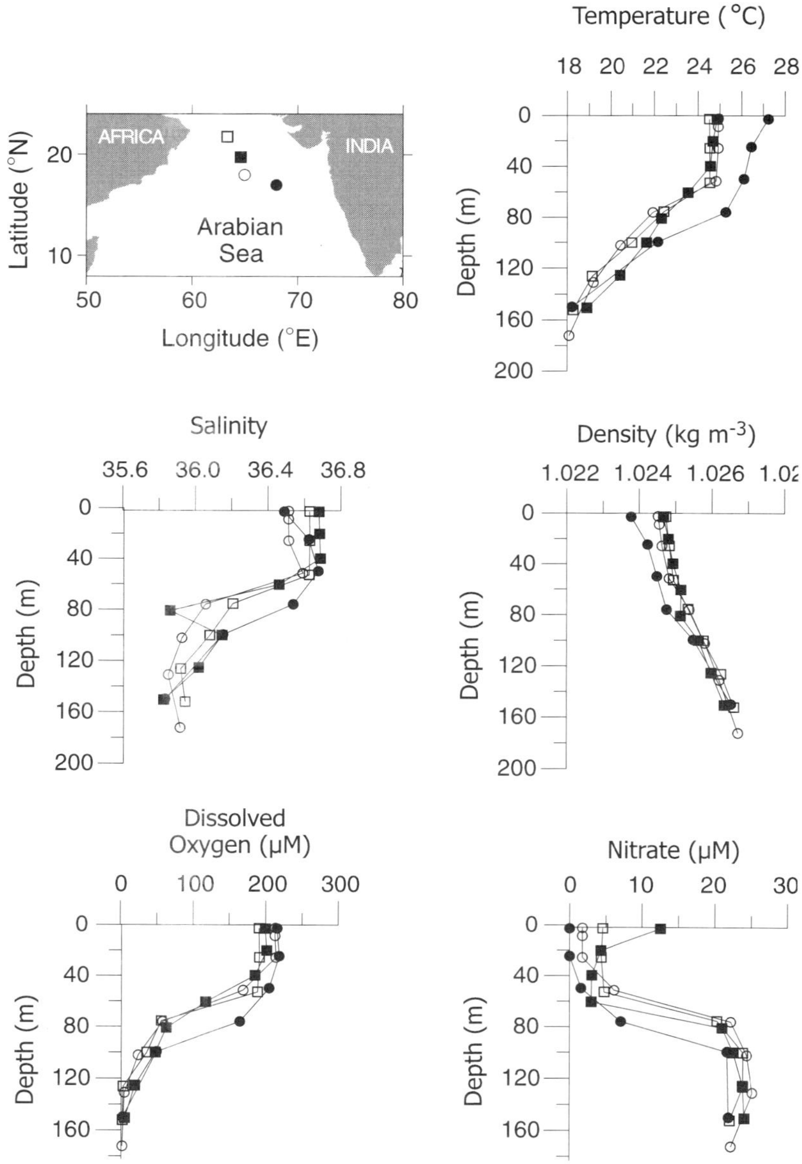

Fig. 2 Vertical profiles of temperature, salinity, density, dissolved oxygen and nitrate in the northern Arabian Sea during the northeast monsoon (SS161)

however, peaks in DMS (9.7 nM of DMS at 41 m) and $DMSP_t$ ($DMSP_t$ maximum of 31.3 nM) were found around a depth of 40 m (Fig. 4). Further, DMS and $DMSP_t$ levels were detectable up to 150–175 m. This occurrence of DMS at deeper depths is unique to the central regions of the Indian Ocean (because of deep mixed layers) and contrasts with that in the Arabian Sea and the Bay of Bengal.

Table 1 lists DMS and $DMSP_t$ variations as averages in the surface mixed layer. In the coastal (southwest coast of India) Arabian Sea the highest values occurred during the SWM and the least were during the intermonsoons. In the Bay of Bengal the concentrations of DMS and $DMSP_t$ near the coast are almost same as the open ocean. The DMS and $DMSP_t$ concentrations in the Central Indian Ocean are nearly similar to those in the Bay of Bengal. On the whole, the Arabian Sea (coastal + open ocean) surface mixed layer contained more than double DMS and $DMSP_t$ (7.8 nM and 43.6 nM,

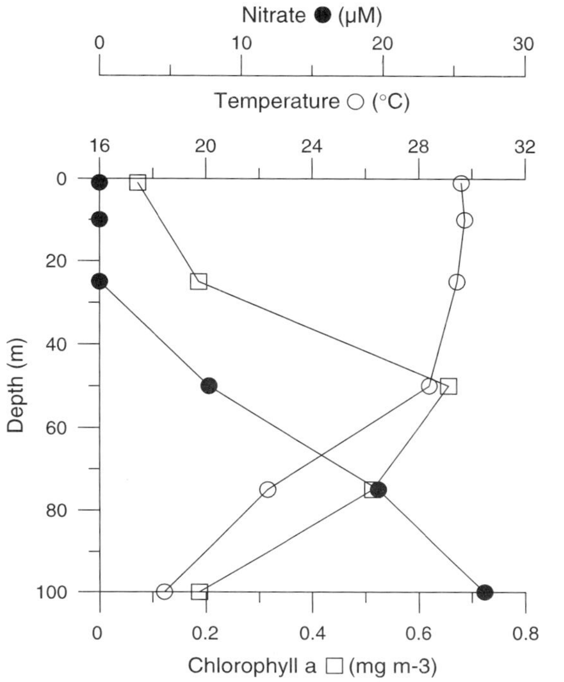

Fig. 3 Profiles of temperature, nitrate and chlorophyll (at 9.94°N, 72.44°E, SK158) showing subsurface chlorophyll maximum

respectively) than the Bay of Bengal (2.8 nM and 12.6 nM, respectively) and central Indian ocean (2.7 nM and 10.1 nM, respectively) and followed the same trend as that of the primary production (Behrenfeld and Falkowski, 1997).

Spatial variations

Figure 5 depicts the variation of DMS in the Arabian Sea, along a section near 15°N (off Goa), during January, July, October and December of 1998. Very clear coastal to open ocean trends were observed. In January (winter), the DMS concentrations varied between 6 and 12 nM within 50 m of the water column near the coast but between 200 and 400 km away from the shore the concentrations ranged between 2 and 4 nM before increasing to 10 nM further offshore. During summer (July 1998), DMS values were higher near the bottom in coastal waters (29 nM), and decreased towards the open ocean. Features similar to those in summer were seen in October 1998 when DMS concentration levels near the coast varied between 15 and 40 nM while in offshore waters these were low. In December 1998 coastal DMS concentrations were around 30 nM but decreased offshore. Thus in the eastern Arabian Sea high DMS concentrations were found closer to the coast.

Data obtained through periodic sampling along a small transect (~13 km, Fig 1b) off the Candolim

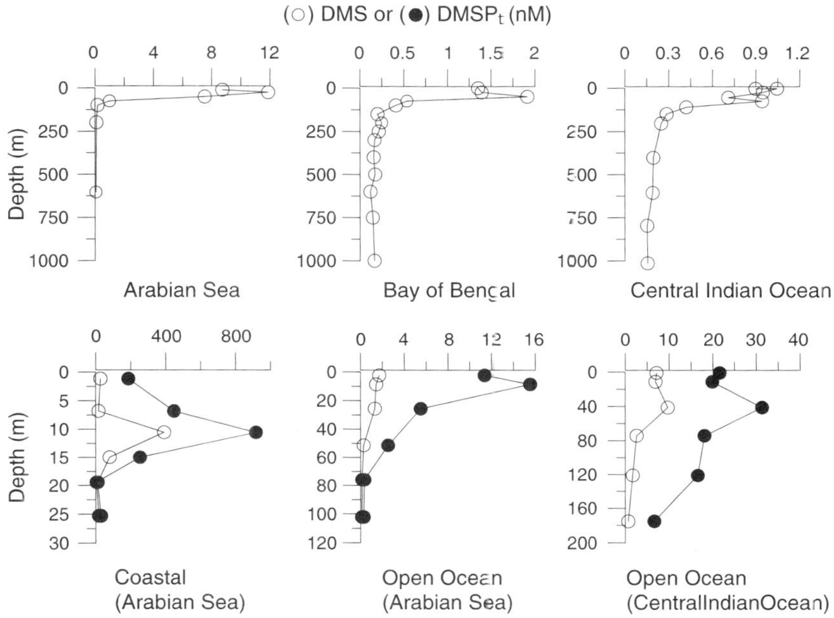

Fig. 4 Typical vertical profiles of DMS and $DMSP_t$ in different regions of the Indian Ocean

Table 1 Averages and ranges (in parenthesis) of DMS and $DMSP_t$ in the surface mixed layer (defined by 0.125 kg m^{-3} density difference) in different seasons and in different regions of the Indian Ocean

Area	Season	Coastal		Open ocean	
		DMS (nM)	DMSP (nM)	DMS (nM)	DMSP (nM)
Arabian Sea	SWM	32.9 (0.3–525.6)	81.1 (2.0–915.9)	1.6 (0.5–3.7)	NM
	NEM	9.0 (1.8–51.0)	NM	2.3 (0.2–11.9)	9.8 (3.6–16.7)
	FIM	5.8 (0.2–64.4)	26.1 (0.5–160.1)	1.4 (0.9–2.2)	9.4 (4.9–16.1)
Bay of Bengal	SWM	3.3 (1.2–7.2)	12.9 (3.1–23.3)	2.8 (0.2–11.1)	13.3 (1.5–30.9)
	FIM	NM	NM	2.8 (0.8–5.5)	10.9 (4.1–19.5)
Central Indian Ocean	NEM	–	–	2.7 (0.2–13.9)	10.1 (1.3–35.9)

NM: Not measured

SWM, NEM and FIM refer to southwest, northwest and fall intermonsoon seasons, respectively

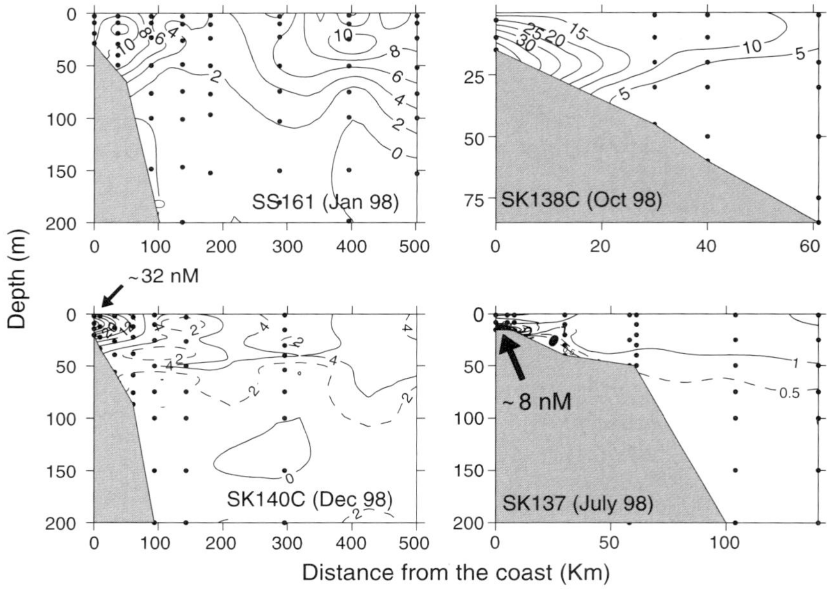

Fig. 5 Longitudinal variations in DMS off Goa in different seasons

coast (Goa) enabled us to examine the extent of seasonal variability in DMS and $DMSP_t$ in close proximity to the coast. During the pre-monsoon (March) of 2001 (Fig. 6a) concentrations of DMS and $DMSP_t$ varied, respectively, from 0.2 to 0.9 nM with an average value of 0.6 nM and between 0.5 and 2.9 nM with a mean of 1.4 nM. These concentrations changed in SW monsoon as upwelled water with low oxygen and high nitrate dominate the coastal regions (Naqvi et al. 2000). DMS and $DMSP_t$ are, in general, higher at the surface because of intense biological activities at surface. Very high photosynthetic fixation and the presence of traces of nitrate seem to have aided higher production of $DMSP_t$ (9–113 nM) and DMS (13–132 nM). The DMS levels found in these coastal waters on 29 September 2000 (Fig. 6b) are about 100 times more than that occurred in March 2001. On the other hand, the highest value for DMS of 525 nM was found very close to the coast in November 1999, where DMS levels varied from 6 nM to 525 nM and $DMSP_t$ varied between 8 nM and 339 nM (Fig. 6c). Coinciding with this trend, extremely high $DMSP_t$ values were found in the same season in the neighbouring Zuari estuary (~419 nM) and in western continental marginal waters off Mangalore (~916 nM). Thus, DMS varied from 0.2 nM in the pre-monsoon season to 525 nM in the fall intermonsoon (or post-SW monsoon) season exhibiting a variability by 2500 times in coastal waters of Goa. Such behaviour could be occurring in at least a few other places

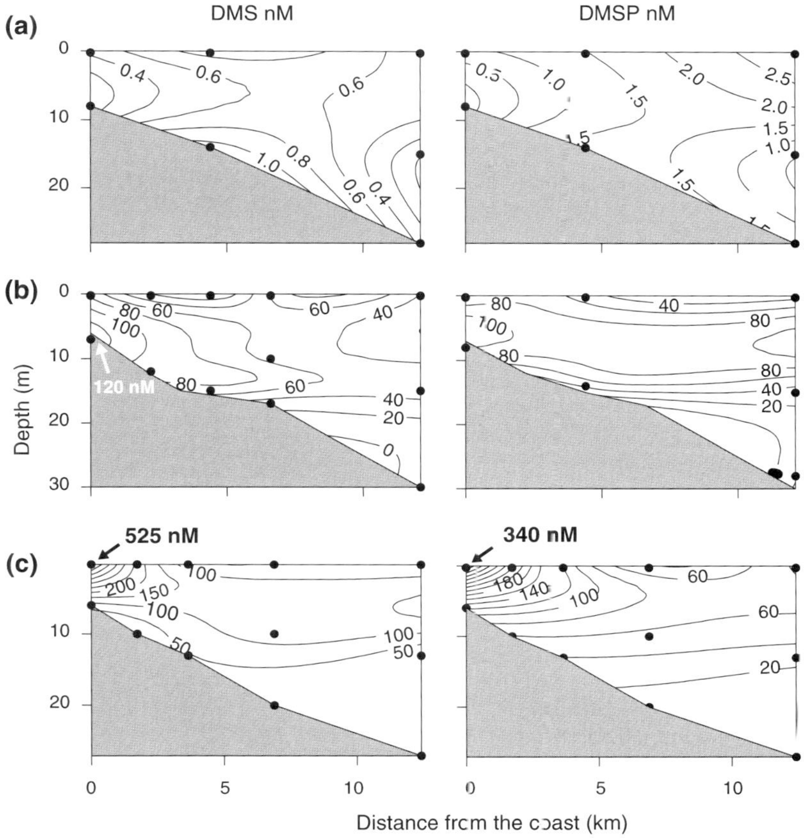

Fig. 6 Time series observations of DMS (left panels) and $DMSP_t$ (right panels) off Candolim in Goa on (**a**) 28/3/2001, (**b**) 29/9/2000 and (**c**) 11/11/1999

along the west coast of India since the oceanographic conditions are similar.

Zonal variations of DMS in the Bay of Bengal are shown in Fig. 7. Near Paradip (86.87°E, 19.98°N), DMS concentrations varied between 0.5 and 3.3 nM in coastal waters but showed little change with increasing distance from the coast. Similar features were noticed off Chennai as well. Here, DMS varied from 0.5 to 3.2 nM. Thus the zonal variability trends of DMS were contrasting between the Arabian Sea (with clear inshore–offshore gradients) and the Bay of Bengal (with no significant variation).

In the central Indian Ocean a different scenario was found. Figure 8 depicts the variation of DMS in the central Indian Ocean observed during the INDOEX cruises of 1998 and 1999. High DMS concentrations (0.1–13.9 nM) were observed between 5°S and 15°S in 1999. These were observed as pockets between surface and 75 m with detectable concentrations down to 100–150 m. While away from these latitudes viz. between 0° and 5°S, and 15°S and 20°S DMS concentrations, in the same depth range, as mentioned above varied between 1 and 5 nM. DMS levels were lower in 1998 with strong latitudinal gradients.

Thus DMS in the Indian Ocean exhibits very high spatial variability in which higher concentrations were observed near to the coast and lower concentrations in the open Ocean. Further, DMS concentrations (MLD average) showed a gradual increase from the central Indian Ocean (2.7 nM) and the Bay of Bengal (2.8 nM) to the Arabian Sea (7.8 nM).

Discussion

DMS at sea surface and its emission

Variations in surface DMS

The envisaged indirect global cooling by DMS products depends on its emission from the ocean.

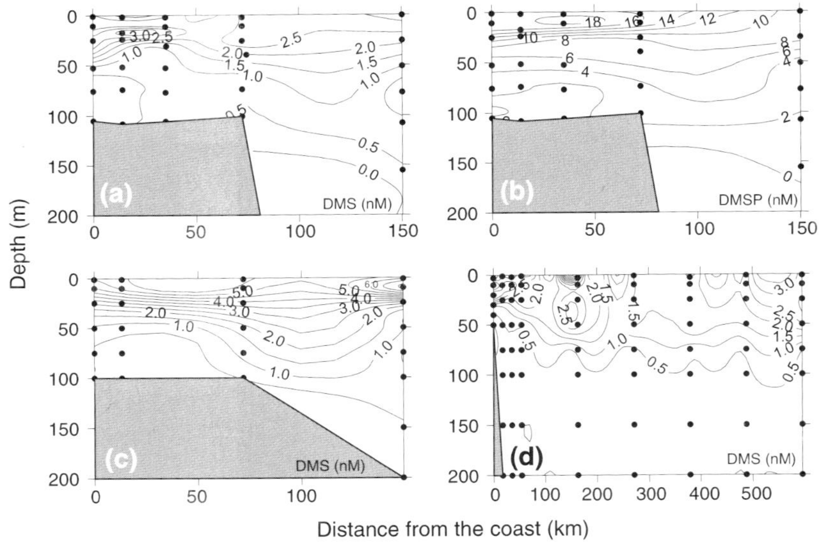

Fig. 7 Variations in DMS and $DMSP_t$ off Paradip (**a**) and (**b**) during SK147A and (**c**) during SK147B and off Chennai (**d**) during SK147B

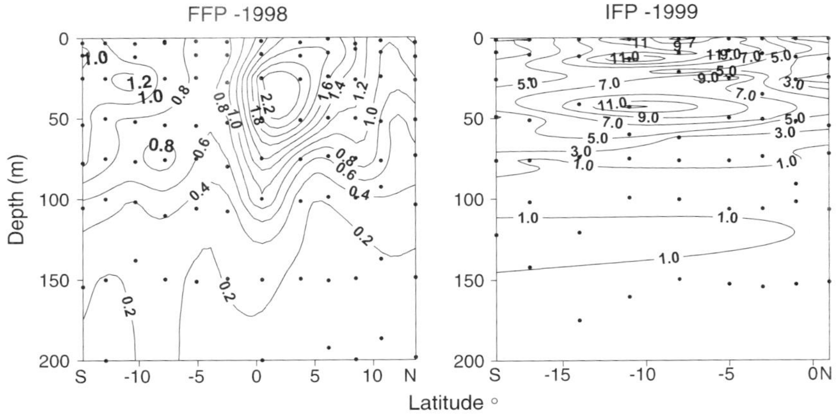

Fig. 8 Variations in DMS (nM) in the central Indian Ocean in the winter season of 1998 (FFP; SK133) and 1999 (IFP; SK141) of the INDOEX project

Variations in surface DMS in the Indian Ocean are shown in Fig. 9. During the northeast monsoon the surface DMS in coastal waters of the Arabian Sea (central and southern regions of the west coast of India) varied between 0.7 nM and 31.7 nM with an average of 6.3 nM, while in the open ocean it was between 0.3 nM and 10.9 nM with a mean of 3.3 nM. The average DMS concentrations in the coastal region were nearly twice that of the open ocean. During the SW monsoon the surface DMS in coastal waters varied between 0.5 nM and 525 nM (the highest value is not shown in Fig. 9) with an average value of 10.3 nM. The highest value was found in near-shore waters of Goa due to very high primary production driven by coastal upwelling and the associated biological processes including hypoxia. In contrast, the intermonsoon, which is a slack period with respect to primary production but rich in bacterial production, experienced low concentrations of DMS. The coastal surface DMS varied between 0.9 nM and 10.7 nM with an average concentration of 2.6 nM while the open ocean surface DMS varied between 0.9 nM and 1.7 nM (average of 1.3 nM) in this season. The mean DMS concentrations at surface were greater in the SW monsoon (10.3 nM) followed by winter (4.3 nM) and intermonsoon (2.1 nM). The annual surface DMS average, in general, for the coastal Arabian Sea was 8.2 nM whereas it was

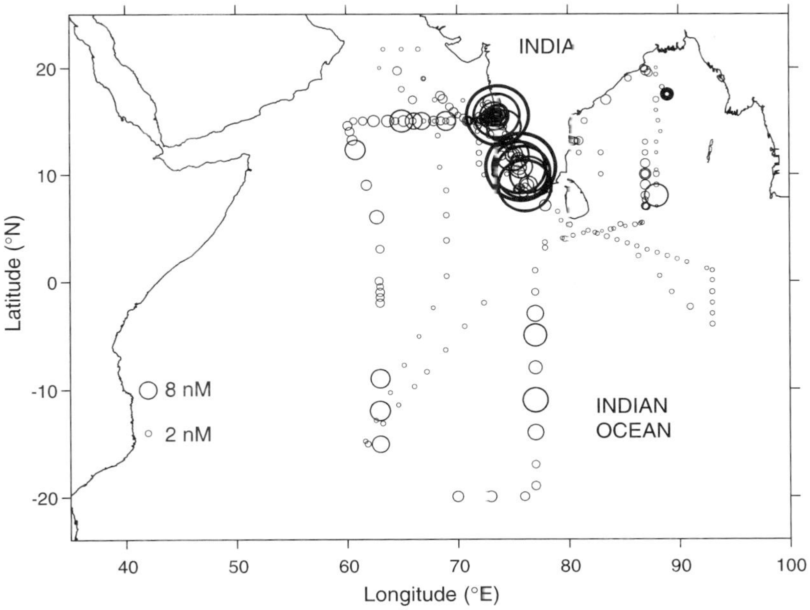

Fig. 9 Variations in surface DMS concentration in the Indian Ocean. The maximum concentration of 525 nM found in the coastal waters of Goa is not shown here

3 nM for the open Arabian Sea. Thus the surface DMS in the Arabian Sea shows very high spatial and temporal variability. The overall surface DMS average for the Arabian Sea was 6.3 nM.

On the other hand, surface DMS concentration in the coastal waters (off the Paradip in the north and Chennai in the south) in the western Bay of Bengal varied from 1.2 to 3.9 nM with an average of 2.6 nM during the SW monsoon, when the central Bay waters contained the surface DMS concentrations of 0.2–11.1 nM having a mean of 2.7 nM. Thus the surface DMS concentrations during the SW monsoon are almost the same in the coastal and open waters of the Bay of Bengal. During intermonsoon, the southern and central Bay of Bengal exhibited surface DMS concentrations between 1.4 and 4.7 nM with a mean of 2.9 nM. The overall average surface DMS concentration in the Bay of Bengal was 2.7 nM. In the central Indian Ocean surface DMS ranged from 0.2 to 11.9 nM during the northeast monsoon with an average concentration of 2.2 nM. All the observations made in the central Indian Ocean were in winter as part of the INDOEX project. The surface DMS in the Arabian Sea was more than double to that in the Bay of Bengal and nearly thrice than that in the Central Indian Ocean. In general the surface DMS concentrations in coastal waters were higher than in the open ocean. The overall average surface DMS for the entire Indian Ocean covered in the present study was 4.2 nM.

Barnard et al. (1982) reported a surface DMS mean of 2.8 nM for the Atlantic Ocean. Wong et al. (2005) in a time series study in the subarctic northeast Pacific Ocean (an HNLC region) found that the mean surface DMS was low in winter (2 nM) in comparison to summer (10 nM). High variability in surface DMS concentrations is found in the southern North Sea with minimal concentrations of 0.13 nM in winter to a maximum of 25 nM in May, which coincided with large blooms of *Phaeocystis pouchetti* (Turner et al. 1996). In the Arctic Ocean surface DMS was found to vary between 0.04 and 12 nM with maximal values occurring along the ice edge (Leck and Persson 1996). Surface DMS concentrations have been found to range between 0.002 and 4.63 nM with a mean of 2.6 nM in the South China Sea, and from 1.8 to 5.7 nM (a mean of 3.4 nM) for the East China Sea, where the highest concentrations occurred in shelf waters and DMS showed strong correlation with chlorophyll (Yang, 2000). Uher et al. (2000) found highly variable surface DMS concentrations in the European western continental margin. In

September 1994 it ranged between 0.6 and 33.4 nM (mean of 2.8 nM) while in July 1995 it varied from 2.9 to 38.5 nM (mean of 7.2 nM). It is apparent from the above comparisons that the Indian Ocean surface DMS average is nearly double that of the Atlantic, the Pacific and the other seas in the Northern hemisphere and is also more than the surface DMS concentrations reported for the China Sea. Our values are, however, comparable to the July 1995 average for the European western continental margin. In addition we also compared our data with the data set for the Indian Ocean from the Pacific Marine Environmental Laboratory (PMEL) database. The database server is the first of its kind to hold 31,059 surface seawater DMS measurements with contributions from more than 50 researchers for different parts of the worlds Ocean. From this server (http:saga.pmel.noaa/dms/) data were downloaded for the Indian Ocean region, which basically covered central and western Arabian Sea and Central Indian Ocean over different seasons. The surface DMS over this coverage varied between 0.76 nM and 12.05 nM with an average of 3.7 nM. Our surface DMS average for the entire Indian Ocean is comparable with that of the PMEL database. A bit higher average surface DMS for the present study might have resulted from the fact that this study covers a lot of the Indian coastline, which is influenced by intense biological activity driven by monsoons (Naqvi et al. 2000).

Wind

Wind is an important driving force in air–sea interaction processes that largely determines the extents of exchanges of heat and materials. The Indian Ocean covered in the present study experienced seasonally variable winds. During the northeast monsoon wind speeds along the eastern coastal Arabian Sea were between 0.7 and 10.4 m s^{-1} with an average speed of 4.2 m s^{-1} while in the open ocean they varied from 1.4 to 9 m s^{-1} with a mean of 4.9 m s^{-1}. During the SW monsoon the winds along the coastal Arabian Sea of India were weaker and varied between 0.5 and 8 m s^{-1} with an average of 3.1 m s^{-1}. During the intermonsoon periods, they ranged between 1.8 and 12.5 m s^{-1} and 3.7 and 6.3 m s^{-1} with average wind speeds of 5.8 and 5.2 m s^{-1} for the coastal and the open ocean regions, respectively. Thus maximal wind speeds in the Arabian Sea were encountered during the intermonsoons along the west coast of India. Along the east coast of India, winds varied between 2.2 and 8.9 m s^{-1} with an average speed of 5.5 m s^{-1} during the SW monsoon while in the open bay winds varied from 4 m s^{-1} to 18.2 m s^{-1} (in association with a low-pressure system), with an average speed of 9.3 m s^{-1}. During the intermonsoons, however, winds in the open bay varied between 1.6 and 8.6 m s^{-1} (an average of 3.8 m s^{-1}). In the central Indian Ocean the winds showed a variation from 1.1 to 12.7 m s^{-1} (with a mean of 5.7 m s^{-1}) in winter. These wind speeds were higher in 1999, than in 1998, because of the development of stormy conditions associated with the Inter Tropical Convergence Zone. The measured wind speeds were in general agreement with the climatological winds recorded in Hastenrath and Lamb (1979). The observed wind speeds in the eastern Arabian Sea, however, are much weaker than those reported for the western parts (~18 m s^{-1}) in the southwest monsoon.

Sea-to-air flux of DMS

In DMS flux calculations it is assumed that DMS in air is negligible compared to that in seawater and therefore the concentration difference used in the flux calculation is essentially equal to DMS concentration in seawater. Figure 10 depicts variations in surface DMS flux in the coastal and open ocean areas of the Indian Ocean. The DMS fluxes from the coastal waters of India are lower than that from the open Ocean. Higher DMS fluxes in the central Indian Ocean are associated with higher DMS abundance and high wind speeds in winter of 1999. In the coastal areas the DMS flux varied between 0.04 μmol m^{-2} d^{-1} and 34.4 μmol m^{-2} d^{-1} with an average value of 3.4 μmol m^{-2} d^{-1} while in the open ocean it varied from 0.03 μmol m^{-2} d^{-1} to 41.1 μmol m^{-2} d^{-1} having a mean DMS flux of 6.3 μmol m^{-2} d^{-1}. Table 2 lists the DMS fluxes from the Indian Ocean both on a seasonal and regional basis. Average DMS fluxes were higher in the northeast monsoon than in other seasons both in coastal and open Arabian Sea regions. During the intermonsoon period, open Arabian Sea fluxes were nearly half of that in winter. Comparatively low winds were found when the measurements were made in coastal waters of the Arabian Sea in summer monsoon. The annual coastal mean DMS flux was 3.4 μmol m^{-2} d^{-1} while the open ocean mean flux was 3.8 μmol m^{-2} d^{-1} resulting in an average flux from the Arabian Sea of 3.5 μmol m^{-2} d^{-1}. The fluxes estimated here for the Arabian Sea have to be considered to be low since our coverage is limited to central and eastern parts and excluded regions of strong winds during the southwest monsoon. Fluxes in the Bay of Bengal were higher in the SW monsoon than in intermonsoon. Higher fluxes (average of 11.1 μmol m^{-2} d^{-1}) during the BOBMEX time series experiment resulted from the enhanced production of surface DMS due to increased biological activity triggered by turbulent conditions and strong winds associated with the low-pressure system. A large variability in flux at this time-series location is seen in Fig. 10. In the central Indian Ocean the DMS flux varied between 0.06 and 29.1 μmol m^{-2} d^{-1} with an average value of 3.5 μmol m^{-2} d^{-1}. Thus the Bay of Bengal exhibited the maximal DMS fluxes during the present study (due to the prevailing low pressure systems) with nearly three times to that in the Arabian Sea and in the central Indian Ocean. The overall DMS flux for the entire Indian Ocean is estimated to be 5.42 μmol m^{-2} d^{-1}. Table 3 shows a comparison between the DMS flux from the Indian Ocean (this study) with those estimated for other regions. It is evident that flux from the Indian Ocean is higher than in many regions of the world Oceans, e.g., the north Pacific, the Atlantic, the North Sea and the Arctic Polar Ocean, but is lower than that reported for the northeast Pacific Ocean. The only DMS estimate for the Indian Ocean, available from the earlier studies, is that of Nguyen et al. (1990) near Amsterdam Island situated near 38°S, which is nearly one-half of that in the present work. However, our flux estimate agrees well with that made by Yang (2000) for the South China Sea. In the present study an area of around 16.1×10^6 km^2 was covered for DMS and $DMSP_t$ study, which is about 21.5% of the total Indian Ocean area or around 4.5% of the global oceanic area. The total flux of DMS from the study area was estimated to be 1.02×10^{12} g S y^{-1}. This works out to contribute 4.1–6.3% to the global DMS emission, which has a range of $16–25 \times 10^{12}$ g S y^{-1}. One of the

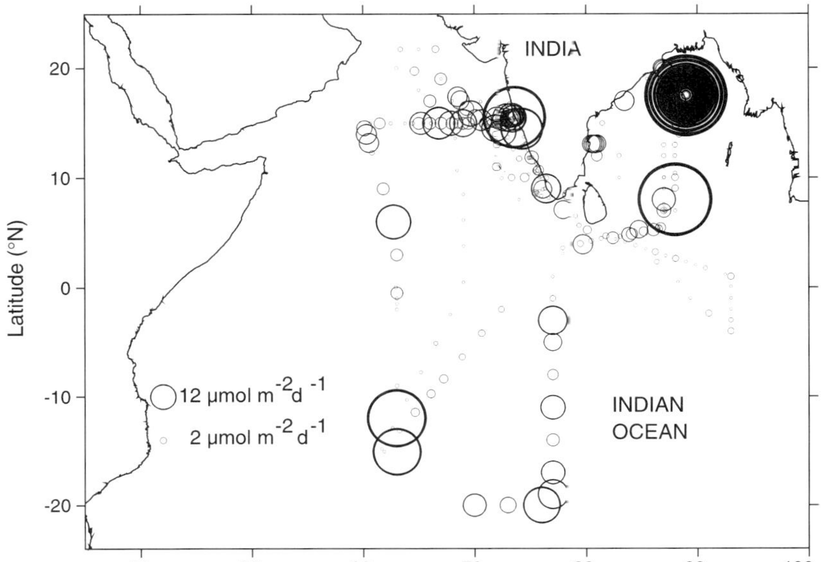

Fig. 10 Variations in DMS flux in the Indian Ocean. Time series station (SK147A & B) in the Bay of Bengal depicts wide variations in the DMS flux

Table 2 Sea-to-air fluxes of DMS (μmol S m^{-2} d^{-1}) in the Indian Ocean*

Area/season	Coastal ocean			Open ocean			Total		
	Min.	Max.	Mean	Min.	Max.	Mean	Min.	Max.	Mean
Arabian Sea									
NE monsoon	0.05	34.4	4.7	0.03	16.8	4.2	0.03	34.4	4.3
SW monsoon	0.04	31	2.6	–	–	–	0.04	31	2.6
Intermonsoon	0.1	13.1	3.9	0.4	3.1	1.6	0.1	13.1	3.13
Average	0.04	34.4	3.4	0.03	16.8	3.8	0.03	34.4	3.5
Bay of Bengal									
SW monsoon	0.18	9.03	3.7	0.4	41.1	11.5	0.18	41.1	11.1
Intermonsoon	–	–	–	0.16	10.9	1.63	0.16	10.9	1.63
Average				0.16	41.1	11.02	0.16	41.1	9.63
Central Indian Ocean									
NE monsoon				0.06	29.1	3.5	0.06	29.1	3.5
Average				0.06	29.1	3.5	0.06	29.1	3.5
Indian Ocean	0.04	34.4	3.4	0.03	41.1	6.3	0.03	41.1	5.42

*Wherever data are not available for coastal/open ocean regions or seasons the available values were assumed to represent the same in other regions/seasons in deriving average fluxes

Table 3 Comparison of DMS fluxes from the Indian Ocean (this study) with that from other oceanic regions

Source	Area	Average flux, μmol S m^{-2} d^{-1}
Wong et al. (2005)	Northeast Pacific Ocean	16
Sharma et al. (1999)	North Pacific Ocean	2.3
Berresheim et al. (1991)	Western North Atlantic Ocean	3.4
Turner et al. (1996)	North Sea	3.84
Leck and Persson (1996)	Arctic Polar Ocean (summer)	2.0
Nguyen et al. (1990)	Amsterdam Island (Indian Ocean, 38°S)	2.1
Yang (2000)	South China Sea	5.5
Yang et al. (2000)	East China Sea	3.4
Present study	Arabian Sea	3.5
	Bay of Bengal	9.63
	Central Indian Ocean	3.5
	Indian Ocean (total)	5.42

limitations of the study is the lack of coverage of certain important areas in the Indian Ocean such as the intense upwelling zones off the coast of Somali and Oman, the east coast of Africa, the west coast of the Indonesian Islands and the south Indian Ocean. Considering the fact that this study covered only 21.5% of the total Indian Ocean, the Indian Ocean could be an important source for DMS sulphur to the atmosphere. It should, however, be cautioned that the flux trends observed among various basins in the Indian Ocean studied here may be different from that expected since we occupied the Bay of Bengal (BOBMEX) and Central Indian Ocean (INDOEX) under turbulent conditions but could not cover high-wind regimes in the Arabian Sea during the southwest monsoon.

Conclusions

Significant spatial and temporal variability in DMS was found in the north and central Indian Ocean. Very high DMS and DMSP of 526 nM and 916 nM, respectively, were found in coastal waters of the Arabian Sea. These are among the highest values found in the world oceans. In the Arabian Sea the lowest DMS and $DMSP_t$ concentrations were observed during the intermonsoons. Lower DMS and $DMSP_t$ concentrations were found to occur in the Bay of Bengal compared to that in the Arabian Sea which might have been due to relatively less biological production and trophic diversity in the bay, as also can be seen from generally lower $DMSP_t$ levels in the bay. Average surface DMS is the highest in the Arabian Sea whereas the average flux is the highest from the Bay of Bengal. DMS from the study area was found to contribute 4–6% to the total global emission, which is comparable to other regions of the world. Export fluxes of sulphur compounds from the Indian Ocean are important since this is one of the most turbulent regions, with regular occurrence of monsoons.

Acknowledgement We thank the Director of NIO, for encouragement and support. Thanks are due to the Departments of Ocean Development and Science and Technology, New Delhi, for financial support and for ship time on board ORV Sagar Kanya and FORV Sagar Sampada. The first author acknowledges the CSIR for support through a fellowship. We would like to acknowledge the various team members, and Dr. V.V.S.S. Sarma in particular, for their help on board. Special thanks are due to Dr. E Saltzman (University of Miami) and Dr. Gillian Malin (University of East Anglia) for their advice in regard to dimethyl sulphide analysis. Suggestions by the two anonymous reviewers helped to improve this manuscript substantially. This is NIO contribution number 4213.

References

Barnard WR, Andreae MO, Watkins WE, Bingemer H, Georgii HW (1982) The flux of dimethylsulfide from the oceans to the atmosphere. J Geophys Res 87C:8787–8793

Behrenfeld MJ, Falkowski PG (1997) Photosynthetic rates derived from satellite-based chlorophyll concentration. Limnol Oceanogr 42(1):1–20

Berresheim H, Andreae MO, Iverson RL, Li SM (1991) Seasonal variations of dimethylsulfide emissions and atmospheric sulfur and nitrogen species over the western north Atlantic Ocean. Tellus 43:353–372

Bhattathiri PMA, Pant A, Sawant S, Gauns M, Matondkar SGP, Mohanraju R (1996) Phytoplankton production and chlorophyll distribution in the eastern and central Arabian Sea in 1994–95. Curr Sci 71:857–862

Charlson RJ, Lovelock JE, Andreae MO, Warren SC (1987) Oceanic phytoplankton, atmospheric sulphur, cloud albedo and climate. Nature 326:655–661

Charlson RJ, Anderson TL, McDuff RE (1992) The sulfur cycle. In: Butcher SS, Charlson RJ, Orians GH, Wolfe GV (eds) Global biogeochemical cycles. Academic Press, London, pp 285–300

Dacey JWH, Wakeham SG (1986) Oceanic dimethyl sulfide: production during zooplankton grazing on phytoplankton. Science 233:1314–1316

Hastenrath S, Lamb PJ (1979) Climatic atlas of the Indian Ocean. University of Wisconsin Press, Wisconsin 19 + 97 charts

Hatton AD, Malin G, Liss PS (1999) Distribution of biogenic sulphur compounds during and just after the southwest monsoon in the Arabian Sea. Deep Sea Res II 46:617–632

Iverson RL, Nearhoof FL, Andreae MO (1989) Production of dimethyl-sulfonium propionate and dimethylsulfide by phytoplankton in estuarine and coastal waters. Limnol Oceanogr 34:53–67

Kettle AJ et al (1999) A global database of sea surface dimethyl sulphide (DMS) measurements and a procedure to predict sea surface DMS as a function of latitude, longitude and month. Global Biogeochem Cycles 13:399–444

Kirst GO, Thiel CH, Wolff J, Nothnagel M, Wanzek R, Ulmke (1991) Dimethyl-sulfoniopropionate (DMSP) in ice-algae and its possible biological role. Mar Chem 35:381–388

Kumar MD, Shenoy DM, Sarma VVSS, George MD, Dandekar M (2002) Export fluxes of dimethylsulfoniopropionate and its break sown gases at the air-sea interface. Geophys Res Lett 29(2):10.1029/2001GL013967. 8-1 to 8-3

Leck C, Persson C (1996) Seasonal and short-term variability in dimethyl sulfide, sulfur dioxide and biogenic sulfur and sea salt aerosol particles in the arctic marine boundary layer during summer and autumn. Tellus 48B:272–299

Liss PS, Malin G, Turner SM (1993) Production of DMS by marine phytoplankton. In: Restelli G, Angeletti G (eds) Dimethyl sulphide: Oceans, atmosphere and climate. Kluwer Academic, Dordrecht, pp 1–14

Lovelock JE, Maggs RJ, Rasmussen RA (1972) Atmospheric dimethyl sulphide and the natural sulphur cycle. Nature 237:452–453

Moret I, Gambaro A, Piazza R, Barbante C, Andreoli C, Corami F, Scarponi G (2000) The seasonal variations of dimethyl sulphide and carbon disulphide in surface waters of the Venice lagoon. Mar Chem 71:283–295

Naqvi SWA, Jayakumar RA, Narvekar PV, Naik H, Sarma VVSS, D'scuza W Joseph S, George MD (2000) Increased marine production of N_2O due to intensifying anoxia on the Indian continental shelf. Nature 408:346–349

Nguyen BC, Mihalopoulos N, Belviso S (1990) Seasonal variation of atmospheric dimethylsulphide at Amsterdam Island in the southern Indian Ocean. J Atmos Chem 11:123–141

Nguyen BC, Mihalopoulos N, Putaud JP, Gaudry A, Gallet L, Keene WC, Galloway JN (1992) Covariations in oceanic dimethyl sulfide, its oxidation products and rain acidity at Amsterdam Island in the Southern Indian Ocean. J Atmos Chem 15:39–53

Saltzman ES, King DB, Holmen K, Leck C (1993) Experimental determination of the diffussion coefficient of dimethylsulfide in water. J Geophys Res 98C:16481–16486

Sharma S, Barrie LA, Plummer D, McConnell JC, Brickell PC, Levasseur M, Gosselin M, Bates TS (1999) Flux estimation of oceanic dimethyl sulfide around North America. J Geophys Res 104D:21327–21342

Shenoy DM, Kumar MD, Sarma VVSS (2000) Controls of dimethyl sulphide in the Bay of Bengal during BOBMEX Pilot Cruise 1998. Proc Indian Acad Sci (Earth Planet Sci) 109:279–283

Shenoy DM, Joseph S, Dileep Kumar M, George MD (2002) Control and inter-annual variability of dimethyl sulfide in Indian Ocean. J Geophys Res 107: 10.1029/2001JD000371

Shenoy DM, Patil JS (2003) Temporal variations in dimethylsulphonio-propionate and dimethyl sulphide in the Zuari estuary, Goa (India). Mar Environ Res 56/3:387–402

Stefels J (2000) Physiological aspects of the production and conversion of DMSP in marine algae and higher plants. J Sea Res 43:183–197

Sunda W, Kieber DJ, Kiene RP, Huntsman S (2002) An antioxidant function for DMSP and DMS in marine algae. Nature 418:317–320

Turner SM, Malin G, Bagander LE, Leck C (1990) Interlaboratory calibration and sample analysis of dimethyl sulfide in water. Mar Chem 29:47–62

Turner SM, Malin G, Nightingale GD, Liss PS (1996) Seasonal variation of dimethyl sulphide in the North Sea and an assessment of fluxes to the atmosphere. Mar Chem 54:243–262

Uher G, Schebeske G, Barlow RG, Cummings DG, Mantoura RFC, Rapsomanikis SR, Andreae MO (2000) Distribution and air-sea gas exchange of dimethyl sulphide at the European western continental margin. Mar Chem 69:277–300

UNESCO (1994) Protocols for the Joint Global Ocean Flux Study (JGOFS). Core Measurements, Scientific committee on Oceanic Research Manual and Guides 29:170

Vairavamurthy A, Andreae MO, Iverson RL (1985) Biosynthesis of dimethylsulfide and dimethylpropiothetin by Hymenomonas carterae in relation to sulfur source and salinity variations. Limnol Oceanogr 30:59–70

Wolfe GV, Steinke M (1996) Grazing-activated production of dimethyl sulfide (DMS) by two clones of *Emiliania huxleyi*. Limnol Oceanogr 41:1151–1160

Wolfe GV, Steinke M, Kirst GO (1997) Grazing-activated chemical defence in a unicellular marine alga. Nature 387:894–897

Wong CS, Wong SE, Richardson WA, Smith GE, Arychuk MD, Page JS (2005) Temporal and spatial distribution of dimethylsulfide in the subarctic northeast Pacific Ocean: a high-nutrient-low-chlorophyll region. Tellus 57B:317–331

Yang GP (2000) Spatial distributions of dimethylsulfide in the South China Sea. Deep Sea Res I 47:177–192

Yang GP, Cong XD, Zhang ZB, Zhao W (2000) Dimethylsulfide in the South China Sea. Chin J Oceanol Limnol 18:162–168

Biogeochemistry (2007) 83:293–309
DOI 10.1007/s10533-007-9089-z

RESEARCH ARTICLE

Gaining integrated understanding of *Phaeocystis* spp. (Prymnesiophyceae) through model-driven laboratory and mesocosm studies

Stuart J. Whipple · Bernard C. Patten · Peter G. Verity · Marc E. Frischer · Jeremy D. Long · Jens C. Nejstgaard · Jon T. Anderson · Anita Jacobsen · Aud Larsen · Joaquin Martinez-Martinez · Stuart R. Borrett

Received: 20 April 2006 / Accepted: 7 September 2006 / Published online: 8 March 2007

Abstract Knowledge of the complex life cycle of *Phaeocystis* is a key to understanding its role in marine ecosystems and global biogeochemistry. An existing life cycle model was modified and used to integrate understanding of the *Phaeocystis* life cycle. In model-driven research, models expose gaps in our understanding, empirical studies ensue, and feedback improves understanding. Following this scheme, three facets of the life cycle model were examined here. With four exceptions, the empirical studies described have been presented in other literature citations. The first facet involved testing for the existence of a process or producing its description. These studies included: demonstration of in vitro colony division in *Phaeocystis pouchetii*, description of in vitro change in colony shape for *P. pouchetii* associated with senescence, determining which *P. pouchetii* life stage is vulnerable to viral infection and lysis, and an experiment designed to determine whether the sediment could be a source of new *Phaeocystis* colonies to overlying waters; results suggested that more-detailed investigation of benthic particles as a physical substrate for colony formation is warranted. The second facet involved investigation of process rate quantification or process control parameters. Process rate quantification included measurements of colony division rate and growth rate using mesocosm-derived colonies. Process control experiments included testing diatom frustule enhancement of *P. pouchetii* colony formation from solitary cells, and investigation of mesozooplanktonic suppression and microzooplanktonic enhancement of *Phaeocystis globosa* colony formation by planktonic grazer infochemicals. The third facet

S. J. Whipple · P. G. Verity · M. E. Frischer
Skidaway Institute of Oceanography,
Savannah, GA, USA

S. J. Whipple (✉) · B. C. Patten
Institute of Ecology, University of Georgia, Athens,
GA, 30602, USA
e-mail: sjwhipple@earthlink.net

J. D. Long
Marine Science Center, Northeastern University,
Nahant, MA, USA

J. C. Nejstgaard · A. Jacobsen · A. Larsen
Department of Biology, University of Bergen, Bergen,
Norway

J. T. Anderson
Estuarine Research Center, Morgan State University,
St. Leonard, MD, USA

J. Martinez-Martinez
Plymouth Marine Laboratory,
Plymouth, Devon, UK

S. R. Borrett
Computational Learning Laboratory,
Stanford University, Stanford, CA, USA

pertained to the molecular identification of genetic differences between single cells and colonies of *P. globosa*. These studies were designed to provide insight to the question of control factors involved in the transition between single cell and colonial life stages. The life cycle model provided a ready place to incorporate new insights and understanding from empirical studies into an existing model, and can be used to improve simulation models of the direct and indirect effects of *Phaeocystis* on global biogeochemistry.

Keywords Conceptual model · Life cycle · Mesocosm · Model-driven research · *Phaeocystis*

Introduction

The genus *Phaeocystis* contains species that may play a significant role in global biogeochemistry (Smith Jr et al. 1991; Lancelot et al. 1994; Wassmann et al. 2005). They occur over extensive areas (Stefansson and Olafsson 1991; Lancelot and Rousseau 1994), supply significant portions of global dimethyl sulphide (DMS; Gabric et al. 1999; Ayers and Gillett 2000), and play a key role in planktonic ecosystems (Verity and Smetacek 1996; Lancelot et al. 1998). Life cycle processes of *Phaeocystis* may have significant effects on ecosystem biogeochemical processes including nutrient uptake and regeneration by life forms of *Phaeocystis*, nutrient release by viral lysis, nutrient release associated with zooplankton grazing, and nutrient release by senescent colonies. Details of the current state of knowledge of the *Phaeocystis* life cycle are comprehensively reviewed in Rousseau et al. (in press, this volume). In order to integrate understanding of the *Phaeocystis* life cycle, a conceptual model was constructed (Whipple et al. 2005a), although the existence of some life stages was uncertain and the controls on many of the processes were poorly understood. This exercise highlighted experiments which could provide a better understanding of the role of the life cycle in species dynamics and ecosystem processes.

Such enhanced understanding will be incorporated into a developing conceptual model (Whipple et al. 2005a), and also into an ecosystem model in which the *Phaeocystis* life cycle is only one set of biotic compartments (Whipple et al. unpubl.). In model-driven research, models expose gaps in our understanding of the *Phaeocystis* life cycle embedded in its ecosystem; empirical studies to address these gaps ensue, and the feedback from iteration of this cycle improves understanding. This paper presents examples of how a life cycle model of *Phaeocystis* has been used to direct research on the *Phaeocystis* life cycle and its ecology, and demonstrates how insight from these studies integrated into a conceptual model context enhances and integrates such empirical findings.

Model description

The *Phaeocystis* life cycle, which has been modified from the model presented in Whipple et al. (2005a), advances clockwise around Fig. 1 from the upper left with Diploid Solitary Cells, as the starting point. These cells flow to and from New Colonies, which is the critical transition stage between solitary and colonial life forms.

After new colonies cells first divide, they transfer to the growing small colonies compartment. Maturation continues through the size classes: small → medium → large via mitosis and mucus production of colony cells. There is also back-transfer from larger to smaller size classes by colony division (Whipple et al. 2005b).

When colonies begin to lose cells, growing colonies are transferred to corresponding size-based senescent colonies. Diploid or haploid solitary cells are released from these to complete the *Phaeocystis* life cycle. Of the four colony-forming *Phaeocystis* species, haploid flagellates have been documented only for *Phaeocystis globosa* (Eilertsen 1989; Jacobsen 2002; Medlin and Zingone in press, this volume; Rousseau et al. in press, this volume). The other life cycle input to diploid solitary cells comes from transformation of new colonies cells. Syngamy has not been documented in any *Phaeocystis* species; however, it has been included in the model since it is hypothesized as necessary to create diploid solitary cells from haploid cells prior to colony formation in *P. globosa* (Rousseau et al. in press, this volume). Since this

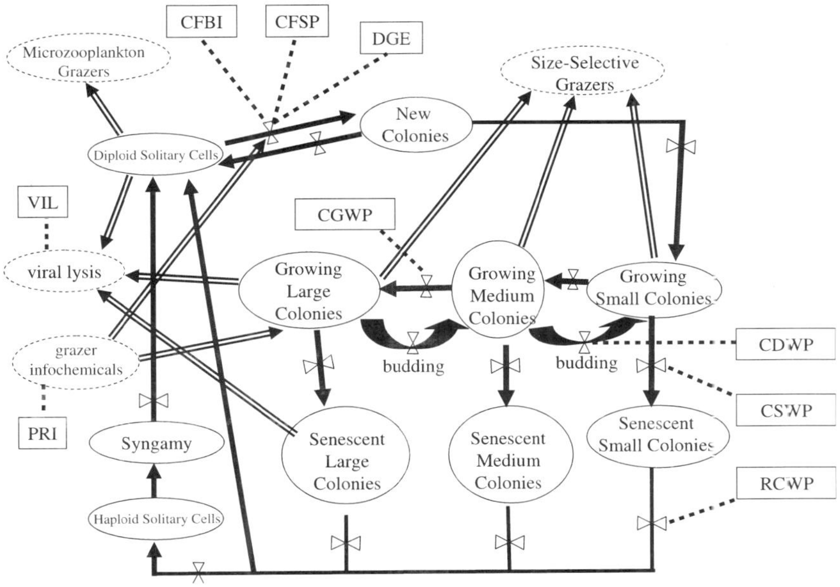

Fig. 1 *Phaeocystis* life cycle diagram. Solid-line ovals: *Phaeocystis* life cycle stages; dashed-line ovals: other ecosystem components. Solid arrows: life cycle transitions; hollow-line arrows: inputs to or outputs from *Phaeocystis*. ⋈ symbols: rate control for life cycle transitions. Rectangles with acronyms: empirical studies with dashed-lines pointing to an associated life cycle stage or life cycle transition

Names and acronyms for experiments described:

CDWP	Colony division in well plates
CFSP	Colony formation with seed particles
CFBI	Colony formation with benthic involvement
CGWP	Colony growth in well plates
CSWP	Colony senescence in well plates
DGE	Differential gene expression
PRI	*Phaeocystis* response to infochemicals
RCWP	Release of colony cells in well plates
VIL	Viral infection and lysis

model was developed to represent the genus *Phaeocystis*, haploid life stages are included, but may only occur in certain species (Whipple et al. 2005a).

Hollow-line arrows in Fig. 1 represent hypothesized flows to or from non-*Phaeocystis* components of the larger ecosystem. In some cases these serve as controls on life cycle flows, and in other cases they are boundary inputs or outputs of the *Phaeocystis* life cycle. Three main outputs from the *Phaeocystis* life cycle are represented. Two are grazing on solitary cells (upper left), and grazing on all size classes of growing colonies (upper right). The third is viral lysis (left center), which is an output process usually emphasized in viral studies. However, infection, an input process, must occur for viruses to complete their life cycle and have an effect on *Phaeocystis* cells. The last process is the input of infochemicals from zooplankton grazers (lower left) into *Phaeocystis* solitary cell-to-colony transformations or colony cells.

Case studies

The conceptual model of this paper serves two functions by: (1) providing a context for integrating what is known, and (2) defining observations and experiments to eliminate unknowns (Whipple et al. 2005a). Figure 1 graphically depicts how the *Phaeocystis* life cycle model and empirical studies conducted to address unknowns may be interrelated; each experiment within our research group described below is denoted by an acronym

Table 1 Description of the *Phaeocystis* life cycle processes and the studies associated with them that are discussed in this paper

Acronym	Name used in Fig. 1	Process description	Source of data
CDWP	Colony division in well plates	In vitro observations of colony division in *P. pouchetii*	Whipple et al. (2005b)
CFBI	Colony formation with benthic involvement	In vitro observation of colonies derived from coarse surficial material collected from shallow fjord sediments	this study
CFSP	Colony formation with seed particles	In vitro observation of colony formation from solitary cells in *P. pouchetii* with diatom frustules	Nejstgaard et al. (in press)
CGWP	Colony growth in well plates	In vitro observations of colony growth in *P. pouchetii*	Whipple et al. (2005b)
CSWP	Colony senescence in well plates	In vitro observations of colony senescence in *P. pouchetii*	this study
DGE	Differential gene expression	Differential gene expression in solitary versus colonial cells in *P. globosa*	this study
PRI	Phaeocystis response to infochemicals	Experiments to determine *P. globosa* response to presence of micro- and mesozooplankton grazers or their infochemicals	Long et al. (submitted); Long (2004)
RCWP	Release of colony cells in well plates	In vitro observations of colony cell release in *P. pouchetii*	this study
SIM	Species identification in mesocosms	18S rDNA probe assessment of *Phaeocystis* species growing in 2003 Norwegian mesocosm	this study
VIL	Viral infection and lysis	In vitro observations of viral infection and lysis for cells from intact and disrupted colonies in *P. pouchetii*	Jacobsen et al. (2005)

(e.g., VIL = viral infection and lysis; left center) indicating where it relates to the life cycle. The nine processes and their associated studies considered in this paper are shown in their life-cycle context in Fig. 1. Descriptions of these processes, their acronyms, and data sources are shown in Table 1; an entry describing verification of the *Phaeocystis* species found in the mesocosm studies (SIM) is also included. Stages or processes involved in the phenomena under investigation are described in relation to the life cycle model. Parenthesized acronyms for the Table 1 experiments appear in the subheadings of this section. In four case studies, detailed methods and results are presented; in other cases, a brief summary of methods and results will be given based on published or submitted literature sources, which may be consulted for further details. The Discussion section considers broader implications of our empirical findings for both *Phaeocystis* life cycle dynamics itself, and potential revisions of the life cycle model.

Mesocosm experiment methods

Many of the observations and experiments discussed below were performed in the context of mesocosm studies conducted in Raunefjorden at the Norwegian National Mesocosm Center, located at the marine biological field station at the University of Bergen in western Norway. Studies were conducted between 4 March and 24 March 2002 and 27 February and 2 April 2003 (Nejstgaard et al. 2006); these studies will be referred to as 2002 and 2003 mesocosm studies in the remainder of the paper. The main results of these mesocosm studies will not be presented in this manuscript; they are provided in detail in Nejstgaard et al. (2006). The purpose of this section is to provide sufficient background on the mesocosm studies so that experiments conducted in their context can be understood and interpreted.

In each study, three transparent polyethylene enclosures (4.5 m deep, 2 m diameter, ca. 11 m^3, 90% PAR) were filled with unfiltered fjord water from 5 m depth. An airlift system completely mixed the enclosures approximately five times per day. To allow for the introduction of new species, to avoid pH changes, and to replace water removed for sampling, 10% of the water in each enclosure was renewed daily with fjord water (Nejstgaard et al. 2006). Two of the three mesocosms were amended with nitrate and phosphate that corresponded to an initial enrichment of 16 μM nitrate and 1 μM phosphate. Daily addition of 1.6 μM nitrate and 0.1 μM phosphate replaced nutrients removed by the 10% water renewal and sampling. In the 2002 experiment, based on low nutrient concentrations measured on March 12, both nutrient enriched mesocosms were augmented with additional nutrients, 8 μM nitrate and 0.5 μM phosphate on the evening of March 12. The third mesocosm in each experiment, which was left unamended, served as a control (Nejstgaard et al. 2006). Detailed description of mesocosm set-up, sampling, analytical methods, and results are described in Nejstgaard et al. (2006).

Species identification in mesocosms (SIM)

During the 2002 and 2003 mesocosm studies many of the *Phaeocystis* colonies were observed to be perfectly spherical, suggesting that there may be *P. globosa* rather than *P. pouchetii* present in western Norwegian fjords. During the second mesocosm experiment (2003) an effort was made to determine the diversity of *Phaeocystis* species present in these mesocosm studies.

Methods

Throughout the 2003 mesocosm study, water (50–500 ml) was collected every third day onto 0.8 μm Supor® membrane filters (Pall corp.) and stored at −80°C until analysis. Following enumeration of *Phaeocystis* sp. single cells and colonies, two samples were chosen for molecular determination of the composition of the *Phaeocystis* sp. community. Gene sequence analysis of *Phaeocystis* sp. small subunit ribosomal RNA (rRNA) genes from samples collected from a fertilized enclosure early in the experiment when only single cells were present (5 March 2003; single cells 15 ml^{-1}, colonies 0 ml^{-1}) and samples collected during the peak of the bloom (20 March 2003; single cells 990 ml^{-1}, colonies 41 ml^{-1}) were examined. Total DNA was extracted from filters using the Ultraclean™ soil DNA isolation (MoBio Laboratories, Inc.) as previously described (Allen et al. 2005). Small subunit rRNA genes from *Phaeocystis* sp. were specifically polymerase chain reaction (PCR) amplified using a universal 18S rRNA targeted primer UnivF-15 (5′ ctg cca gta gtc ata tgc; Frischer et al. 2002) and a *Phaeocystis* genus-specific primer PHAEO1 (5′ cgg tcg agg tgg act cgt; Lange et al. 1996). Amplification reactions were facilitated using an Applied Biosystems 9,700 thermocycler in 20 μl volumes. Each amplification reaction consisted of 8 μl Eppendorf ® MasterMix® (Eppendorf North America), 8.4 μl PCR grade distilled water, 3 μl each primer (100 ng μl^{-1}), and 3 μl of undiluted or 1:10 diluted DNA extract. Purified DNA from the extracted samples was not quantified. Reaction conditions consisted of an initial denaturation step (94°C, 3 min) followed by 35 amplification cycles (94°C, 15 s; 53°C, 15 s; 72°C, 30 s) followed by a 7 min final extension at 72°C. PCR amplicons were examined on an agarose gel to insure that only a single band was produced. Single bands were cloned into the bacterial plasmid vector PCR®2.1 using the TOPO TA Cloning® kit (Invitrogen, Carlsbad, CA) and sequenced in both directions as previously described (Frischer et al. 2002).

Results

A total of 17 clones was sequenced and the identity of each cloned sequence was confirmed to be *Phaeocystis pouchetii*, indicating that it is unlikely that any other species of *Phaeocystis* were present in these studies.

Testing for the existence of a process or producing a description of a process

Colony formation with benthic involvement (CFBI)

Benthic involvement in the life cycle of various *Phaeocystis* species has been discussed for many years in the literature. Kayser (1970) describes his observation of benthic cells in culture as follows: "It is interesting to observe that single cells of *Phaeocystis pouchetii* exhibit a tendency to attach themselves to solid surfaces... We may assume that in the open sea single cells are attached to solid surfaces on the sea floor or to motile particles and constantly supply the free water with single cells or young colonies" (Kayser 1970, pp. 209–210). Rousseau et al. (1994, p. 26) concluded, there is "...absolutely no evidence for a truly differentiated benthic stage."

Rousseau et al. (in press, this volume) describe *P. globosa* diploid flagellates as forming new colonies within a day after adhesion to a surface (Cariou et al. 1994; Kayser 1970; Kornmann 1955; Rousseau et al. 1994). Non-living particles (Rousseau et al. 1994), culture vessel walls (Cariou et al. 1994; Kayser 1970), and diatoms (Weisse et al. 1986; Boalch 1987; Rousseau et al. 1994) have been reported as adhesion sites (Rousseau et al. in press, this volume). Attachment of cells to surfaces led to the assumption that a benthic stage, acting as an overwintering form, exists in the natural environment (Kayser 1970; Rousseau et al. in press, this volume). The general conclusion from these reports is that the diploid flagellate described in these studies is not a truly differentiated benthic form (Rousseau et al. in press, this volume).

Empirical work conducted during the 2003 mesocosm study tested whether *P. pouchetii* solitary cells or colonies would be resuspended from fjord sediment incubated with filtered water in controlled environmental chambers.

Methods

Benthic substrate, which consisted of medium-grained sands with shells and biotic remains, was hand-collected by SCUBA divers from 12 m depth in a coastal embayment near Bergen, Norway on March 14, 2003. The sediments were kept isolated from the overlying water during collection and experimental preparation, and were overlaid with filtered seawater during incubation. A set of 20 cc aliquots of drained but damp sediments were placed in the bottom of triplicate 250 ml PC bottles and gradually overlaid with 200 ml of the following water types: artificial seawater; autoclaved fjord water; 0.8 μm filtered water from mesocosm #1 (control: no nutrients added); 0.8 μm filtered water from mesocosm #2 (NO_3 and PO_4 added). The bottles were incubated under dim light (1.7 μmol m^{-2} s^{-1}, measured using a Biospherical Instruments QSL−100 meter with a spherical light collector; 12:12 L:D) at 3°C in a walk-in cold room and sampled at day 0 (start of incubation) and at day 7. The abundance, diameter, and cell density of *P. pouchetii* colonies were measured microscopically.

Results

Phaeocystis pouchetii colonies was observed in all sediment-water treatments after seven days of incubation (Fig. 2). The highest number of colonies developed in bottles containing filtered water from mesocosm 2 (M2) treatments, which had received daily nitrogen and phosphorus additions (Fig. 2a). M2-water samples also contained colonies with the largest diameter, and highest number of cells per colony (Figs. 2b,c).

Existence of colony division in P. pouchetii (CDWP) Using in vitro observations of mesocosm-derived *P. pouchetii* colonies in well plates in 2002, Whipple et al. (2005b) reported that 10–12% of wells contain colonies that divided at least once. They also observed that division of colonies was only observed after many hours had elapsed. This may indicate that colonies undergo a period of colony enlargement by cell division and mucus production before they divide to produce daughter colonies (Whipple et al. 2005b). This pattern of colony multiplication was also reported for *P. globosa* (Verity et al. 1988).

Identification of molecular markers that distinguish between single cells and colonies A first-order hypothesis-generating question associated with the multiform life cycle of *Phaeocystis* sp. is the identification of cellular and genetic differences between single and colonial cells. To begin to address this question in the context of the complex *Phaeocystis* life cycle, a differential display PCR approach was used to identify genes whose expression was regulated in colonial versus single cells. These studies were conducted with

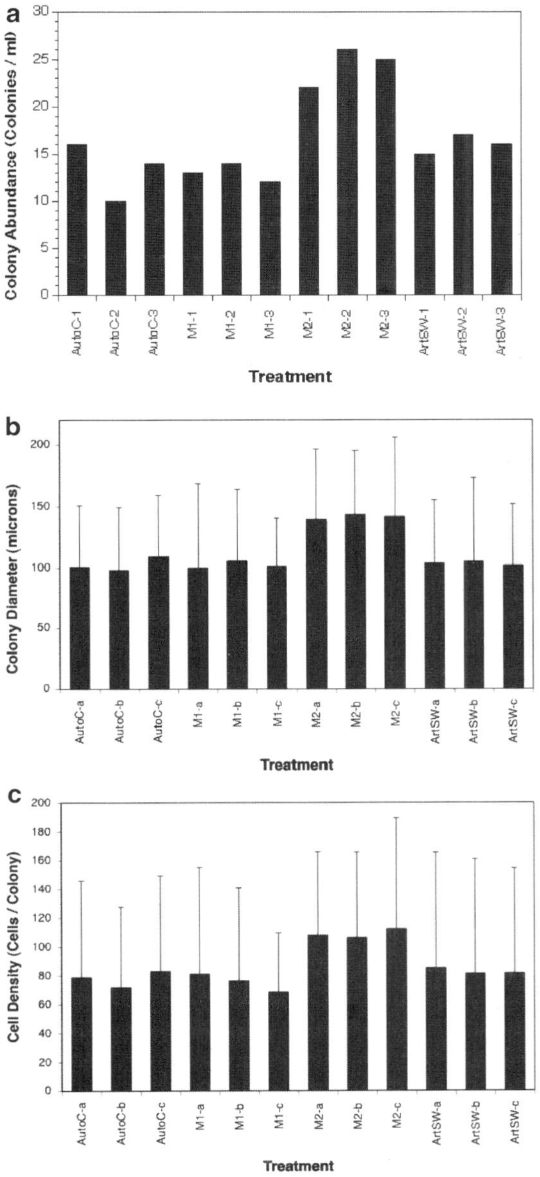

Fig. 2 Results of surficial material collected from shallow fjord sediments experiment. Treatment key: ArtSW#: artificial seawater; AutoC#: autoclaved fjord water; M1- #: 0.8 μm filtered water from mesocosm #1 (= no nutrients added); M2- #: 0.8 μm filtered water from mesocosm #2 (NO_3 and PO_4 added). (**a**) Abundance of *P. pouchetii* colonies (colonies ml^{-1}); histograms represent colony estimates for each replicate bottle. (**b**) Colony diameter (μm); error bars represent standard deviations of estimates within each bottle calculated from triplicate counts. (**c**) Cell density (cells $colony^{-1}$); error bars represent standard deviations of estimates within each bottle calculated from triplicate counts

laboratory cultures of *P. globosa* (CCMP 1528, Provasoli-Guillard National Center for Culture of Marine Phytoplankton, USA).

Methods

Three independent cultures of *P. globosa* CCMP 1528 were grown for one week at 20°C, 14:10 light:dark, in L1-Si media (Guillard and Hargraves, 1993). Under these conditions both single cells and colonies were produced in each culture. Single cells were separated from colonies by reverse osmosis with a 10 μm mesh sieve and hand pump. Cells and colonies were separately collected onto 25 mm, 5 μm Durapore filters (#SVLP02500, Millipore, Bedford, MA). Approximately 7.5×10^6 single cells and 5×10^4 small colonies (ca. 1 mm) were immobilized per filter. RNA was extracted from each sample using the Qiagen RNeasy Plant Mini Kit (Cat# 74903, Qiagen, Valencia, CA) following the manufacturer's recommended procedures (Qiagen, Valencia, CA). Purified RNA extracts of replicate samples from each culture were combined (6 extractions for each cell type). RNA was quantified by fluorometry using RiboGreen (Molecular Probes, Inc.) and purified RNA was stored at −80°C until analysis.

Differentially expressed genes were identified by differential display following cDNA synthesis from ca. 1 μg of total RNA using the Advantage® cDNA PCR kit (Cat# K1905–1, Clontech Laboratories, Inc., Palo Alto, CA). Differential display analysis was facilitated using reagents and procedures associated with the commercially available delta differential display kit® (Cat# K1810–1, Clontech Laboratories, Inc., Palo Alto, CA). Amplification of cDNA was accomplished with all 90 unique combinations of 10 arbitrary 25-mer primers (P primers) and 9 Oligo(dT) primers (T primers) supplied in the delta differential display kit. Differential display amplification reactions were conducted in 20 μl volumes with 2 μl 10X Klen Taq PCR reaction buffer, 14.2 μl PCR grade distilled water, 0.2 μl dNTP mix (5 mM each), 0.4 μl 50X Advantage® KlenTaq, 0.2 μl ^{33}P dATP, and 1 μl of a 1:40 or a 1:10 dilution of cDNA template. Amplification conditions were as

Table 2 Differentially expressed gene fragments from *P. globosa* colonial and single cells

Cell type	No. of bands	Con[a]	Partially regulated		Fully regulated	
			Up[a] reg	Down reg[a]	On	Off
Cells	1365	1193	32	32	117	16
Colonies	1254	1193	7	7	7	15

[a] Con, constitutively expressed; up reg, upregulated; down reg, downregulate

recommended by Clontech and consisted of three low-stringency PCR cycles (1st cycle: 94°C, 5 min; 40°C, 5 min; 68°C, 5 min), second and third cycles (94°C, 30 s; 40°C, 30 s; 68°C, 5 min), and 25 additional high-stringency cycles (94°C, 20 s; 60°C, 30 s; 68°C, 2 min) followed by a dwell cycle for 7 min at 68°C. Amplification was performed in a dedicated Amplitron thermocycler (Barnstead/Thermolyne, Dubuque, IA). Following confirmation of PCR amplification of cDNA by agarose gel electrophoresis, 5 μl of PCR product was mixed with gel loading dye (2 ml 10× TAE, 7.5 ml 40% sucrose, 830 μl 6× bromophenol blue, 100 ml distilled water) and 3 μl was loaded onto a 7M urea, 6% acrylamide gel (553 × 381 × 4 mm) with 1× TBE buffer and electrophoretically separated at 1500V, 50W for 2.5–3 h. The gel was transferred to filter paper, covered with plastic wrap and dried on a vacuum gel dryer (Fisher Scientific, Pttsburgh, PA) for 1 h at 80°C. After drying, bands were detected by autoradiography in a film cassette with a ^{33}P intensifying screen. Films were exposed for 4–6 days at −80°C. All bands were manually assigned into three primary categories; constitutively expressed, differentially expressed, or fully regulated. Bands categorized as constitutively expressed were those that appeared at approximately the same density in both colonial and single cells. Bands identified as differentially expressed were those whose autoradiographic density was substantially different between colonial and single cells. Fully regulated bands were those that were either present or absent in one cell type compared to the other.

Following identification of differentially expressed genes, the seven genes that were present in colonial cells but not in single cells were cloned and sequenced. Differentially expressed PCR bands were excised from the dried gel as recommended by Clontech. Briefly, bands were cut from the gel/filter paper, placed in 40 μl of DNase- and RNase-free water, covered with two drops of sterile mineral oil and heated at 100°C for 5 min. The filter paper was then discarded. The PCR product was re-amplified with the same primers originally used to amplify the band of interest. Following amplification, PCR amplicons were examined on an agarose gel to insure that only a single band was produced. Single bands were cloned into the bacterial plasmid vector PCR®2.1 using the TOPO TA Cloning® kit (Invitrogen, Carlsbad, CA) and sequenced in both directions using the M13 priming sites as previously described (Allen et al., 2005). After sequence assembly and editing using the Vector NTI Suite 8 software package (InforMax, Bethesda, MD), putative gene identities were determined using BLASTn and BLAST searches for short nearly identical matches (NCBI, http://www.ncbi.nih.gov)

Results

A total of 1,365 expressed gene fragments derived from single cells and 1,254 expressed gene fragments in colonial cells were observed in these studies (Table 2). Of these expressed fragments, 117 unique fragments were observed in single cells that were not observed in colonial cells and seven fragments that were expressed in colonial cells but absent in single cells were detected. Of the 172 differential expressed genes observed in flagellated cells, 87% of them (149) were either up regulated or turned on compared to colonial cells. In comparison, only 23% (14) of the 61 differentially expressed genes observed in colonial cells were upregulated or turned on relative to single cells.

The seven gene fragments that were observed only in colonial cells were cloned and sequenced,

Table 3 Differentially expressed genes associated with *P. globosa* colonies

Clone	Size (bp)	Genbank accession no.	Closest blast hit Genbank accession no.	Identity	E-value	Predicted function
PgC-3	490 bp	DQ886384	AAP94719	*Emiliania huxleyi* NADH Dehydrogenase subunit I	4×10^{-14}	Electron transport
PgC-10	342 bp	DQ886385	ZP00777402	UDP-3-0-acyl N-acetylglucosamine deacetylase (bacterial)	1×10^{-37}	Lipid assembly
PgC-19	213 bp	DQ886386	AY741371	*Emiliania huxleyi* strain CCMP 373 rRNA gene	9×10^{-50}	Protein synthesis
PgC-13	384 bp	DQ886387	XP639957	Hypothetical protein DDB01862282(*Dictyostelium discoideum*)	2×10^{-6}	Similar to putative stress responsive protein Srg6
PgC-5	261 bp	DQ886388	AF100332	Dendrobium grex Madame Thong-IN putative 21D7 protein (ovg29)	6×10^{-5}	Differential expressed gene during orchid floral transition
PgC-6	486 bp	DQ886389	AF107586	Dendrobium grex Madame Thong-IN putative cell control protein (otg4)	4×10^{-4}	Differentially expressed gene during orchid flora transition gene involved in cell division

although one did not produce useable sequence information (Table 3). The size of these fragments ranged from 213 to 490 bp with an average of 362 ± 114 bp. Of the genes that were putatively identified, one was involved in energy production (PgC-3) with closest similarity to the nicotinamide adenine dinucleotide (NADH) dehydrogenase subunit I. PgC-10 was most closely related to a bacterial UDP-3-0-acyl N-acetylglucosamine deacetylase coding gene involved in lipid assembly, and PgC-19 was most similar to the large subunit ribosomal gene from *Emiliania huxleyi*. Interestingly, the remaining three fragments appear to be related to stress proteins or genes associated with flowering in higher plants. PgC-13 was most similar to a hypothetical stress response protein from the amoeba *Dictyostelium discoideum*. Both PgC-5 and PgC-6 appeared to be related to genes associated with flowering in an Orchid species. The sequences of these gene fragments have been deposited in Genbank with accession numbers DQ886384-9.

Determination of life stage vulnerable to viral infection in P. pouchetii (VIL) Viruses and the processes of viral infection and lysis have been characterized for *P. pouchetii* and *P. globosa* (Bratbak et al. 1998a, b; Brussaard et al. 2004, 2005; Jacobsen 2000, 2002; Jacobsen et al. 1996; Veldhuis et al. 2005). Determining which stages of the life cycle are vulnerable to viral infection and lysis is critical for understanding the effects of viruses on *Phaeocystis* life cycle dynamics and ecology. Most previous work with *P. pouchetii* (Bratbak et al. 1998b; Hamm et al. 1999; Jacobsen 2000; Jacobsen et al. 1996) and *P. globosa* (Brussaard et al. 2004; Brussaard et al. 2005) indicated that solitary cells are vulnerable to viral infection and lysis while colony cells are not. However, in one set of experiments, Baudoux and Brussaard (2005) showed that colonial *P. globosa* cells can be infected by the PgV virus. To determine which life stages of *P. pouchetii* are vulnerable to viral infection and lysis, Jacobsen et al. (2005) conducted a set of experiments during our 2003 mesocosm study. In one set of experiments *P. pouchetii* solitary cells were incubated with different amounts of virus additions and a control containing no viruses (Jacobsen et al. 2005). In a second set of experiments, colony cells that were detached from their colony matrix were incubated with different amounts of virus additions and a control containing no viruses; for comparison solitary flagellate cells were also included in this experiment. Colonies were shaken strongly in order to detach the cells from the mucus; this method is effective in producing colonial cells free of their mucus matrix because, in contrast to

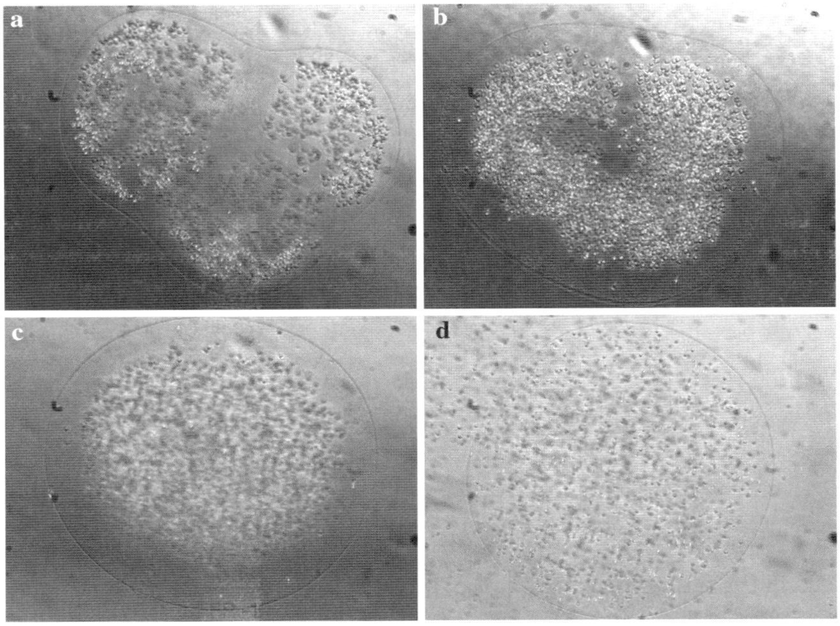

Fig. 3 A time sequence of a single *P. pouchetii* colony isolated and maintained in a well plate from the 2003 mesocosms (see Mesocosm Experiment Methods); data format: longest linear dimension and date and time of photo (**a**) 880 mm ; 26 March 2003, 11:33 (**b**) 920 mm ; 27 March 2003, 11:57 (**c**) 930 mm ; 28 March 2003, 09:41 (**d**) 890 mm ; 29 March 2003, 09:19; Cells within colony skin and in well plate observed to be motile at this time

P. globosa, colonies of *P. pouchetii* easily rupture (Jacobsen et al. 2005).

Intact colonies incubated with viruses showed no evidence of viral infection or lysis (Jacobsen et al. 2005). In experiments where colonial cells were detached from their colonies, detached colonial cells showed no evidence of cell lysis, whereas the solitary flagellated cells were completely lysed during the experimental period (Jacobsen et al. 2005).

Colony shape changes accompany colony senescence in P. pouchetii (CSWP; RCWP) The senescence of *Phaeocystis* colonies has been observed under natural, mesocosm, and laboratory conditions (Davidson and Marchant 1992; Lancelot and Rousseau 1994; Rousseau et al. 1994; Veldhuis et al. 1986; Verity et al. 1988). Solitary-cell release from senescent colonies at bloom termination has also been documented (Kornmann 1955; Rousseau et al. 1994; Verity et al. 1988). However, there is little detailed understanding of general colony-senescence and bloom-termination processes. In an in vitro study of *P. pouchetii*, colonies were isolated in well plates from the 2003 mesocosms, observed over a three-day period, and digital micrographs were taken to record colony morphology and measurement of dimensions.

During this study, a single colony was photographed over time that shows a change in morphology from the typical lobular *P. pouchetii* shape to a spherical shape (Fig. 3a–d). The time interval required for the observed shape change was two days. Approximately one day after the observed shape change, the colony was observed to contain motile cells and had also released some cells into the medium (Fig. 3d). While the possibility exists that spherical colonies were *P. globosa*, we found no evidence for the presence of *P. globosa* in these studies (see section in this manuscript entitled Species Identification in Mesocosms). Further investigation is needed to document the observed senescence phenomenon in other contexts, and to rigorously describe and quantify the observed morphological changes.

Quantification or investigation of a process or control parameter

Process rate quantification

Division (CDWP) and growth (CGWP) of P. pouchetii colonies Whipple et al. (2005b) reported that in vitro median colony division rates of mesocosm-derived *P. pouchetii* colonies ranged from 0.21–0.28 divisions day^{-1}. They also found that these colonies required an average of

3.5–4.9 days to complete one division; culture data from Cariou (described in Rousseau et al., 1994) indicated that *P. globosa* colony division occurred after 4–5 days, indicating similarity to *P. pouchetii* (Whipple et al. 2005b). Whipple et al. (2005b) reported that median growth rates of colonies, measured as maximum linear dimension, ranged from near zero to 7 μm h^{-1}.

Control parameters

Effect of diatom frustules on colony formation rate in P. pouchetii (CFSP) The transition from solitary cells to colonial forms is a critical event in the *Phaeocystis* life cycle. Many blooms consist primarily of colonial life stages (but see Wassmann et al. 2005). Even after nearly a century of study, no definitive set of conditions has been established as necessary for colony formation from solitary cells. Diatom frustules have been considered as particle nuclei for colony formation, including a number of observations of small *Phaeocystis* colonies attached to the setae of diatom frustules (especially *Chaetoceros*) (Boalch 1987; Lancelot and Rousseau 1994; Rousseau et al. 1994). Rousseau et al. (1994) stated that "...experimental work under controlled laboratory conditions (Rousseau and Davies, unpublished data), gives strong evidence that any microscopic particles, either biological (e.g. diatoms), organic or mineral (sand, glass wool) may act as substrate for colony development" (Rousseau et al. 1994, p. 35). Peperzak et al. (2000b) suggested that, in addition to providing a substrate that counteracts local turbulence, diatoms may influence colony formation by secreting vitamin B1, a possible growth stimulator, or depleting vitamin B12, a postulated growth inhibitor. These observations have led to the hypothesis that diatom frustules might enhance colony formation rate.

During the 2002 mesocosm study, an experiment was conducted to determine if in vitro *P. pouchetii* colony formation rates would be enhanced by the presence of diatom frustules added to culture medium; detailed methods and results appeared in Nejstgaard et al. (in press). Colony formation rates were estimated using water from a nutrient-amended mesocosm that was filtered to remove *Phaeocystis* colonies but allow solitary cells to pass. Diatomaceous earth was added to colonies in treatment well plates, while control well plates received only filtered mesocosm water (Nejstgaard et al. 2006). In all the experiments, new colonies increased approximately linearly over time. The colony formation rate for the filtered mesocosm water ranged from 1.3–1.9 colonies ml^{-1} d^{-1}; rates for water with diatomaceous earth ranged from 2.6–2.8 colonies ml^{-1} d^{-1}. Linear regression slopes for the filtered water versus water with diatomaceous earth were significantly different (Nejstgaard et al. 2006). The general conclusion of Nejstgaard et al. (2006) was that diatomaceous earth enhanced the in vitro colony formation rate for *P. pouchetii*.

Infochemical effect on colony formation due to mesozooplankton and microzooplankton (PRI). Long and colleagues conducted a series of experiments to test the ability of *P. globosa* solitary cells to respond to grazer-associated signals by altering colony formation or colony size. Colony formation was assessed when *P. globosa* encountered either microzooplankton or mesozooplankton grazers directly, only the chemical signals from these grazers, or no grazer cues (Long 2004; Long et al. submitted). In initial experiments with mixed zooplankton, the three dominant grazers were adult copepods of taxa *Acartia tonsa* and *Pseudodiaptomus pelagicus* (size ca. 1mm) and the heterotrophic dinoflagellate *Noctiluca scintillans*; the size of *P. globosa* colonies was about 50 μm in diameter. In other experiments, *A. tonsa* and *Euplotes* spp. were used alone as separate grazer treatments.

Two types of containers, both of which consisted of connecting compartments separated by a membrane and allowed for chemical exchange between connecting compartments while preventing algae and grazers from moving between compartments were used. *P. globosa* cells were added to both compartments while micro- or mesozooplankton grazers were added to only one compartment in grazer treatments. One set of containers was 1 L plastic bottles with sides removed and covered with Nitex mesh® (1 μm) using superglue. These mesh bottles were placed inside 2 l glass beakers that allowed for chemical exchange across 260 cm^2 of mesh. The second set of containers, called 'communication chambers', was constructed from two 0.64 cm thick, clear, acrylic, capped tubes (each 8.89 cm in diameter and approximately 7.94 cm in length, for a final volume of 500 ml each) that screwed into each other and were separated by a 0.8 μm polycarbonate membrane filter (Long 2004; Long et al., submitted).

Phaeocystis globosa cells were added to both compartments while grazers were added to only one in grazer treatments. This allowed for simultaneous assessment of the direct and indirect effects of grazers on colony formation compared to grazer-free controls (Long 2004; Long et al., submitted). To enhance exchange of chemical signals between experimental chambers, mesh bottles were placed in bench top orbital shakers and rotated at 50 rpm; each bottle was also gently lifted twice a day until half of its volume was displaced from one chamber to the other (Long 2004; Long et al., submitted). Additional details of the setup of the experiments, sampling methods, and results can be found in Long (2004) and Long et al. (submitted).

In the initial mixed zooplankton experiments, Long et al. (submitted) found that crustacean mesozooplankton, dominated by adult copepods *A. tonsa* and *P. pelagicus*, significantly suppressed colony formation. In other experiments, Long et al. (submitted) found that infochemical cues alone from the mesozooplankton *A. tonsa* alone suppressed colony formation. Long et al. (submitted) also observed that the colony suppression effect of *A. tonsa* was density dependent. In experiments with the microzooplankton *Euplotes* spp., Long et al. (submitted) reported that infochemical cues from this ciliate enhanced *P. globosa* colony formation.

Discussion

Solitary cell transformation to colonies

Diatom frustules enhance *P. pouchetii* colony formation in vitro (Nejstgaard et al. 2006), a process that may therefore accelerate the in situ transformation of solitary motile cells into new colonies. This is the critical step in the formation of colony-based *Phaeocystis* blooms, and thus is essential in the construction of quantitative ecosystem models of *Phaeocystis*. Future research must elaborate details of the control of colony formation rate by diatom frustules, such as dose response, species differences (*Phaeocystis* or diatoms), and how this control might change over the course of a bloom. One method by which these results could be represented in mechanistic models is to include a colony formation rate algorithm that varies colony formation rate with respect to the density of diatom frustules in the *Phaeocystis* environment.

The results of chemical cue experiments point to a potentially important role of consumer organisms in influencing the *Phaeocystis* life cycle. Tang (2003) found that grazing by microzooplankton resulted in an increase in the proportion of cells in the colony life stages, and increased the mean size of colonies; Long et al. (submitted) also found that infochemical cues from ciliates (*Euplotes* spp.) enhanced colony formation in *P. globosa*. In contrast, Long et al. (submitted) found that chemical signals from the mesozooplankton *Acartia tonsa* suppressed colony formation in *P. globosa*. Because of the different responses observed for meso- versus microzooplankton grazers, Long et al. (submitted) postulated that *P. globosa* might be able to detect grazing, identify the grazer type, and respond with enhancement or suppression of colony formation or growth of existing colonies. This type of induced response by *Phaeocystis* implies a sophisticated control mechanism for colony formation where chemical signals are received, processed, and a response is made dependent on the information contained in the signal. Further empirical work might include identification of chemicals in the signal, and determination of specificity of the signal for grazer taxa. The Tang (2003) and Long et al. (submitted) results would allow for control structures to be added to mechanistic models involving detection of grazers, identification of sender, and appropriate response of *Phaeocystis* solitary cells or colonies to grazers. In addition, the results of Long et al. (submitted), that the colony suppression effect of *A. tonsa* is density-dependent, gives the first data for producing a dose-response algorithm for the effects of these grazer-derived infochemicals in plankton models containing the *Phaeocystis* life cycle stages. These sort of controls demonstrate one aspect of the biocomplexity that organism semiotics can generate in ecosystem models.

Experimental results using coarse surficial material collected from shallow fjord sediments indicate that benthic particles may function as sites for colony formation. The colonies that appeared in the overlying water during the incubation were derived from cells or very small colonies associated with the sediments. It is inferred that these cells or small colonies were naturally associated with these sediments, however the possibility cannot be excluded that they were artificially mixed into the sediments at time of collection. These studies were not designed to prove or disprove the occurrence or existence of a true benthic life cycle stage, which has not been found since its original hypothesized existence (Kornmann 1955; Kayser 1970). Rather, the most parsimonious explanation is that these coarse sediments functioned as attachment sites for the initiation of new colonies, very much like diatoms and diatom frustules in the water column (Rousseau et al. 1994; Jacobsen 2002; Nejstgaard et al. 2006). Colonies were present in the fjord throughout the mesocosm study period (Nejstgaard et al. 2006). These results suggest that shallow sediments underlying illuminated waters during the period of *Phaeocystis* colony growth may serve as colony formation sites. The ecological contribution of colonies derived from such a mechanism must await a thorough seasonal study that includes sampling before and after the period of colony appearance in the water column. This finding suggests that some form of benthic involvement should be included in *Phaeocystis* life cycle models; these experiments do not provide evidence for a distinct benthic life cycle stage of *P. pouchetii*. Mechanisms that provide for benthic sediments providing colony formation sites could be provisionally included in models of *Phaeocystis* life cycles. Empirical studies to provide details of these sediment mechanisms are required to further elaborate conceptual or mechanistic models.

Comparison of gene expression profiles between single and colonial *Phaeocystis* cells indicated that a larger array of genes were expressed during the single stage life cycle phase than during the colonial life cycle stage, implying increased biochemical complexity associated with single cells. While the differential display analyses conducted as part of these studies in no way represent a comprehensive characterization of molecular differences between the colonial and single cell life cycles, intriguingly several of the genes that were expressed in colonial *Phaeocystis* cells and not in flagellated single cells appear to be related to stress, cell signaling, and complex developmental transitions in higher plants. Furthermore, the presence of genes in *Phaeocystis* most closely associated with phylogenetically diverse organisms including bacteria, amoeba, and higher plants, in addition to the closely related haptophyte *Emiliania huxleyi*, suggest that the genome of *Phaeocystis* is likely composed of a complex mosaic of genes and regulatory elements similar to other microalgae, including diatoms, whose genomes have recently been fully sequenced (Montsant et al. 2005).

The studies described here also demonstrate the tractability of *Phaeocystis* sp. to genomic and transcriptomics studies in the future. With relatively little modification it was possible to adapt commercially available RNA purification kits for work with plant cells to produce RNA suitable for constructing cDNA libraries from both single cells and colonies of *Phaeocystis globosa*. Several previous reports have demonstrated that genomic and plastid DNA suitable for molecular studies is also relatively easily obtained from a variety of *Phaeocystis* species (Zingone et al. 1999; Lange et al. 2002). As is occurring in all fields of biology, access to molecular information from individual organisms and communities of organisms is yielding tremendous insights into the physiology, life history, and ecology of organisms including microalgae for which molecular information is available (Allen 2005, Walker et al. 2005). Undoubtedly, important new insights into the life history of *Phaeocystis* will be gained when complete genomic and comprehensive gene expression data become available.

Colony proliferation

Documentation of colony division in vitro provides evidence for two types of colony proliferation in *P. pouchetii*. First, colonies may form from solitary cells and grow by mitosis and secretion of extracellular mucus to form the large colonies observed in most blooms. Relationships between cell number and biomass has been documented for *P. globosa* (Rousseau et al. 1990) and *P. pouchetii* (Verity et al. in press); such data would enhance the usefulness of *P. pouchetii* colony size measurements. Second, existing colonies may undergo division and may also increase in size after division by mitosis and secretion of extracellular mucus (Whipple et al. 2005b). The latter pathway allows the number of colonies to increase faster than would be possible if all colonies were produced from solitary cells, because daughter colonies produced by colony division start out with hundreds of cells, and may attain a large size much faster than colonies established with a solitary cell. With the possibility of colony division, colonies that form from solitary cells constitute a source population from which large numbers of multicellular colonies may be derived by successive colony growth and colony division stages. Modeling this mode of colony proliferation could provide insight into its possible effects on *Phaeocystis* bloom dynamics. Further work will be needed to determine the extent to which this pattern is observed in natural *Phaeocystis* blooms.

Colony senescence and solitary cell release

Observation of spherical colonies with supporting genetic data, identifying them with high confidence as *P. pouchetii*, indicates that colony shape alone should be confirmed with other methods to unambiguously identify the species of *Phaeocystis* colonies observed in the field. The observed in vitro morphological changes of colonies of *P. pouchetii* suggest that there may be many changes in colony cell activities and physiology that would be required to perform this morphological change during colony senescence. It is not known in detail what changes the colony cells undergo during senescence; however, this colony shape change observation points to a period where major changes may be occurring. Further investigation of the colony cell characteristics during this shape-change period might provide further insight into the senescence processes in *P. pouchetii*. Qualitative shape changes, as described in these results, cannot usefully be incorporated into *Phaeocystis* life cycle models; however, if details of cellular physiology during senescence were to be discovered by further study of this colony stage, they would likely prove useful for model development.

Colony or solitary cell losses: viral infection and lysis

If blooms form predominantly from colonies, *P. pouchetii* can escape viral lysis by forming colonies. Until colonies undergo senescence and release solitary flagellate cells, which would be predicted to be vulnerable to viruses, *P. pouchetii* colony cells would be protected from viral infection. Transparent exopolymer particles (TEP), which appear to bind viruses (Brussaard et al. 2005), introduce another control that might be especially significant during senescence of *Phaeocystis* colonies at the end of blooms.

Jacobsen et al. (2005) report that detached colonial cells of *P. pouchetii* appear to be non-infectable. This suggests that the colony cells may have or acquire some inherent resistance to viral infection that does not involve the colony integument. Because the detached colonies would be surrounded by colony mucus that contains TEP, the PgV particles might have been bound to TEP, reducing their ability to infect the detached cells. This potential confounding factor should be addressed in future experiments. Jacobsen (2002) proposed that a colony integument pore size too small to admit virus particles might be a mechanism for prevention of viral infection of colonies. Results reported in Jacobsen et al. (2005) indicate that another mechanism might be responsible for the lack of colony cell lysis observed for *P. pouchetii*. These experiments were not designed to elucidate the specific viral infection mechanisms, and further investigation is needed to confirm these preliminary observations and investigate mechanisms of lysis resistance (Jacobsen et al., 2005). Ruardij et al. (2005) incorporated viral lysis into a simulation model of *P. globosa*; if the results of Jacobsen et al. (2005) were incorporated into a simulation model of *P. pouchetii*, viral lysis would only involve solitary cells and once cells were in their colonial form they would be invulnerable to viruses.

Summary and conclusions

Several compartments, process flows, and their controls that were included in Whipple et al. (2005a) were empirically investigated to fill gaps in our understanding. Two major aspects of the conceptual model were involved.

The first involved testing for the existence of a life cycle process or producing a description of such a process. Five examples discussed here include (**a**) demonstration of in vitro colony division in *P. pouchetii* (CDWP); (**b**) description of in vitro change in colony shape for *P. pouchetii* associated with senescence (CSWP); (**c**) determining which life stage of *P. pouchetii*, solitary cells or colonies, is vulnerable to viral infection and lysis (VIL); (**d**) demonstration of potential benthic involvement in the *P. pouchetii* life cycle (CFBI); (**e**) demonstration of differential gene expression in solitary versus colonial cells of *P. globosa* (DGE).

The second facet involved quantification of processes or regulatory controls. Two examples of process rate quantification for *P. pouchetii* are: (**a**) in vitro colony division rate (CDWP) and (**b**) in vitro colony growth rate (CGWP). There were also two examples of investigating process controls. The first was diatom frustule enhancement of colony formation from solitary cells in *P. pouchetii* (CFSP), and the second was suppression of colony formation in *P. globosa* by infochemicals released by mesozooplanktonic grazers, and enhancement of colony formation by infochemicals released by microzooplankton grazers (PRI).

The significance of investigating these features of *Phaeocystis* ecology in the context of a *Phaeocystis* life cycle model is that new insights and understanding are readily accommodated in a pre-existing organized scheme of established facts and hypotheses. This new understanding, made coherent by the life cycle model, can improve ecosystem simulation models (e.g., Canziani and Hallam 1996; Verity 2000; Lancelot et al. 2005; Ruardij et al. 2005) designed to receive the life cycle information and relate it to the broader ecosystem. From this perspective, *Phaeocystis* is a definitive example of a complex adaptive system (CAS) (Waldrop 1992; Kauffman 1993, 1995) incorporated within its broader ecosystem, also seen as a complex adaptive system. Using models, conceptual or ecosystem, to drive empirical research that is then incorporated into new model structures can ultimately provide better insight into the direct and indirect effects of *Phaeocystis* populations on such important contemporary problems as global biogeochemical cycles, as well as illuminate the roles of *Phaeocystis* in its organized ecosystems.

Acknowledgements The authors would like to thank T. Sorlie, A. Aadnesen, and H. Gjertsen of the University of Bergen Marine Biological Station for assistance during mesocosm experiments. We thank current and former Skidaway Institute of Oceanography staff: M. Zirbel, T. Walters, J. Danforth, K. Kohlberg, G. Smalley, D. Heaton, J. Brofft, and R. Hristov. We also thank current and former Georgia Institute of Technology student and faculty members: T. Barsby and M. Hay. Funding was provided by US National Science Foundation grant OPP-00-83381. This paper is a contribution to the SCOR WG 120 "*Phaeocystis*, major link in the biochemical cycling of climate-relevant elements".

References

Allen AE (2005) Defining the molecular basis for energy balance in marine diatoms under fluctuating environmental conditions. J Phycol 41:1073–1076

Allen AE, Booth MG, Verity PG, Frischer ME (2005) Influence of nitrate availability on the distribution and abundance of heterotrophic bacterial nitrate assimilation genes in the Barents Sea during summer. Aquat Microb Ecol 39:247–255

Ayers GP, Gillett RW (2000) DMS and its oxidation products in the remote marine atmosphere. J Sea Res 43:275–286

Baudoux A-C, Brussaard CPD (2005) Characterization of different viruses infecting the marine harmful algal bloom species *Phaeocystis globosa*. Virology 341:80–90

Boalch GT (1987) Recent blooms in the western English Channel. Rapp P-v Reun Cons int Explor Mer 187:94–97

Bratbak G, Jacobsen A, Heldal M (1998a) Viral lysis of *Phaeocystis pouchetii* and bacterial secondary production. Aquat Microb Ecol 16:11–16

Bratbak G, Jacobsen A, Heldal M, Nagasaki K, Thingstad F (1998b) Viral production in *Phaeocystis pouchetii* and its relation to host cell growth and nutrition. Aquat Microb Ecol 16:1–9

Brussaard CPD, Kuipers B, Veldhuis MJW (2005) A mesocosm study of *Phaeocystis globosa* population dynamics I. Regulatory role of viruses in bloom control. Harmful Algae 4:859–874

Brussaard CPD, Short SM, Frederickson CM, Suttle CA (2004) Isolation and phylogenetic analysis of novel viruses infecting the phytoplankton *Phaeocystis globosa* (Prymnesiophyseae). Appl Environ Microbiol 70:3700–3705

Canziani GA, Hallam TG (1996) A mathematical model for *Phaeocystis* sp. dominated plankton community dynamics. I The basic model Nonlinear World 3:19–76

Cariou V, Casotti R, Birrien JL, Vaulot D (1994) The initiation of *Phaeocystis* colonies. J Plankton Res 16:458–470

Davidson AT, Marchant HJ (1992) The biology and ecology of *Phaeocystis* (Prymnesiophyceae). In: Round FE, Chapman DJ (eds) Progress in Phycological Research. Biopress Ltd, Bristol, UK pp 1–45

Eilertsen HC (1989) *Phaeocystis pouchetii* (Hariot) Lagerheim, a key species in Arctic marine ecosystems: Life history and physiology. Rapp P-v Reun Cons int Explor Mer 188:131

Frischer ME, Hansen AS, Wyllie JA, Wimbush J, Murray J, Nierzwicki-Bauer SA (2002) Specific amplification of the 18S rRNA gene as a method to detect zebra mussel (*Dreissena polymorpha*) larvae in plankton samples. Hydrobiologia 487:33–44

Gabric AJ, Matrai PA, Vernet M (1999) Modelling the production of dimethylsulfide during the vernal blooms in the Barents Sea. Tellus Ser B Chem Phys Meteorol 51:919–937

Guillard RRL, Hargraves PE (1993) *Stichochrysis immobilis* is a diatom, not a chrysophyte. Phycologia 32:234–236

Hamm CE, Simson DA, Merkel R, Smetacek V (1999) Colonies of *Phaeocystis globosa* are protected by a thin but tough skin. Mar Ecol Prog Ser 187:101–111

Jacobsen A (2000) New aspects of bloom dynamics of *Phaeocystis pouchetii* (Haptophyta) in Norwegian waters. PhD Dissertation. Department of Fisheries and Marine Biology (45pp). University of Bergen, Bergen, Norway

Jacobsen A (2002) Morphology, relative DNA content and hypothetical life cycle of *Phaeocystis pouchetii* (Prymnesiophyceae); with special emphasis on the flagellated cell type. Sarsia 87:338–349

Jacobsen A, Bratbak G, Heldal M (1996) Isolation and characterization of a virus infecting *Phaeocystis pouchetii* (Prymnesiphyceae). J Phycol 32:923–927

Jacobsen A, Martínez-Martínez J, Verity P, Frischer ME, Sandaa RA, Larsen A (2005) Are colonies or colonial cell of *Phaoecystis pouchetii* (Prymnesiophyceae) suseptible to virus infection? American society of limnology and oceanography, summer meeting. ASLO, Santiago de Compostela, Spain

Kauffman SA (1993) The origins of order. Self-organization and selection in evolution. Oxford University Press, New York

Kauffman SA (1995) At home in the Universe. The search for laws of self-organization and complexity. Oxford University Press, New York

Kayser H (1970) Experimental-ecological investigations on *Phaeocystis poucheti* (Haptophyceae). Helgol wiss Meeresunters 20:195–212

Kornmann PV (1955) Beobachtungen an *Phaeocystis*-Kulturen. Helgol wiss Meeresunters 5:218–233

Lancelot C, Keller MD, Rousseau V, Smith WO, Mathot S (1998) Autecology of the marine Haptophyte *Phaeocystis* sp. In: Anderson DM, Cembella AD, Hallegraeff GM (eds) Physiological ecology of harmful algal blooms. Springer-Verlag, Berlin pp 209–224

Lancelot C, Rousseau V (1994) Ecology of *Phaeocystis*: the key role of colony forms. In: Green JC, Leadbeater BSC (eds) The haptophyte algae. Clarendon Press, Oxford pp 229–245

Lancelot C, Spitz Y, Gypens N, Ruddick K, Becquevort S, Rousseau V, Lacroix G, Billen G (2005) Modelling diatom and *Phaeocystis* blooms and nutrient cycles in the Southern Bight of the North Sea: the MIRO model. Mar Ecol Prog Ser 289:63–78

Lancelot C, Wassmann P, Barth H (1994) Ecology of *Phaeocystis*-dominated ecosystems. J Mar Syst 5:1–4

Lange M, Chen Y-Q, Medlin LK (2002) Molecular genetic delineation of *Phaeocystis* species (Prymnesiophyceae) using coding and non-coding regions of nuclear and plastid genomes. Eur J Phycol 37:77–92

Lange M, Laure G, Vaulot D, Simon N, Amann RI, Ludwig W (1996) Identification of the class Prymnesiophyceae and the genus *Phaeocystis* with ribosomal RNA-targeted nucleic acid probes detected by flow cytometry. J Phycol 32:868–872

Long JD (2004) Plasticity of consumer-prey interactions in the sea: chemical signaling, learned aversion, and ecological consequences. PhD Dissertation. Department of Biology (119pp). Georgia Institute of Technology, Atlanta, GA, USA. URL link: http://www.etd.gatech.edu/theses/available/etd-11182004–164652/unrestricted/long_jeremy_d_200412_phd.pdf

Long JD, Smalley GW, Barsby T, Anderson JT, and Hay ME (submitted) Chemical cues induce consumer-specific defences in a bloom-forming marine phytoplankton. Proceedings of the National Academy of Science

Medlin LK, Zingone A (in press) A review: the genus *Phaeocystis* and its species. Biogeochemistry

Montsant A, Jabbari K, Maheswari U, Bowler C (2005) Comparative genomics of the pennate diatom *Phaeodactylum tricornutum*. Plant Physiol 137:500–513

Nejstgaard JC, Frischer ME, Verity PG, Anderson JT, Jacobsen A, Zirbel MJ, Larsen A, Martínez-Martínez J, Sazhin AF, Walters T, Bronk DA, Whipple SJ, Borrett SR, Patten BC, Long JD (2006) Plankton development and trophic transfer in sea water enclosures added nutrients and *Phaeocystis pouchetii*. Mar Ecol Prog Ser 321:99–121

Peperzak L, Colijn F, Vrieling EG, Gieskes WWC, Peeters JCH (2000a) Observations of flagellates in colonies of *Phaeocystis globosa* (Prymnesiophyceae); a hypothesis for their position in the life cycle. J Plankton Res 22:2181–2203

Peperzak L, Gieskes WWC, Duin R, Colijn F (2000b) The vitamin B requirement of *Phaeocystis globosa* (Prymnesiophyceae). J Plankton Res 22:1529–1537

Rousseau V, Jacobsen A, Verity P, Whipple S (in press) The life cycle of *Phaeocystis*: state of knowledge and presumptive role in ecology. Biogeochemistry

Rousseau V, Mathot S, Lancelot C (1990) Calculating carbon biomass of Phaeocystis sp. from microscopic observations Mar Biol 107:305–314

Rousseau V, Vaulot D, Casotti R, Cariou V, Lenz J, Gunkel J, Baumann M (1994) The life cycle of *Phaeocystis* (Prymnesiophyceae): evidence and hypotheses. J Mar Syst 5:23–39

Ruardij P, Veldhuis MJW, Brussaard CPD (2005) Modeling the bloom dynamics of the polymorphic phytoplankter *Phaeocystis globosa*: impact of grazers and viruses. Harmful Algae 4:941–963

Smith Jr WO, Codispoti LA, Nelson DM, Manley T, Buskey EJ, Niebauer HJ, Cota GF (1991) Importance of *Phaeocystis* blooms in the high-latitude ocean carbon cycle. Nature 352:514–516

Stefansson U, Olafsson J (1991) Nutrients and fertility of Icelandic waters. Rit Fiskideildar 12:1–56

Tang KW (2003) Grazing and colony size development in *Phaeocystis globosa* (Prymnesiophyceae): the role of a chemical signal. J Plankton Res 25:831–842

Veldhuis MJW, Brussaard CPD, Noordeloos AAM (2005) Living in a *Phaeocystis* colony: a way to be a successful algal species. J Sea Res 53:841–858

Veldhuis MJW, Colijn F, Venekamp LAH (1986) The spring bloom of *Phaeocystis pouchetii* (Haptophyceae) in Dutch coastal waters. Neth J Sea Res 20:37–48

Verity PG (2000) Grazing experiments and model simulations of the role of zooplankton in *Phaeocystis* food webs. J Sea Res 43:317–343

Verity PG, Smetacek V (1996) Organism life cycles, predation, and the structure of marine pelagic ecosystems. Mar Ecol Prog Ser 130:277–293

Verity PG, Villareal TA, Smayda TJ (1988) Ecological investigations of blooms of colonial *Phaeocystis pouchetii*. II. The role of life-cycle phenomena in bloom termination. J Plankton Res 10:749–766

Verity PG, Whipple SJ, Nejstgaard JC, and Alderkamp A-C (in press) Colony size, cell density, carbon and nitrogen contents of Phaeocystis pouchetii from western Norway. J Plankton Res

Waldrop MM (1992) Complexity. The emerging science at the edge of order and chaos. Simon and Schuster, New York

Walker TL, Collet C, Purton S (2005) Algal transgenics in the genomic era. J Phycol 41:1077–1093

Wassmann P, Ratkova T, Reigstad M (2005) The contribution of single and colonial cells of *Phaecystis pouchetii* to spring and summer blooms in the north-eastern North Atlantic. Harmful Algae 4:823–840

Weisse T, Grimm N, Hickel W, Martens P (1986) Dynamics of *Phaeocystis pouchetii* blooms in the Wadden Sea of Sylt (German Bight, North Sea). Estuar Coast Shelf Sci 23:171–182

Whipple SJ, Patten BC, Verity PG (2005a) Life cycle of the marine alga *Phaeocystis*: a conceptual model to summarize literature and guide research. J Mar Syst 57:83–110

Whipple SJ, Patten BC, Verity PG (2005b) Colony growth and evidence for colony multiplication in *Phaeocystis pouchetii* (Prymnesiophyceae) isolated from mesocosm blooms. J Plankton Res 27:495–501

Zingone A, Chretiennot-Dinet M-J, Lange M, Medlin L (1999) Morphological and genetic characterization of *Phaeocystis cordata* and *P. jahnii* (Prymnesiophyceae), two new species from the Mediterranean Sea. J Phycol 35:1322–1337

Biogeochemistry (2007) 83:311–330
DOI 10.1007/s10533-007-9090-6

REVIEW PAPER

Current understanding of *Phaeocystis* ecology and biogeochemistry, and perspectives for future research

Peter G. Verity · Corina P. Brussaard · Jens C. Nejstgaard · Maria A. van Leeuwe · Christiane Lancelot · Linda K. Medlin

Received: 1 June 2006 / Accepted: 13 September 2006 / Published online: 16 March 2007

Abstract The phytoplankton genus *Phaeocystis* has well-documented, spatially and temporally extensive blooms of gelatinous colonies; these are associated with release of copious amounts of dimethyl sulphide (an important climate-cooling aerosol) and alterations of material flows among trophic levels and export from the upper ocean. A potentially salient property of the importance of *Phaeocystis* in the marine ecosystem is its physiological capability to transform between solitary cell and gelatinous colonial life cycle stages, a process that changes organism biovolume by 6–9 orders of magnitude, and which appears to be activated or stimulated under certain circumstances by chemical communication. Both life-cycle stages can exhibit rapid, phased ultradian growth. The colony skin apparently confers protection against, or at least reduces losses to, smaller zooplankton grazers and perhaps viruses. There are indications that *Phaeocystis* utilizes chemistry and/or changes in size as defenses against predation, and its ability to create refuges from biological attack is known to stabilize predator–prey dynamics in model systems. Thus the life cycle form in which it occurs, and particularly associated interactions with viruses, determines whether *Phaeocystis* production flows through the traditional "great fisheries" food chain, the more regenerative microbial food web, or is exported from the mixed layer of the ocean.

Despite this plethora of information regarding the physiological ecology of *Phaeocystis*, fundamental interactions between life history traits and system ecology are poorly understood. Research summarized here, and described in the various papers in this special issue, derives from a central question: how do physical (light, temperature, particle distributions, hydrodynamics), chemical (nutrient resources, infochemistry,

P. G. Verity (✉)
Skidaway Institute of Oceanography,
10 Ocean Science Circle, Savannah, GA, 31411, USA
e-mail: Peter.Verity@skio.usg.edu

C. P. Brussaard
Royal Netherlands Institute for Sea Research,
P.O. Box 59, 1790 AB Den Burg, The Netherlands

J. C. Nejstgaard
UNIFOB AS, Department of Biology,
University of Bergen, Bergen High Technology
Centre, 5020 Bergen, Norway

M. A. van Leeuwe
University of Groningen, Biological Centre,
P.O. Box 14, 9750 AA Haren, The Netherlands

C. Lancelot
Ecologie des Systèmes Aquatiques, Université Libre
de Bruxelles, Campus de la Plaine – C.P. 221,
Bd. du Triomphe, 1050 Bruxelles, Belgium

L. K. Medlin
Alfred Wegener Institute for Polar and Marine
Research, Am Handelshafen 12, 27570 Bremerhaven,
Germany

allelopathy), biological (grazers, viruses, bacteria, other phytoplankton), and self-organizational mechanisms (stability, indirect effects) interact with life-cycle transformations of *Phaeocystis* to mediate ecosystem patterns of trophic structure, biodiversity, and biogeochemical fluxes? Ultimately the goal is to understand and thus predict why *Phaeocystis* occurs when and where it does, and the bio-feedbacks between this keystone species and the multitrophic level ecosystem.

Keywords Biocomplexity · Plankton life cycles · *Phaeocystis* · Viruses · Zooplankton

Introduction

The assembly of phytoplankton communities is a continuous process in which instantaneous external assessments, e.g., species abundances, diversity, and bulk constituents, reflect the outcome of a host of physical, chemical, and biological processes operating at a hierarchy of temporal and spatial scales. The absolute abundance of a given taxon increases according to growth, immigration, physical concentration, and other mechanisms, whereas losses including grazing, lysis, sedimentation, dilution, and emigration act to stabilize population fluctuations. In this manner, many species generally compete yet coexist in a water mass of seemingly similar properties. That one species becomes numerically dominant, much less forms a near monospecific bloom, is a remarkable phenomenon, especially given that many loss processes are density dependent in nature. To do so regularly, over large spatial and long temporal scales, is indicative of a life-history strategy more successful than the competition. It may also indicate that one species is more genetically diverse than another, and that diversity enables it to exist over these scales as one suite of genotypes gives way to another suite of genotypes to perpetuate the bloom. From this perspective, each suite of genotypes conceivably thrives in a discrete set of environmental parameters.

The genus *Phaeocystis* is one such taxon. Whereas it contains at least six species (Medlin and Zingone this volume; Rousseau et al. this volume), it is ecologically acknowledged that the blooms are typically caused by the two cold water species, *P. pouchetii* (high boreal and arctic waters) and *P. antarctica* (Southern Ocean), and by *P. globosa* (primarily cold north temperate waters). What is unusual in marine ecosystems is that such blooms are primarily a life cycle event: these *Phaeocystis* species transform from a tiny single-celled morphotype into a colonial stage in which hundreds or thousands of cells are embedded within a gelatinous matrix and protected behind an elastic but solute-permeable community membrane. Something about this transformation and/or the characteristics of the colony life-cycle stage (Lancelot and Rousseau 1994; Lancelot et al. 1998) permit *Phaeocystis* to out-compete other phytoplankton, which belong to a variety of algal classes with diverse strategies for competing for light and nutrients, but without a functionally comparable gelatinous house. At sufficiently high concentrations, colony blooms have been associated with a variety of ecosystem changes and negative effects on fisheries and fish farming (Lancelot et al. 1987; Schoemann et al. 2005; Nejstgaard et al. this volume), such that blooms of *Phaeocystis* colonies are considered as harmful algal blooms (HABs, e.g., Anderson et al. 1998, Veldhuis and Wassmann 2005). Less is known about the occurrence and significance of solitary cell stages of *Phaeocystis*, primarily because these small nanoplankton require specific careful methods for quantification. However, blooms of solitary cells have been reported on occasion, e.g., Ratkova and Wassmann (2002) and Wassmann et al. (2005), and thus the temporal and spatial distribution of *Phaeocystis* may have been undersampled. Solitary cells may also be important to over-wintering strategies (see below) but relevant quantitative studies are lacking. Nevertheless, given that colonies derive from solitary cells, knowledge of their dynamics and mechanisms influencing their occurrence are important to understanding the success of this genus (Reigstad and Wassmann, this volume).

Thus the causes of life-cycle alterations in *Phaeocystis* may be under environmental controls, whether abiotic or biotic, and elucidation of these controls will shed light on the fundamental processes that regulate ecosystem organization and biogeochemical flows. Research attempting to explain these life-cycle transformations from the perspective of resource limitation or physiological capability have been underway for almost a century, but have failed to provide conclusive evidence of why *Phaeocystis* is so successful in sequestering resources and dominating its competitors. Emphasis of studies on shifts in life cycle dominance in situ has increasingly focused on influences by other organisms, either direct causes such as viral- or grazer-mediated reduction of abundance of specific stages, or indirect causes such as stage transformation induced by allelopathy or infochemicals. Because of the large change in effective organism size when it transforms from solitary cells to colonies or vice-versa, *Phaeocystis* can function in reality as a dual species: a nanoplanktonic solitary cell, a micro- to macroplanktonic colony, or both concurrently. But whereas we understand some of the evolutionary costs and benefits of these two stages (see Verity and Medlin 2003; Veldhuis et al. 2005 for discussions), and some of the factors influencing biomass accumulation of these stages, we do not yet understand how they interact to make the keystone phytoplankton that is *Phaeocystis*. If we could, the expected insights into planktonic ecosystem structure and function might be remarkable.

This was the rationale and motivation for the formation of the scientific committee on ocean research (SCOR) working group #120 ("Marine Phytoplankton and Global Climate Regulation: The *Phaeocystis* spp. Cluster as a Model"). The working group's mission during the period 2000–2005 was to define research needs on *Phaeocystis*, and to provide a logistical mechanism to integrate results into a revised conceptual model of this taxon's role in the transfer of elements relevant to climate regulation, such as carbon and sulfur, and the biogeochemical processes that influence such a role. These goals were a natural development from previous symposia on *Phaeocystis* held in Belgium (1991) and the Netherlands (1999), and resulting in special publications in the *Journal of Marine Systems* (1994, Volume 5) and *Journal of Sea Research* (2000, Volume 43), respectively. As such, the field of interest was broad, encompassing genes to climate, with a goal of generating sufficient information to place *Phaeocystis* not only in the context of biodiversity but also in the framework of global international programs such as the surface ocean lower atmosphere study (SOLAS: http://www.uea.ac.uk/env/solas) and integrated marine biogeochemistry and ecosystem research (IMBER: http://www.igbp.kva.se/obe). The ultimate objective was to improve predictability in the essentially chaotic plankton world. The following section summarizes progress regarding key aspects of the fitness, ecology, and biogeochemistry of *Phaeocystis* that occurred during the working group's organizational lifetime, and which are described in more detail in papers in this special issue of *Biogeochemistry*.

New perspectives on *Phaeocystis* and its role in upper ocean biogeochemistry

Phaeocystis-dominated ecosystems include large and biogeochemically important portions of the global ocean, which are trophically unusual and poorly understood. Life-cycle alternations caused by system stresses on *Phaeocystis*, whether abiotic or biotic, restructure the entire food web and change ecosystem characteristics ranging from nutrient utilization to trophic energy transfer to vertical export of biogeochemically salient elements. Evidence supports the notion that *Phaeocystis* is so successful because, through a combination of life cycle stages, a range of colony sizes, and perhaps grazer-activated defenses, it can reduce mortality more than competing phytoplankton. Interactions between *Phaeocystis* and other organisms, especially zooplankton, may involve sophisticated signaling, and attempts to accurately model and predict the complexity of these ecosystems requires quantitative information on the feedbacks between *Phaeocystis* and its environment. For these reasons, among many, *Phaeocystis* is a model organism from which to begin the study of complexity in marine pelagic ecosystems.

Taxonomy

The genus *Phaeocystis* is thought to have originated 75 million years ago. There are currently six recognized species: *P. antarctica* Karsten, *P. cordata* Zingone et Chretiennot-Dinet, *P. globosa*

Scherffel, *P. jahnii* Zingone, *P. pouchetii* (Hariot) Lagerheim, and *P. scrobiculata* Moestrup (Medlin et al. 1994; Zingone et al. 1999; Medlin and Zingone, this volume; Rousseau et al. this volume), and several as yet undescribed species are thought to exist. However, one or more of these may represent species complexes. For example, the existence of different internal transcribed spacer (ITS) alleles in a single strain of *Phaeocystis globosa* was interpreted as evidence of hybridization between cryptic species within the *P. globosa* morphotype (Lange et al. 2002). Normally ITS regions should be homogenized by concerted evolution but this was not the case in *P. globosa*. In hybrids of land-plant species, both parental copies are present in the offspring and a continued crossing of the F1 generation will eventually lead to the loss of one of the parental types. These regions were not evolving too rapidly for homogenization to take place based on relative rates tests made on the 18S ribosomal deoxyribonucleic acid (rDNA; Lange et al. 1996). Cryptic species within *Phaeocystis globosa* had been inferred earlier in this species based on differences in DNA content determined by flow cytometry among strains of the species (Vaulot et al. 1994). Thus there seems to be substantial inter- and intraspecific variation in this genus.

Gaebler et al. (this volume) attempted to trace the biogeographical history of *Phaeocystis* strains in the Southern Ocean by quantifying the genetic diversity within populations of *P. antarctica* from each of the three major gyres in the Antarctic continental waters and calculating the gene flow between them. They described methods to quantify genetic diversity, and preliminary results for *P. antarctica* in comparison to two other colonial species: *P. globosa* and *P. pouchetii*. Results showed differences in the AFLP patterns between isolates of *P. antarctica* from different regions, and that a wide variety of microsatellite motifs could be obtained from the three *Phaeocystis* species. Initial results (Lange et al. 2002) support the ITS analysis that the populations from the Antarctic circumpolar current (ACC) are seeding the continental gyres.

Lange et al. (2002) used sequence variation to compare 22 isolates representing a global distribution of the genus. The most conserved 18S rRNA analysis suggested that an undescribed unicellular *Phaeocystis* sp. (isolate PLY559) is a sister taxon to the Mediterranean unicellular *Phaeocystis jahnii*; this clade branched prior to the divergence of all other *Phaeocystis* species, including the colonial ones. *Phaeocystis globosa* and *pouchetii* have multiple different ITS copies, suggestive of cryptic species that are still able to hybridize. A molecular clock was constructed that estimates the divergence of the cold-water forms from the warm-water forms to be about 30 Mya and the divergence of *P. antarctica* and *P. pouchetii* to be about 15 Mya. Medlin and Zingone (this volume) have reviewed the morphological information on the validly described *Phaeocystis* species and presented some new information on undescribed species. The combined molecular and morphological data implied that the number of species in the genus is still underestimated, and that cryptic or pseudo-cryptic diversity requires a sound assessment in future research on this genus.

Life cycles

The complex life cycle of *Phaeocystis* is the key to understanding its role in marine ecosystems (Verity et al. 1988b; Rousseau et al. 1994; Peperzak et al. 2000). However, despite almost a century of study, the number and role of the different cell types involved in the *Phaeocystis* life cycle is still uncertain. Rousseau et al. (this volume) provided a detailed review of the different morphotypes of the six recognized *Phaeocystis* species, and found four different cell types. There are two types of scaly flagellates: one produces star-forming filaments and has been reported for all species except *P. jahnii*, while the other, lacking filaments and stars, has been found in *P. globosa* and *P. jahnii*. The three colony-forming species, *P. globosa*, *P. pouchetii*, and *P. antarctica*, share three cell types in common: a flagellate with scales and filaments, a colonial cell, and a flagellate devoid of scales and filaments. In contrast, only flagellates with scales and filaments have been observed in the non-colony-forming species, *P. scrobiculata* and *P. cordata*. The authors provided evidence for a haploid–diploid life cycle in *P. globosa* that includes sexuality during colony bloom formation, and two types of vegetative reproduction.

The high genetic diversity recovered in the preliminary genetic analysis done by Gaebler et al. (this volume) would suggest that sexual reproduction is quite common.

Different hypotheses have been proposed to explain the selective pressures that led to the evolution of colony formation in *Phaeocystis*, including as a strategy to reduce population mortality by grazing or viral lysis (Verity and Medlin 2003; Brussaard et al. this volume; Nejstgaard et al. this volume). If cells in colonies have lower per capita losses than freely dispersed cells, and also have equivalent or even higher intrinsic growth rates (Veldhuis et al. 2005), then colonies would appear to provide a distinct life-cycle advantage. Nevertheless, there remain numerous questions regarding the ecological roles of the different life-cycle stages, before their significance can be completely understood (Rousseau et al. this volume). However, there is universal agreement that these life-cycle transitions may have large direct and indirect effects on trophic dynamics and biogeochemistry of planktonic ecosystems. In coccolithophorids, another member of the Prymnesiophyta, life-cycle changes are likely related to niche partitioning (Noel et al. 2004). The haploid holococcolith stage thrives in oligotrophic, deeper oceanic waters, whereas the diploid heterococcolith stage lives in more-eutrophic coastal water. To evaluate the hypothesis of the effect of life cycle transitions on trophic dynamics, a better understanding of the occurrence, transitions, and causal mechanisms of the life cycle is needed.

Yet another life-cycle issue is how *Phaeocystis* survives the long polar winter, i.e., several months of darkness. It would be advantageous in such an environment for photosynthetic solitary cells to be mixotrophic (capable of also ingesting bacteria or other small particles), or for mixotrophy to be induced by decreasing photoperiod or some other signal of the impending months of darkness. Phagotrophy of bacteria and phytoplankton by other prymnesiophytes related to *Phaeocystis* has been documented (Nygaard and Tobiesen 1993; Tillmann 1998), but to the authors' knowledge it is unknown whether *Phaeocystis* has such capability. It is known that with the induction of low temperature and low light, the expression of photosynthetic genes is increased in sea-diatoms (Mock and Valentin 2004), but whether or not this is a universal response of other polar phytoplankton has not been investigated. Interestingly, *Phaeocystis* may serve as a source of photosynthetic potential for its own predators. A new undescribed dinoflagellate, related to *Karenia* and *Karlodinium*, has been found in the Antarctic with kleptoplastids from *Phaeocystis* (Gast et al. 2006). Cultures of this heterotrophic dinoflagellate are maintained on single cells of *Phaeocystis*, from which it extracts its plastids by a yet as unknown mechanism to become photosynthetic. The dinoflagellate has also been seen inside *Phaeocystis* colonies, but it is unknown if the dinoflagellate prefers colonial or unicellular stages as its plastid source. The dinoflagellate is quite abundant in Antarctic waters, and this relationship represents another unusual relationship in understanding the biological role of *Phaeocystis* and its predators.

To provide a framework for the details of cell biology, genetics, and chemical signaling as controls on *P. pouchetii* life-cycle transitions, Whipple et al. (2005a) constructed a 15-compartment conceptual life cycle that includes solitary cell types and size classes of growing and senescent colonies. Whipple et al. (2005b; this volume) described how this conceptual approach was used to direct integrated research into various facets of *P. pouchetii* ecology: appearance of colonies from fjord sediments; in vitro colony division; in vitro changes in colony shape of *P. pouchetii* associated with senescence; determining which *P. pouchetii* life stage is vulnerable to viral infection and lysis; colony division rate and growth rates; enhancement by diatoms frustules of *P. pouchetii* colony formation from solitary cells; and investigation of mesozooplankton suppression and microzooplankton enhancement of *P. globosa* colony formation by grazer infochemicals.

Although life-cycle alternations may be the key to unraveling the biofeedbacks that occur between *Phaeocystis* and the ecosystem, many questions remain unanswered. For example, heteromorphic life cycles such as those exhibited by *Phaeocystis* function in macroalgae as adaptations to reduce grazing pressure (Lubchenco and Cubit 1980; Slocum 1980). Do life-cycle transformations such as development of motility and emigration of cells from colonies (Verity et al. 1988b) happen randomly, are they driven only by physical (light, temperature) or chemical (nutrient) stimuli, or are they also under biological control such as grazing (Tang 2003), or perhaps intra- or interspecies signaling? In any case, these transformations are best viewed as responses to stressors [grazing, viruses, nutrient and light limitation, prolonged darkness (polar winter)]. By changing its life-cycle morphology, *Phaeocystis* changes the size structure of the ecosystem and also essentially disperses offspring to another future where the stressor may be avoided or reduced.

Environmental regulation

The distribution patterns of *Phaeocystis* species can partly be explained by its specific light and nutrient requirements, which are thoroughly reviewed in Schoemann et al. (2005). The genus occurs under a wide range of light intensities. In temperate regions, the ability of *Phaeocystis* to grow at low light levels allows bloom development early in the season (Peperzak et al. 1998). Massive colony blooms in Antarctic waters early in spring are only terminated once nutrients become depleted. In those areas, high biomass of *Phaeocystis* cells may also be found under the sea ice, where light intensities are often <5 μmol quanta m^{-2} s^{-1} (van Hilst and Smith 2002).

The concentrations of the macronutrients PO_4 and NO_3 largely control biomass development of *Phaeocystis* in temperate areas (Lancelot et al. 1987; Verity et al. 1988a; Veldhuis et al. 1991). Early studies demonstrated that this reflected the ability of *Phaeocystis* to both utilize organic phosphate (Admiraal and Veldhuis 1987) and be competitive at high nitrate levels (Lancelot et al. 1998). Gieskes et al. (this volume) studied the extensive data set on *Phaeocystis* that is provided by the continuous plankton recorder (CPR), a sampling device towed behind ships-of opportunity throughout the North Atlantic since 1948 (Reid et al. 2003). *Phaeocystis* abundance has mainly been restricted to neritic regions and limited to springtime, with the southeastern North Sea as a hotspot. Abundance and length of seasonal occurrence of *Phaeocystis* in the North Sea was not correlated with the increasing eutrophication during the period of data collection. This is apparently because *Phaeocystis* is not only influenced by anthropogenic nutrient enrichment in river discharge: nutrients in the Channel/Atlantic water entering through the Dover Straits may be equally or more important, although this region is also influenced by discharge from the Seine and Somme rivers. Only very close to the Dutch coast was phytoplankton biomass related significantly to Rhine river run-off in spring; elsewhere in the North Sea, Atlantic water incursion overruled anthropogenic nutrient enrichment. As discussed later, the relative importance of *Phaeocystis* blooms compared to diatoms in the Belgian coastal zone is related to the combined effects of riverine nutrient load and the North Atlantic oscillation (NAO: Breton et al. 2006).

In the Southern Ocean, algal growth is partly controlled by iron when this micronutrient is available in limiting concentrations, but variable-light climate also plays an important role. Thus a complex interaction between iron and light limitation has been described (Stefels and van Leeuwe 1998; van Leeuwe and Stefels 1998; Arrigo and Tagliabue 2005). Recently, experiments were designed to study photoacclimation in *P. antarctica* under more natural light conditions (van Leeuwe and Stefels, this issue). Cells were exposed to a variable light regime that mimicked vertical mixing of the water column. Iron limitation was invoked by adding ethylenediaminetetraacetic acid (EDTA) to natural seawater. Periodic exposure to high irradiance induced rapid conversion of the light-harvesting pigment, diadinoxanthin, into the photoprotective pigment, diatoxanthin. Under iron limitation, cells showed characteristics of high light acclimation. An expansion of the photoprotective pigment pool coincided with weaker diurnal patterns of pigment cycling. These results supported prior conclusions that photoacclimation can be successful under conditions of iron limitation.

DiTullio et al. (this volume) studied the pigment composition of freshly isolated strains of *P. antarctica*. The experiments were performed in EDTA-free media thus utilizing the natural organic Fe binding ligands present in the seawater, and applying ecologically relevant Fe concentrations (i.e., subnanomolar). Ratios of the pigments 19′-hexanoyloxyfucoxanthin: chlorophyll *a* increased while ratios of fucoxanthin: chlorophyll *a* decreased under increasing Fe-limited conditions. Thus DiTullio and colleagues confirmed the findings by van Leeuwe and Stefels. The ratios retrieved in the laboratory experiments mimic those typically measured in *P. antarctica* spring blooms in the Ross Sea, which stresses the importance of Fe limitation in this region.

At low light intensities, the iron demand of cells increases in relation with enlargement of the photosynthetic apparatus (Raven 1990; van Leeuwe and de Baar 2000). Sedwick et al. (this volume) studied the iron requirements of *P. antarctica* in the laboratory, and recorded a relatively high iron requirement (a half-saturation constant of 0.45 nM dissolved iron) at low light levels (at ca. 20 μmol photons m^{-2} s^{-1}). Their results showed that dissolved iron availability plays a primary role in regulating blooms of colonial *P. antarctica* in the southern Ross Sea during summer. Sedwick and colleagues also performed field studies in the Ross Sea, which showed that in late spring iron was still in ample supply and *Phaeocystis* growth was controlled by the ambient light conditions rather than by iron availability. It appears that in situ cellular iron demand was <0.45 nM Fe. The authors related this low iron requirement to the relatively high ambient light levels.

Annual patterns of *P. antarctica* in the Southern Ocean may thus be determined by the interplay between variable micronutrient concentrations and irradiance availability. In contrast, patterns in high latitudes in the northern hemisphere, where iron is not typically limiting, are primarily determined by nutrient conditions and large-scale hydrographical phenomena.

Characterization of organic matter and biodegradability

The gross biochemical composition of *Phaeocystis* differs to a certain extent from other phytoplankton. For example, colonies appear to have a specific carbohydrate signature, including a rather unusual, or even unique, nuclear magnetic resonance (NMR) spectra of the mucus (van Boekel 1992; Janse et al. 1996). A combination of sterols and fatty acids can be used as a biomarker of *Phaeocystis*, and a set of fatty acids, sterols, and pigments is diagnostic for the class of algae to which *Phaeocystis* belongs (Prymnesiophyceae), although individually they are widespread among phytoplankton. However, the fatty acids 20:5w3 (eicosapentaenoic acid, EPA) and 22:6w3 (docosahexaenoic acid, DHA), which are essential for the growth of multicellular animals, occur in only trace amounts in *Phaeocystis*, whereas they are common in other phytoplankton (Nichols et al. 1991). Moreover, based upon the fatty acid and lipid contents, solitary cells are more nutritious than colonies.

Phaeocystis is well known for producing extravagant amounts of particulate and dissolved organic carbon, especially during the colonial stage of its life cycle. Alderkamp et al. (2005; this volume) reviewed the characteristics and dynamics of the organic matter compartments produced by *Phaeocystis*. When nutrients become limiting and *Phaeocystis* blooms reach a stationary growth phase, excess energy is stored as carbohydrates that leads to an increase in C/N and C/P ratios of *Phaeocystis* organic material. At the end of a bloom, deterioration of colonies, autolysis of cells, grazing of colonies and cells, and lysis of cells that are virally infected release dissolved organic matter (DOM) rich in glucan and mucopolysaccharides. Whereas this organic matter is potentially readily degradable by heterotrophic bacteria (Brussaard et al. 2005b; Ruardij et al. 2005), the high C/N and C/P ratios of *Phaeocystis* organic matter may lead to nutrient limitation of microbial degradation. Furthermore, carbohydrate hydrogel formation may physically retard the degradation potential of the *Phaeocystis* organic material, and also accelerate its sedimentation. The production of dissolved organic matter and subsequent formation of transparent exopolymers (TEP) may influence microbial population dynamics directly through bacterial colonization and indirectly through scavenging of predators and viruses (Brussaard et al. 2005b).

The relationship between *Phaeocystis* photosynthesis, DOC release, and microbial dynamics is thus complex and likely mediated by nutrient availability (Thingstad and Billen 1994). For example, bacterial degradation of primary production in high-nutrient low-chlorophyll regions can be limited by iron availability (Arrieta et al. 2004). Iron supply can either directly stimulate microbial activity, or do so indirectly by enhancing DOC production of phytoplankton upon their release from iron limitation. Becquevort et al. (this volume) investigated indirect effects of Fe supply on bacterial degradation by measuring bacterial response to organic substrates derived from *P. antarctica* cultures set up in high and low Fe-amended treatments using Antarctic seawater. Iron addition influenced the relative abundance of colonies versus single cells of *Phaeocystis* and caused a decrease in the C/N ratio of *Phaeocystis* material. In turn this altered microbial community composition and cellular activity levels, such that more *Phaeocystis*-derived organic matter was degraded in high iron compared to low-iron bioassays. The authors concluded that by increasing bacterial degradation and thereby preventing the accumulation of dissolved organic carbon, the positive effect of Fe supply on the carbon biological pump via enhanced phytoplankton export may partly be counteracted.

Another aspect of *Phaeocystis* colonies that is under-appreciated is that they provide sites for temporary settlement of other plankton. For example, Sazhin et al. (this volume) enumerated the consortia of organisms associated with colonies during blooms of *P. globosa* in the eastern English Channel and *P. pouchetii* in mesocosms in western Norway. In both environments, mass development of the small needle-shaped diatom, *Pseudonitzschia* species, occurred on *Phaeocystis* colonies, and comprised up to 70% as much carbon as contained in the *Phaeocystis* cells inside the colonies at the end of the bloom. The authors proposed that the diatoms use the organic substance of the colonial matrix for growth.

Grazing

Despite almost a century of qualitative observations and two decades of quantitative studies, there remains considerable uncertainty about the extent to which *Phaeocystis*, especially colonies, provide a good food for, and are ingested significantly by, zooplankton (reviews by Weisse et al. 1994; Schoemann et al. 2005; Nejstgaard et al. this volume). Reported feeding rates are very variable and span the full range from non-detectable to close to physiological maxima for the respective grazer and prey size. A variety of filtering or suspension-feeding organisms decrease or stop their feeding when *Phaeocystis* colonies are present, including zooplankton and various bivalves (Weisse et al. 1994; Smaal and Twisk 1997; Nejstgaard et al. this volume). Active rejection of colonies has also been observed (Kamermans 1994) and mechanical disturbance of the gills of bivalves and the mouthparts of zooplankton caused by colony mucus has been hypothesized. Widespread speculation attends the notion that *Phaeocystis* may be unpalatable, the wrong size, noxious, toxic, or combinations of these (see Weisse et al. 1994 for earlier review), at least during some bloom phases when investigated in situ (Estep et al. 1990). Colony blooms themselves are offered as evidence of reduced mortality for whatever reason; in contrast, regions and seasons with regular blooms are also ecosystems with great fisheries. Hamm et al. (1999) argued that the colony membrane may be largely responsible for minimizing a variety of loss processes including grazing, viruses, cell lysis, and sedimentation. In this regard, intact small colonies of *P. cf. pouchetii* with living cells have been observed in copepod fecal pellets from the Barents Sea (P.G. Verity, E. Arashkevitch, P. Wassmann, unpubl.). Seuront et al. (this volume) offered another mechanism that protects *Phaeocystis* against grazing. They investigated the distribution patterns of chlorophyll *a* and seawater excess viscosity over the course of a phytoplankton spring bloom dominated by *Phaeocystis globosa*. Positive and negative correlations were measured during the bloom and after the formation of foam in the turbulent zone. The authors suggested that the biologically induced increase in seawater viscosity might be a competitive advantage to *Phaeocystis* as a potential anti-predator strategy. In addition, through this unique environmental engineering strategy, *P. globosa* can create a favorable, turbulent-free physical habitat that protects colony integrity.

In marine systems, both proto- and metazoan zooplankton use species-specific contact chemoreception and diffusable chemical cues in analyzing particles they encounter (Friedman and Strickler 1975; Spero 1985; Verity 1988). If grazing is selective and varies in response to prey ratio, abundance, or palatability, then investment in mortality reduction by prey is rewarded. Considerable evidence illustrates that many proto- and metazoan zooplankton are indeed selective feeders (e.g., Alcaraz et al. 1980; Verity 1991; Sanders and Wickham 1993). Hence, theory would suggest that phytoplankton such as *Phaeocystis* would evolve defenses against predation. But relatively little is known about production of secondary chemicals or metabolites that function as feeding deterrents in *Phaeocystis* as well as for other phytoplankton (see reviews by Pohnert 2004; Yoshida et al. 2004). *Phaeocystis* has been shown to have toxic effects on several organisms and cell types (see Nejstgaard et al. this volume for a review), and dimethylsulphoniopropionate (DMSP) and its cleavage products have been suggested to function as grazing deterrents in the closely related *Emiliania huxleyi* (e.g., Wolfe 2000; Strom et al. 2003). Further, lectins are known to function as infochemicals signaling the presence of certain prey to predators so that the predators can recognize and select their prey item (Wooten and Roberts 2006). *Phaeocystis* is known to possess lectins in the wall of its colonial stage (van Rijssel et al. 1997; Hamm et al. 1999), but perhaps these lectins function in the opposite way by discouraging the predator rather than signaling its presence to the predator.

Phytoplankton vary widely in their physiology and internal chemistry as a function of growth conditions, and may change their physiology considerably and irreversibly when isolated from nature into culture conditions (Edvardsen and Paasche 1998). Likewise it is also known that many zooplankton may show strong variations in feeding rates in response to chemical and physiological cues (e.g., Dutz et al. 2005; Koski et al. 2005; Dutz and Koski 2006; Long and Hay 2006). Nejstgaard et al. (this volume) critically reviewed the evidence for and against *Phaeocystis* as a nutritious ration for herbivores and omnivores. Based on available data, they concluded that much of the reported variation in feeding on *Phaeocystis* especially between field and laboratory studies may (a) be due to methodological problems (including use of nonspecific bulk pigments in field investigations, and zooplankton not having alternative prey available in laboratory experiments), and (b) reflect physiological differences between *Phaeocystis* species, strains, and cell types within a strain, and/or induced changes in algae during culture. This study showed that this may have resulted in an overestimation of feeding rates in laboratory studies of as much as fivefold compared to field estimates based on specific cell counts. They further posited that colony size provides a (partial) refuge from predation by small copepods (such as *Temora* spp. and *Acartia* spp.) at least for *P. globosa*, while colonies of *P. pouchetii* appear less protected from large copepods (such as *Calanus*), and that chemical signaling between the predator and its prey may be one of the major factors controlling grazing on *Phaeocystis*. From presently available data, it is unclear whether colonies of *P. pouchetii* are less protected from copepod grazing compared to similarly sized colonies of *P. globosa* due to differences in the colony physiology. This should be tested in future experiments. For smaller colonies, size may also be less important for those protozooplankton whose feeding mechanisms permit engulfment of larger prey, e.g., heterotrophic dinoflagellates (Stelfox-Widdicombe et al. 2004).

Allelopathy and toxicity

The effects of anti-metabolites on phytoplankton were recognized over 40 years ago (Johnston 1963). Marine plankton themselves also produce allelochemicals that have antagonistic effects. Gauthier and Aubert (1981) presented several examples within and between bacteria, phytoplankton, and zooplankton. Pratt (1966) reported allelopathy between the diatom *Skeletonema costatum* and the flagellate *Olisthodiscus luteus*. Where investigated, release of allelochemicals by marine microalgae has been observed (Uchida 1977; Chan et al. 1980; Van Alstyne 1986; Targett and Ward 1991), although the identity of the allelochemicals was seldom known. Arguments have also been presented in support of contact inhibition in microalgae (Costas et al. 1993).

Phaeocystis has long been suspected to negatively impact co-occurring organisms, including allelochemical and toxic effects, and recently it has been identified as a harmful algal bloom (HAB) species, at least in European waters (Veldhuis and Wassmann 2005). Van Rijssel et al. (this volume) used a bioassay to quantify the potential impact of living *P. pouchetii*, collected during mesocosm blooms in western Norway, on other living cells. Red blood cells were used as model targets representing unprotected cells, and their lysis were quantified by monitoring dissolved hemoglobin. Hemolytic activity was best correlated with *P. pouchetii* abundance compared to other phytoplankton, including diatoms, and was enhanced by increased temperature and light. In comparison to other harmful algae, living *P. pouchetii* colonies were highly hemolytic and the mechanism was apparently quite different from the hemolytic harmful algal bloom species studied so far. Whereas it can be argued that red blood cells may not mimic the responses of marine phytoplankton, the lytic response of the chemicals to the membranes is likely the same. Long et al. (subm.) reached similar conclusions that *Phaeocystis* possesses allelopathic properties from experiments with both algal cultures and natural plankton communities during *Phaeocystis* blooms in mesocosms. Results such as these support the notion that *Phaeocystis* may use plant chemistry in its competitive interactions with other phytoplankton, as well as a means of adjusting to grazing pressure, as discussed earlier. These antagonistic effects may help account for the prodigious, near-monospecific colony blooms regularly observed in high latitudes. It seems increasingly likely that "allelochemistry ... may help explain complexity in biological systems" (Lewis 1986).

Viruses

Viruses are now universally recognized as potentially being a major contributor to phytoplankton and bacterial mortality. Given a life cycle that includes solitary cell and gelatinous colony stages, and their role in food web dynamics and biogeochemical fluxes, viral infection of *Phaeocystis* may affect not only host population dynamics, but also ecosystem structure and function. Viral infection and lysis have been described in *P. pouchetii* and *P. globosa* (Bratbak et al. 1998a, b; Jacobsen 2002; Brussaard et al. 2005a). Among other issues, determining which stages of the life cycle are vulnerable to viral infection and lysis is critical for understanding the effects of viruses on *Phaeocystis* life-cycle dynamics and ecology. Brussaard et al. (this volume) reviewed the role of viruses with respect to *Phaeocystis* bloom dynamics, and offered a discussion on potential factors affecting viral control. The colony morphotype provides protection against infection (Brussard et al. 2005a; Jacobsen et al. 2005, subm.), with the importance of viral infection being inversely related to colony size (Ruardij et al. 2005). Degradation of the colonial matrix and the resulting formation of transparent exopolymeric particles (TEP) also indirectly affect the process of viral infection by acting to scavenge viruses from the water column (Brussaard et al. 2005b). It becomes more and more obvious that viral lysis of *Phaeocystis* may be significant to biogeochemical processing in the water column by affecting nutrient remineralization from *Phaeocystis* biomass, stimulating bacterial productivity, and increasing release of climatically active organic sulfur compounds.

Sedimentation and vertical export

Aggregation and accompanying sedimentation is a physical process that can provide a significant fate for *Phaeocystis* blooms in situ, in addition to trophic losses caused by grazing and viral lysis. Understanding of vertical export of organic material or cells comprising *Phaeocystis* blooms thus helps to constrain water column biogeochemistry and reflects physical water column and pelagic food web structures. For example the fluxes through the water column not only depend on grazing pressure per se but also the (meso-)zooplankton present. TEP formation and coagulation associated with colony degradation may favor export. The importance of sedimentation to the fate of blooms, and the significance of vertical export of POC associated with *Phaeocystis* blooms is hard to define for all species and cases, mainly due to the extraordinary plasticity in the ecology and life cycle of the species (Wassmann 1994; Schoemann et al. 2005).

Reigstad and Wassmann (this volume) have made an attempt to quantify the various export mechanisms and determined cell and colony sedimentation rates and their contribution to the vertical POC directly. Their compilation of data, ranging from polar to sub-Arctic and boreal regions, suggests that the overall contribution of *Phaeocystis* species to vertical export of biogenic matter is not significant. Colony blooms are often associated with strong vertical export in the upper photic zone, but between 40 m and 100 m, the contribution of *Phaeocystis* cells to total export declines to only ~3% at 100 m depth. By comparison, vertical export of diatoms is significantly larger. Other processes that favor export of *Phaeocystis*-derived carbon, such as grazing and fecal pellet production, downwelling, and sinking mucilagenous matter, are difficult to quantify, but even considering colony mucilage, the contribution of *Phaeocystis* apparently increases only slightly to <5% of POC export.

Not withstanding these results, episodic sedimentation events may still be of local importance (e.g., DiTullio et al. 2000). Such events are hard to monitor, e.g., they can be associated with grazing by highly mobile krill (Hamm et al. 2001), and other difficulties remain. Reigstad and Wassmann (this volume) note how difficult it is to quantify the carbon content of *Phaeocystis* colonies and also refer to the problems caused by fragmentation of colonies. Given the known links between *Phaeocystis* physiology (Brussaard et al. 2005b; Mari et al. 2005; Alderkamp et al., this volume), it seems likely that bloom export is coupled to both nutrient regime and zooplankton dynamics. While N limitation would favor retention of fragmented colony material in upper euphotic zone (Mari et al. 2005), P limitation may induce rapidly colony disruption resulting in heavy and sticky mucus aggregates (thought to sink rapidly). Again, it is difficult to quantify carbon export along this pathway, since this pool is not visible as pigments or cells using microscopy, and perhaps not even picked up accurately by traditional POC measurements (Reigstad and Wassmann, this volume).

Additional information on sedimentation rates can be derived from tracer studies. Schmidt et al. (this volume) and Lalande (2006) show, however, that interpretation of tracers is not straightforward in the presence of *Phaeocystis*. Particulate thorium-234 (^{234}Th) removal from the water column and presence in sediment traps can be used as indication of vertical export of biogenic matter (Buesseler et al. 2006) However, the occurrence of *Phaeocystis* can temporarily delay the export of particulate ^{234}Th from surface waters, e.g., sediment trap POC fluxes were much lower than large-volume POC fluxes in such waters (Lalande 2006). Schmidt et al. (this volume) suggests that large flocculation negatively affects the scavenging of ^{234}Th onto biological particles. When large colonies make up a significant fraction of the vertical flux, ^{234}Th clearly underestimated particle fluxes.

DMS(P) and the sulfur cycle

Spring blooms dominated by *Phaeocystis* have been associated with considerable production of dimethyl sulphide (DMS). It is thought that the atmospheric oxidation products of DMS result in cooling of the Earth's atmosphere and thereby reduce the effects of greenhouse gases such as CO_2. Given the global importance of DMS for the biogeochemical sulfur cycle and climate, the observed correlations between *Phaeocystis* blooms and DMS in the atmosphere are significant (Turner et al. 1995; Kwint and Kramer 1996). The main direct source for DMS is dimethylsulphoniopropionate (DMSP), which is produced by several phytoplankton including *Emiliania huxleyi* and *Phaeocystis* spp. (Keller et al. 1989; Liss et al. 1994). DMSP has properties in common with the compatible solute glycine betaine, and production of DMSP and DMS may be related to physiological stress (Stefels 2000; Sunda et al. 2002). *Phaeocystis* has an extracellularly located, membrane-bound DMSP-lyase enzyme that cleaves DMSP into equimolar amounts of DMS and acrylate (Stefels and Dijkhuizen 1996). It has long been speculated that the production of acrylate may function as a microbial inhibitor (Sieburth 1960). But ambient concentrations of acrylate during *Phaeocystis* blooms are inadequate to expect bacterial inhibition. However, acrylate concentrations inside the colonial matrix are much higher (Noordkamp et al. 2000), providing the potential to function as a grazing deterrent. Wolfe et al. (1997) reported that strains of the related genus *Emiliania*, which contained low levels of DMSP-lyase were preferred by grazers over strains with high DMSP-lyase levels. The mechanism apparently was lyase-induced release of high concentrations of acrylate in grazer food vacuoles. Given the very high concentrations of acrylate in the colony (Noordkamp et al. 1998) and also high DMSP-lyase activity in cells (Stefels and van Boekel 1993), DMS production and grazing-activated DMSP cleavage by *Phaeocystis* is thought to contribute to the avoidance of healthy colonies by protozoan and metazoan zooplankton (Wolfe et al. 2000). It has also recently been shown that copepods react strongly to DMS in the feeding current (Steinke et al. 2006), although it is not clear whether this is a cue to trigger ingestion of the *Phaeocystis* cell or its micropredator by the copepod (see discussion in Nejstgaard et al. this volume).

Stefels et al. (this volume) reviewed the functional relationships between physical and chemical ecosystem parameters, biological productivity, and the production and consumption of DMSP and DMS. They made specific attempts to quantify these functional relationships, and described the various factors controlling DMSP production (species composition, abiotic factors), the conversion of DMSP to DMS (by algal and bacterial enzymes), the fate of DMSP-sulfur (e.g., grazing, microbial consumption, sedimentation), and factors controlling DMS removal from the water column (microbial consumption, photo-oxidation, emission to the atmosphere). The authors concluded with specific recommendations for future studies needed to improve accuracy of model predictions linking ecosystem structure and function to DMS dynamics. One of these recommendations is the subdivision of the algal realm into six groups, one of which (Phytoflagellate B) is based on the features of *Phaeocystis*. Although it cannot be excluded that a few other prymnesiophytes with high DMSP-lyase activity fall within this same category, *Phaeocystis* is thought to be the main contributor. This outstanding position is indicative of the global importance of this single genus. In this regard, DiTullio et al. (this volume) described rapid sedimentation of DMS and DMSP at the end of spring blooms of *P. antarctica* in the Ross Sea. Shenoy and Kumar (this volume) provided a detailed overview of spatial and temporal patterns of organic sulfur compounds in the Indian Ocean, where *Phaeocystis globosa* blooms have recently been reported to occur (Madhupratap et al. 2000; Schoemann et al. 2005).

Modeling

The search for organizational pattern in plankton and its causes is not new. The realization is growing that there is more to the organization of nature than can be revealed in empirical data. Thus there must be theory, and there must be modeling. The role of model simulation in understanding system complexity and in decision-making is recognized even in the business world (Schrage and Peters 1999). Models offer the ability to test conceptual understanding of how components of a given system are linked, and to simulate complex interactions in a quantifiable repeatable manner. Modeling, however, succeeds only if new insights subsequently lead to validation or rejection of hypotheses upon more detailed examination. Thus information extracted from model runs is strongly linked to the chosen structure (trophic resolution) of the model and to the parameterizations of the interactions between the components. When properly validated, models can be useful in both hindcast and forecast modes and used for guiding decision-making in environmental matters.

Life-cycle, food web and ecosystem models have all been formulated to facilitate understanding of the ecology of *Phaeocystis* and the impact of their blooms on marine food web structures and biogeochemical cycles. Environmental application of these models included the control of *Phaeocystis* blooms in eutrophic coastal waters (Gypens et al. 2004; Lancelot et al. 2006) and their contribution to the oceanic biological pump of atmospheric CO_2 (Gypens et al. 2004; Pasquer et al. 2005; Tagliabue and Arrigo 2005).

Conceptual models for the life cycle of *Phaeocystis* (Rousseau et al. 1994, this volume; Whipple et al. 2005a, b; this volume) have been essential to progress in understanding *Phaeocystis* ecosystems by providing the basis for subsequent experimental studies and for best trophic resolution in mathematical models. Mechanistic models including different life forms of *Phaeocystis* as the dominant primary producer and their various micro- and meso-grazers have been developed to mimic *Phaeocystis* bloom dynamics in mesocosms and investigate the role of life history attributes at the population level. These include several types of food web models such as individual-based models that track cells and colonies as they grow and divide (Canziani and Hallam 1996), food web models that investigate trophic interactions (Verity 2000), and a combination of these two (Ruardij et al. 2005). The latter model is the most sophisticated, describing the dynamics of *Phaeocystis* free-living and colonial cells including 50 colony classes (2–80,000 cells per colony). Simulations obtained with these different models suggest critical processes determining *Phaeocystis* life form dominance and the exchanges between them. These are the specific growth rate of colony cells compared to solitary cells, their light and nutrient controls (e.g., Ruardij et al. 2005), the mechanisms controlling colony lysis, as well as top–down controls (including virus) specific to the different *Phaeocystis* life forms.

Several ecosystem models with *Phaeocystis* and diatoms as dominant primary producers but various trophic resolution, have been implemented to describe seasonal cycles of ecosystem constituents and elemental fluxes in areas where *Phaeocystis* blooms have been recorded: the Barents Sea (Wassmann and Slagstad 1993), the southern North Sea (Gypens et al. 2004; Lancelot et al. 2005; Lacroix et al. 2006), and the Southern Ocean (Arrigo et al. 2003; Pasquer et al. 2005; Tagliabue and Arrigo 2005, 2006).

Mechanisms controlling *Phaeocystis* dominance over diatoms in the iron-limited Southern Ocean, and the influence of ecosystem structure on oceanic capability of absorbing atmospheric CO_2, were investigated by two ice-ocean ecosystem models applied to the Ross Sea: the coupled ice atmosphere ocean (CIAO; Arrigo et al. 2003; Tagliabue and Arrigo 2005, 2006) and the seawter microbial community (SWAMCO-4; Pasquer et al. 2005) models. These models differed in their trophic resolution, notably in the description of *Phaeocystis* life forms with only free-living cells in CIAO, while both colony forms and free-living cells were explicitly described in SWAMCO-4. Simulations all suggested a similar substantial uptake of atmospheric CO_2 associated with *Phaeocystis* blooms under control of light and iron availability, but explanatory mechanisms for *Phaeocystis* success were contrasting, due to different *Phaeocystis* life-cycle resolutions. In CIAO simulations, the success of *Phaeocystis* over diatoms was explained by better photosynthetic performance of *Phaeocystis* cells at low light and low iron, while iron availability determined the height of the bloom and the related CO_2 sink (Tagliabue and Arrigo 2005). In contrast, the explicit description of colony forms in SWAMCO-4 simulations emphasized the key iron-sequestering role played by the *Phaeocystis* colony polysaccharidic matrix, which deprived diatoms of iron (Pasquer et al. 2005).

Mechanisms regulating the diatom–*Phaeocystis* succession and the magnitude of colony blooms in the eutrophic coastal waters of the eastern Channel and Southern North Sea were investigated with the MIRO model implemented in a multi-box (Gypens et al. 2004; Lancelot et al. 2005, 2006) and a tridimensional (Lacroix et al. 2006) structure. MIRO was very similar to SWAMCO-4, including two *Phaeocystis* stages (free-living cells and colonies) as well as their interrelations, trophic fates, and associated carbon and nutrient fluxes. Interpretation of MIRO simulations suggested a key role of *Phaeocystis* colony affinity for low PO_4 in determining the success and magnitude of *Phaeocystis* blooms along the NO_3 rich coastal zone, as already proposed from field observations (Veldhuis and Admiraal 1987; Veldhuis et al. 1991). Historical reconstruction of eutrophication in the area indicated that a well-balanced nitrate and phosphate enrichment as observed between 1960 and 1989 was more beneficial to diatoms than to *Phaeocystis*, while the contemporary decrease in phosphate loads while high nitrate is maintained is beneficial to *Phaeocystis* (Lancelot et al. 2006). The latter study concluded that future management of nutrient emission reduction aiming at decreasing

Phaeocystis blooms in the southern North Sea without impacting on diatoms should decrease nitrate loads. Further MIRO simulations showed that: (i) a significant sink of atmospheric CO_2 was associated with *Phaeocystis* blooms in the southern North Sea (Gypens et al. 2004), (ii) the coastal ecosystem had a low nutrient retention and elimination capacity, (iii) the trophic efficiency of the planktonic system was low, and (iv) both were modulated by meteorological forcing (Lancelot et al. 2005).

Beside mechanistic models, inductive models built on cause–effect relationships such as Regression Analysis (Breton et al. 2006) and fuzzy logic (Chen and Mynett 2004) have been developed as early-warning tools for predicting the occurrence, timing, and magnitude of undesirable *Phaeocystis* blooms in the southern North Sea. Chen and Mynett (2004) demonstrated that the integration of a fuzzy cellular automata model in a three-dimensional numerical model describing the light and nutrient fields was a promising approach to forecast possible *Phaeocystis* blooms in Dutch coastal waters. Statistical analysis of 14 years of diatom and *Phaeocystis* records in Belgian coastal waters identified synergy between climate driven by the North Atlantic oscillation (NAO) and human-induced river-based nitrate inputs (Breton et al. 2006), and predicted that elevated *Phaeocystis* blooms during years of middle NAO while high NAO years would favor diatom blooms.

Concluding thoughts on the complexity of *Phaeocystis* ecosystems and recommendations for future research

Much of the research on this enigmatic phytoplankton species cluster during the 20th century was conducted from two mutually reinforcing perspectives: that phytoplankton patterns in time and space could be understood essentially from a perspective of physiological ecology, and that plankton, being microscopic, were necessarily rudimentary in their physiology, behavior, interactions, and integration of environmental stimuli. We now know better. For example, appreciation is spreading that evolutionary strategies which result in, e.g., 10% reduction in mortality are every bit as effective selective forces as those that enhance growth by 10% (Verity and Smetacek 1996). Single-celled plankton are no longer viewed as biologically simple creatures, merely because they bear little resemblance to complex human structures: auto- and heterotrophic unicellular plankton are fully capable of assessing their environment with multiple sensors and communicating with one another using a host of chemicals. It should not be surprising therefore that *Phaeocystis* is one of these complex organisms.

The papers deriving from the SCOR symposium on *Phaeocystis* and the arguments marshaled here infer that it is essential to distinguish between factors affecting production and standing stocks, in which the nature, type, and amount of resources may be essential, versus factors affecting time/space distributions, diversity, community composition, and energy transfer, in which predation, morphology, chemical ecology, and life cycles may be more influential. *Phaeocystis* is one of the few recognized key species or trophic engineers (*sensu* Lawton and Jones 1995) whose morphologies and transformations result in their contributing the bulk of biogenic fluxes or fish production in specific ecosystems (Verity and Smetacek 1996). Population oscillations represent complex ecosystem behavior influenced by component organisms, and evidence clearly indicates that *Phaeocystis* reorganizes these complex ecosystems (Fig. 1). Thus biocomplexity drives biogeochemistry, but the fundamental mechanisms behind these patterns are only now becoming appreciated. Through chemistry and/or size, the life cycle of *Phaeocystis* finds refuge from predation and viral infection, a process thought to stabilize ecosystems (see Strom and Loukos 1998). Diversification at a given trophic level promotes diversification at the levels above (through more food options) and below (through resource utilization), and is considered responsible for adaptive radiation throughout the course of evolution (Stanley 1973; Butterfield 1997). Interestingly, the strategy for success apparently used by *Phaeocystis* (gelatinous morphology, heteromorphic life cycle, size changes, and allelochemistry) is very successful in freshwater ecosystems. Is this because grazing pressure is greater in freshwater compared to marine systems? Is *Phaeocystis* so

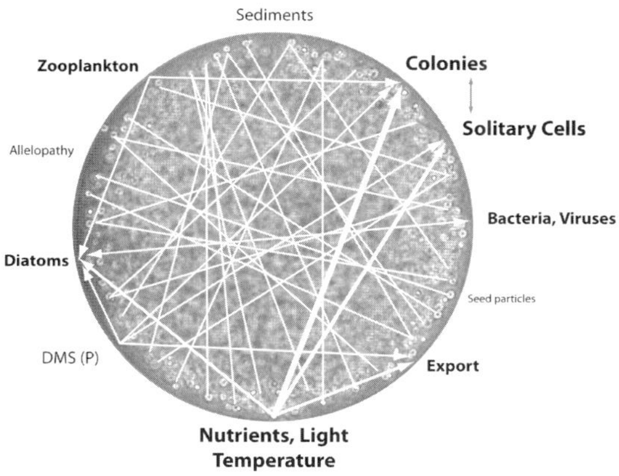

Fig. 1 A diorama of a *Phaeocystis* ecosystem, in which physical, chemical, geological, and biological components interact to varying degrees to influence food web structure, biogeochemical flows, and ecosystem function. For simplicity, only well-documented components are illustrated. Current extent of our knowledge base is qualitatively related to font size. Arrow magnitude and direction indicate major known flows; increasingly smaller and more numerous arrows imply unknown connections, the importance of indirect flows among ecosystem compartments, and overall system complexity

ecologically successful because it is one of the few marine phytoplankton to combine these behavioral and physiological capabilities? Ultimately, taxa such as *Phaeocystis* may help unravel similarities and differences in the biocomplexity of freshwater versus marine pelagic ecosystems, and help us better understand and predict aquatic ecosystem structure and function.

Recent studies including those conducted under the aegis of SCOR working group 120 have advanced understanding of *Phaeocystis* ecosystems, as reviewed above and described elsewhere in this special issue. However, there remain several salient topics where data and information needs are foremost. For example, organelle-specific or complete genomes of several phytoplanktons have been sequenced (*Prochlorococcus*: Dufresne et al. 2003; *Emiliania*: Puerta et al. 2004; *Ostreococcus*: Derelle et al. 2006) and this information has identified metabolic pathways previously unsuspected, e.g., a complete urea cycle in *Thalassiosira pseudonana* (Armbrust et al. 2004). Thus it would likely prove very useful to sequence *Phaeocystis* in the future (see the Joint Genome Institute website http://genome.jgi-psf.org/euk_home.html for eukaryotes planned and already sequenced), and such knowledge would enable answers to questions such as the identity of genes responsible for life cycle transformations. In this regard, determination of protocols and culture conditions under which either growth stage can be maintained reliably in culture, and induced to transform to the other stage, would permit a vast array of experiments aimed at quantifying evolutionary advantages and disadvantages of each life-cycle stage, their susceptibility to mortality, and role in biogeochemistry. Comparisons of genomic architecture would also enhance the basis for comparing adaptations in the field. In general, quantitative understanding of controls and causes, and accurate prediction of switches between colonies and solitary cells, would advance the field tremendously.

Another important gap remains. *Phaeocystis* is suspected of producing large pools of dissolved organic matter, and the magnitude of this extracellular release likely vary with each life stage, their physiological status, and environmental stresses. Moreover, its composition and lability influences the microbial food web, production of transparent exopolymers, and thus particle coagulation and sedimentation. Thus we need to learn more about this pool, its sources, turnover, and sinks, to determine the specific role of *Phaeocystis* in biogeochemical cycles.

Future needs are not limited to empirical studies, and in fact substantial progress likely will not be achieved without continued model development. These include the need for the larger modeling community to share and combine models, to agree upon minimum formats or base models for widespread use, and to use common parameterizations. Where datasets permit, model intercomparisons should be conducted. Experimentalists clearly need to be intimately involved in the entire process, from model conceptualization to simulations, especially regarding *Phaeocystis* (Whipple et al. 2005a). While empiricists and modelers often speak different languages (Anderson 2005; Flynn 2005), the effort to design and implement such models can integrate research and teach us more about ecosystems than the outcomes or predictions of the models themselves.

Acknowledgments This special issue of *Biogeochemistry* represents the culmination of extensive efforts by many scientists involved in various aspects of *Phaeocystis* research. Integration of their activities would not have been possible without the support of the scientific committee on ocean research (SCOR) (www.jhu.edu/scor, accessed 6/30/06). SCOR working group #120 was devoted to *Phaeocystis* studies; chair and co-chair were W.W.C. Gieskes and S. Belviso, respectively. We thank E. Urban at SCOR and the organizers and participants of the *Phaeocystis* workshops held in 2002 (University of East Anglia, UK), 2004 (Savannah, GA, USA), and 2005 (University of Groningen, The Netherlands). Participation by the senior author in the research and SCOR activities reported here was provided by USA National Science Foundation grant OPP-00-83381 and Department of Energy grant FG02-98ER62531. We thank G. Malin, J. Stefels, and W.W.C. Gieskes for improvements to an earlier draft, and A. Boyette for drafting the figure.

References

Admiraal W, Veldhuis MJW (1987) Determination of nucleosides and nucleotides in seawater by HPLC: application to phosphatase activity in cultures of the alga *Phaeocystis pouchetii*. Mar Ecol Prog Ser 36:277–285

Alcaraz M, Paffenhöfer G-A, Strickler JR (1980) Catching the algae: a first account of visual observations of filter-feeding copepods. In: Kerfoot WC (ed) Evolution and ecology of zooplankton communities. Univ. Press of New England, Hanover, NH, pp 241–248

Alderkamp A-C, Buma AGJ, van Rijssel M The carbohydrates of *Phaeocystis* and their degradation in the microbial food web. Biogeochemistry. doi:10.1007/s10533-007-9078-2

Alderkamp A-C, Nejstgaard JC, Verity PG, Zirbel MJ, Sazhin AF, van Rijssel M (2006) Dynamics in carbohydrate composition of *Phaeocystis pouchetii* colonies during spring blooms in mesocosms. J Sea Res 55:169–181

Anderson TR (2005) Plankton functional type modeling: running before we can walk? J Plankton Res 27:1073–1081

Anderson DM, Cembella AD, Hallegraeff GM (1998) Physiological ecology of harmful algal blooms. Springer-Verlag, Berlin

Armbrust EV et al (2004) The genome of the diatom *Thalassiosira pseudonana*: ecology, evolution, and metabolism. Science 306:79–86

Arrieta JM, Weinbauer MG, Lute C, Herndl GJ (2004) Response of bacterioplankton to iron fertilization in the Southern Ocean. Limnol Oceanogr 49:799–808

Arrigo KR, Tagliabue A (2005) Iron in the Ross Sea, Part II: impact of discrete iron addition strategies. J Geophys Res 110, C03010. DOI:10.1029/2004JC002568

Arrigo KR, Worthen DL, Robenson DH (2003) A coupled ocean-ecosystem model of the Ross Sea: 2. Iron regulation of phytoplankton taxonomic variability and primary production. J Geophys Res 108(C7):3231. DOI:10:1029/2001JC000856

Becquevort S, Lancelot C, Schoemann V Experimental study on the role of Fe in the bacterial degradation of organic matter derived from *Phaeocystis antarctica*. Biogeochemistry. doi:10.1007/s10533-007-9079-1

Bratbak G, Jacobsen A, Heldal M (1998a) Viral lysis of *Phaeocystis pouchetii* and bacterial secondary production. Aquat Microb Ecol 16:11–16

Bratbak G, Jacobsen A, Heldal M, Nagasaki K, Thingstad F (1998b) Virus production in *Phaeocystis pouchetii* and its relation to host cell growth and nutrition. Aquat Microb Ecol 16:1–9

Breton E, Rousseau V, Parent J-Y, Ozer J, Lancelot C (2006) Hydroclimatic modulation of diatom/*Phaeocystis* blooms in nutrient-enriched Belgian coastal waters (North Sea). Limnol Oceanogr 51:1401–1409

Brussaard CPD, Bratbak G, Baudoux A-C, Ruardij P *Phaeocystis* and its interaction with viruses. Biogeochemistry. doi:10.1007/s10533-007-9096-0

Brussaard CPD, Kuipers B, Veldhuis MJW (2005a) A mesocosm study of *Phaeocystis globosa* population dynamics. I. Regulatory role of viruses in bloom control. Harmful Algae 4:859–874

Brussaard CPD, Mari X, Van Bleijswijk JDL, Veldhuis MJW (2005b) A mesocosm study of *Phaeocystis globosa* population dynamics. II. Significance for the microbial community. Harmful Algae 4:875–893

Buesseler KO, Benitez-Nelson CR, Moran SB, Burd A, Charette M, Cochran JK, Coppola L, Fisher NS, Fowler SW, Gardner WD, Guo LD, Gustafsson O, Lamborg C, Masque P, Miquel JC, Passow U, Santschi PH, Savoye N, Stewart G, Trull T (2006) An assessment of particulate organic carbon to thorium-234 ratios in the ocean and their impact on the application of 234Th as a POC flux proxy. Mar Chem 100:213–233

Butterfield NJ (1997) Plankton ecology and the Proterozoic–Phanerozoic transition. Paleobiology 23:247–262

Canziani GA, Hallam TG (1996) A mathematical model for a *Phaeocystis* sp. dominated plankton community dynamics. I. The basic model. Nonlin World 3:19–76

Chan AT, Andersen RJ, LeBlanc MJ, Harrison PJ (1980) Algal plating as a tool for investigating allelopathy among marine microalgae. Mar Biol 59:7–13

Chen Q, Mynett A (2004) Predicting algal blooms in the Dutch coast by integrated numerical and fuzzy cellular automata. In: Liong S-Y, Phoon K-K, Babovic V (eds) Proceedings of the 6th international conference on hydroinformatics. World Scientific Publishing Company, Singapore, pp 502–510

Costas E, Aguilera A, Gonzalez-Gil S, Lopez-Rodas V (1993) Contact inhibition: also a control for cell proliferation in unicellular algae? Biol Bull 184:1–5

Derelle E et al (2006) Genome analysis of the smallest free-living eukaryote *Ostreococcus tauri* unveils many unique features. Proc Natl Acad Sci 103:11647–11652

DiTullio GR, Garcia N, Riseman SF, Sedwick PN Effects of iron concentration on the pigment composition of *Phaeocystis antarctica* in the Ross Sea. Biogeochemistry. doi:10.1007/s10533-007-9080-8

DiTullio GR, Grebmeier JM, Arrigo KR, Lizotte MP, Robinson DH, Leventer A, Barry JB, VanWoert ML, Dunbar RB (2000) Rapid and early export of *Phaeo-*

cystis antarctica blooms in the Ross Sea, Antarctica. Nature 404:595–598

Dufresne A et al (2003) Genome sequence of the cyanobacterium *Prochlorococcus marinus* SS120, a nearly minimal oxyphototrophic genome. Proc Natl Acad Sci 100:10020–10025

Dutz J, Klein Breteler WCM, Kramer G (2005) Inhibition of copepod feeding by exudates and transparent exopolymer particles (TEP) derived from a *Phaeocystis globosa* dominated phytoplankton community. Harmful Algae 4:929–940

Dutz J, Koski M (2006) Low grazing vulnerability in flagellated, solitary cells of the prymnesiophyte *Phaeocystis globosa*. Limnol Oceanogr 51:1230–1238

Edvardsen B, Paasche E (1998) Bloom dynamics and physiology of *Prymnesium* and *Chrysochromulina*. In: Anderson DM, Cembella AD, Hallegraef GM (eds) The physiological ecology of harmful algal blooms. Springer Verlag, Heidelberg, pp 193–208

Estep KW, Nejstgaard JC, Skjoldal HR, Rey F (1990) Predation by copepods upon natural populations of *Phaeocystis pouchetii* as a function of the physiological state of the prey. Mar Ecol Prog Ser 67:235–249

Flynn KJ (2005) Castles built on sand: dysfunctionality in plankton models and the inadequacy of dialogue between biologists and modelers. J Plankton Res 27:1205–1210

Friedman MM, Strickler JR (1975) Chemoreception and feeding in the calanoid copepods. Proc Nat Acad Sci USA 72:4185–4188

Gaebler S, Hayes PK, Medlin LK Methods used to reveal genetic diversity in the colony forming prymnesiophyte *Phaeocystis antarctica* – preliminary results. Biogeochemistry. doi:10.1007/s10533-007-9084-4

Gast RJ, Moran D, Dennett MR, Beaudoin DJ, Caron DA (2006) Studies in protistan diversity in Antarctica. Abstract 57th Annual meeting of the International Society of Protozoologists, June 20–24 (p 24). Lisbon, Portugal

Gauthier MJ, Aubert M (1981) Chemical telemediators in the marine environment. In: Duursma EK, Dawson E (eds) Marine organic chemistry: evolution, composition, interactions, and chemistry of organic matter in seawater. Elsevier, Amsterdam, pp 225–257

Gieskes WWC, Leterme SC, Peletier H, Edwards M, Reid PC Annual variation of *Phaeocystis* colonies Atlantic-wide since 1948, and interpretation of long-term changes in '*Phaeocystis* hotspot' North Sea. Biogeochemistry. doi:10.1007/s10533-007-9082-6

Gypens N, Lancelot C, Borges AV (2004) Carbon dynamics and CO_2 air-sea exchanges in the eutrophied coastal waters of the Southern Bight of the North Sea: a modelling study. Biogeosciences 1:147–157

Hamm CE, Simson DA, Merkel R, Smetacek V (1999) Colonies of *Phaeocystis globosa* are protected by a thin but tough skin. Mar Ecol Prog Ser 187:101–111

Hamm C, Reigstad M, Riser CW, Muhlebach A, Wassmann P (2001) On the trophic fate of *Phaeocystis pouchetii*. VII. Sterols and fatty acids reveal sedimentation of *Phaeocystis*—derived organic matter via krill fecal strings. Mar Ecol Prog Ser 209:55–69

Jacobsen A (2002) Morphology, relative DNA content, and hypothetical life cycle of *Phaeocystis pouchetii* (Prymnesiophyceae); with special emphasis on the flagellated cell type. Sarsia 87:338–349

Jacobsen A, Martinez-Martinez J, Verity P, Frischer ME, Sandaa R-A, Larsen A (2005) Are colonies or colonial cells of *Phaeocystis pouchetii* (Prymnesiophyceae) susceptible to virus infection? American Society of Limnology and Oceanography, Summer Meeting, June 19–24, Santiago de Compostela, Spain

Jacobsen A, Larsen A, Martínez-Martínez J, Frischer ME, Verity PG Are colonies and colonial cells of *Phaeocystis pouchetii* (Haptophyta) susceptible to viral infection? Aquat Microb Ecol (Subm.)

Janse I, van Rijssel M, van Hall PJ, Gerwig GJ, Gottschal JC, Prins RA (1996) The storage glucan of *Phaeocystis globosa* (Prymnesiophyceae) cells. J Phycol 32:382–387

Johnston R (1963) Antimetabolites as an aid to the study of phytoplankton nutrition. J Mar Biol Assoc UK 43:409–425

Kamermans P (1994) Nutritional value of solitary cells and colonies of *Phaeocystis* sp. for the bivalve *Macoma balthica* (L.). Ophelia 39:35–44

Keller MD, Bellows WK, Guillard RRL (1989) Dimethyl sulfide production in marine phytoplankton. In: Saltzman ES, Cooper WJ (eds) Biogenic sulfur in the environment. American Chemical Society, Wash DC, pp 167–182

Koski M, Dutz J, Klein-Breteler WCM (2005) Selective grazing of *Temora longicornis* in different stages of a *Phaeocystis globosa* bloom – a mesocosm study. Harmful Algae 4:915–927

Kwint RLJ, Kramer KJM (1996) Annual cycle of the production and fate of DMS and DMSP in a marine coastal system. Mar Ecol Prog Ser 134:217–224

Lacroix G, Ruddick R, Park Y, Gypens N, Lancelot C (2007) Validation of the 3D biogeochemical model MIRO&CO with field nutrient and phytoplankton data and MERIS-derived surface chlorophyll a images. J Mar Syst 64:66–88

Lalande C (2006) Vertical export of biogenic in the Barents and Chukchi Seas. PhD. Thesis, University of Knoxville, Tennessee, USA

Lancelot C, Rousseau V (1994) Ecology of *Phaeocystis*: the key role of colony forms. In: Green JC, Leadbeater BSC (eds) The haptophyte algae. Clarendon Press, Oxford, pp 229–245

Lancelot C, Billen G, Sournia A, Weisse T, Colijn F, Veldhuis M, Davies A, Wassman P (1987) *Phaeocystis* blooms and nutrient enrichment in the continental coastal zones of the North Sea. Ambio 16:38–46

Lancelot C, Gypens N, Billen G, Garnier J, Roubeix V (2007) Testing an integrated river-ocean mathematical tool for linking marine eutrophication to land use: the *Phaeocystis* dominated Belgian coastal zone (Southern North Sea) over the past 50 years. J Mar Syst 64:216–228

Lancelot C, Keller MD, Rousseau V, Smith WO Jr, Mathot S (1998) Autecology of the marine haptophyte *Phaeocystis* sp. In: Anderson DM, Cembella AD, Hallegraeff GM (eds) Physiological ecology of harmful algal blooms, NATO ASI series, vol G41. Springer-Verlag, Berlin, pp 209–224

Lancelot C, Spitz Y, Gypens N, Ruddick K, Becquevort S, Rousseau V, Lacroix G, Billen G (2005) Modelling diatom–*Phaeocystis* blooms and nutrient cycles in the Southern Bight of the North Sea: the MIRO model. Mar Ecol Prog Ser 289:63–78

Lange M, Chen Y-Q, Medlin LK (2002) Molecular genetic delineation of *Phaeocystis* species (Prymnesiophyceae) using coding and non-coding regions of nuclear and plastid genomes. Eur J Phycol 37:77–92

Lange M, Guillou L, Vaulot D, Simon N, Amann RI, Ludwig W, Medlin LK (1996) Identification of the class Prymnesiophyceae and the genus *Phaeocystis* with ribosomal RNA-targeted nucleic acid probes detected by flow cytometry. J Phycol 32:858–868

Lawton JH, Jones CG (1995) Linking species and ecosystems: organisms as ecosystem engineers. In: Jones CG, Lawton JH (eds) Linking species and ecosystems. Chapman and Hall, NY, pp 141–150

Lewis WM Jr (1986) Evolutionary interpretations of allelochemical interactions in phytoplankton algae. Am Nat 127:184–194

Liss PS, Malin G, Turner SM, Holligan PM (1994) Dimethyl sulfide and *Phaeocystis*: a review. J Mar Syst 5:41–53

Long JD, Hay ME (2006) When intraspecific exceeds interspecific variance: effects of phytoplankton morphology and growth phase on copepod feeding and fitness. Limnol Oceanogr 51:988–996

Long JD, Anderson JT, Nejstgaard JC, Verity PG, Hay ME Allelopathy of a bloom-froming marine phytoplankton, *Phaeocystis*, in mesocosm blooms and laboratory cultures. Aquat Microb Ecol (Subm.)

Lubchenco J, Cubit J (1980) Heteromorphic life histories of certain marine algae as adaptations to variations in herbivory. Ecology 61:676–687

Madhupratap M, Sawant S, Gauns M (2000) A first report on a bloom of the marine prymnesiophycean, *Phaeocystis globosa*, from the Arabian Sea. Ocean Acta 23:83–90

Mari X, Rassoulzadegan F, Brussaard CPD, Wassmann P (2005) Dynamics of transparent exopolymeric particles (TEP) production by *Phaeocystis globosa* under N- or P-limitation: a controlling factor of the export/retention balance. Harmful Algae 4:895–914

Medlin LK, Lange M, Baumann MEM (1994) Genetic differentiation among three colony-forming species of *Phaeocystis*: further evidence for the phylogeny of the Prymnesiophyta. Phycologia 33:199–212

Medlin L, Zingone A A Review: the genus *Phaeocystis* and its species. Biogeochemistry. doi:10.1007/s10533-007-9087-1

Mock T, Valentin K (2004) Photosynthesis and cold acclimation: molecular evidence from a polar diatom. J Phycol 40:732–741

Nejstgaard JC, Tang KW, Steinke M, Dutz J, Koski M, Antajan E, Long JD Zooplankton grazing on *Phaeocystis*: a quantitative review and future challenges. Biogeochemistry (in press)

Nichols PD, Skerrat JH, Davidson A, Burton H, McMeekin TA (1991) Lipids of cultured *Phaeocystis pouchetii*: signatures for food web, biogeochemical and environmental studies in Antarctica and the Southern Ocean. Phytochem 30:3209–3214

Noel MH, Kawachi M, Inouye I (2004) Induced dimorphid life cycle of a coccolithophorid *Calytrosphaera sphaeroidea* (Prymnesiophyceae). J Phycol 40:112–129

Noordkamp DJB, Gieskes WWC, Gottschal JC, Forney LJ, van Rijssel M (2000) Acrylate in *Phaeocystis* colonies does not affect the surrounding bacteria. J Sea Res 43:287–296

Noordkamp DJB, Schotten M, Gieskes WWC, Forney LJ, Gottschal JC, van Rijssel M (1998) High acrylate concentrations in the mucus of *Phaeocystis globosa* colonies. Aquat Microb Ecol 16:45–52

Nygaard K, Tobiesen A (1993) Bacterivory in algae: a survival strategy during nutrient limitation. Limnol Oceanogr 38:273–279

Pasquer B, Laruelle G, Becquevort S, Schoemann V, Gosse H, Lancelot C (2005) Linking ocean biogeochemical cycles and ecosystem structure and function: results of the complex Swamco-4 model. J Sea Res 53:93–108

Peperzak L, Colijn F, Gieskes WWC, Peeters JCH (1998) Development of the diatom–*Phaeocystis* spring bloom in the Dutch coastal zone of the North Sea: the silicon versus the daily irradiance threshold hypothesis. J Plankton Res 20:517–537

Peperzak L, Duin RMN, Colijn F, Gieskes WWC (2000) Growth and mortality of flagellates and non-flagellate cells of *Phaeocystis globosa* (Prymnesiophyceae). J Plankton Res 22:107–119

Pohnert G (2004) Chemical defense strategies of marine organisms. In: Schulz SE (ed) Topics in current chemistry. Springer-Verlag GmBH, Berlin, pp 179–219

Pratt DM (1966) Competition between *Skeletonema costatum* and *Olisthodiscus luteus* in Narragansett Bay and in culture. Limnol Oceanogr 11:447–455

Puerta MVS, Bachvaroff TR, Delwiche CF (2004) The complete mitochondrial genome sequence of the haptophyte *Emiliania huxleyi* and its relation to heterokonts. DNA Res 11:1–10

Ratkova T, Wassmann P (2002) Seasonal variation and spatial distribution of phyto- and protozooplankton in the central Barents Sea. J Mar Syst 38:47–75

Raven JA (1990) Predictions of Mn and Fe use efficiencies of phototrophic growth as a function of light availability for growth and of C assimilation pathway. New Phytol 116:1–18

Reid PC, Colebrook JM, Matthews JBL, Aiken J (2003) Continuous plankton recorder: concepts and history, from plankton indicator to undulating recorders. Prog Oceanogr 58:117–173

Reigstad M, Wassmann P Does *Phaeocystis* spp. contribute significantly to vertical export of biogenic matter? Biogeochemistry. doi:10.1007/s10533-007-9093-3

Rousseau V, Chrétiennot-Dinet M-J, Jacobsen A, Verity P, Whipple S The life cycle of *Phaeocystis*: state of knowledge and presumptive role in ecology. Biogeochemistry doi:10.1007/s10533-007-9085-3

Rousseau V, Vaulot D, Casotti R, Cariou V, Lenz J, Gunkel JJ, Baumann M (1994) The life cycle of

Phaeocystis (Prymnesiophyceae): evidence and hypotheses. J Mar Syst 5:23–39

Ruardij P, Veldhuis MJW, Brussaard CPD (2005) Modeling the bloom dynamics of the polymorphic phytoplankter *Phaeocystis globosa*: impact of grazers and viruses. Harmful Algae 4:941–963

Sanders RW, Wickham SA (1993) Planktonic protozoa and metazoa: predation, food quality, and population control. Mar Microb Food Webs 7:197–223

Sazhin AF, Felipe Artigas L, Nejstgaard JC, Frischer ME Colonization of *Phaeocystis* species by pennate diatoms and other protists: an important contribution to colony biomass. Biogeochemistry. doi:10.1007/s10533-007-9086-2

Schmidt S, Belviso S, Wassmann P, Thouzeau G, Stefels J, Reigstad M Vernal sedimentation trends in north Norwegian fjords: temporary anomaly in ^{234}Th particulate fluxes related to *Phaeocystis pouchetii* proliferation. Biogeochemistry. doi:10.1007/s10533-007-9094-2

Schoemann V, Becquevort S, Stefels J, Rousseau V, Lancelot C (2005) *Phaeocystis* blooms in the global ocean and their controlling mechanisms: a review. J Sea Res 53:43–66

Schrage M, Peters T (1999) Serious play: how the world's best companies simulate to innovate. Harvard Business School Press, Cambridge, MD

Sedwick PN, Garcia N, Riseman SF, Marsay CM, DiTullio GR Evidence for high iron requirements of colonial *Phaeocystis antarctica* in the Ross Sea. Biogeochemistry. doi:10.1007/s10533-007-9081-7

Seuront L, Lacheze C, Doubell MJ, Seymour JR, Mitchell JG The influence of *Phaeocystis globosa* bloom dynamics on microscale spatial patterns of phytoplankton biomass and bulk-phase seawater viscosity. Biogeochemistry. doi:10.1007/s10533-007-9097-z

Shenoy DM, Dileep Kumar M Variability in abundance and fluxes of dimethyl sulphide in the Indian Ocean. Biogeochemistry. doi:10.1007/s10533-007-9092-4

Sieburth JMcN (1960) Acrylic acid, an "antibiotic" principle in *Phaeocystis* blooms in Antarctic waters. Science 132:676–677

Slocum CJ (1980) Differential susceptibility to grazers in two phases of an intertidal alga: advantages of heteromorphic generations. J Exp Mar Biol Ecol 46:99–110

Smaal AC, Twisk F (1997) Filtration and absorption of *Phaeocystis* cf. *globosa* by the mussel *Mytilus edulis* L. J Exp Mar Biol Ecol 209:33–46

Spero HJ (1985) Chemosensory capabilities in the phagotrophic dinoflagellate *Gymnodinium fungiforme*. J Phycol 21:181–184

Stanley SM (1973) An ecological theory for the sudden origin of multicellular life in the late Precambrian. Proc Nat Acad Sci USA 70:1486–1489

Stefels J (2000) Physiological aspects of the production and conversion of DMSP in marine algae and higher plants. J Sea Res 43:183–197

Stefels J, Dijkhuizen L (1996) Characteristics of DMSP-lyase in *Phaeocystis* sp. (Prymnesiophyceae). Mar Ecol Prog Ser 131:307–313

Stefels J, Van Boekel WHM (1993) Production of DMS from dissolved DMSP in axenic cultures of the marine phytoplankton species *Phaeocystis* sp. Mar Ecol Prog Ser 97:11–18

Stefels J, van Leeuwe MA (1998) Effects of iron and light stress on the biogeochemical composition of Antarctic *Phaeocystis* sp. (Prymnesiophyceae). I. Intracellular DMSP concentrations. J Phycol 34:486–495

Stefels J, Steinke M, Turner S, Malin G, Belviso S Environmental constraints on the production of the climatically active gas dimethylsulphide (DMS) and implications for ecosystem modeling. Biogeochemistry. doi:10.1007/s10533-007-9091-5

Steinke M, Stefels J, Stamhuis E (2006) Dimethyl sulfide triggers search behavior in copepods. Limnol Oceanogr 51:1925–1930

Stelfox-Widdicombe CE, Archer SD, Burkill PH, Stefels J (2004) Microzooplankton grazing in *Phaeocystis* and diatom-dominated waters in the southern North Sea in spring. J Sea Res 51:37–51

Strom SL, Loukos H (1998) Selective feeding by protozoa: model and experimental behaviors and their consequences for population stability. J Plankton Res 20:831–846

Strom S, Wolfe G, Holmes J, Stecher H, Shimeneck C, Lambert S, Moreno E (2003) Chemical defense in microplankton. I. Feeding and growth rates of heterotrophic protests on the DMS-producing phytoplankter *Emiliania huxleyi*. Limnol Oceanogr 48:217–229

Sunda W, Kieber DJ, Kiene RP, Huntsman S (2002) An antioxidant function for DMSP and DMS in marine algae. Nature 418:317–320

Tagliabue A, Arrigo KR (2005) Iron in the Ross Sea: 1. Impact on CO_2 fluxes via variation in phytoplankton functional group and non-Redfield stoichiometry. J Geophys Res 110, C03009. DOI: 10.1029/2004JC002531

Tagliabue A, Arrigo KR (2006) Processes governing the supply of iron to phytoplankton in stratified seas. J Geophys Res 111, C06019. DOI: 10.1029/2005JC003363

Tang KW (2003) Grazing and colony size development in *Phaeocystis globosa* (Prymnesiophyceae): the role of a chemical signal. J Plankton Res 25:831–842

Targett NM, Ward JE (1991) Bioactive microalgal metabolites: mediation of subtle ecological interactions in phytophagous suspension-feeding marine invertebrates. Bioorganic Mar Chem 4:91–118

Thingstad F, Billen G (1994) Microbial degradation of *Phaeocystis* material in the water column. J Mar Syst 5:55–66

Tillmann U (1998) Phagotrophy by a plastidic haptophyte, *Prymnesium patelliferum*. Aquat Microb Ecol 14:155–160

Turner SM, Nightingale PD, Broadgate W, Liss PS (1995) The distribution of dimethyl sulfide and dimethylsulphoproprionate in Antarctic waters and sea ice. Deep-Sea Res II 42:1059–1080

Uchida T (1977) Excretion of a diatom inhibitory substance by *Prorocentrum micans* Ehrenberg. Jap J Ecol 27:1–4

Van Alstyne KL (1986) Effects of phytoplankton taste and smell on feeding behavior of the copepod *Centropages hamatus*. Mar Ecol Prog Ser 34:187–190 van Boekel WHM (1992) *Phaeocystis* colony mucus components and the importance of calcium ions for stability. Mar Ecol Prog Ser 87:301–305

van Hilst CM, Smith WO Jr (2002) Photosynthesis/irradiance relationships in the Ross Sea, Antarctica, and their control by phytoplankton assemblage composition and environmental factors. Mar Ecol Prog Ser 226:1–12

van Leeuwe MA, de Baar HJW (2000) Photoacclimation by the Antarctic flagellate *Pyramimonas* sp. (Prasinophyceae) in response to iron limitation. Eur J Phycol 35:295–303

van Leeuwe MA, Stefels J (1998) Effects of iron and light stress on the biogeochemical composition of Antarctic *Phaeocystis* sp. (Prymnesiophyceae). II. Pigment composition. J Phycol 34:496–503

van Leeuwe MA, Stefels J Photosynthetic responses in *Phaeocystis antarctica* towards varying light and iron conditions. Biogeochemistry. doi:10.1007/s10533-007-9083-5

van Rijssel M, Alderkamp A-C, Nejstgaard JC, Sazhin AF, Verity PG Haemolytic activity of living *Phaeocystis pouchetii* during mesocosm blooms. Biogeochemistry. doi:10.1007/s10533-007-9095-1

van Rijssel M, Hamm CE, Gieskes WWC (1997) *Phaeocystis globosa* (Prymnesiophyceae) colonies: hollow structures built with small amounts of polysaccharides. Eur J Phycol 32:185–192

Vaulot D, Birrien J-L, Marie D, Casotti R, Veldhuis MJW, Kraaij GW, Chretiennot-Dinet M-J (1994) Morphology, ploidy, pigment composition, and genome size of cultured strains of *Phaeocystis* (Prymnesiophyceae). J Phycol 30:1022–1035

Veldhuis MJW, Admiraal W (1987) Influence of phosphate depletion on the growth and colony formation of *Phaeocystis pouchetii*. Mar Biol 95:47–54

Veldhuis MJW, Wassmann P (2005) Bloom dynamics and biological control of a high biomass HAB species in European coastal waters: a *Phaeocystis* case study. Harmful Algae 4:805–809

Veldhuis MJW, Brussaard CPD, Noordeloos AAM (2005) Living in a *Phaeocystis* colony: a way to be a successful algal species. Harmful Algae 4:841–858

Veldhuis MJW, Colijn F, Admiraal W (1991) Phosphate utilization in *Phaeocystis pouchetii* (Haptophyceae). Mar Biol 12:53–62

Verity PG (1988) Chemosensory behavior in marine planktonic ciliates. Bull Mar Sci 43:772–782

Verity PG (1991) Feeding in planktonic protozoans: evidence for non-random acquisition of prey. Mar Microb Food Webs 5:69–76

Verity PG (2000) Grazing experiments and model simulations of the role of zooplankton in *Phaeocystis* food webs. J Sea Res 43:317–343

Verity PG, Medlin LK (2003) Observations on colony formation by the cosmopolitan phytoplankton genus *Phaeocystis*. J Mar Syst 43:153–164

Verity PG, Smetacek V (1996) Organism life cycles, predation, and the structure of marine pelagic ecosystems. Mar Ecol Prog Ser 130:277–293

Verity PG, Villareal TA, Smayda TJ (1988a) Ecological investigations of blooms of colonial *Phaeocystis pouchetii*. I. Abundance, biochemical composition, and metabolic rates. J Plankton Res 10:219–248

Verity PG, Villareal TA, Smayda TJ (1988b) Ecological investigations of blooms of colonial *Phaeocystis pouchetii*. II. The role of life cycle phenomena in bloom termination. J Plankton Res 10:749–766

Wassmann P (1994) Significance of sedimentation for the termination of *Phaeocystis* blooms. J. Mar Syst 5:81–100

Wassmann P, Ratkova T, Reigstad M (2005) The contribution of single and colonial cells of *Phaeocystis pouchetii* to spring and summer blooms in the north-eastern North Atlantic. Harmful Algae 4:823–840

Wassmann P, Slagstad D (1993) Seasonal and annual dynamics of particulate carbon flux in the Barents Sea, a model approach. Polar Biol 13:363–372

Weisse T, Tande K, Verity P, Hansen F, Gieskes W (1994) The trophic significance of *Phaeocystis* blooms. J Mar Syst 5:67–79

Whipple SJ, Patten BC, Verity PG (2005a) Life cycle of the marine alga *Phaeocystis*: a conceptual model to summarize literature and guide research. J Mar Syst 57:83–110

Whipple SJ, Patten BC, Verity PG (2005b) Colony growth and evidence for colony multiplication in *Phaeocystis pouchetii* (Prymnesiophyceae) isolated from mesocosm blooms. J Plankton Res 27:495–501

Whipple SJ, Patten BC, Verity PG, Nejstgaard JC, Long JD, Anderson JT, Jacobsen A, Larsen A, Martinez-Martinez J, Borrett SR. Gaining integrated understanding of *Phaeocystis* spp. (Prymnesiophyceae) through model-driven laboratory and mesocosm studies. Biogeochemistry. doi:10.1007/s10533-007-9089-2

Wolfe GV (2000) The chemical defense ecology of marine unicellular plankton: constraints, mechanisms, and impacts. Biol Bull 198:225–244

Wolfe GV, Levasseur M, Cantin G, Michaud S (2000) DMSP and DMS dynamics and microzooplankton grazing in the Labrador Sea: application of the dilution technique. Deep-Sea Res I 47:2243–2264

Wolfe GV, Steinke M, Kirst GO (1997) Grazing-activated chemical defense in a unicellular marine alga. Nature 387:894–897

Wooten EC, Roberts EC (2006) Biochemical recognition of prey by planktonic protozoa. Abstract 57th Annual meeting of the International Society of Protozoologists, June 20–24, 2006 (p 57). Lisbon, Portugal

Yoshida T, Hairston NG, Ellner SP (2004) Evolutionary trade-off between defense against grazing and competitive ability in a simple unicellular alga, *Chlorella vulgaris*. Proc Royal Soc Lond Ser B, Biol Sci 271:1947–1953

Zingone A, Chretiennot-Dinet M-J, Lange M, Medlin L (1999) Morphological and genetic characterization of *Phaeocystis cordata* and *P. jahnii* (Prymnesiophyceae), two new species from the Mediterranean Sea. J Phycol 35:1322–1337